THIN SHELL THEORY

V.V. NOVOZHILOV

THIN SHELL THEORY

TRANSLATED FROM THE SECOND RUSSIAN EDITION BY

P. G. LOWE

Engineering Laboratory, Cambridge

EDITED BY

J. R. M. RADOK

Professor of Mathematics, University of Adelaide

SECOND AUGMENTED AND REVISED EDITION

Springer-Science+Business Media, B.V.

ISBN 978-94-017-5354-8 ISBN 978-94-017-5352-4 (eBook)
DOI 10.1007/978-94-017-5352-4

Originally published by P. Noordhoff, Ltd. Groningen, The Netherlands in 1964.
Softcover reprint of the hardcover 2nd edition 1964

AUTHOR'S PREFACE

TO THE RUSSIAN 1951 EDITION

At the present time, the theory of thin shells (curved plates) is one of the more active branches of the theory of elasticity which is receiving everywhere a great deal of attention.

This is explained by the fact that thin-walled shell constructions combine light weight with high strength, as a result of which they have found wide application in naval, aeronautical and boiler engineering as well as in reinforced concrete roof designs. It must be emphasized that the practical possibilities of the utilization of shells have by no means been exhausted and we are all the time witnesses to the perfection of different types of construction; on the one hand this takes the form of an extension of the range of employment of shells, on the other hand, of a more thorough analysis of their properties (i.e. by improvement of the methods of stress analysis).

Soviet scientists have contributed extensively to the theory of shells and it may be said, without exaggeration, that during the past 15–20 years a large number of valuable and profound ideas in this theory have arisen in our country.

This book is devoted to the analysis of stresses and deformations in shells; this problem is treated here as a linear one, i.e., displacements are assumed to be small in comparison with the thickness of the shell and the strains are not to exceed the limit of proportionality. Linear shell theory is of considerable practical importance, since the assumptions on which it is based are in many problems close to the true conditions.

Non-linear shell theory is of equal practical importance and so is the associated stability theory of shells. However, since they differ essentially from the linear theory as regards their formulation as well as their methods of solution, these theories must be treated individually, and they deserve special monographs.

This book is mainly based on the work of the Russian school in the theory of shells which includes also certain results obtained by the author. It is divided into four chapters the first of which is devoted to the general (moment) theory of thin shells, the second to the membrane (momentless) theory, the third and the fourth to the moment theory of two types of shells which are important in industrial applications, namely, cylindrical shells and shells of revolution.

In order to ease for the reader the burden of understanding the subject and applying it independently to practical problems, the book has been provided with many examples taken from different fields of application. The second chapter dealing with the membrane theory contains especially many of them. The author has there aimed at maximum simplicity and physical clarity (to the prejudice even of mathematical generality and rigour) having in mind that for the engineer the problem of the membrane state of stress of shells is of paramount interest.

ПРЕДИСЛОВИЕ К ПЕРЕВОДУ НА АНГЛИЙСКИЙ ЯЗЫК КНИГИ „ТЕОРИЯ ТОНКИХ ОБОЛОЧЕК"

Эта книга посвящена линейной теории оболочек и освещает данный вопрос с единой точки зрения, суть которой состоит в широком использовании комплексного преобразования. Данный прием (идея которого восходит еще к ученым Цюрихской школы) обобщен здесь на любые задачи линеаризированной теории оболочек.

Мне кажется, что такое изложение имеет ряд преимуществ, поскольку оно делает уравнения более обозримыми, облегчает их преобразование, а иногда и упрощает их интегрирование. Вместе с тем данный путь, разумеется, отнюдь не является единственным. Можно указать хотя бы на прекрасные книги, написанные двумя другими русскими учеными — В. З. Власовым и А. А. Гольденвейзером, посвященные тому же вопросу, но излагающие его в другом духе.

Необходимо далее отметить, что данная книга была написана около десяти лет назад и не учитывает, естественно, те результаты, которые были получены в этот период. Тем не менее она не слишком устарела, поскольку в линейной теории оболочек за это время принципиальных изменений не произошло.

Я должен, в заключение, принести свою искреннюю благодарность профессору Бруклинского политехнического института J. R. M. Radok, взявшему на себя большой труд по редактированию перевода моей книги на английский язык. Надеюсь, что издание этого перевода хотя бы в некоторой мере будет способствовать налаживанию деловых и дружеских отношений между специалистами по теории упругости Советского Союза и других стран.

Ленинград,
30 октября 1958г.

(В. В. Новожилов)

Я с удовлетворением узнал, что перевод моей книги "Теория тонких оболочек" вторично издается на английском языке. Пользуюсь случаем передать свой привет и наилучшие пожелания её читателям.

Ленинград, апрель, 1964г. В. В. Новожилов

PREFACE TO THE ENGLISH TRANSLATION OF "THE THEORY OF THIN SHELLS"

This book is devoted to the linear theory of shells and it elucidates the inherent problems from a unified point of view which essentially consists of the wide utilization of complex transformations. The underlying method (the idea for which originated from the Zürich school) has been generalized here to include all problems of the linearized theory of shells.

I believe that such a study has a number of advantages, since it makes the equations more accessible, facilitates their transformation and sometimes simplifies their integration. At the same time, this approach is, of course, not at all the only one. I must mention here the fine books, written by two other Russian Scientists – V. Z. Vlasov and A. L. Gol'denweizer which are devoted to the same problem, but each giving his account in a different spirit.

It should also be mentioned that this book has been written almost ten years and that it naturally does not include those results which have been obtained during this period. Nevertheless it has not become obsolete, since during this time no principal changes have occurred in the linear theory of shells.

In conclusion, I must express my sincere gratitude to Professor J. R. M. Radok of the Polytechnic Institute of Brooklyn for taking on himself the sizable task of editing the English translation of my book. I hope that the publication of this translation will in some measure contribute to professional and social interchanges between specialists in the theory of elasticity of the Soviet Union and of other countries.

Leningrad, 30 October 1958 V. V. Novozhilov

PREFACE TO THE SECOND ENGLISH TRANSLATION

I have learnt with satisfaction that a second edition of the English translation of my Book "Theory of thin shells" is to be published. I am using this opportunity to send my regards and best wishes to its readers.

Leningrad, April 1964 V. V. Novozhilov

TRANSLATOR'S PREFACE

In this translation a sentence by sentence rendering of the original has been aimed at. Any alterations or additions included have been recorded as footnotes. However a number of typographical corrections have been made in the text and figures without special mention. In this connection thanks are due to professor Novozhilov for supplying us with a list of errata. Further the sections in the translation have been numbered consecutively throughout the book rather than chapter by chapter as was used in the Russian original. In addition Auther and Subject Indices have been added.

My personal thanks are due, in particular, to the University of Sydney Authorities for facilities enabling me to complete the work of translation.

London, August 1959 P. G. LOWE

CONTENTS

Chapter I

Chapter II

CHAPTER III

Chapter IV

CHAPTER I

THE GENERAL THEORY OF THIN ELASTIC SHELLS

§ 1. Fundamental concepts and assumptions

In the theory of elasticity, the term shell is applied to bodies bounded by two curved surfaces, the distance between the surfaces being small in comparison with the other dimensions. The locus of points which lie at equal distances from these two surfaces defines the middle surface of the shell.

Drawing now at any arbitrary point of the middle surface the perpendicular to the middle surface, the length intercepted between the surfaces of the shell determines its thickness which will be designated by δ. The thickness may vary in magnitude; however, in what follows, it will be assumed almost exclusively that the shell has constant thickness, as is mostly the case in practice.

If the shell has no other boundary besides the above two surfaces, it will be called complete. Every complete shell can be divided into parts which will be referred to as incomplete shells. It will be assumed that each edge of an incomplete shell is a plane curve and that the cuts are made perpendicular to the middle surface. The middle surface, thickness and edges completely define the geometry of a shell.

It follows from the above discussion that shells bear the same relation to plates as curved beams to straight beams. Therefore, in the literature, shells are frequently referred to as curved plates. This last expression may prove useful when applied to incomplete shells, but it is of little value when applied to complete shells. Thus, for example, a complete thin-walled sphere may hardly be referred to as a curved plate. In contrast, the name thin-walled sphere for an incomplete shell is also insufficiently suggestive, so that there, at least, the term curved plate could serve some purpose. However, in general, only a remote connection exists between these terms and, for the sake of brevity, it is proposed here to refer to all bodies in the class under discussion as shells.

The theory of shells forms a part of the theory of elasticity concerned with the study of the deformation of elastic bodies under the influence of given loads. For this purpose, it will be assumed that the material of the shell is isotropic and obeys Hooke's Law and that the displacements at a point are small in comparison with the thickness of the shell. If this last limitation is not imposed, the governing differential equations become non-linear and their solution encounters great difficulties. For a large range of practical problems, however, these limitations do not impose severe restrictions on the usefulness of the theory. It should be noted here that the results obtained below must always be interpreted in the light of these restrictions.

There are two different classes of shells — thick shells and thin shells. A shell will be called thin, if the maximum value of the ratio δ/R (where R is the radius of curvature of the middle surface) can be neglected in comparison with unity. Correspondingly, shells will be called thick whenever such terms cannot be neglected. This division of shells into two classes is not fully defined as long as it is not accompanied by a statement regarding what value of the above ratio can be neglected in comparison with unity. For ordinary technical calculations, it is admissible to have a relative error of 5%, and for thin shells this corresponds to

$$\max\left(\frac{\delta}{R}\right) \leqslant \frac{1}{20}.$$

Hence shells for which this inequality is violated will be regarded as thick shells.

It is clearly suggested by the above discussion that for thin shells some terms in the basic formulae may be neglected, while for thick shells these terms must be retained. Therefore, in correspondence with this subdivision of shells, the theory will be divided into a theory of shells of arbitrary thickness and into a theory of thin shells. Obviously the theory of thin shells will be considerably simpler than that of shells of the other type.

It will be noted here that shells possess many useful properties arising from their elastic nature and that by suitable design they can be made to support large loads even when they are very thin. In this connection, the designer may gain a great deal by considering a plate as a slightly curved shell in the same way was as he may replace a beam by a shallow arch.

This property of shells is readily utilized in constructions which are strong and adaptable to a broad range of applications — such as occur in aircraft, ships and reinforced concrete design — in short, everywhere where light weight is a vitally important factor. For a large number of practical applications, the thicknesses of shells lie in the range

$$\frac{1}{1000} \leqslant \frac{\delta}{R} \leqslant \frac{1}{50},$$

i.e., in the range of thin shells, so that the theory presented in this book is of great practical interest.

As has already been indicated, there is an analogy between plates and shells, suggesting for the latter the alternate term of curved plates. This is the basis for the adoption into a theory of shells of a number of methods from the theory of plates. Therefore a short summary of the historical development of the theory of plates will now be presented.

In the classical theory of plates, there are two fundamental methods for the solution of problems. The first method was proposed by A. Cauchy [39] and S. D. Poisson [50], the second by G. Kirchhoff [46]. The method of Cauchy and Poisson is based on the expansion of the displacements and stresses in the plate in power series of z, i.e., of the distance of a given point from the middle plane of the plate. Retaining the least possible number of terms in these series, one obtains the equations derived by Mlle. Sophie Germain *. By retaining more terms, it is possible to obtain more exact solutions of the problem, these solutions tending to the exact ones as the number of terms becomes infinite. Thus the method of Cauchy and Poisson gives a universal approach to the theory of plates. However, there arose a dispute which developed in two directions:

Firstly, B. de St. Venant [40, page 725] contested the legality of the method, asserting that the corresponding series will, as a rule, diverge. (It should be stated here that the region and nature of the convergence of these series has hitherto remained unclarified.)

Secondly, arguments arose regarding the formulation of the relevant boundary conditions: whether there were to be four or five. The method of Cauchy and Poisson was not able to completely resolve this basic question. (At least, this was the opinion of the scientists of the last century.)

Therefore, the other method which was proposed by Kirchhoff [46] and which introduced physical meaning into the theory of plates has remained in use until to-day. Kirchhoff based his reasoning on several assumptions, analogous to those used in the theory of beams. His assumptions may be reduced to the following statements:

a) The straight fibres of a plate which are perpendicular to the middle surface before deformation remain so after deformation and do not change their length.

b) The normal stresses acting on planes parallel to the middle surface may be neglected in comparison with the other stresses.

The method of Kirchhoff excels over the method of Cauchy-Poisson

* See Todhunter and Pearson 'History of Elasticity.' Vol. I. § 414 p. 209

by its great clarity and physical meaning: It introduced into the basic theory a simplification which had a definite physical interpretation and which was a completely obvious extension of the well proven theory of beams. The introduction of the concepts of internal forces and moments produced a further link between the theory of plates and the theory of beams and eventually it cleared up the problem of the boundary conditions for plates which, as has already been mentioned, remained for a long time a subject for discussion. However, at this stage, it should be pointed out that Kirchhoff's method is insufficient in one respect, namely that it is only approximate and cannot be developed into a more accurate theory. In contrast, the theory of Cauchy-Poisson has the advantage that, once the problem of the convergence of its series has been resolved, it would, in principle, permit the definition of the exact solution.

The two methods in the theory of plates described above can be effectively employed in the development of a theory of shells. A theory of shells, founded on the hypotheses of Kirchhoff, was first presented in the work of G. Aron [37]; however, his development was not strictly correct. These inaccuracies were noticed and corrected by A. E. H. Love in his paper [48] in which he formulated a theory of shells in analogy to the theory of plates of Kirchhoff. A version of the theory of shells, proposed by Love, was given in one of the chapters of his book [49] in a somewhat different form from that in his paper [48].

In spite of its popularity, the development given by Love is not free from inadequacies, in that it is inconsistent with regard to small terms: some are retained and others which are of the same order of magnitude are rejected. These insufficiencies were already noted a long time ago and several authors, using Love's theory, have added certain small terms for the sake of consistency, others have omitted small terms which had been retained. Hence there have originated different versions of the formulae of shell theory which, however, only differ from the prototype — the theory of Love — by terms which are small.

The deficiencies in the formulation given by Love and also the lack of a discussion of the important question, namely, how to write the relations between the forces, moments and the displacements of the middle surface, has for a long time precluded the presentation of the basic theory in a generally acceptable form.

All these deficiencies have been studied in the publications of the Russian School in the theory of shells, led by the late Academician B. G. Galerkin. Although the works of Galerkin [93, 94, 128] relate to thick shells, he was able by his original methods to obtain all

all the formulae of the theory of shells from the equations of the general theory of elasticity. He thus played an important part in the development of a mathematically rigorous theory for thin shells. The equations of the theory of thin shells were for the first time deduced in this way by A. I. Lur'e [15, 17] who, however, was not able to give the necessary criterion for simplifying his formulae. This criterion was given in the papers [30, 28] where the errors introduced by Kirchhoff into the theory of thin shells were elucidated and where it was proved that these errors are of order δ/R in comparison with unity. In this way it was established that the above mentioned assumptions are sufficient for the construction of the theory of thin shells.

Valuable additions to the general theory of shells are given in the works of A. L. Gol'denweizer [6, 5] in which, for the first time, are formulated the conditions for the compatibility of strain. In addition, Gol'denweizer showed there the possibility of identically satisfying the equations of equilibrium by means of four stress functions, his result being analogous to that obtained by Lur'e [17]. In the work [32] which forms the basis of this book the present writer showed that it was possible to represent the equations of the theory of shells in complex form. In this way the equations assumed a more compact form, simplifying their study and, in particular, the understanding of their transformations and of their solution. These transformations are essentially a generalisation of the transformation discovered by E. Meissner [154]. An extension of the work in [32] is given in the article by Kh. M. Mushtari [24].

A statement should still be made about the work of the Russian School devoted to the simplification of the equations of shell theory, applicable to a definite range of problems, for example, the calculation of edge effects and the study of shallow shells. Mention should be made of the works of Kh. M. Mushtari [20, 21], S. Feinberg [36], V. Z. Vlasov [3, 4] and U. N. Rabotnov [34]. Particular attention has in recent times been given to the theory of shallow shells by V. Z. Vlasov [4] and a number of his pupils [80, 81], etc.

A classification of solutions of problems in the theory of shells has been given by Gol'denweizer [8, 9] and Mushtari [25] in whose publications may be found the basis of a general principle for simplifications of the equations of the theory of shells, admissible in various concrete cases.

This summarises the contribution of Russian work to the development of the general theory of shells.

Side by side with the development of a theory of shells based on an analogy with Kirchhoff's theory of plates, attempts have been made to construct more complicated shell theories with correspondingly higher degrees of accuracy. The earliest such attempt is due to A. Basset [38] whose work is less known than A. E. Love's [48], although it is in many respects not less interesting. In particular, Basset was the first to pay attention to the fact that in shell theory the error inherent in the Kirchhoff hypothesis is, generally speaking, more essential than in plate theory. Basset's idea has been developed further in [30].

The work of F. Krauss [47] and N. Kil'chevskii [12] is concerned with a shell theory which is analogous to the plate theory due to Cauchy and Poisson (cf. above).

Recently, in the U.S.A., much attention has been given to a shell theory developed by E. Reissner [37]. This is the simplest of all theories which are not based on the Kirchhoff hypothesis [37, 17a].

Since only problems of thin shells are to be considered, this book will be based on the two fundamental hypotheses a) and b), given earlier. Adoption of these hypotheses is equivalent to a reduction of the problem of the deformation of a shell to the study of the deformation of the middle surface in the same way as the introduction of the hypothesis regarding plane sections in the bending of a beam reduces that problem to a study of the bending of the neutral axis.

Thus, the above hypotheses are essentially an extension of the ideas underlying the elementary theory of beams. It would appear that a simpler approach to the study of shells is impossible, if one has in mind a theory which is to apply to all problems; on the basis of the above assumption regarding normals remaining normals and the simplifications with their inherent errors, one is thus justified in calling the theory of this book the technical theory of shells.

§ 2. Some fundamental results from the theory of surfaces

In this chapter, a summary of certain fundamental relations from the theory of surfaces will be given. It will be assumed that the reader has some knowledge of differential geometry and vector analysis.

Every smooth surface may be determined in cartesian coordinates by three equations:

$$x = f_1(\alpha_1, \alpha_2),$$
$$y = f_2(\alpha_1, \alpha_2), \tag{2.1}$$
$$z = f_3(\alpha_1, \alpha_2),$$

in which f_1, f_2, f_3 are definite, continuous, single-valued functions of the two variable parameters α_1, α_2.

The ranges of these parameters will be restricted, so that every point of the surface corresponds to one and only one pair of values of α_1, α_2.

Then these parameters may be called curvilinear coordinates of the surface under consideration.

Letting the parameter α_2 in (2.1) take successively different constant values and varying α_1, one obtains on the surface a family of curves which will be named the coordinate lines α_1. In a similar manner, giving α_1 different constant values and letting α_2 vary, a second family of curves, the coordinate lines α_2 are found. In view of the above limitations on the range of the variables, only one curve of each of these two families of curves will pass through each point of the surface. Hence, every point of the surface can be considered as the intersection of two coordinate lines α_1 and α_2 (see Fig. 1).

For each surface there exist an infinite number of possible choices for the functions f_1, f_2, f_3 and depending on their choice the form of the coordinate lines on the surface will change. However, not all of these possible alternatives will be equally convenient. As a rule, the most expedient of these alternatives will be that curvilinear system of coordinates in which the formulae for the surface attain their simplest form. Such a system is that in which the two families of coordinate curves are

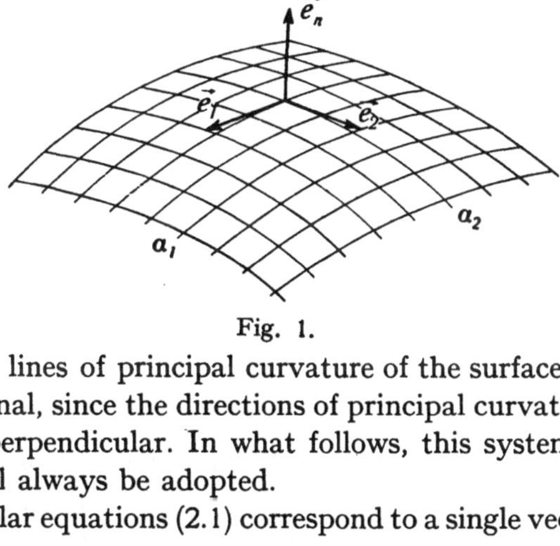

Fig. 1.

simultaneously lines of principal curvature of the surface. Such a system will be orthogonal, since the directions of principal curvature on a surface are mutually perpendicular. In what follows, this system of curvilinear coordinates will always be adopted.

The three scalar equations (2.1) correspond to a single vector equation

$$\mathbf{r} = \mathbf{r}(\alpha_1, \alpha_2) \tag{2.2}$$

the projections of which on the axes x, y, z are given by (2.1).

Consider the derivatives of $\mathbf{r}$ with respect to α_1 and α_2

$$\begin{aligned} \mathbf{r}_{\alpha_1} &= \frac{\partial \mathbf{r}}{\partial \alpha_1}, \\ \mathbf{r}_{\alpha_2} &= \frac{\partial \mathbf{r}}{\partial \alpha_2}, \end{aligned} \tag{2.3}$$

where the vectors $\mathbf{r}_{\alpha_1}$, $\mathbf{r}_{\alpha_2}$ will at each point of the surface be tangential to the α_1, α_2 lines respectively. As a result, it is seen that an increase of one of the curvilinear coordinates corresponds to a shift of the vector $\mathbf{r}$ on the surface in the direction corresponding to this coordinate line. Hence the vector

$$d\mathbf{s} = \frac{\partial \mathbf{r}}{\partial \alpha_1} d\alpha_1 + \frac{\partial \mathbf{r}}{\partial \alpha_2} d\alpha_2 \tag{2.4}$$

is equal in magnitude and direction to the segment, joining the points (α_1, α_2) and $(\alpha_1 + d\alpha_1, \alpha_2 + d\alpha_2)$ of the surface.

The square of the length of this segment is given by

$$|d\mathbf{s}|^2 = |\mathbf{r}_{\alpha_1}|^2 d\alpha_1^2 + 2(\mathbf{r}_{\alpha_1}, \mathbf{r}_{\alpha_2}) d\alpha_1 d\alpha_2 + |\mathbf{r}_{\alpha_2}|^2 d\alpha_2^2, \tag{2.5}$$

where the symbol $|\ |^2$ indicates the scalar product of the vector with itself, which is equal to the square of the length of the vector. Similarly, the symbol $(\ ,\ )$ represents the scalar product of two different vectors.

Introduce now the notation

$$\begin{aligned} |\mathbf{r}_{\alpha_1}|^2 = A_1^2 = \left(\frac{\partial x}{\partial \alpha_1}\right)^2 + \left(\frac{\partial y}{\partial \alpha_1}\right)^2 + \left(\frac{\partial z}{\partial \alpha_1}\right)^2, \\ |\mathbf{r}_{\alpha_2}|^2 = A_2^2 = \left(\frac{\partial x}{\partial \alpha_2}\right)^2 + \left(\frac{\partial y}{\partial \alpha_2}\right)^2 + \left(\frac{\partial z}{\partial \alpha_2}\right)^2. \end{aligned} \tag{2.6}$$

On the basis of the assumed orthogonality of the curvilinear coordinates, one finds then

$$(\mathbf{r}_{\alpha_1}, \mathbf{r}_{\alpha_2}) = 0. \tag{2.7}$$

Hence the formula (2.5) may be written in the form

$$ds^2 = A_1^2 d\alpha_1^2 + A_2^2 d\alpha_2^2. \tag{2.8}$$

Thus, in the particular cases when only one curvilinear coordinate is varied, one has

$$\begin{aligned} ds_1 &= A_1 d\alpha_1, \\ ds_2 &= A_2 d\alpha_2, \end{aligned} \tag{2.9}$$

where ds_1, ds_2 are the increases of the arc length along the coordinate lines α_1, α_2, corresponding to the increase of the curvilinear coordinates by $d\alpha_1$, $d\alpha_2$ respectively. The quantities A_1, A_2 are called the Lamé parameters.

In future, all vectors which are functions of the points of the surface will be given in terms of their projections on the directions of the tangents to the coordinate lines α_1 and α_2 and on the normal to the surface at the point considered. Since these reference axes change their direction for a passage from one point of the surface to another, it is necessary to investigate the differentiation of these vectors with respect to α_1, α_2.

In view of the fact that every vector, referred to a given system, can be written in the form

$$\mathbf{T} = T_{\alpha_1}\mathbf{e}_1 + T_{\alpha_2}\mathbf{e}_2 + T_n\mathbf{e}_n, \tag{2.10}$$

where $T_{\alpha_1}, T_{\alpha_2}, T_n$ are scalars (projections) and $\mathbf{e}_1, \mathbf{e}_2, \mathbf{e}_n$ the orthogonal unit vectors

$$\begin{aligned} \mathbf{e}_1 &= \frac{1}{A_1}\mathbf{r}_{\alpha_1}, \\ \mathbf{e}_2 &= \frac{1}{A_2}\mathbf{r}_{\alpha_2}, \\ \mathbf{e}_n &= [\mathbf{e}_1.\mathbf{e}_2], \end{aligned} \tag{2.11}$$

the study of the differentiation rule for vectors reduces to that for the unit vectors (2.11).

In the last formula of (2.11), the symbol [.] denotes the vector product. In future, the system $\mathbf{e}_1$, $\mathbf{e}_2$, $\mathbf{e}_n$ will be assumed to be right-handed. In addition, it will be assumed that $\mathbf{e}_n$ is directed to the convex side of the surface (or to the side of the centres of negative curvature, if the signs of the radii of curvature are not equal). These conditions do not contradict each other, since the choice of the positive directions for α_1 and α_2 can be arranged according to convenience.

A start will now be made with the study of the rules for the differentiation of the unit vectors $\mathbf{e}_1$, $\mathbf{e}_2$, $\mathbf{e}_n$. Consider two neighbouring points on the coordinate line α_1 and construct at each of these points the normals $\mathbf{e}_n$ (Fig. 2). These normals intersect at the point O. The corresponding principal radius of curvature of the surface will be denoted by R_1.

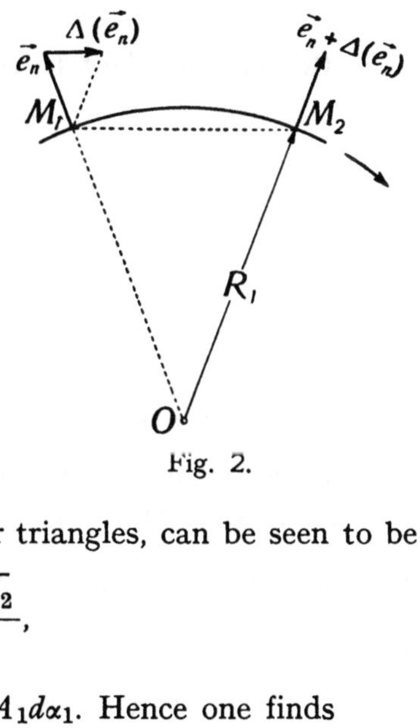

Fig. 2.

It is seen from Fig. 2 that the increase of the unit vector $\mathbf{e}_n$ for passage from the point M_1 to the infinitely near point M_2 is to sufficient accuracy parallel to the chord M_1M_2, i.e., in the limit it is parallel to the unit vector $\mathbf{e}_1$. The magnitude of this increase, on the basis of similar triangles, can be seen to be

$$\frac{|\Delta(\mathbf{e}_n)|}{|\mathbf{e}_n|} = \frac{\overline{M_1M_2}}{R_1},$$

where the length $|\mathbf{e}_n| = 1$ and $\overline{M_1M_2} = A_1 d\alpha_1$. Hence one finds

$$\frac{\partial \mathbf{e}_n}{\partial \alpha_1} = \frac{A_1}{R_1}\mathbf{e}_1. \tag{2.12}$$

In a similar manner, one may calculate the derivative of $\mathbf{e}_n$ with respect to α_2

$$\frac{\partial \mathbf{e}_n}{\partial \alpha_2} = \frac{A_2}{R_2} \mathbf{e}_2, \tag{2.13}$$

where R_2 denotes the second principal radius of curvature of the surface at the point considered.

Note that in deducing the formulae (2.12), (2.13) it had to be assumed that for the passage from the point M_1 to the infinitely near point M_2 the unit vector $\mathbf{e}_n$ changes in such a way that it remains to second order of smallness coplanar and rotates about the corresponding centre of curvature of the surface. This is one of the fundamental properties of lines of principal curvature. If the given unit vector had been shifted along any other line on the surface, it would be impossible to assume that it remains coplanar. The proof of the above stated property of the lines of curvature is known in the theory of surfaces as the 'Theorem of Rodrigues'.

Consider now the evaluation of the derivatives of the unit vectors $\mathbf{e}_1$, $\mathbf{e}_2$. First it will be noted that

$$\frac{\partial \mathbf{r}_{\alpha_1}}{\partial \alpha_2} = \frac{\partial \mathbf{r}_{\alpha_2}}{\partial \alpha_1}, \tag{2.14}$$

since this is equivalent to the identity

$$\frac{\partial^2 \mathbf{r}}{\partial \alpha_2 \partial \alpha_1} = \frac{\partial^2 \mathbf{r}}{\partial \alpha_1 \partial \alpha_2}.$$

Hence

$$\frac{\partial A_1 \mathbf{e}_1}{\partial \alpha_2} = \frac{\partial A_2 \mathbf{e}_2}{\partial \alpha_1}. \tag{2.15}$$

Further, consider the derivative $\dfrac{\partial \mathbf{e}_1}{\partial \alpha_1}$ and study its components along the axes $\mathbf{e}_1$, $\mathbf{e}_2$, $\mathbf{e}_n$. In the direction $\mathbf{e}_1$, this derivative has obviously no projection, since the length of the vector $\mathbf{e}_1$ is constant and its change is perpendicular to $\mathbf{e}_1$. As regards the projection of $\dfrac{\partial \mathbf{e}_1}{\partial \alpha_1}$ on $\mathbf{e}_2$, it is given by

$$\left(\mathbf{e}_2, \frac{\partial \mathbf{e}_1}{\partial \alpha_1}\right) = \frac{\partial(\mathbf{e}_2, \mathbf{e}_1)}{\partial \alpha_1} - \left(\mathbf{e}_1, \frac{\partial \mathbf{e}_2}{\partial \alpha_1}\right) = -\left(\mathbf{e}_1, \frac{\partial \mathbf{e}_2}{\partial \alpha_1}\right), \tag{2.16}$$

since $(\mathbf{e}_1, \mathbf{e}_2) = 0$ (the lines α_1 and α_2 being orthogonal).

Substituting on the right-hand side of this equality for the derivative $\frac{\partial \mathbf{e}_2}{\partial \alpha_1}$ its value given by (2.15), i.e.,

$$\frac{\partial \mathbf{e}_2}{\partial \alpha_1} = \frac{1}{A_2} \frac{\partial A_1 \mathbf{e}_1}{\partial \alpha_2} - \frac{1}{A_2} \frac{\partial A_2}{\partial \alpha_1} \mathbf{e}_2, \tag{2.17}$$

one finds

$$\left(\mathbf{e}_2, \frac{\partial \mathbf{e}_1}{\partial \alpha_1}\right) = -\frac{1}{A_2}\left(\mathbf{e}_1, \frac{\partial A_1 \mathbf{e}_1}{\partial \alpha_2}\right) + \frac{1}{A_2} \frac{\partial A_2}{\partial \alpha_1} (\mathbf{e}_1, \mathbf{e}_2) =$$

$$= -\frac{A_1}{A_2}\left(\mathbf{e}_1, \frac{\partial \mathbf{e}_1}{\partial \alpha_2}\right) - \frac{1}{A_2} \frac{\partial A_1}{\partial \alpha_2} (\mathbf{e}_1, \mathbf{e}_1). \tag{2.18}$$

However, $(\mathbf{e}_1, \mathbf{e}_2) = 1$ and $\left(\mathbf{e}_1, \frac{\partial \mathbf{e}_1}{\partial \alpha_2}\right) = 0$, as has been indicated above, and therefore, finally,

$$\left(\mathbf{e}_2, \frac{\partial \mathbf{e}_1}{\partial \alpha_1}\right) = -\frac{1}{A_2} \frac{\partial A_1}{\partial \alpha_2}. \tag{2.19}$$

There still remains to evaluate the projection of $\frac{\partial \mathbf{e}_1}{\partial \alpha_1}$ on $\mathbf{e}_n$. This is given by

$$\left(\mathbf{e}_n, \frac{\partial \mathbf{e}_1}{\partial \alpha_1}\right) = \frac{\partial (\mathbf{e}_n, \mathbf{e}_1)}{\partial \alpha_1} - \left(\mathbf{e}_1, \frac{\partial \mathbf{e}_n}{\partial \alpha_1}\right) = -\left(\mathbf{e}_1, \frac{\partial \mathbf{e}_n}{\partial \alpha_1}\right), \tag{2.20}$$

since $(\mathbf{e}_1, \mathbf{e}_n) = 0$.

By use of (2.12), one has

$$\left(\mathbf{e}_1, \frac{\partial \mathbf{e}_n}{\partial \alpha_1}\right) = -\frac{A_1}{R_1} (\mathbf{e}_1, \mathbf{e}_1) = -\frac{A_1}{R_1}. \tag{2.21}$$

Thus all three components of the derivative $\frac{\partial \mathbf{e}_1}{\partial \alpha_1}$ have been found.

In a similar fashion, one may proceed with the determination of the second derivatives of the same unit vector, i.e., of $\frac{\partial \mathbf{e}_1}{\partial \alpha_2}$. On the basis of considerations similar to those above, its projection on the direction $\mathbf{e}_1$ will be equal to zero. Using (2.15), one obtains

$$\left(\mathbf{e}_2, \frac{\partial \mathbf{e}_1}{\partial \alpha_2}\right) = -\frac{1}{A_1} \frac{\partial A_1}{\partial \alpha_2} (\mathbf{e}_1, \mathbf{e}_2) + \frac{1}{A_1}\left(\mathbf{e}_2, \frac{\partial A_2 \mathbf{e}_2}{\partial \alpha_1}\right) = \frac{1}{A_1} \frac{\partial A_2}{\partial \alpha_1}, \tag{2.22}$$

and finally

$$\left(\mathbf{e}_n, \frac{\partial \mathbf{e}_1}{\partial \alpha_2}\right) = \frac{\partial(\mathbf{e}_n, \mathbf{e}_1)}{\partial \alpha_2} - \left(\mathbf{e}_1, \frac{\partial \mathbf{e}_n}{\partial \alpha_2}\right) = -\left(\mathbf{e}_1, \frac{\partial \mathbf{e}_n}{\partial \alpha_2}\right). \tag{2.23}$$

However, by (2.13),

$$\left(\mathbf{e}_1, \frac{\partial \mathbf{e}_n}{\partial \alpha_2}\right) = (\mathbf{e}_1, \mathbf{e}_2) \frac{A_2}{R_2} = 0, \tag{2.24}$$

and therefore the derivative $\dfrac{\partial \mathbf{e}_1}{\partial \alpha_2}$ has no projection on the normal.

Similarly, the projections of the derivatives $\dfrac{\partial \mathbf{e}_2}{\partial \alpha_1}$, $\dfrac{\partial \mathbf{e}_2}{\partial \alpha_2}$ can be calculated. The results, however, can be written by analogy with those above. One thus finds

$$\begin{aligned}
\frac{\partial \mathbf{e}_1}{\partial \alpha_1} &= -\frac{1}{A_2}\frac{\partial A_1}{\partial \alpha_2}\mathbf{e}_2 - \frac{A_1}{R_1}\mathbf{e}_n, \\
\frac{\partial \mathbf{e}_1}{\partial \alpha_2} &= +\frac{1}{A_1}\frac{\partial A_2}{\partial \alpha_1}\mathbf{e}_2, \\
\frac{\partial \mathbf{e}_2}{\partial \alpha_2} &= -\frac{1}{A_1}\frac{\partial A_2}{\partial \alpha_1}\mathbf{e}_1 - \frac{A_2}{R_2}\mathbf{e}_n, \\
\frac{\partial \mathbf{e}_2}{\partial \alpha_1} &= \frac{1}{A_2}\frac{\partial A_1}{\partial \alpha_2}\mathbf{e}_1, \\
\frac{\partial \mathbf{e}_n}{\partial \alpha_1} &= \frac{A_1}{R_1}\mathbf{e}_1, \\
\frac{\partial \mathbf{e}_n}{\partial \alpha_2} &= \frac{A_2}{R_2}\mathbf{e}_2.
\end{aligned} \tag{2.25}$$

In what follows, these formulae permit differentiation of vectors with given projections on to the "local" axes of coordinates $\mathbf{e}_1$, $\mathbf{e}_2$, $\mathbf{e}_n$ (i.e., on axes which depend on the position of the particular point on the surface). These equations lead immediately to important differential relations between the parameters A_1, A_2 and the principal radii of curvature R_1, R_2. These formulae can be obtained from the conditions that the order of differentiation of mixed second derivatives of vectors may be reversed.

For example, it follows from the identity

$$\frac{\partial^2 \mathbf{e}_n}{\partial \alpha_1 \partial \alpha_2} = \frac{\partial^2 \mathbf{e}_n}{\partial \alpha_2 \partial \alpha_1}$$

that

$$\frac{\partial}{\partial \alpha_2}\left(\frac{A_1}{R_1}\mathbf{e}_1\right) = \frac{\partial}{\partial \alpha_1}\left(\frac{A_2}{R_2}\mathbf{e}_2\right)$$

or

$$\frac{\partial}{\partial \alpha_2}\left(\frac{A_1}{R_1}\right)\mathbf{e}_1 + \frac{A_1}{R_1}\,\frac{\partial \mathbf{e}_1}{\partial \alpha_2} = \frac{\partial}{\partial \alpha_1}\left(\frac{A_2}{R_2}\right)\mathbf{e}_2 + \frac{A_2}{R_2}\,\frac{\partial \mathbf{e}_2}{\partial \alpha_1}. \quad (2.26)$$

Replacing $\frac{\partial \mathbf{e}_1}{\partial \alpha_2}$ and $\frac{\partial \mathbf{e}_2}{\partial \alpha_1}$ by their values (2.25) and transferring all terms to the left side, one finds

$$\left[\frac{\partial}{\partial \alpha_2}\left(\frac{A_1}{R_1}\right) - \frac{1}{R_2}\,\frac{\partial A_1}{\partial \alpha_2}\right]\mathbf{e}_1 - \left[\frac{\partial}{\partial \alpha_1}\left(\frac{A_2}{R_2}\right) - \frac{1}{R_1}\,\frac{\partial A_2}{\partial \alpha_1}\right]\mathbf{e}_2 = 0. \quad (2.27)$$

Since this vector equation must be satisfied identically, one has thus the two relations

$$\begin{aligned}\frac{\partial}{\partial \alpha_1}\left(\frac{A_2}{R_2}\right) &= \frac{1}{R_1}\,\frac{\partial A_2}{\partial \alpha_1},\\ \frac{\partial}{\partial \alpha_2}\left(\frac{A_1}{R_1}\right) &= \frac{1}{R_2}\,\frac{\partial A_1}{\partial \alpha_2},\end{aligned} \quad (2.28)$$

which are called in the theory of surfaces the conditions of Codazzi. These conditions have been obtained here for the particular case when the parameters α_1, α_2 are principal coordinates of the surface.

Next consider the other identity

$$\frac{\partial^2 \mathbf{e}_1}{\partial \alpha_1 \partial \alpha_2} = \frac{\partial^2 \mathbf{e}_1}{\partial \alpha_2 \partial \alpha_1},$$

which, using (2.25), can be reduced to the form

$$\left[\frac{\partial}{\partial \alpha_1}\left(\frac{1}{A_1}\,\frac{\partial A_2}{\partial \alpha_1}\right) + \frac{\partial}{\partial \alpha_2}\left(\frac{1}{A_2}\,\frac{\partial A_1}{\partial \alpha_2}\right) + \frac{A_1 A_2}{R_1 R_2}\right]\mathbf{e}_2 + \left[\frac{\partial}{\partial \alpha_2}\left(\frac{A_1}{R_1}\right) - \frac{1}{R_2}\,\frac{\partial A_1}{\partial \alpha_2}\right]\mathbf{e}_n = 0. \quad (2.29)$$

Again, in order that this vector equation will be satisfied identically, each of the expressions in square brackets must vanish separately. This gives two further differential relations, only the first of which is new, while the second coincides with one of the relations (2.28) above. Equating the first square brackets above to zero, one obtains

$$\frac{\partial}{\partial \alpha_1}\left(\frac{1}{A_1}\frac{\partial A_2}{\partial \alpha_1}\right)+\frac{\partial}{\partial \alpha_2}\left(\frac{1}{A_2}\frac{\partial A_1}{\partial \alpha_2}\right)=-\frac{A_1A_2}{R_1R_2}. \tag{2.30}$$

In the theory of surfaces this relation is called the condition of Gauss.

There still remains to consider one further identity, namely

$$\frac{\partial^2 \mathbf{e}_2}{\partial \alpha_1 \partial \alpha_2}=\frac{\partial^2 \mathbf{e}_2}{\partial \alpha_2 \partial \alpha_1}.$$

It will be left to the reader to verify by a reasoning quite analogous to the above that no new differential relations between A_1, A_2, R_1, R_2 arise from this equality.

It follows from the above that the four functions A_1, A_2, R_1, R_2 of the two parameters α_1, α_2, if they are chosen arbitrarily, will not, in general, define a surface. They may be considered as Lamé parameters and principal radii of curvature of the surface only in the case that they satisfy the conditions of Gauss and Codazzi. If these conditions are satisfied, these parameters determine a definite surface except for its position in space.

In conclusion, attention will be drawn to the fact that the parameter

$$K=\frac{1}{R_1R_2} \tag{2.31}$$

is called the Gaussian curvature of a surface; points of a surface are subdivided into elliptic ($K > 0$), parabolic ($K = 0$) and hyperbolic ($K < 0$) points. A surface, all points of which are elliptic, will be called a surface of positive Gaussian curvature. If all points are parabolic, one has a surface of zero Gaussian curvature, while a surface all points of which are hyperbolic is said to have negative Gaussian curvature.

It is shown in the theory of surfaces that the properties of surfaces in bending (without change of length of an arbitrary curve drawn on the surface) are closely linked to the sign of their Gaussian curvature. The conditions under which a surface cannot be free from bending differ for surfaces of positive, zero and negative curvatures.

The formulae (2.25), (2.28), (2.30) of the theory of surfaces are sufficient for the development of the theory of shells and they will be frequently used in the following pages.

§ 3. The law of variation of displacements across the thickness of the shell

Let the position of a point on the middle surface of the shell be determined by the curvilinear coordinates α_1, α_2 which will be assumed to be the principal coordinates, so that the coordinates of the lines are simultaneously lines of principal curvature of the middle surface. Further, as the result of deformation, let the arbitrary point O on the middle surface undergo the displacement $\mathbf{\Delta}$. This vector will be a function of α_1, α_2, that is, a function of the particular point on the middle surface, and its knowledge is equivalent to determining the deformation of the middle surface. Let the projections of the displacement $\mathbf{\Delta}$ on the directions $\mathbf{e}_1$, $\mathbf{e}_2$, $\mathbf{e}_n$ at the point O of the middle surface be denoted by u, v, w. In other words, u will be the projection of the displacement of the middle surface on the direction of the tangent to the line α_1, v the projection of this displacement on the direction of the tangent to the line α_2, w the projection on the normal to the middle surface (at the point O).

At the point O erect the normal to the middle surface and consider a second point O_1 in the interior of the shell and located on this normal at a distance z from the middle surface (z will be assumed to be positive, if the segment OO_1 projects on the positive direction of $\mathbf{e}_n$).

The displacement of the point O_1 will be denoted by $\mathbf{\Delta}_z$ and its projections on $\mathbf{e}_1$, $\mathbf{e}_2$, $\mathbf{e}_n$ by $u_{(z)}$, $v_{(z)}$, $w_{(z)}$. The vector $\mathbf{\Delta}_{(z)}$ will be a function of α_1, α_2 as well as of the third coordinate z, since the displacement of the point O_1 must depend on that of the point O of the middle surface and on its distance from that point.

On the basis of the assumptions of § 1, the quantities $u_{(z)}$, $v_{(z)}$, $w_{(z)}$ will now be expressed in terms of the displacements of the middle surface u, v, w and z, using the following reasoning.

During the process of deformation, let the point O move into the position O' and the point O_1 into the position O_1' (Fig. 3). It follows from the first of the assumptions of § 1 that the segment $O'O_1'$ will be perpendicular to the deformed middle surface and that its length will be equal to OO_1 — namely z. Hence one has the vector equation

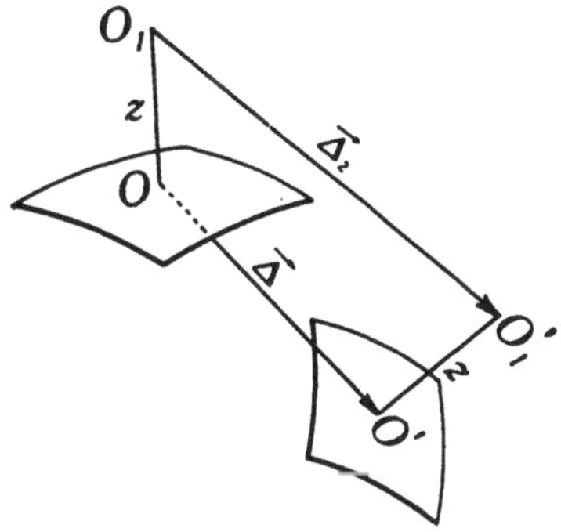

Fig. 3.

$$z\mathbf{e}_n + \mathbf{\Delta}_z = z\mathbf{e}_n' + \mathbf{\Delta}, \tag{3.1}$$

where $\mathbf{e}_n'$ is the unit vector normal to the deformed middle surface. This equation expresses the fact that the two vector sums on either side are equal to one and the same vector OO_1' (Fig. 3).

Solving (3.1) for $\mathbf{\Delta}_z$, it is found that

$$\mathbf{\Delta}_z = \mathbf{\Delta} + (\mathbf{e}_n' - \mathbf{e}_n)z, \tag{3.2}$$

where the unit vector $\mathbf{e}_n'$, neglecting terms involving the product of displacements and their derivatives, may be determined from the relation

$$\mathbf{e}_n' = [\mathbf{e}_1'.\mathbf{e}_2']. \tag{3.3}$$

Here $\mathbf{e}_1'$ and $\mathbf{e}_2'$ are the unit vectors tangential to the lines α_1 and α_2 on the deformed middle surface.

Using vector notation, the equation of the deformed middle surface has then the form

$$\mathbf{R} = \mathbf{r} + \boldsymbol{\Delta} = \mathbf{r} + u\mathbf{e}_1 + v\mathbf{e}_2 + w\mathbf{e}_n. \tag{3.4}$$

This formula expresses the fact that the vector $\mathbf{R}$ to the point α_1, α_2 of the deformed middle surface equals the sum of the vector $\mathbf{r}$, describing the middle surface before deformation, and the displacement vector $\boldsymbol{\Delta}$.

Consider the vector

(3.5)

and introduce instead of the derivatives $\partial\mathbf{e}_1/\partial\alpha_1$, $\partial\mathbf{e}_2/\partial\alpha_2$, $\partial\mathbf{e}_n/\partial\alpha_3$ their values from (2.25). Then

$$\frac{1}{A_1} = \frac{\partial\mathbf{R}}{\partial\alpha_1} = (1 + \varepsilon_1)\mathbf{e}_1 + \omega_1\mathbf{e}_2 - \vartheta\mathbf{e}_n, \tag{3.6'}$$

where

$$\begin{aligned}
\varepsilon_1 &= \frac{1}{A_1}\frac{\partial u}{\partial\alpha_1} + \frac{1}{A_1A_2}\frac{\partial A_1}{\partial\alpha_2}v + \frac{w}{R_1},\\
\omega_1 &= \frac{1}{A_1}\frac{\partial v}{\partial\alpha_1} - \frac{1}{A_1A_2}\frac{\partial A_1}{\partial\alpha_2}u,\\
\vartheta &= -\frac{1}{A_1}\frac{\partial w}{\partial\alpha_1} + \frac{u}{R}.
\end{aligned} \tag{3.7'}$$

Similarly, one has

$$\frac{1}{A_2}\frac{\partial\mathbf{R}}{\partial\alpha_2} = \omega_2\mathbf{e}_1 + (1 + \varepsilon_2)\mathbf{e}_2 - \psi\mathbf{e}_n, \tag{3.6''}$$

where

$$\begin{aligned}
\varepsilon_2 &= \frac{1}{A_2}\frac{\partial v}{\partial\alpha_2} + \frac{1}{A_1A_2}\frac{\partial A_2}{\partial\alpha_1}u + \frac{w}{R_2},\\
\omega_2 &= \frac{1}{A_2}\frac{\partial u}{\partial\alpha_2} - \frac{1}{A_1A_2}\frac{\partial A_2}{\partial\alpha_1}v,\\
\psi &= -\frac{1}{A_2}\frac{\partial w}{\partial\alpha_2} - \frac{v}{R_2}.
\end{aligned} \tag{3.7''}$$

On the basis of (3.6), one may determine also the unit vectors tangential to the coordinate lines α_1, α_2 on the deformed middle surface

$$\mathbf{e}_1' = \frac{1}{A_1'}\frac{\partial \mathbf{R}}{\partial \alpha_1}, \qquad \mathbf{e}_2' = \frac{1}{A_2'}\frac{\partial \mathbf{R}}{\partial \alpha_2}, \tag{3.8}$$

where A_1', A_2', the lengths of the vectors $\partial\mathbf{R}/\partial\alpha_1$, $\partial\mathbf{R}/\partial\alpha_2$, are given by

$$A_1' = \sqrt{\left(\frac{\partial \mathbf{R}}{\partial \alpha_1}, \frac{\partial \mathbf{R}}{\partial \alpha_1}\right)}, \quad A_2' = \sqrt{\left(\frac{\partial \mathbf{R}}{\partial \alpha_2}, \frac{\partial \mathbf{R}}{\partial \alpha_2}\right)}. \tag{3.9}$$

Substituting here from (3.6), one obtains expressions for A_1', A_2' in terms of A_1, A_2, R_1, R_2 and the displacements of the middle surface. Here and in what follows, products of the displacements and their partial derivatives will be neglected as second order quantities, because, as has been stated earlier, one is here concerned only with a linear shell theory.

Enforcing this degree of accuracy, one finds

$$A_1' \approx A_1(1 + \varepsilon_1), \quad A_2' \approx A_2(1 + \varepsilon_2), \tag{3.10}$$

and hence

$$\mathbf{e}_1' = \frac{1}{A_1'}\frac{\partial \mathbf{R}}{\partial \alpha_1} \approx \mathbf{e}_1 + \omega_1\mathbf{e}_2 - \vartheta\mathbf{e}_n, \tag{3.11}$$

$$\mathbf{e}_2' = \frac{1}{A_2'}\frac{\partial \mathbf{R}}{\partial \alpha_2} \approx \omega_2\mathbf{e}_1 + \mathbf{e}_2 - \psi\mathbf{e}_n. \tag{3.12}$$

The last formulae are obtained by dividing (3.6′), (3.6″) by $1 + \varepsilon_1$, $1 + \varepsilon_2$, respectively, and neglecting terms of second and smaller order of magnitude.

One can now express also the unit normal vector to the middle surface in terms of its displacement by substituting (3.11), (3.12) into (3.3). For this purpose, it will be recalled that the rule for the vector product of an orthogonal system of unit vectors is

$$\begin{aligned} [\mathbf{e}_1, \mathbf{e}_2] &= \mathbf{e}_n, \\ [\mathbf{e}_2, \mathbf{e}_n] &= -[\mathbf{e}_n, \mathbf{e}_2] = \mathbf{e}_1, \\ [\mathbf{e}_n, \mathbf{e}_1] &= -[\mathbf{e}_1, \mathbf{e}_n] = \mathbf{e}_2. \end{aligned} \tag{3.13}$$

Substituting (3.11), (3.12) into (3.3) and neglecting second order terms, one obtains

$$\mathbf{e}_n' \approx [\mathbf{e}_1', \mathbf{e}_2'] = [\mathbf{e}_1, \mathbf{e}_2] + [\mathbf{e}_1, (\omega_2\mathbf{e}_1 - \psi\mathbf{e}_n)] + [(\omega_1\mathbf{e}_2 - \vartheta\mathbf{e}_n), \mathbf{e}_2]. \tag{3.14}$$

However, the vector product of a vector by itself vanishes. Hence, using (3.13), one finds

$$\mathbf{e}_n' = \mathbf{e}_n + \vartheta\mathbf{e}_1 + \psi\mathbf{e}_2. \tag{3.15}$$

The formulae (3.11), (3.12) and (3.15) express $\mathbf{e}_i'$ in terms of $\mathbf{e}_i$.

The parameters ω_1, ω_2, ϑ, ψ have simple geometric interpretations which will be explained by means of the vector products

$$[\mathbf{e}_1, \mathbf{e}_1'] = \omega_1\mathbf{e}_n + \vartheta\mathbf{e}_2, \quad [\mathbf{e}_2', \mathbf{e}_2] = \omega_2\mathbf{e}_n + \psi\mathbf{e}_1. \tag{3.16}$$

The magnitude of these products is equal to the sines of the angles by which $\mathbf{e}_1$ and $\mathbf{e}_2$ rotate as a result of the deformation of the middle surface. By (3.7), these sines are small first order quantities and can be identified with the angles of rotation themselves, so that the rotation of $\mathbf{e}_1$ consists of the angle ω_1 about $\mathbf{e}_n$ and the angle ϑ about $\mathbf{e}_2$. Correspondingly, the rotation of $\mathbf{e}_2$ comprises ω_2 about $\mathbf{e}_n$ and ψ about $\mathbf{e}_1$.

The quantities ϑ and ψ have a simple geometrical interpretation. As can be seen by inspection of the expression (3.15), ϑ and ψ are the projections of the vector $\mathbf{e}_n'$, i.e., of the vector normal to the deformed middle surface on the directions $\mathbf{e}_1$ and $\mathbf{e}_2$. Since $\mathbf{e}_n'$ is of unit length and small displacements are assumed, these projections are equal to the angles by which during deformation the normal to the middle surface rotates about the axes $\mathbf{e}_2$ and $\mathbf{e}_1$ respectively. One may likewise say that ϑ and ψ are the angles by which during deformation the tangents to the lines α_1 and α_2 of the middle surface rotate about the axes $\mathbf{e}_2$ and $\mathbf{e}_1$ respectively, since these angles will be equal to the above-mentioned angles of rotation of the normal.

Returning now to (3.2) and replacing in it $\mathbf{e}_n'$ by the expression (3.15), one finds

$$\mathbf{\Delta}_{(z)} = \mathbf{\Delta} + z(\vartheta\mathbf{e}_1 + \psi\mathbf{e}_2). \tag{3.17}$$

Projecting this vector equation on to the directions $\mathbf{e}_1$, $\mathbf{e}_2$, $\mathbf{e}_n$, one arrives at

$$\begin{aligned} u_{(z)} &= u + z\vartheta, \\ v_{(z)} &= v + z\psi, \\ w_{(z)} &= w. \end{aligned} \tag{3.18}$$

These last formulae express the components of displacement of an arbitrary point of the shell in terms of the displacement of its corresponding point on the middle surface.

Thus, in accordance with the initial assumptions, the law of variation of the displacements through the thickness of the shell is linear and the normal displacement $w_{(z)}$ is independent of z.

§ 4. The deformation of a shell and the deformation of its middle surface

Consider a point O on the middle surface of a shell before deformation occurs and next to it a point lying on the line α_1 passing through the point O. Before and after deformation the distance between these points will be given by

(a) $$ds_1 = A_1 d\alpha_1, \tag{4.1}$$

and

(b) $$ds_1' = A_1' d\alpha_1 \tag{4.2}$$

respectively, where A_1' is the Lamé parameter for the deformed middle surface. In the preceding section, this parameter has been determined to first order accuracy by the formulae (3.9, 3.10).

The relative increase of a linear element of the middle surface, having direction α_1, is determined by

$$\frac{ds_1' - ds_1}{ds_1}.$$

Using (4.1), (4.2) and (3.9), it is found that this relative increase is given by

$$\varepsilon_1 = \frac{1}{A_1}\frac{\partial u}{\partial \alpha_1} + \frac{1}{A_1 A_2}\frac{\partial A_1}{\partial \alpha_2} v + \frac{w}{R_1}. \tag{4.3}$$

The formula for the corresponding increase in the direction α_2 may now be written by analogy:

$$\varepsilon_2 = \frac{1}{A_2}\frac{\partial v}{\partial \alpha_2} + \frac{1}{A_1 A_2}\frac{\partial A_2}{\partial \alpha_1} u + \frac{w}{R_2}. \tag{4.4}$$

One further quantity characterizing the deformation of the middle surface has still to be determined, namely, the shear strain defined as the cosine of the angle between the lines α_1, α_2 after deformation. It is clear from this definition that the shear strain will be assumed to be positive, if the angle between the lines α_1, α_2 (a right angle prior to deformation) decreases as the result of the deformation. The cosine of the above angle will be equal to the scalar product of the unit vectors $\mathbf{e}_1'$, $\mathbf{e}_2'$

$$\omega = (\mathbf{e}_1', \mathbf{e}_2'). \tag{4.5}$$

Substituting here the values of the vectors as given by (3.11) and (3.12) and taking into consideration that the mixed products of the unit vectors, i.e., $(\mathbf{e}_1, \mathbf{e}_2)$, $(\mathbf{e}_2, \mathbf{e}_n)$, $(\mathbf{e}_1, \mathbf{e}_n)$ are zero, it is found that

$$\omega = \left(\frac{1}{A_1}\frac{\partial v}{\partial \alpha_1} - \frac{1}{A_1 A_2}\frac{\partial A_1}{\partial \alpha_2} u\right)(\mathbf{e}_2, \mathbf{e}_2) + \left(\frac{1}{A_2}\frac{\partial u}{\partial \alpha_2} - \frac{1}{A_1 A_2}\frac{\partial A_2}{\partial \alpha_1} v\right)(\mathbf{e}_1, \mathbf{e}_1) =$$

$$= \frac{A_2}{A_1} \frac{\partial}{\partial \alpha_1}\left(\frac{v}{A_2}\right) + \frac{A_1}{A_2} \frac{\partial}{\partial \alpha_2}\left(\frac{u}{A_1}\right) = \omega_1 + \omega_2 \tag{4.6}$$

Now let there be drawn inside the shell a surface lying at a distance z from the middle surface (and, in future, to be referred to as a parallel surface) and consider any point on the middle surface and the pair of coordinate lines passing through this point. As the normal to the middle surface passes along these lines, two lines α_1 and α_2 are generated on the parallel surface.

At the point of intersection of these lines consider the triad of unit vectors $\mathbf{e}_1'$, $\mathbf{e}_2'$, $\mathbf{e}_n'$ which are directed along the tangents to the α_1 and α_2 lines and along the normal to the parallel surface respectively. In view of the method of construction of this last surface, $\mathbf{e}_1'$, $\mathbf{e}_2'$ will be parallel to $\mathbf{e}_1$, $\mathbf{e}_2$, and $\mathbf{e}_n'$ parallel to the same straight line as $\mathbf{e}_n$. Hence it is clear that the net of lines α_1, α_2 on the parallel surface is orthogonal and that the normal to the middle surface is at the same time also normal to the parallel surface. Moreover, the lines α_1, α_2 on the parallel surface will themselves be lines of principal curvature, since for infinitely small displacements along either of these lines the unit vector $\mathbf{e}_n'$, coinciding in direction with $\mathbf{e}_n$, will remain coplanar (see § 2).

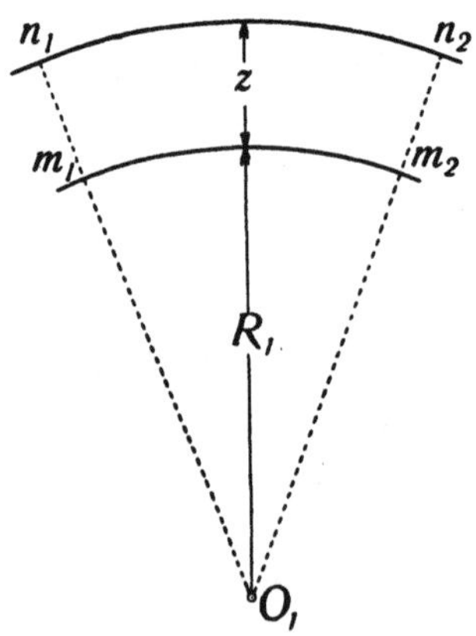

Fig. 4.

At two adjacent points of the middle surface, lying on the line α_1, construct the unit normal vectors. Fig. 4 shows the cross-section of the middle surface, defined by the directions of the above-mentioned two normal vectors, where the points m_1, m_2 belong to the middle surface and n_1, n_2 are the corresponding points of the parallel surface. In this figure, O_1 is the centre of curvature of the middle surface for the cross-section drawn in the manner stated above.

If the radius of curvature of this cross-section of the middle surface is equal to R_1, then it is clear from Fig. 4 that the corresponding radius of curvature of the parallel surface will be given by

$$R_1^{(z)} = R_1 + z. \tag{4.7}$$

Further, if the length of the arc m_1m_2 of the middle surface is equal to $ds_1 = A_1 d\alpha_1$, then the length of the arc n_1n_2 of the parallel surface will be given by

$$ds_1^{(z)} = A_1\left(1 + \frac{z}{R_1}\right) d\alpha_1;$$

this follows from the similarity of the sectors $m_1m_2O_1$ and $n_1n_2O_1$ of Fig. 4.

It follows from the above discussion that the principal radii of curvature of the parallel surface and its Lamé parameters will be given by

$$\begin{aligned} A_1^{(z)} &= A_1(1 + z/R_1), \quad & R_1^{(z)} &= R_1 + z, \\ A_2^{(z)} &= A_2(1 + z/R_2), \quad & R_2^{(z)} &= R_2 + z. \end{aligned} \tag{4.8}$$

On the basis of the simplifying assumptions, when the points of the middle surface undergo displacements u, v, w, the corresponding points of the parallel surface receive the displacements $u_{(z)}$, $v_{(z)}$, $w_{(z)}$, determined by (3.18). Further, the radii of curvature and the Lamé parameters of the parallel surface are known. Therefore, the change in length and shear strain of this last surface may now be determined by using the formulae (4.3), (4.4) and (4.6) and replacing in them u, v, w by $u_{(z)}$, $v_{(z)}$, $w_{(z)}$ and A_1, A_2, R_1, R_2 by $A_1^{(z)}$, $A_2^{(z)}$, $R_1^{(z)}$, $R_2^{(z)}$. Performing these substitutions in (4.3) and using (4.8), one finds

$$\varepsilon_1^{(z)} = \frac{1}{A_1(1 + z/R_1)} \frac{\partial u_{(z)}}{\partial \alpha_1} + $$

$$+ \frac{1}{A_1A_2(1 + z/R_1)(1 + z/R_2)} \frac{\partial A_1(1 + z/R_1)}{\partial \alpha_2} v_{(z)} + \frac{w_{(z)}}{R_1 + z}. \tag{4.9}$$

However, by the second of the Codazzi conditions (2.28),

$$\frac{\partial A_1(1 + z/R_1)}{\partial \alpha_2} = \frac{\partial A_1}{\partial \alpha_2} + z \frac{\partial}{\partial \alpha_2}\left(\frac{A_1}{R_1}\right) = (1 + z/R_2)\frac{\partial A_1}{\partial \alpha_2},$$

as a result of which (4.9) may be rewritten in the form

$$\varepsilon_1^{(z)} = \frac{1}{1 + z/R_1}\left(\frac{1}{A_1}\frac{\partial u_{(z)}}{\partial \alpha_1} + \frac{1}{A_1A_2}\frac{\partial A_1}{\partial \alpha_2} v_{(z)} + \frac{w_{(z)}}{R_1}\right). \tag{4.10}$$

Substituting now the values of $u_{(z)}$, $v_{(z)}$, $w_{(z)}$ from (3.18), one finds

$$\varepsilon_1^{(z)} = \frac{1}{1 + z/R_1}(\varepsilon_1 + z\kappa_1), \tag{4.11}$$

where

$$\kappa_1 = \frac{1}{A_1}\frac{\partial \vartheta}{\partial \alpha_1} + \frac{1}{A_1A_2}\frac{\partial A_1}{\partial \alpha_2}\psi \tag{4.12}$$

and ε_1 is defined by (4.3). The parameter $\varepsilon_1^{(z)}$ is the change in length of the linear element of the parallel surface in the direction α_1.

In a similar manner, it can be shown that the change in length in the direction α_2 is given by

$$\varepsilon_2^{(z)} = \frac{1}{1 + z/R_2}(\varepsilon_2 + z\kappa_2), \tag{4.13}$$

where

$$\kappa_2 = \frac{1}{A_2}\frac{\partial \psi}{\partial \alpha_2} + \frac{1}{A_1 A_2}\frac{\partial A_2}{\partial \alpha_1}\vartheta. \tag{4.14}$$

Consider next the formula for the shear strain (4.6) for the purpose of obtaining from it the corresponding shear strain on the parallel surface. From (4.8) and the Codazzi condition (2.28), one finds

$$\omega^{(z)} = \frac{1}{1 + z/R_1}\left[\frac{1}{A_1}\frac{\partial v_{(z)}}{\partial \alpha_1} - \frac{1}{A_1 A_2}\frac{\partial A_1}{\partial \alpha_2}u_{(z)}\right] + $$

$$+ \frac{1}{1 + z/R_2}\left[\frac{1}{A_2}\frac{\partial u_{(z)}}{\partial \alpha_2} - \frac{1}{A_1 A_2}\frac{\partial A_2}{\partial \alpha_1}v_{(z)}\right], \tag{4.15}$$

or, after substituting for $u_{(z)}$, $v_{(z)}$ from (3.18),

$$\omega^{(z)} = \frac{1}{1 + z/R_1}(\omega_1 + z\tau_1) + \frac{1}{1 + z/R_2}(\omega_2 + z\tau_2), \tag{4.16}$$

where

$$\begin{aligned} \omega_1 &= \frac{1}{A_1}\frac{\partial v}{\partial \alpha_1} - \frac{1}{A_1 A_2}\frac{\partial A_1}{\partial \alpha_2}u, \quad \tau_1 = \frac{1}{A_1}\frac{\partial \psi}{\partial \alpha_1} - \frac{1}{A_1 A_2}\frac{\partial A_1}{\partial \alpha_2}\vartheta, \\ \omega_2 &= \frac{1}{A_2}\frac{\partial u}{\partial \alpha_2} - \frac{1}{A_1 A_2}\frac{\partial A_2}{\partial \alpha_1}v, \quad \tau_2 = \frac{1}{A_2}\frac{\partial \vartheta}{\partial \alpha_2} - \frac{1}{A_1 A_2}\frac{\partial A_2}{\partial \alpha_1}\psi. \end{aligned} \tag{4.17}$$

The formula (4.16) contains four separate functions of α_1, α_2, namely ω_1, ω_2, τ_1, τ_2. However, it is possible to reduce it to a form involving only two separate functions of the curvilinear coordinates of the middle surface. In order to show that this is possible, the existence of the identity

$$\tau_1 + \frac{\omega_2}{R_1} = \tau_2 + \frac{\omega_1}{R_2} \tag{4.18}$$

will be noted which is easily verified by substituting in it for τ_1, τ_2, ω_1, ω_2

their expressions in terms of u, v, w and by performing the differentiation by use of the Codazzi conditions.

Introducing now, in addition to $\omega = \omega_1 + \omega_2$, the new notation

$$\tau = \tau_1 + \frac{\omega_2}{R_1} = \tau_2 + \frac{\omega_1}{R_2} \tag{4.19}$$

and reducing the terms of (4.16) to a common denominator, one finds

$$\omega^{(z)} = \frac{1}{(1 + z/R_1)(1 + z/R_2)} \left[\omega + 2z\tau + \frac{z^2}{R_1 R_2} (\tau_1 R_1 + \tau_2 R_2) \right]. \tag{4.20}$$

However, by (4.19),

$$\tau_1 R_1 + \tau_2 R_2 = (R_1 + R_2)\tau - \omega, \tag{4.21}$$

and therefore, finally,

$$\omega^{(z)} = \frac{1}{(1+z/R_1)(1+z/R_2)} \left\{ \left(1 - \frac{z^2}{R_1 R_2}\right)\omega + 2\left[1 + \left(\frac{1}{R_1} + \frac{1}{R_2}\right)\frac{z}{2}\right] z\tau \right\}. \tag{4.22}$$

The formulae (4.11), (4.13) and (4.22) determine the change in length and in the shear of a surface lying at a distance z from the middle surface of the shell. With their help the study of the deformations of a shell is reduced to that of the deformation of the middle surface.

The formulae (4.11), (4.13) and (4.22) contain six different functions of transfer to the middle surface, namely:

$$\begin{aligned}
\varepsilon_1 &= \frac{1}{A_1} \frac{\partial u}{\partial \alpha_1} + \frac{1}{A_1 A_2} \frac{\partial A_1}{\partial \alpha_2} v + \frac{w}{R_1}, \\
\varepsilon_2 &= \frac{1}{A_2} \frac{\partial v}{\partial \alpha_2} + \frac{1}{A_1 A_2} \frac{\partial A_2}{\partial \alpha_1} u + \frac{w}{R_2}, \\
\omega &= \frac{A_2}{A_1} \frac{\partial}{\partial \alpha_1} \left(\frac{v}{A_2}\right) + \frac{A_1}{A_2} \frac{\partial}{\partial \alpha_2} \left(\frac{u}{A_1}\right), \\
\kappa_1 &= -\frac{1}{A_1} \frac{\partial}{\partial \alpha_1} \left(\frac{1}{A_1} \frac{\partial w}{\partial \alpha_1} - \frac{u}{R_1}\right) - \frac{1}{A_1 A_2} \frac{\partial A_1}{\partial \alpha_2} \left(\frac{1}{A_2} \frac{\partial w}{\partial \alpha_2} - \frac{v}{R_2}\right), \qquad (4.23) \\
\kappa_2 &= -\frac{1}{A_2} \frac{\partial}{\partial \alpha_2} \left(\frac{1}{A_2} \frac{\partial w}{\partial \alpha_2} - \frac{v}{R_2}\right) - \frac{1}{A_1 A_2} \frac{\partial A_2}{\partial \alpha_1} \left(\frac{1}{A_1} \frac{\partial w}{\partial \alpha_1} - \frac{u}{R_1}\right), \\
\tau &= -\frac{1}{A_1 A_2} \left(\frac{\partial^2 w}{\partial \alpha_1 \partial \alpha_2} - \frac{1}{A_1} \frac{\partial A_1}{\partial \alpha_2} \frac{\partial w}{\partial \alpha_1} - \frac{1}{A_2} \frac{\partial A_2}{\partial \alpha_1} \frac{\partial w}{\partial \alpha_2}\right) +
\end{aligned}$$

$$+\frac{1}{R_1}\left(\frac{1}{A_2}\frac{\partial u}{\partial \alpha_2}-\frac{1}{A_1A_2}\frac{\partial A_1}{\partial \alpha_2}u\right)+\frac{1}{R_2}\left(\frac{1}{A_1}\frac{\partial v}{\partial \alpha_1}-\frac{1}{A_1A_2}\frac{\partial A_2}{\partial \alpha_1}v\right).$$

The geometrical meaning of ε_1, ε_2, ω has been elucidated earlier: they are the relative changes in length and in the shear of the middle surface. In order to interpret the three remaining parameters κ_1, κ_2, τ geometrically, consider a small element of the middle surface bounded by two pairs of adjacent lines α_1, α_2. The knowledge of the changes in length and of the shear at the point $O(\alpha_1, \alpha_2)$ completely characterizes the variation of the dimensions of the given element after deformation.

However, this information is insufficient for the determination of the shape of the element, since, as the result of deformation, the element may undergo bending as well as changes in length and shear and the above-mentioned three parameters do not account for this effect. The bending of the element of the surface may be characterized by changes of its curvatures in the directions α_1, α_2 and by a twist.

Consider the increment of the unit vector tangential to the line α_1 on the deformed middle surface, i.e., the vector $\mathbf{e}_1'$ in the following two cases:

a) as it moves along α_1 over the infinitesimal distance $ds_1' = A_1'd\alpha_1$,

b) as it moves along α_2 over the infinitesimal distance $ds_2' = A_2'd\alpha_2$.

The projection of $\Delta\mathbf{e}_1'$ on the normal to the deformed middle surface in the case a) will be equal to the angle of rotation of $\mathbf{e}_1'$ in the plane of $\mathbf{e}_1'$ and $\mathbf{e}_n'$, i.e., in the plane of the cross-section normal to the deformed surface and tangential to the line α_1, as it is displaced along α_1 by ds_1'. Thus, by the definition of curvature, one can write

$$\lim_{\Delta s_1'\to 0}\left(\frac{\Delta\mathbf{e}_1'}{\Delta s_1'}\,\mathbf{e}_n'\right)=\frac{1}{A_1'}\frac{\partial\mathbf{e}_1'}{\partial\alpha_1}\,\mathbf{e}_n'=-\frac{1}{R_1'},$$

where R_1' is the radius of curvature of the normal section of the deformed middle surface, tangential to the line α_1.

The minus sign in this formula is explained by the fact that the positive direction of the unit normal is in the direction of the centres of negative curvature of the surface (cf. above).

In the case b), the projection of $\Delta\mathbf{e}_1'$ on $\mathbf{e}_n'$ will be equal to the angle of rotation of $\mathbf{e}_1'$ about α_2 (the direction of $\mathbf{e}_2'$) as it moves along this line by a distance ds_2'. This angle can be called the angle of twist of the element of the middle surface about the direction $\mathbf{e}_2'$; hence the quantity τ_{α_2}', defined by

$$\tau_{\alpha_2}'=-\lim_{\Delta s_2'\to 0}\left(\frac{\Delta\mathbf{e}_1'}{\Delta s_2}\,\mathbf{e}_n'\right)=-\frac{1}{A_2'}\frac{\partial\mathbf{e}_1'}{\partial\alpha_2}\,\mathbf{e}_n'$$

can be called the twist of the deformed middle surface about the line α_2.

Reasoning analogously with respect to the vector $\mathbf{e}_1'$, one finds

$$\frac{1}{R_2'} = -\frac{1}{A_2'}\frac{\partial \mathbf{e}_2'}{\partial \alpha_2}\mathbf{e}_n', \quad \tau_{\alpha_1}' = -\frac{1}{A_1'}\frac{\partial \mathbf{e}_2'}{\partial \alpha_1}\mathbf{e}_n',$$

where R_2' is the radius of curvature of the normal section of the middle surface, tangential to the line α_2, and τ_{α_1}' is the twist of the deformed middle surface about the line α_1.

Using the formulae of the preceding two sections, the quantities $R_1', R_2', \tau_{\alpha_1}', \tau_{\alpha_2}'$ are readily expressed in terms of u, v, w. For example, by (3.11), (3.15), (2.25),

$$-\frac{1}{R_1'} = \frac{1}{A_1'}\frac{\partial \mathbf{e}_1'}{\partial \alpha_1}\mathbf{e}_n' = \frac{1}{A_1'}\left(\frac{\partial \mathbf{e}_1}{\partial \alpha_1} + \omega_1\frac{\partial \mathbf{e}_2}{\partial \alpha_1} - \vartheta\frac{\partial \mathbf{e}_n}{\partial \alpha_1} + \right.$$

$$\left. + \mathbf{e}_2\frac{\partial \omega_1}{\partial \alpha_1} - \mathbf{e}_n\frac{\partial \vartheta}{\partial \alpha_1}\right)(\mathbf{e}_n + \vartheta\mathbf{e}_1 + \psi\mathbf{e}_2) =$$

$$= \frac{1}{A_1'}\left[\left(\omega_1\frac{1}{A_2}\frac{\partial A_1}{\partial \alpha_2} - \frac{\vartheta A_1}{R_1}\right)\mathbf{e}_1 + \left(-\frac{1}{A_2}\frac{\partial A_1}{\partial \alpha_2} + \frac{\partial \omega_1}{\partial \alpha_1}\right)\mathbf{e}_2 + \right.$$

$$\left. + \left(-\frac{A_1}{R_1} - \frac{\partial \vartheta}{\partial \alpha_1}\right)\mathbf{e}_n\right](\mathbf{e}_n + \vartheta\mathbf{e}_1 + \psi\mathbf{e}_2).$$

Performing the scalar products and neglecting, as in all such manipulations, products of the displacements and their derivatives, one finds

$$-\frac{1}{R_1'} = -\frac{1}{R_1}(1 - \varepsilon_1) - \frac{1}{A_1}\frac{\partial \vartheta}{\partial \alpha_1} - \frac{1}{A_1 A_2}\frac{\partial A_1}{\partial \alpha_2}\psi = -\frac{1}{R_1}(1 - \varepsilon_1) - \kappa_1.$$

Hence, the increase of curvature of the middle surface of a shell subject to displacements u, v, w is given by

$$\kappa_1' = \frac{1}{R_1'} - \frac{1}{R_1} = \kappa_1 - \frac{\varepsilon_1}{R_1}.$$

Similarly, one finds

$$\kappa_2' = \frac{1}{R_2'} - \frac{1}{R_2} = \kappa_2 - \frac{\varepsilon_2}{R_2},$$

$$\tau_{\alpha_1}' = \tau_{\alpha_2}' = \tau_1 + \frac{\omega_2}{R_1} = \tau_2 + \frac{\omega_1}{R_2} = \tau.$$

It follows from the above that, firstly, the parameters κ_1, κ_2, τ characterize the changes of curvature and twist of the middle surface of a shell as it deforms. Secondly, the two twist parameters $\tau_{\alpha_1}, \tau_{\alpha_2}$ (about the lines α_1 and α_2) are equal to each other and coincide with the parameter τ introduced earlier.

In this context, attention should be drawn to the fact that the variations of the curvatures of the middle surface κ_1', κ_2' involve the parameters

κ_1, κ_2 which depend on the angles of rotation of the tangents to the lines α_1, α_2, caused by the deformation, explicitly as well as the strains of the middle surface ε_1, ε_2. This shows that the extension (or compression) of the middle surface is accompanied by twisting. However, this twist may be neglected in comparison with the twists caused by the rotation of the tangents to the lines, α_1, α_2 which are given by the parameters ϑ, ψ.

On the basis of these observations, one has the following results.

Expanding the denominators in (4.11) and (4.13) in powers of δ/R, one finds

$$\varepsilon_1^{(z)} = \varepsilon_1 + z\kappa_1' - \frac{z^2}{R}\kappa_1' + \ldots, \quad \varepsilon_2^{(z)} = \varepsilon_2 + z\kappa_2' - \frac{z^2}{R_2}\kappa_2' + \ldots \tag{4.24}$$

If the shell is sufficiently thin, one can already neglect the third terms, since the quantities concerned involve the small factor δ/R. Thus, one may write

$$\varepsilon_1^{(z)} \approx \varepsilon_1 + z\kappa_1' = \varepsilon_1\left(1 - \frac{z}{R_1}\right) + z\kappa_1,$$

$$\varepsilon_2^{(z)} \approx \varepsilon_2 + z\kappa_2' = \varepsilon_2\left(1 - \frac{z}{R_2}\right) + z\kappa_2. \tag{4.25}$$

Neglecting here z/R_1, z/R_2 in comparison with unity, one obtains now

$$\varepsilon_1^{(z)} \approx \varepsilon_1 + z\kappa_1, \quad \varepsilon_2^{(z)} \approx \varepsilon_2 + z\kappa_2. \tag{4.26}$$

However, the transition from (4.25) to (4.26) is obviously equivalent to writing

$$\kappa_1' \approx \kappa_1, \quad \kappa_2' \approx \kappa_2, \tag{4.27}$$

i.e., to the identification of the parameters κ_1, κ_2 with the changes of curvature of the middle surface of the shell during deformation.

Next, the following question will be asked. How is one to write the above formulae in those cases, where $\varepsilon_1^{(z)}$ and $\varepsilon_2^{(z)}$ vanish or are very small and when the distortion of the middle surface is entirely determined by terms in κ_1', κ_2' which depend on ε_1, ε_2? On the basis of (4.25), the answer to this question is that the distribution of the deformations $\varepsilon_1^{(z)}$, $\varepsilon_2^{(z)}$ over the thickness δ will be (exact to terms of order δ/R in comparison with unity) almost uniform, and hence one may assume that $\varepsilon_1^{(z)} \approx \varepsilon_1$, $\varepsilon_2^{(z)} \approx \varepsilon_2$, i.e., one may write $\kappa_1' = \kappa_2' = \kappa_1 = \kappa_2 = 0$.

As a consequence, one can call in future the parameters κ_1, κ_2 the parameters of the changes of curvature of the middle surface of a shell during deformation, or simply the parameters of the change of curvature (although, strictly speaking, this is not entirely correct).

It will be noted from above that the parameters κ_1, κ_2 are positive, if the curvature of the middle surface is increased by deformation.

The torsion τ will be positive, if it leads to positive shear strain in a parallel surface lying with respect to the middle surface on the side of the negative centres of curvature.

For the element of the middle surface, shown in Fig. 5, the torsion will be positive, if, as the sides PP_1 and QQ_1 rotate with respect to each other, the point Q descends, since under these circumstances the angle between the lines α_1, α_2 of the element of the parallel surface will decrease.

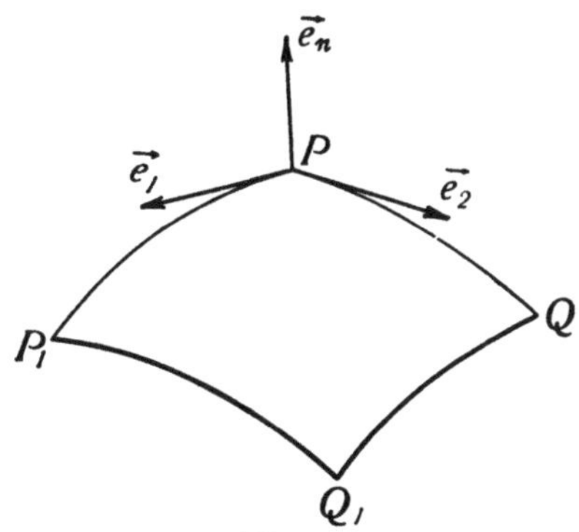

Fig. 5.

Thus one arrives at the conclusion that the deformation of the middle surface of a shell is completely determined by the six parameters ε_1, ε_2, ω, κ_1, κ_2, τ the first three of which characterize the variations of the dimensions of a small element of the middle surface, while the remaining three characterize its distortion.

Correspondingly, these six parameters, in terms of which the displacements of the middle surface are expressed by (4.23), will be called the deformations of the middle surface. If all these parameters are equal to zero, the surface is not deformed, i.e., either its displacements u, v, w are equal to zero or the surface is displaced as a rigid body.

It should be noted that it is not accidental that the number of parameters, describing the deformation of the middle surface, is equal to six. In the theory of surfaces, it may be proved that, in the general case when a surface is given in non-orthogonal curvilinear coordinates, its dimensions and shape are completely determined by six parameters (the coefficients of the first and second quadratic forms).

Hence, the results obtained above conform completely in as much as the coordinate lines α_1, α_2 on the deformed middle surface cease to be orthogonal.

§ 5. The conditions for the compatibility of strain in the middle surface

Consider the two vectors

$$\frac{\partial \mathbf{R}}{\partial \alpha_1} = A_1[(1 + \varepsilon_1)\mathbf{e}_1 + \omega_1 \mathbf{e}_2 - \vartheta \mathbf{e}_n],$$

$$\frac{\partial \mathbf{R}}{\partial \alpha_2} = A_2[\omega_2 \mathbf{e}_1 + (1 + \varepsilon_2)\mathbf{e}_2 - \psi \mathbf{e}_n] \qquad (5.1)$$

and differentiate the first with respect to α_2, the second with respect to α_1, using (2.25). Then one finds

$$\frac{\partial^2 \mathbf{R}}{\partial\alpha_2\partial\alpha_1} = \left(\frac{\partial A_1(1+\varepsilon_1)}{\partial\alpha_2} - \omega_1\frac{\partial A_2}{\partial\alpha_1}\right)\mathbf{e}_1 + \left[(1+\varepsilon_1)\frac{\partial A_2}{\partial\alpha_1} + \right.$$

$$\left. + \frac{\partial A_1\omega_1}{\partial\alpha_2} - \frac{A_1A_2}{R_2}\vartheta\right]\mathbf{e}_2 - \left(\frac{\partial A_1\vartheta}{\partial\alpha_2} + \frac{A_1A_2}{R_2}\omega_1\right)\mathbf{e}_n, \quad (5.2)$$

$$\frac{\partial^2 \mathbf{R}}{\partial\alpha_1\partial\alpha_2} = \left[(1+\varepsilon_2)\frac{\partial A_1}{\partial\alpha_2} + \frac{\partial A_2\omega_2}{\partial\alpha_1} - \frac{A_1A_2}{R_1}\psi\right]\mathbf{e}_1 +$$

$$+ \left[\frac{\partial A_2(1+\varepsilon_2)}{\partial\alpha_1} - \omega_2\frac{\partial A_1}{\partial\alpha_2}\right]\mathbf{e}_2 - \left(\frac{\partial A_2\psi}{\partial\alpha_1} + \frac{A_1A_2}{R_1}\omega_2\right)\mathbf{e}_n.$$

These vectors are equal, whence follow the three equalities

$$\frac{\partial A_1\varepsilon_1}{\partial\alpha_2} - \frac{\partial A_2\omega}{\partial\alpha_1} - \varepsilon_2\frac{\partial A_1}{\partial\alpha_2} = -A_2\left(\frac{\partial\omega_1}{\partial\alpha_1} + \frac{A_1}{R_1}\right)\psi,$$

$$\frac{\partial A_2\varepsilon_2}{\partial\alpha_1} - \frac{\partial A_1\omega}{\partial\alpha_2} - \varepsilon_1\frac{\partial A_2}{\partial\alpha_1} = -A_1\left(\frac{\partial\omega_2}{\partial\alpha_2} + \frac{A_2}{R_2}\right)\vartheta,$$

$$\frac{\partial A_1\vartheta}{\partial\alpha_2} - \frac{\partial A_2\psi}{\partial\alpha_1} = A_1A_2\left(\frac{\omega_2}{R_1} - \frac{\omega_1}{R_2}\right),$$

where $\omega = \omega_1 + \omega_2$.

Next, consider the vector $\mathbf{e}_n'$ of (3.14) and find its second derivatives $\partial^2\mathbf{e}_n'/\partial\alpha_1\partial\alpha_2$, $\partial^2\mathbf{e}_n'/\partial\alpha_2\partial\alpha_1$, using (4.15), (4.12), (4.14) and (2.25):

$$\frac{\partial^2 \mathbf{e}_n'}{\partial\alpha_2\partial\alpha_1} = \left[\frac{\partial}{\partial\alpha_2}\left(\frac{A_1}{R_1}\right) + \frac{\partial}{\partial\alpha_2}(A_1\kappa_1) - \tau_1\frac{\partial A_2}{\partial\alpha_1}\right]\mathbf{e}_1 +$$

$$+ \left(\frac{1}{R_1}\frac{\partial A_2}{\partial\alpha_1} + \kappa_1\frac{\partial A_2}{\partial\alpha_1} + \frac{\partial A_1\tau_1}{\partial\alpha_2} - \vartheta\frac{A_1A_2}{R_1R_2}\right)\mathbf{e}_2 -$$

$$- \left[\frac{A_1A_2}{R_2}\tau_1 + \frac{\partial}{\partial\alpha_2}\left(\frac{A_1}{R_1}\vartheta\right)\right]\mathbf{e}_n, \quad (5.4)$$

$$\frac{\partial^2 \mathbf{e}_n'}{\partial\alpha_1\partial\alpha_2} = \left(\frac{1}{R_2}\frac{\partial A_1}{\partial\alpha_2} + \kappa_2\frac{\partial A_1}{\partial\alpha_2} + \frac{\partial A_2\tau_2}{\partial\alpha_1} - \psi\frac{A_1A_2}{R_1R_2}\right)\mathbf{e}_1 +$$

$$+ \left[\frac{\partial}{\partial\alpha_1}\left(\frac{A_2}{R_2}\right) + \frac{\partial}{\partial\alpha_1}(A_2\kappa_2) - \tau_2\frac{\partial A_1}{\partial\alpha_2}\right]\mathbf{e}_2 -$$

$$- \left[\frac{A_1A_2}{R_1}\tau_2 + \frac{\partial}{\partial\alpha_1}\left(\frac{A_2}{R_2}\psi\right)\right]\mathbf{e}_n.$$

Equating these vectors and using for this purpose (2.28) and (4.18), one finds

$$\frac{\partial}{\partial\alpha_2}(A_1\varkappa_1) - \kappa_2\frac{\partial A_1}{\partial\alpha_2} - \frac{\partial A_2\tau}{\partial\alpha_1} - \tau\frac{\partial A_2}{\partial\alpha_1} + \frac{\omega}{R_1}\frac{\partial A_2}{\partial\alpha_1} = -\frac{A_2}{R_2}\left(\frac{\partial\omega_1}{\partial\alpha_1} + \frac{A_1}{R_1}\psi\right),$$

$$\frac{\partial}{\partial\alpha_1}(A_2\kappa_2) - \kappa_1\frac{\partial A_2}{\partial\alpha_1} - \frac{\partial A_1\tau}{\partial\alpha_2} - \tau\frac{\partial A_1}{\partial\alpha_2} + \frac{\omega}{R_2}\frac{\partial A_1}{\partial\alpha_2} = -\frac{A_1}{R_1}\left(\frac{\partial\omega_2}{\partial\alpha_2} + \frac{A_2}{R_2}\vartheta\right), \tag{5.5}$$

$$\frac{A_1A_2}{R_2}\tau_1 + \frac{\partial}{\partial\alpha_2}\left(\frac{A_1}{R_1}\vartheta\right) = \frac{A_1A_2}{R_1}\tau_2 + \frac{\partial}{\partial\alpha_1}\left(\frac{A_2}{R_2}\psi\right).$$

The right-hand sides of the first two equations coincide, apart from factors, with the right-hand sides of the equation (5.3). Hence one has

$$\frac{\partial}{\partial\alpha_2}(A_1\kappa_1) - \kappa_2\frac{\partial A_1}{\partial\alpha_2} - \frac{\partial A_2\tau}{\partial\alpha_1} - \tau\frac{\partial A_2}{\partial\alpha_1} + \frac{\omega}{R_1}\frac{\partial A_2}{\partial\alpha_1} - \frac{1}{R_2}\left(\frac{\partial A_1\varepsilon_1}{\partial\alpha_2} - \frac{\partial A_2\omega}{\partial\alpha_1} - \varepsilon_2\frac{\partial A_1}{\partial\alpha_2}\right) = 0, \tag{5.6'}$$

$$\frac{\partial}{\partial\alpha_1}(A_2\kappa_2) - \kappa_1\frac{\partial A_2}{\partial\alpha_1} - \frac{\partial A_1\tau}{\partial\alpha_2} - \tau\frac{\partial A_1}{\partial\alpha_2} + \frac{\omega}{R_2}\frac{\partial A_1}{\partial\alpha_2} - \frac{1}{R_1}\left(\frac{\partial A_2\varepsilon_2}{\partial\alpha_1} - \varepsilon_1\frac{\partial A_2}{\partial\alpha_1} - \frac{\partial A_1\omega}{\partial\alpha_2}\right) = 0. \tag{5.6''}$$

The third equation (5.5) becomes an identity after one has performed the differentiations using (4.15) and (2.28).

Further, one can consider the mixed second derivatives of the vectors $\mathbf{e}_1'$, $\mathbf{e}_2'$ and obtain from the equalities

$$\frac{\partial^2\mathbf{e}_1'}{\partial\alpha_1\partial\alpha_2} = \frac{\partial^2\mathbf{e}_1'}{\partial\alpha_2\partial\alpha_1}, \qquad \frac{\partial^2\mathbf{e}_2'}{\partial\alpha_1\partial\alpha_2} = \frac{\partial^2\mathbf{e}_2'}{\partial\alpha_2\partial\alpha_1}$$

six relations analogous to (5.6). Certain of these relations, after some transformations, become identities, others lead again to the relations (5.6). However, one new differential relation is obtained which is connected with the deformation of the middle surface, namely

$$\frac{\kappa_1}{R_2} + \frac{\kappa_2}{R_1} + \frac{1}{A_1A_2}\left\{\frac{\partial}{\partial\alpha_1}\frac{1}{A_1}\left[A_2\frac{\partial\varepsilon_2}{\partial\alpha_1} + \frac{\partial A_2}{\partial\alpha_1}(\varepsilon_2 - \varepsilon_1) - \right.\right.$$

$$-\tfrac{1}{2}A_1\frac{\partial\omega}{\partial\alpha_2}-\frac{\partial A_1}{\partial\alpha_2}\omega\Big]+\frac{\partial}{\partial\alpha_2}\frac{1}{A_2}\times \tag{5.6'''}$$

$$\times\left[A_1\frac{\partial\varepsilon_1}{\partial\alpha_2}+\frac{\partial A_1}{\partial\alpha_2}(\varepsilon_1-\varepsilon_2)-\tfrac{1}{2}A_2\frac{\partial\omega}{\partial\alpha_1}-\frac{\partial A_2}{\partial\alpha_1}\omega\right]\Bigg\}=0.$$

The three differential relations (5.6) which link the six parameters $\varepsilon_1, \varepsilon_2, \omega, \kappa_1, \kappa_2, \tau$, characterizing the deformation of the middle surface of the shell, have been derived in an analogous manner as in § 2 the Gauss-Codazzi relations. In essence, they are the Gauss-Codazzi relations for the deformation of the middle surface, simplified with the aid of analogous relations for the middle surface before deformation.

The equalities (5.6) must be fulfilled whatever are the displacements u, v, w of the middle surface, i.e., they must become identities if the deformations of the middle surface are replaced by the expressions (4.23). The reader will readily verify that this is true by performing this operation, taking into consideration (2.28) and (2.30).

The relations (5.6) which have been derived first by A. L. Gol'denweizer [5] play in the theory of shells the same role as the St. Venant relations in the theory of elasticity. They are, firstly, the conditions of continuity for the deformation of the middle surface and, secondly, the conditions of compatibility, the fulfillment of which ensures the possibility of determining displacements from given strain components $\varepsilon_1, \varepsilon_2, \omega, \kappa_1, \kappa_2, \tau$ of a shell.

§ 6. Forces and moments

As is usual in the study of the theory of elasticity, the investigation of the geometrical properties of the deformations of shells will be followed by that of the corresponding stresses.

Let a space element be separated from a shell by means of four cuts, normal to its middle surface and tangential to the lines α_1, $\alpha_1 + d\alpha_1$, α_2, $\alpha_2 + d\alpha_2$, the height of which will be finite and equal to the thickness of the shell (Fig. 6).

Let

σ_{11}, σ_{22} be the normal stresses, acting on the faces of this element,

σ_{12}, σ_{21} the shear stresses on these faces, acting parallel to the middle surface,

σ_{13}, σ_{23} the shear stresses on these faces, acting normal to the middle surface.

As is usual in the theory of elasticity, stresses acting on faces the outward normals to which coincide with the positive directions of α_1, α_2

will be assumed to be positive whenever they project positively on $\mathbf{e}_1$, $\mathbf{e}_2$, $\mathbf{e}_n$.

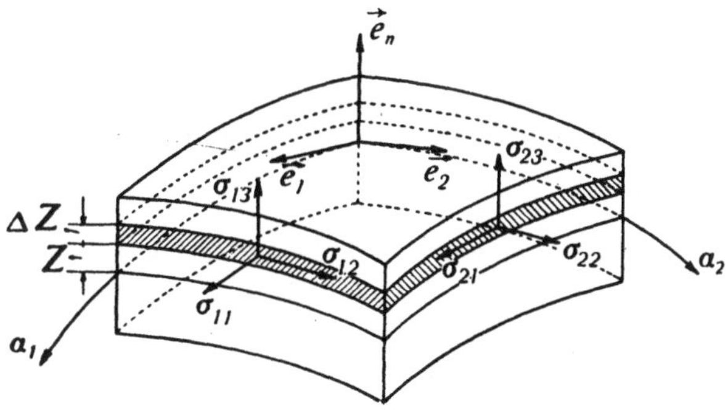

Fig. 6.

As regards the stresses acting on faces for which the external normal is in the opposite direction to α_1 or α_2, they will be assumed to be positive when they project on the negative directions of $\mathbf{e}_1$, $\mathbf{e}_2$, $\mathbf{e}_n$. In Fig. 6, all stresses shown are positive.

Since the basic assumptions have led to a completely defined law of variation of the deformations through the thickness of the shell, it may now be assumed that the law of variation of the stresses σ_{11}, σ_{22}, σ_{12}, σ_{21} is known.

As in the theory of plates, it is convenient to introduce instead of these stresses, statically equivalent forces and bending moments. Consider that face of the element in Fig. 6 which is perpendicular to the line α_1 (Fig. 7).

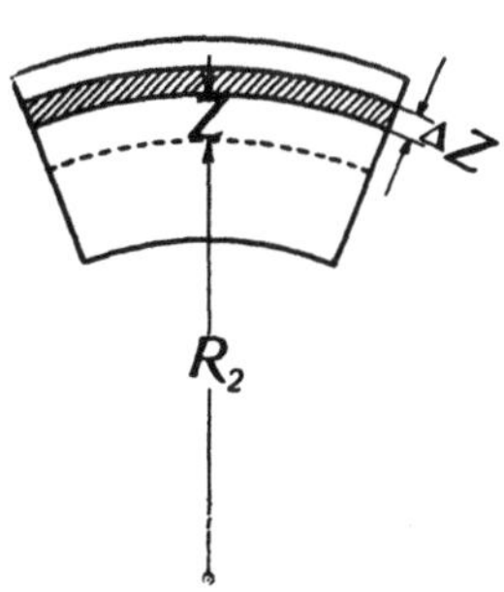

Fig. 7.

On the shaded area of this face, there will act the normal stress σ_{11} and the shear stresses σ_{12}, σ_{13}. Let the length of the arc of the middle surface be equal to $ds_2 = A_2 d\alpha_2$, then the length of the shaded region belonging to the parallel surface will be given by

$$ds_2^{(z)} = A_2(1 + z/R_2)\, d\alpha_2.$$

Hence the resultants of the stresses σ_{11}, σ_{12}, σ_{13}, acting on the entire area of the element of the shell will be equal to

$$S_{11} = A_2 d\alpha_2 \int_{-\delta/2}^{+\delta/2} \sigma_{11}(1 + z/R_2)\, dz,$$

$$S_{12} = A_2 d\alpha_2 \int_{-\delta/2}^{+\delta/2} \sigma_{12}(1 + z/R_2)\, dz,$$

$$S_{13} = A_2 d\alpha_2 \int_{-\delta/2}^{+\delta/2} \sigma_{13}(1 + z/R_2)\, dz.$$

These resultants act on the element of the arc of the line α_2 of the middle surface which is equal to $ds_2 = A_2 d\alpha_2$. Hence one finds for the forces per unit length of the given line

$$\begin{aligned} T_1 &= \frac{S_{11}}{A_2 d\alpha_2} = \int_{-\delta/2}^{+\delta/2} \sigma_{11}(1 + z/R_2)\, dz, \\ T_{12} &= \frac{S_{12}}{A_2 d\alpha_2} = \int_{-\delta/2}^{+\delta/2} \sigma_{12}(1 + z/R_2)\, dz, \\ N_1 &= \frac{S_{13}}{A_2 d\alpha_2} = \int_{-\delta/2}^{+\delta/2} \sigma_{13}(1 + z/R_2)\, dz. \end{aligned} \tag{6.1}$$

In an analogous manner, the following quantities may be introduced:

$$\begin{aligned} T_2 &= \int_{-\delta/2}^{+\delta/2} \sigma_{22}(1 + z/R_1)\, dz, \\ T_{21} &= \int_{-\delta/2}^{+\delta/2} \sigma_{21}(1 + z/R_1)\, dz, \\ N_2 &= \int_{-\delta/2}^{+\delta/2} \sigma_{23}(1 + z/R_1)\, dz \end{aligned} \tag{6.2}$$

which are the forces per unit length of the line α_1 of the middle surface. Consider next the moments of the stresses σ_{11}, σ_{12}, acting on one of the faces of the element of the shell about the directions $\mathbf{e}_2$, $\mathbf{e}_1$. This leads to the two new quantities

$$M_1 = \int_{-\delta/2} z(1 + z/R_2)\sigma_{11}\, dz, \tag{6.3}$$

$$M_{12} = \int_{-\delta/2}^{+\delta/2} z(1 + z/R_2)\sigma_{12} dz$$

which are the bending and twisting moments per unit length of the line α_2.

Analogously, by considering the second face of the element of the shell which has acting on it stresses σ_{22}, σ_{21}, one arrives at the quantities

$$M_2 = \int_{-\delta/2}^{+\delta/2} z(1 + z/R_1)\sigma_{22} dz,$$
$$M_{21} = \int_{-\delta/2}^{+\delta/2} z(1 + z/R_1)\sigma_{21} dz, \quad (6.4)$$

the bending and twisting moments per unit length of the line α_1.

The positive directions of the forces and moments given by the above formulae are shown in Figs. 8 and 9.

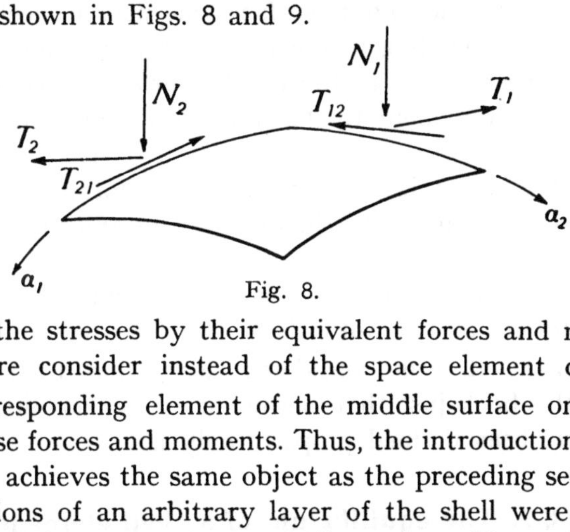

Fig. 8.

Replacing the stresses by their equivalent forces and moments, one may in future consider instead of the space element cut from the shell the corresponding element of the middle surface on the sides of which act these forces and moments. Thus, the introduction of the forces and moments achieves the same object as the preceding sections. There, the deformations of an arbitrary layer of the shell were expressed in terms of the deformation of the middle surface. Here, the stresses in the shell are replaced by their resultants, likewise acting in the middle surface.

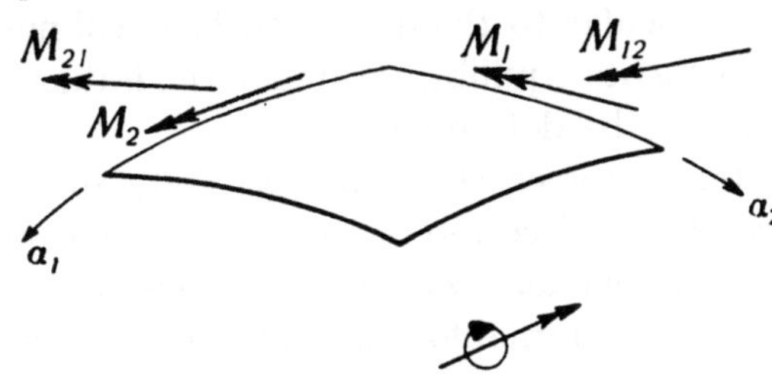

Fig. 9.

However, it is essential to note that in the previous sections, when investigating the deformations of the layers of the shell parallel to the middle surface, the shear displacements corresponding to the stresses σ_{13}, σ_{23} were neglected, since the assumption regarding normals remaining normals during deformation is obviously equivalent to this simplification. Thus it might be thought that one should neglect N_1, N_2, and consequently assume that the shearing stresses are zero. However, this is not correct, since these forces play an essential role in the equilibrium equations. The hypothesis which requires normals to remain normals is only applied to determine the law of deformation of a fibre of the shell, parallel to the middle surface. In the study of the equilibrium of forces this condition is not used.

Hence, the first hypothesis of Kirchhoff should, strictly speaking, be formulated somewhat differently from the way in which this was done in § 1, — namely, instead of the point "a" in that section it should be stated that for the determination of the deformations of the fibres of the shell, parallel to the middle surface, one may neglect the shearing displacements corresponding to the stresses σ_{13}, σ_{23} and the strains in the direction perpendicular to the middle surface. This situation is, of course, not at all equivalent to the categorical statement that normals remain normals.

This remark also applies to the customary formulation of the law relating to plane sections in the theory of beams and to that of the basic hypotheses in the theory of plates.

In view of the fact that these classical definitions are used in all textbooks without, to the author's knowledge, ever referring to this misunderstanding, it was considered acceptable to pursue at the beginning of this book just that method of exposition; that is the reason why they have been defined properly only now.

§ 7. The equilibrium of a shell element

The small element shown in Fig. 6 which has been separated from a shell by four cuts perpendicular to the middle surface will now be considered and the conditions for its equilibrium under the influence of all external and internal loads will be determined.

Firstly the external forces will comprise body forces acting on the element and, secondly, surface forces (stresses) acting on the upper and lower boundaries of the element, which are sections of the curved surfaces bounding the shell. The internal forces will be stresses on the side faces of the element.

In the preceding section, all the internal stresses were transferred to the boundaries of the section of the middle surface corresponding to the considered element of the shell and they were replaced by statically equivalent forces and moments. An analogous operation will now be executed for the external forces by replacing them by statically equivalent stresses distributed over the middle surface. The middle surface will thus be loaded by forces as well as moments. However, in practical problems, the latter can be neglected.

In fact, in view of the smallness of the thickness of the shell, it may be assumed that the body forces (weight and inertia forces) are uniformly distributed throughout the thickness of the shell, so that the moments about the middle surface vanish at least to a term of order δ/R in comparison with unity. As regards the external forces applied to the surfaces bounding the shell, they will in all practical cases act normal to the middle surface (fluid or gas pressure) and hence again the moment transferred to the middle surface will vanish. Thus, for all practical purposes, it may be assumed that, after transfer of the external forces to the middle surface, the latter will only be stressed by distributed forces.

Let the external force, acting on a unit area of the middle surface, be denoted by $\mathbf{q}$ and the components of this vector in the directions $\mathbf{e}_1, \mathbf{e}_2, \mathbf{e}_n$ by q_1, q_2, q_n.

Now, instead of considering the equilibrium of an element of the shell, one may study the equilibrium of the corresponding element of the middle surface (Fig. 10).

The sides of this element will then be acted upon by the following internal forces and moments:

On the side OC:

the force $$-\mathbf{T}^{(1)} = -(T_1\mathbf{e}_1 + T_{12}\mathbf{e}_2 + N_1\mathbf{e}_n)A_2 d\alpha_2$$

and the moment
$$-\mathbf{M}^{(1)} = -(M_1\mathbf{e}_2 - M_{12}\mathbf{e}_1)A_2 d\alpha_2;$$

On the side ED:

the force $$\mathbf{T}^{(1)} + \frac{\partial \mathbf{T}^{(1)}}{\partial \alpha_1} d\alpha_1$$

and the moment $$\mathbf{M}^{(1)} + \frac{\partial \mathbf{M}^{(1)}}{\partial \alpha_1} d\alpha_1;$$

On the side EO:

the force $$-\mathbf{T}^{(2)} = -(T_2\mathbf{e}_1 + T_{21}\mathbf{e}_2 + N_2\mathbf{e}_n)A_1 d\alpha_1$$

and the moment $$-\mathbf{M}^{(2)} = -(M_{21}\mathbf{e}_2 - M_2\mathbf{e}_1)A_1 d\alpha_1;$$

On the side DC:
the force

$$\mathbf{T}^{(2)} + \frac{\partial \mathbf{T}^{(2)}}{\partial \alpha_2} d\alpha_2$$

and the moment

$$\mathbf{M}^{(2)} + \frac{\partial \mathbf{M}^{(2)}}{\partial \alpha_2} d\alpha_2.$$

In addition, the element will be subject to the external surface loading

$$\mathbf{q} A_1 A_2 d\alpha_1 d\alpha_2 = (q_1 \mathbf{e}_1 + q_2 \mathbf{e}_2 + q_n \mathbf{e}_n) A_1 A_2 d\alpha_1 d\alpha_2.$$

In these formulae, and likewise in the following calculations, changes in the dimensions and in the shape of the element of the middle surface arising from its deformation have been neglected, since their inclusion would have been equivalent to the calculation of products of the deriva-

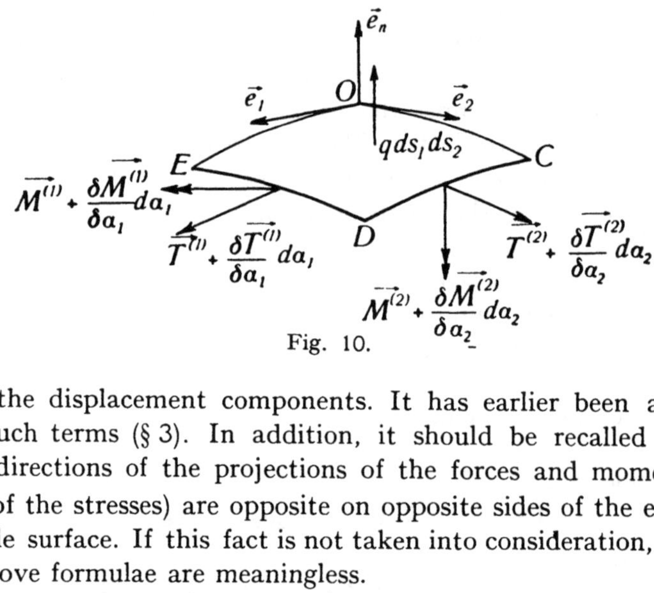

Fig. 10.

tives of the displacement components. It has earlier been agreed to neglect such terms (§ 3). In addition, it should be recalled that the positive directions of the projections of the forces and moments (and likewise of the stresses) are opposite on opposite sides of the element of the middle surface. If this fact is not taken into consideration, the signs in the above formulae are meaningless.

On the basis of this discussion, the condition that the resultant vector of all the forces acting on the element of the middle surface is to vanish can be written in the form

$$-\mathbf{T}^{(1)} + \mathbf{T}^{(1)} + \frac{\partial \mathbf{T}^{(1)}}{\partial \alpha_1} d\alpha_1 - \mathbf{T}^{(2)} + \mathbf{T}^{(2)} + \frac{\partial \mathbf{T}^{(2)}}{\partial \alpha_2} d\alpha_2 + \\ + \mathbf{q} A_1 A_2 d\alpha_1 d\alpha_2 = 0, \tag{7.1}$$

or, after cancellations,

$$\frac{\partial \mathbf{T}^{(1)}}{\partial \alpha_1} d\alpha_1 + \frac{\partial \mathbf{T}^{(2)}}{\partial \alpha_2} d\alpha_2 + \mathbf{q} A_1 A_2 d\alpha_1 d\alpha_2 = 0. \tag{7.2}$$

Replacing here $\mathbf{T}^{(1)}$, $\mathbf{T}^{(2)}$ and $\mathbf{q}$ by their earlier expressions and using the rules (2.25) for differentiation of unit vectors, one finds after dividing by $d\alpha_1 d\alpha_2$

$$\left[\frac{\partial A_2 T_1}{\partial \alpha_1} + \frac{\partial A_1 T_{21}}{\partial \alpha_2} + \frac{\partial A_1}{\partial \alpha_2} T_{12} - \frac{\partial A_2}{\partial \alpha_1} T_2 + \frac{A_1 A_2}{R_1} N_1 + A_1 A_2 q_1\right] \mathbf{e}_1 +$$

$$+ \left[\frac{\partial A_2 T_{12}}{\partial \alpha_1} + \frac{\partial A_1 T_2}{\partial \alpha_2} + \frac{\partial A_2}{\partial \alpha_1} T_{21} - \frac{\partial A_1}{\partial \alpha_2} T_1 + \frac{A_1 A_2}{R_2} N_2 + A_1 A_2 q_2\right] \mathbf{e}_2 +$$

$$+ \left[\frac{\partial A_2 N_1}{\partial \alpha_1} + \frac{\partial A_1 N_2}{\partial \alpha_2} - A_1 A_2 \left(\frac{T_1}{R_1} + \frac{T_2}{R_2}\right) + A_1 A_2 q_n\right] \mathbf{e}_n = 0. \tag{7.3}$$

In order that this vector equation will be satisfied, the terms in the three square brackets must vanish separately, since a vector can only be identically zero if each of its components satisfy this condition.

Hence the following three differential equations are obtained:

$$\begin{aligned}
&\frac{1}{A_1 A_2}\left[\frac{\partial A_2 T_1}{\partial \alpha_1} + \frac{\partial A_1 T_{21}}{\partial \alpha_2} + \frac{\partial A_1}{\partial \alpha_2} T_{12} - \frac{\partial A_2}{\partial \alpha_1} T_2\right] + \frac{N_1}{R_1} + q_1 = 0,\\
&\frac{1}{A_1 A_2}\left[\frac{\partial A_2 T_{12}}{\partial \alpha_1} + \frac{\partial A_1 T_2}{\partial \alpha_2} + \frac{\partial A_2}{\partial \alpha_1} T_{21} - \frac{\partial A_1}{\partial \alpha_2} T_1\right] + \frac{N_2}{R_2} + q_2 = 0,\\
&\frac{1}{A_1 A_2}\left[\frac{\partial A_2 N_1}{\partial \alpha_1} + \frac{\partial A_1 N_2}{\partial \alpha_2}\right] - \frac{T_1}{R_1} - \frac{T_2}{R_2} + q_n = 0.
\end{aligned} \tag{7.4}$$

They express the condition that the resultant vector of all internal and external forces acting on an element of the shell is to vanish.

Next, the conditions for the vanishing of the resultant moment acting on the same element will be determined. On the basis of the earlier results, the geometric sum of all the interior moments acting on the sides of an element of the middle surface will be given by

$$\mathbf{M}^{(1)} + \frac{\partial \mathbf{M}^{(1)}}{\partial \alpha_1} d\alpha_1 - \mathbf{M}^{(1)} + \mathbf{M}^{(2)} + \frac{\partial \mathbf{M}^{(2)}}{\partial \alpha_2} d\alpha_2 - \mathbf{M}^{(2)} =$$

$$= \frac{\partial \mathbf{M}^{(1)}}{\partial \alpha_1} d\alpha_1 + \frac{\partial \mathbf{M}^{(2)}}{\partial \alpha_2} d\alpha_2. \tag{7.5}$$

Substituting here for $\mathbf{M}^{(1)}$ and $\mathbf{M}^{(2)}$ and taking into consideration the differentiation rule for the unit vectors, one finds

$$\frac{\partial \mathbf{M}^{(1)}}{\partial \alpha_1} d\alpha_1 + \frac{\partial \mathbf{M}^{(2)}}{\partial \alpha_2} d\alpha_2 =$$

$$= \left\{ - \left[\frac{\partial A_2 M_{12}}{\partial \alpha_1} + \frac{\partial A_1 M_2}{\partial \alpha_2} + \frac{\partial A_2}{\partial \alpha_1} M_{21} - \frac{\partial A_1}{\partial \alpha_2} M_1 \right] \mathbf{e}_1 + \right.$$

$$+ \left[\frac{\partial A_2 M_1}{\partial \alpha_1} + \frac{\partial A_1 M_{21}}{\partial \alpha_2} + \frac{\partial A_1}{\partial \alpha_2} M_{12} - \frac{\partial A_2}{\partial \alpha_1} M_2 \right] \mathbf{e}_2 +$$

$$\left. + \left[\frac{M_{12}}{R_1} - \frac{M_{21}}{R_2} \right] A_1 A_2 \mathbf{e}_n \right\} d\alpha_1 d\alpha_2. \tag{7.6}$$

Consider now the moment about the point O (Fig. 10) of all the forces, acting on the element of the middle surface. It will be noted that the moment due to the surface loading will have a higher order of smallness than the remaining moments, since it is calculated by multiplying the vector $\mathbf{q}$, firstly, by the area of the element (a second order quantity) and, secondly, by the moment arm (a first order quantity). On the other hand, the remaining moments involve only factors which are of second order of smallness. On this basis, the moment due to the vector $\mathbf{q}$ may be neglected as a third order quantity.

Next, the moment of the forces will be determined. It is easily seen that the components of the forces in the plane of the middle surface give, to a high order accuracy, only moments about the axis $\mathbf{e}_n$, while the components of the forces normal to the middle surface lead only to moments about the axes $\mathbf{e}_1$ and $\mathbf{e}_2$. For this purpose, multiplying the forces by the appropriate moment arms, one may neglect the differential forces, since they are of second order and lead to third order terms.

Taking all this into consideration, one obtains the following expression for the moment of all the forces about the point O [retaining, as in (7.6), only second order terms]:

$$\{N_2 \mathbf{e}_1 - N_1 \mathbf{e}_2 + (T_{12} - T_{21}) \mathbf{e}_n\} A_1 A_2 d\alpha_1 d\alpha_2. \tag{7.7}$$

The vanishing of the resultant moment of all forces acting on the element of the middle surface leads now to the requirement that the sum of the expressions (7.6) and (7.7) must vanish. Thus, after dividing by $d\alpha_1 d\alpha_2$, one arrives at the three differential equations:

$$\frac{1}{A_1 A_2} \left[\frac{\partial A_2 M_1}{\partial \alpha_1} + \frac{\partial A_1 M_{21}}{\partial \alpha_2} + \frac{\partial A_1}{\partial \alpha_2} M_{12} - \frac{\partial A_2}{\partial \alpha_1} M_2 \right] - N_1 = 0,$$

$$\frac{1}{A_1 A_2} \left[\frac{\partial A_2 M_{12}}{\partial \alpha_1} + \frac{\partial A_1 M_2}{\partial \alpha_2} + \frac{\partial A_2}{\partial \alpha_1} M_{21} - \frac{\partial A_1}{\partial \alpha_2} M_1 \right] - N_2 = 0, \tag{7.8}$$

$$T_{12} - T_{21} + \frac{M_{12}}{R_1} - \frac{M_{21}}{R_2} = 0.$$

The six equations (7.4) and (7.8) are the conditions for the equilibrium of a small element of the shell.

It will now be shown that the last of these equations, i.e., the equation

$$T_{12} - T_{21} + \frac{M_{12}}{R_1} - \frac{M_{21}}{R_2} = 0 \tag{7.9}$$

is identically satisfied. In fact, replacing in this equation the forces and moments by their expressions from § 6, one finds

$$T_{12}-T_{21}+\frac{M_{12}}{R_1}-\frac{M_{21}}{R_2} = \int_{-\delta/2}^{+\delta/2} (1+z/R_1)(1+z/R_2)(\sigma_{12}-\sigma_{21})\,dz \equiv 0. \tag{7.10}$$

Thus, independently of the assumptions adopted for the theory of shells, the equality (7.9) will be fulfilled, because it is implied by the symmetry of the stress tensor.

The above six equations of equilibrium involve ten unknowns, i.e., the problem of the theory of shells is 'internally' statically indeterminate (statically indeterminate, even when only an infinitely small element of the shell is being considered). Its solution is therefore impossible unless several of the unknowns entering into the equilibrium equations are related through an elastic law to the deformations.

In conclusion, it should be emphasized that the deduced static equations for shells involve only an approximation, arising from the assumption that the displacements of the shell are to be small in comparison with its thickness. An approximation, based on the initial assumptions, does not enter into these equations, because these assumptions were not used in their deduction.

§ 8. Functions which satisfy the homogeneous statical equations identically

Simultaneously and independently of each other, Lur'e [17] and Gol'denweizer [6] proved that, if the components of the surface loading on a shell

$$q_1 = q_2 = q_n = 0,$$

i.e., if the shell is only loaded at the edges, then all the ten unknown forces and moments, entering into the system of equations (7.4), (7.8), can be expressed in terms of four arbitrary functions a, b, c, χ of α_1 and α_2 in such a way that the system is identically satisfied. This result is a generalization of a well known idea in the plane theory of elasticity (due to Airy) to shell problems, except that, instead of a single auxiliary function, the present, more complicated problems require four auxiliary functions. The proof of this statement is based on the fact that the equations of equilibrium of the shell element may be transformed into a form which is similar to the conditions of compatibility of the strain of the middle surface.

Introduce the notation [cf. the identity (7.9)]

$$S = T_{12} - \frac{M_{21}}{R_2} = T_{21} - \frac{M_{12}}{R_1}, \quad H = \frac{M_{12} + M_{21}}{2} \tag{8.1}$$

and eliminate from the equations (7.4) the forces N_1, N_2 by substituting for them their values as given by (7.8). Then, taking into consideration the above notation and using for the differentiations the conditions of Codazzi, the equations (7.4) may be written in the form

$$\begin{gathered}
\frac{\partial A_2 T_1}{\partial \alpha_1} + \frac{\partial A_1 S}{\partial \alpha_2} + \frac{\partial A_1}{\partial \alpha_2} S - \frac{\partial A_2}{\partial \alpha_1} T_2 + \\
+ \frac{1}{R_1}\left[\frac{\partial A_2 M_1}{\partial \alpha_1} - \frac{\partial A_2}{\partial \alpha_1} M_2 + 2\frac{\partial A_1 H}{\partial \alpha_2} + 2\frac{R_1}{R_2}\frac{\partial A_1}{\partial \alpha_2} H\right] = - A_1 A_2 q_1, \\
\frac{\partial A_2 S}{\partial \alpha_1} + \frac{\partial A_1 T_2}{\partial \alpha_2} + \frac{\partial A_2}{\partial \alpha_1} S - \frac{\partial A_1}{\partial \alpha_2} T_1 + \\
+ \frac{1}{R_2}\left[\frac{\partial A_1 M_2}{\partial \alpha_2} - \frac{\partial A_1}{\partial \alpha_2} M_1 + 2\frac{\partial A_2 H}{\partial \alpha_1} + 2\frac{R_2}{R_1}\frac{\partial A_2}{\partial \alpha_1} H\right] = - A_1 A_2 q_2, \\
\frac{T_1}{R_1} + \frac{T_2}{R_2} - \frac{1}{A_1 A_2}\left\{\frac{\partial}{\partial \alpha_1}\frac{1}{A_1}\left[\frac{\partial A_2 M_1}{\partial \alpha_1} + \frac{\partial A_1 H}{\partial \alpha_2} + \frac{\partial A_1}{\partial \alpha_2} H - \frac{\partial A_2}{\partial \alpha_1} M_2\right] + \right. \\
\left. + \frac{\partial}{\partial \alpha_2}\frac{1}{A_2}\left[\frac{\partial A_2 H}{\partial \alpha_1} + \frac{\partial A_1 M_2}{\partial \alpha_2} + \frac{\partial A_2}{\partial \alpha_1} H - \frac{\partial A_1}{\partial \alpha_2} M_1\right]\right\} = q_n.
\end{gathered} \tag{8.2}$$

It will now be assumed that the problem is homogeneous, i.e., that

$$q_1 = q_2 = q_n = 0,$$

and the three arbitrary functions a, b, c of the variables α_1, α_2 will be introduced which are to be related to the unknowns in the equations by means of the formulae

$$\begin{aligned}
T_1 &= \kappa_2(a, b, c), \\
T_2 &= \kappa_1(a, b, c), \\
S &= -\tau\,(a, b, c), \\
M_1 &= -\varepsilon_2\,(a, b, c), \\
M_2 &= -\varepsilon_1\,(a, b, c),
\end{aligned} \tag{8.3}$$

$$H = \tfrac{1}{2}\omega\,(a, b, c),$$

where $\varepsilon_1(a, b, c), \ldots \tau(a, b, c)$ are understood to be the expressions for the strains in the middle surface, defined by (4.23) with u, v, w replaced by a, b, c respectively. By substituting now (8.3) in (8.2), it may be verified that this leads to expressions which are identical with the conditions of compatibility of strain for the middle surface, i.e., with equations (5.1). But these last equations would be identically satisfied by the deformations of the middle surface whatever the displacements u, v and w may be. Hence clearly the equations (8.2) will also be identically satisfied by the expressions (8.3) involving the arbitrary auxiliary functions a, b and c.

Express now the forces T_{12}, T_{21} and the moments M_{12}, M_{21} in terms of these functions. On the basis of (8.1) and (8.3), one may write

$$\begin{aligned} M_{12} + M_{21} &= \omega(a, b, c), \\ T_{12} - \frac{M_{21}}{R_2} &= -\tau(a, b, c), \\ T_{21} - \frac{M_{12}}{R_1} &= -\tau(a, b, c). \end{aligned} \tag{8.4}$$

By the first of these formulae

$$\begin{aligned} M_{12} &= \tfrac{1}{2}\omega(a, b, c) + \Phi, \\ M_{21} &= \tfrac{1}{2}\omega(a, b, c) - \Phi, \end{aligned} \tag{8.5}$$

where Φ is a function which from the point of view of satisfying the static equations has the same arbitrary character as a, b, c.

Substituting now for M_{12}, M_{21} in the last two of the formulae (8.4) their expressions (8.5), one finds

$$\begin{aligned} T_{12} &= -\tau(a, b, c) + \frac{1}{2R_2}\,\omega(a, b, c) - \Phi/R_2, \\ T_{21} &= -\tau(a, b, c) + \frac{1}{2R_1}\,\omega(a, b, c) + \Phi/R_1. \end{aligned} \tag{8.6}$$

There still remains the determination of the normal shearing forces N_1, N_2 in terms of the same four functions. This may be done by substituting from the above formulae in the first two of the equations (7.8). After some manipulations, involving the Gauss-Codazzi relations, one obtains

$$\begin{aligned} N_1 &= \frac{1}{R_2}\,\vartheta(a, c) - \frac{1}{A_2}\,\frac{\partial\chi}{\partial\alpha_2}, \\ N_2 &= \frac{1}{R_1}\,\psi(a, c) + \frac{1}{A_1}\,\frac{\partial\chi}{\partial\alpha_1}, \end{aligned} \tag{8.7}$$

where $\vartheta(a, c)$, $\psi(a, c)$ are the quantities obtained by replacing in (3.16) the displacements u, v, w by a, b, c, while

$$\chi = \Phi - \frac{1}{2A_1A_2}\left[\frac{\partial A_2 a}{\partial \alpha_2} - \frac{\partial A_2 b}{\partial \alpha_1}\right].$$

All the ten forces and moments, characterizing the state of stress of the shell, have now been expressed in terms of the four functions a, b, c, Φ and this has been done in such a manner that by substitution of these expressions into the equations of equilibrium of the element of the shell (8.2) (for $q_1 = q_2 = q_n = 0$) these equations are satisfied identically.

§ 9. The strain energy for a shell

The discussion so far has dealt with the geometry of shells and the equilibrium conditions for its elements, and thus it did not depend on the elastic properties of its material. For this reason, these properties have not been considered so far. However, since the equations of equilibrium obtained in § 7 for an element of the shell are statically indeterminate, the problem cannot be solved until the elastic properties of the shell have been established.

It is known that the elastic nature of a body is completely specified by the expression for its potential energy of deformation. This expression will now be deduced for thin shells made from a material which is homogeneous, isotropic and obeys Hooke's law. For this purpose, a start will be made with the generalized formula of the general theory of elasticity

$$V = \tfrac{1}{2}\int(\sigma_{11}e_{11} + \sigma_{22}e_{22} + \sigma_{33}e_{33} + \sigma_{12}e_{12} + \sigma_{13}e_{13} + \sigma_{23}e_{23})\,dO, \tag{9.1}$$

where σ_{11}, σ_{22}, σ_{33} are the normal stresses on three mutually perpendicular faces of an element, cut from the elastic body,
e_{11}, e_{22}, e_{33} are the corresponding strains,
σ_{12}, σ_{13}, σ_{23} are the shear stresses on these faces,
e_{12}, e_{13}, e_{23} are the corresponding shear deformations,
dO is the volume of the element,

and the integration in (9.1) extends over the entire volume of the body under consideration.

This formula will be applied to an element separated from a shell by four neighbouring cuts perpendicular to the lines α_1, α_2 of the middle surface and by two cuts parallel to the middle surface (Fig. 6), the volume of which will be equal to the product of its three sides (since they are perpendicular to each other)

$$dO = A_1A_2(1 + z/R_1)(1 + z/R_2)\,d\alpha_1 d\alpha_2 dz. \tag{9.2}$$

The hypotheses of Kirchhoff are equivalent to neglecting the stress σ_{33} in comparison with the remaining stresses and the strains e_{13}, e_{23} in comparison with the remaining strains (§ 1). Hence it follows from these hypotheses that one may write

$$V = \tfrac{1}{2}\iiint(\sigma_{11}e_{11}+\sigma_{22}e_{22}+\sigma_{12}e_{12})A_1A_2(1+z/R_1)(1+z/R_2)\,d\alpha_1 d\alpha_2 dz, \tag{9.3}$$

where the first integral extends through the thickness of the shell (i.e., from $z = -\delta/2$ to $z = +\delta/2$) and the remaining integrals are taken over its entire middle surface.

By Hooke's law, one has for homogeneous and isotropic bodies

$$\begin{aligned}\sigma_{11} &= \frac{E}{1+\mu}\left(e_{11} + \frac{\mu}{1-2\mu}e\right),\\ \sigma_{22} &= \frac{E}{1+\mu}\left(e_{22} + \frac{\mu}{1-2\mu}e\right),\\ \sigma_{33} &= \frac{E}{1+\mu}\left(e_{33} + \frac{\mu}{1-2\mu}e\right),\\ \sigma_{12} &= \frac{E}{2(1+\mu)}e_{12},\end{aligned} \tag{9.4}$$

where

E is Young's modulus,

μ is Poisson's ratio,

$e = e_{11} + e_{22} + e_{33}$.

Neglecting in (9.4) the stress σ_{33} in comparison with the remaining stresses, one finds

$$\frac{1+\mu}{E}\sigma_{33} = e_{33} + \frac{\mu e}{1-2\mu} = \frac{1}{1-2\mu}[(1-\mu)e_{33} + \mu(e_{11} + e_{22})] = 0,$$

and hence

$$e_{33} = -\frac{\mu}{1-\mu}(e_{11} + e_{22}). \tag{9.5}$$

Introducing this value for the strain e_{33} into the first two of the formulae (9.4), one obtains

$$\begin{aligned}\sigma_{11} &= \frac{E}{1-\mu^2}(e_{11} + \mu e_{22}),\\ \sigma_{22} &= \frac{E}{1-\mu^2}(e_{22} + \mu e_{11}),\end{aligned} \tag{9.6}$$

or, since in this case e_{11}, e_{22}, e_{33} are identical with the strains $\varepsilon_1^{(z)}$, $\varepsilon_2^{(z)}$, $\omega^{(z)}$ considered in § 1, one may write

$$\sigma_{11} = \frac{E}{1-\mu^2}\,[\varepsilon_1{}^{(z)} + \mu\varepsilon_2{}^{(z)}],$$

$$\sigma_{22} = \frac{E}{1-\mu^2}\,[\varepsilon_2{}^{(z)} + \mu\varepsilon_1{}^{(z)}], \tag{9.7}$$

$$\sigma_{12} = \frac{E}{2(1+\mu)}\,\omega^{(z)},$$

where $\varepsilon_1{}^{(z)}$, $\varepsilon_2{}^{(z)}$, $\omega^{(z)}$ are given in terms of the deformations of the middle surface ε_1, ε_2, ω, κ_1, κ_2, τ and the coordinate z by the formulae (4.11), (4.13), (4.22).

After substituting these expressions in (9.7) and then from (9.7) into (9.3), the integration with respect to z can be carried out. In order to simplify the result, the integrand in (9.3) will now be expanded in a series of powers of z, retaining terms of up to and including the second order.

To this order of accuracy the formulae for $\varepsilon_1{}^{(z)}$, $\varepsilon_2{}^{(z)}$, $\omega^{(z)}$ assume the form

$$\varepsilon_1{}^{(z)} = \varepsilon_1 + z\left(\kappa_1 - \frac{\varepsilon_1}{R_1}\right) - \frac{z^2}{R_1}\left(\kappa_1 - \frac{\varepsilon_1}{R_1}\right),$$

$$\varepsilon_2{}^{(z)} = \varepsilon_2 + z\left(\kappa_2 - \frac{\varepsilon_2}{R_2}\right) - \frac{z^2}{R_2}\left(\kappa_2 - \frac{\varepsilon_2}{R_2}\right),$$

$$\omega^{(z)} = \omega + 2z\left[\tau - \left(\frac{1}{R_1} + \frac{1}{R_2}\right)\frac{\omega}{2}\right] -$$

$$- z^2\left(\frac{1}{R_1} + \frac{1}{R_2}\right)\left[\tau - \frac{R_1{}^2 + R_2{}^2}{R_1 + R_2}\,\frac{\omega}{R_1R_2}\right].$$

Introducing (4.25) into (9.7) and into (9.3) (replacing there e_{11}, e_{22}, e_{33}) and rejecting powers of z higher than the second, one finds

$$V = \frac{E}{2(1-\mu^2)}\int\int\left\{\int_{-\delta/2}^{+\delta/2}(Q_0 + Q_1 z + Q_2 z^2)\,dz\right\}A_1A_2\,d\alpha_1\,d\alpha_2, \tag{9.8}$$

where

$$Q_0 = (\varepsilon_1 + \varepsilon_2)^2 - 2(1-\mu)\left(\varepsilon_1\varepsilon_2 - \frac{\omega^2}{4}\right),$$

$$Q_2 = (\kappa_1 + \kappa_2)^2 - 2(1-\mu)(\kappa_1\kappa_2 - \tau^2) -$$

$$- 2\left(\frac{1}{R_1} - \frac{1}{R_2}\right)(\varepsilon_1\kappa_1 - \varepsilon_2\kappa_2) - (1-\mu)\left(\frac{1}{R_1} + \frac{1}{R_2}\right)\omega\tau + \tag{9.9}$$

$$+ \left(\frac{1}{R_1} - \frac{1}{R_2}\right)\left(\frac{\varepsilon_1{}^2}{R_1} - \frac{\varepsilon_2{}^2}{R_2}\right) + \frac{1-\mu}{2}\left(\frac{1}{R_1{}^2} - \frac{1}{R_1R_2} + \frac{1}{R_2{}^2}\right)\omega^2.$$

The value of Q_1 will be of no interest, since

$$\int_{-\delta/2}^{+\delta/2} Q_1 z\, ds = Q_1 \int_{-\delta/2}^{+\delta/2} z\, dz = 0.$$

After carrying out the integration with respect to z in (9.8), one finds

$$V = \frac{E\delta}{2(1-\mu^2)} \iint \left[(\varepsilon_1 + \varepsilon_2)^2 - 2(1-\mu)\left(\varepsilon_1\varepsilon_2 - \frac{\omega^2}{4}\right)\right] A_1 A_2 d\alpha_1 d\alpha_2 +$$

$$+ \frac{E\delta^3}{24(1-\mu^2)} \iint \Big\{ (\kappa_1 + \kappa_2)^2 - 2(1-\mu)(\kappa_1\kappa_2 - \tau^2) -$$

$$- 2\left(\frac{1}{R_1} - \frac{1}{R_2}\right)(\varepsilon_1\kappa_1 - \varepsilon_2\kappa_2) -$$

$$- (1-\mu)\left(\frac{1}{R_1} + \frac{1}{R_2}\right)\omega\tau + \left(\frac{1}{R_1} - \frac{1}{R_2}\right)\left(\frac{\varepsilon_1^2}{R_1} - \frac{\varepsilon_2^2}{R_2}\right) +$$

$$+ \frac{1-\mu}{2}\left(\frac{1}{R_1^2} - \frac{1}{R_1 R_2} + \frac{1}{R_2^2}\right)\omega^2 \Big\} A_1 A_2 d\alpha_1 d\alpha_2, \quad (9.10)$$

where the integration extends only over the middle surface of the shell.

Since in this theory of shells there is an admissible error of order δ/R in comparison with unity, replacement of (9.1) by (9.3) introduces into the potential energy an error of the stated order. Hence, in (9.10), terms of this order may not be rejected and it must be discovered which terms are to be retained or rejected. This process will be more difficult, since (9.10) contains the quantities ε_1, ε_2, ω and κ_1, κ_2, τ, where each group has different dimensions (namely — the first group contains dimensionless quantities, while those in the second group have the dimension $[L]^{-1}$).

Introduce therefore instead of κ_1, κ_2, τ the parameters

$$\varepsilon_1' = \frac{\delta}{2}\kappa_1, \quad \varepsilon_2' = \frac{\delta}{2}\kappa_2, \quad \omega' = \tau\delta,$$

which have simple physical interpretations; namely, ε_1', ε_2' are the strains in the extreme fibres of the shell, caused by the bending κ_1, κ_2, and ω' is the shear strain in the outer layer of the shell, arising from the twist τ.

These last interpretations follow from (4.11), (4.13) and (4.22) after neglecting there the ratios z/R_1, z/R_2 in comparison with unity and letting $z = \dfrac{\delta}{2}$.

The quantities ε_1', ε_2', ω' are now dimensionally similar to ε_1, ε_2, ω. Introducing these quantities into (9.10) instead of κ_1, κ_2, τ, this formula may be reduced to the form

$$V = \frac{E\delta}{2(1-\mu^2)} \iint \Big\{ (\varepsilon_1 + \varepsilon_2)^2 - 2(1-\mu)\left(\varepsilon_1\varepsilon_2 - \frac{\omega^2}{4}\right) + \\
+ \frac{1}{3}\left[(\varepsilon_1' + \varepsilon_2')^2 - 2(1-\mu)\left(\varepsilon_1'\varepsilon_2' - \frac{\omega'^2}{4}\right)\right] - \\
- \frac{1}{3}\left(\frac{\delta}{R_1} - \frac{\delta}{R_2}\right)(\varepsilon_1\varepsilon_1' - \varepsilon_2\varepsilon_2') - \frac{1-\mu}{12}\left(\frac{\delta}{R_1} + \frac{\delta}{R_2}\right)\omega\omega' + \\
+ \frac{1}{12}\left(\frac{\delta}{R_1} - \frac{\delta}{R_2}\right)\left(\frac{\delta}{R_1}\varepsilon^2 - \frac{\delta}{R_2}\varepsilon_2^2\right) + \\
+ \frac{1-\mu}{24}\left(\frac{\delta^2}{R_1^2} - \frac{\delta^2}{R_1R_2} + \frac{\delta^2}{R_2^2}\right)\omega^2 \Big\} A_1A_2\,d\alpha_1\,d\alpha_2. \tag{9.11}$$

The integrand above is now a quadratic form in the six dimensionless parameters ε_1, ε_2, ω, ε_1', ε_2', ω'. Comparing its three groups of terms

$$\text{(I)} \quad (\varepsilon_1 + \varepsilon_2)^2 - 2(1-\mu)\left(\varepsilon_1\varepsilon_2 - \frac{\omega^2}{4}\right) + \\
+ \frac{1}{3}\left[(\varepsilon_1' + \varepsilon_2')^2 - 2(1-\mu)\left(\varepsilon_1'\varepsilon_2' - \frac{\omega'^2}{4}\right)\right],$$

$$\text{(II)} \quad -\frac{1}{3}\left(\frac{\delta}{R_1} - \frac{\delta}{R_2}\right)(\varepsilon_1\varepsilon_1' - \varepsilon_2\varepsilon_2') - \frac{1-\mu}{12}\left(\frac{\delta}{R_1} + \frac{\delta}{R_2}\right)\omega\omega',$$

$$\text{(III)} \quad \frac{1}{12}\left(\frac{\delta}{R_1} - \frac{\delta}{R_2}\right)\left(\frac{\delta}{R_1}\varepsilon_1^2 - \frac{\delta}{R_2}\varepsilon_2^2\right) + \frac{1-\mu}{24}\left(\frac{\delta^2}{R_1^2} - \frac{\delta^2}{R_1R_2} + \frac{\delta^2}{R_2^2}\right)\omega^2,$$

it is seen that for arbitrary values of the parameters ε_1', ε_2', ω', ε_1, ε_2, ω the sum of the terms of the second group, if compared to that of the terms of the first group, will be a quantity of order not greater than δ/R, while the sum of the terms of the third group, compared with that of the first group, will be of order δ^2/R^2. This fact is readily established by making all possible assumptions concerning the relative magnitudes of the parameters ε_1, ε_2, ω, ε_1', ε_2', ω (the least favourable of which will be the case when all quantities are of equal order). Hence it follows that, having agreed on the basic assumptions for the theory, one must, in order to be mathematically consistent, neglect the terms (II) and (III) in comparison with the term (I).

Thus the potential energy of deformation in the theory of thin shells must be written in the form

$$V = \frac{E\delta}{2(1-\mu^2)} \iint \left[(\varepsilon_1 + \varepsilon_2)^2 - 2(1-\mu)\left(\varepsilon_1\varepsilon_2 - \frac{\omega^2}{4}\right)\right] A_1A_2 d\alpha_1 d\alpha_2 +$$

$$+ \frac{E\delta^3}{24(1-\mu^2)} \iint [(\kappa_1+\kappa_2)^2 - 2(1-\mu)(\kappa_1\kappa_2 - \tau^2)] A_1A_2 d\alpha_1 d\alpha_2, \quad (9.12)$$

where the double integrals extend over the entire middle surface. In this formula, the first term represents the potential energy of extension and shear, the second that of bending and torsion.

Formula (9.12) is quite analogous to the expression for the potential energy of plates. It contains the deformations of the middle surface which are expressed in terms of the three displacement components u, v, w by (4.23).

§ 10. Relations between the forces, moments and the strains of the middle surface

Consider now the expression for the variational change of the potential energy of deformation which, as is known from the theory of elasticity, is of the form

$$\delta(V) = \iiint [\sigma_{11}\delta(e_{11}) + \sigma_{22}\delta(e_{22}) + \sigma_{33}\delta(e_{33}) + \sigma_{12}\delta(e_{12}) +$$
$$+ \sigma_{13}\delta(e_{13}) + \sigma_{23}\delta(e_{23})]\, dO, \quad (10.1)$$

where $\delta(\ldots)$ is the symbol for the variational increment.

Using the basic assumptions, this formula may be written for the case of thin shells in the form

$$\delta(V) = \iiint [\sigma_{11}\delta(e_{11}) + \sigma_{22}\delta(e_{22}) + \sigma_{12}\delta(e_{12})](1 + z/R_1)$$
$$\times (1 + z/R_2) A_1A_2 d\alpha_1 d\alpha_2 dz, \quad (10.2)$$

where integration extends over the entire volume of the shell.

Introducing here

$$\delta(e_{11}) = \delta(\varepsilon_1^{(z)}) = \frac{1}{1 + z/R_1} [\delta(\varepsilon_1) + z\delta(\kappa_1)],$$

$$\delta(e_{22}) = \delta(\varepsilon_2^{(z)}) = \frac{1}{1 + z/R_2} [\delta(\varepsilon_2) + z\delta(\kappa_2)], \quad (10.3)$$

$$\delta(e_{12}) = \delta(\omega^{(z)}) = \frac{1}{1 + z/R_1} [\delta(\omega_1) + z\delta(\tau_1)] + \frac{1}{1 + z/R_2} [\delta(\omega_2) + z\delta(\tau_2)]$$

and using (6.1), (6.2), (6.3), (6.4), one finds

$$\delta(V) = \iint [T_1\delta(\varepsilon_1) + T_2\delta(\varepsilon_2) + T_{12}\delta(\omega_1) + T_{21}\delta(\omega_2) +$$

$$+ M_1\delta(\kappa_1) + M_2\delta(\kappa_2) + M_{12}\delta(\tau_1) + M_{21}\delta(\tau_2)]A_1A_2 d\alpha_1 d\alpha_2, \quad (10.4)$$

where the integration ranges over the middle surface of the shell.

But, by (4.9),

$$\delta(\tau_1) = \delta(\tau) - \frac{1}{R_1}\delta(\omega_2),$$

$$\delta(\tau_2) = \delta(\tau) - \frac{1}{R_2}\delta(\omega_1), \quad (10.5)$$

$$\delta(\omega) = \delta(\omega_1) + \delta(\omega_2).$$

Hence (10.4) may be reduced to the form

$$\delta(V) = \iint [T_1\delta(\varepsilon_1) + T_2\delta(\varepsilon_2) + S\delta(\omega) + M_1\delta(\kappa_1) + M_2\delta(\kappa_2) +$$

$$+ 2H\delta(\tau)]A_1A_2 d\alpha_1 d\alpha_2, \quad (10.6)$$

where, as in § 8,

$$S = T_{12} - \frac{M_{21}}{R_2} = T_{21} - \frac{M_{12}}{R_1},$$

$$H = \tfrac{1}{2}(M_{12} + M_{21}). \quad (10.7)$$

On the basis of (9.12) and (10.6), relations between the quantities T_1, T_2, S, M_1, M_2, H and the deformations of the middle surface may be readily established. For this purpose, one has to determine the variational increase of the potential energy starting from (9.12) and then to compare the coefficients for variational changes of the deformations in this expression with the corresponding coefficients in (10.6).

In this manner, the following relations are obtained:

$$T_1 = \frac{E\delta}{1-\mu^2}(\varepsilon_1 + \mu\varepsilon_2), \qquad T_2 = \frac{E\delta}{1-\mu^2}(\varepsilon_2 + \mu\varepsilon_1),$$

$$M_1 = \frac{E\delta^3}{12(1-\mu^2)}(\kappa_1 + \mu\kappa_2), \quad M_2 = \frac{E\delta^3}{12(1-\mu^2)}(\kappa_2 + \mu\kappa_1), \quad (10.8)$$

$$S = \frac{E\delta}{2(1+\mu)}\omega, \qquad H = \frac{E\delta^3}{12(1+\mu)}\tau.$$

The last two of the formulae (10.8), written out in full, represent the three equations (8.1)

$$
\begin{aligned}
T_{12} - \frac{M_{21}}{R_2} &= \frac{E\delta}{2(1+\mu)}\,\omega, \\
T_{21} - \frac{M_{12}}{R_1} &= \frac{E\delta}{2(1+\mu)}\,\omega, \\
M_{12} + M_{21} &= \frac{E\delta^3}{6(1+\mu)}\,\tau,
\end{aligned}
\tag{10.9}
$$

which by their very nature are insufficient to express each of the quantities T_{12}, T_{21}, M_{12}, M_{21} uniquely in terms of ω and τ.

One finds from these equations

$$
\begin{gathered}
T_{12} = \frac{E\delta}{2(1+\mu)}\left(\omega + \frac{\delta^2}{6R_2}\,\tau\right) - \Phi/R_2, \\
T_{21} = \frac{E\delta}{2(1+\mu)}\left(\omega + \frac{\delta^2}{6R_1}\,\tau\right) + \Phi/R_1, \\
M_{12} = \frac{E\delta^3}{12(1+\mu)}\,\tau + \Phi, \quad M_{21} = \frac{E\delta^3}{12(1+\mu)}\,\tau - \Phi,
\end{gathered}
\tag{10.10}
$$

where from the point of view of the preceding reasoning the function Φ may be chosen to suit convenience. It is curious that this indeterminateness (as has been noted recently) has no importance for the further study of the theory of shells. It has turned out that, in general, it is not necessary to find in the theory of shells the forces T_{12}, T_{21} and the moments M_{12}, M_{21} separately, since these forces and moments only enter into the equilibrium equations (after elimination of N_1, N_2) and into the boundary conditions in the combinations denoted above by S and H. For the equilibrium equations, this feature was established in § 8, while for the boundary conditions, this fact will be established in the next section.

Thus, for the solution of problems of the theory of shells (independently of whether they will be solved in terms of forces and moments or in terms of displacements), it is not necessary to express T_{12}, T_{21}, M_{12}, M_{21} in terms of the deformations of the middle surface, but only the corresponding quantities S and H. The values of T_{12}, T_{21}, M_{12}, M_{21} are required separately only for the determination of the normal shearing forces N_1, N_2. However, the evaluation of these latter forces is not usually of much practical interest, since in thin shells the shear stresses σ_{13}, σ_{23} corresponding to given forces are always very small (in comparison with the remaining stresses), a fact on which, in essence, the hypotheses of Kirchhoff are based. The little importance attached to these stresses is evident from the fact that no reference to the literature can be given where the question of determining the maximum values of the stresses σ_{13}, σ_{23} for given values of the normal shearing forces N_1, N_2 is discussed.

If for some reason it is desired to evaluate the quantities T_{12}, T_{21}, M_{12}, M_{21} separately, one may proceed to the determination of the function Φ in (10.10), for example, in the following manner: it is found from (6.3) and (6.4) that

$$M_{12} - M_{21} = 2\Phi = \left(\frac{1}{R_2} - \frac{1}{R_1}\right) \int_{-\delta/2}^{+\delta/2} z^2 \sigma_{12} dz; \tag{10.11}$$

substituting here for σ_{12} from the last of the formulae (9.7) and for $\omega^{(z)}$ from (4.22), one has

$$2\Phi = \frac{E}{2(1+\mu)}\left(\frac{1}{R_2} - \frac{1}{R_1}\right) \int_{-\delta/2}^{+\delta/2} \frac{z^2}{(1+z/R_1)(1+z/R_2)} \left\{\left(1 - \frac{z^2}{R_1 R_2}\right)\omega + \right.$$

$$\left. + 2\left[1 + \frac{1}{2}\left(\frac{1}{R_1} + \frac{1}{R_2}\right) z\right] z\tau\right\} dz; \tag{10.12}$$

neglecting now in this last expression (under the integral sign) quantities of order δ/R_1, δ/R_2 in comparison with unity, i.e., letting

$$1 + z/R_1 \approx 1 + z/R_2 \approx 1 - \frac{z^2}{R_1 R_2} \approx 1 + \frac{1}{2}\left(\frac{1}{R_1} + \frac{1}{R_2}\right) z \approx 1,$$

one obtains for Φ the approximate expression

$$\Phi \approx \left(\frac{1}{R_2} - \frac{1}{R_1}\right) \frac{E\delta^3}{48(1+\mu)}\, \omega. \tag{10.13}$$

Introducing this expression into (10.10), one arrives after some analysis at the conclusion that the effect of the introduction of Φ is always insignificant, since it is beyond the limit of the error of the initial assumptions of the theory of thin shells. Using this result, one may set $\Phi = 0$ in (10.10), identifying in this way M_{12} and M_{21} with H.

Exactly this point of view will be adopted in what follows. One thus arrives at the final formulae expressing the forces and moments in terms of the deformation of the middle surface:

$$T_1 = \frac{Eh}{1-\nu^2}(\varepsilon_1 + \nu\varepsilon_2), \qquad T_2 = \frac{Eh}{1-\nu^2}(\varepsilon_2 + \nu\varepsilon_1),$$

$$M_1 = \frac{Eh^3}{12(1-\nu^2)}(\kappa_1 + \nu\kappa_2), \qquad M_2 = \frac{Eh^3}{12(1-\nu^2)}(\kappa_2 + \nu\kappa_1),$$

$$T_{12} = \frac{Eh}{2(1+\nu)}\left(\omega + \frac{h^2}{6R_2}\right)\tau, \qquad T_{21} = \frac{Eh}{2(1+\nu)}\left(\omega + \frac{h^2}{6R_1}\tau\right),$$

$$M_{12} = M_{21} = H = \frac{Eh^3}{12(1+\nu)}\tau.$$

The above development exhausts the question regarding the relations between the forces, moments and the deformations of the middle surface in the theory of thin shells. These are given by (10.8), obtained from the expressions for the potential energy (9.12), simplified in accordance with the accuracy of the initial assumptions of the theory of thin shells.

The attentive reader will no doubt note that there exists another and simpler way of arriving at the relations between the forces, moments and deformations than that which was used above. In fact, using (6.1), (6.2), (6.3), (6.4), (9.7), (4.11), (4.13), (4.22), one may write

$$T_1 = \frac{E}{1-\mu^2}\int_{-\delta/2}^{+\delta/2}\left\{\frac{1+z/R_2}{1+z/R_1}(\varepsilon_1 + z\kappa_1) + \mu(\varepsilon_2 + z\kappa_2)\right\} dz,$$

$$T_2 = \frac{E}{1-\mu^2}\int_{-\delta/2}^{+\delta/2}\left\{\frac{1+z/R_1}{1+z/R_2}(\varepsilon_2 + z\kappa_2) + \mu(\varepsilon_1 + z\kappa_1)\right\} dz,$$

$$T_{12} = \frac{E}{2(1+\mu)}\int_{-\delta/2}^{+\delta/2}\left\{\left(1 - \frac{z^2}{R_1 R_2}\right)\omega + \right.$$
$$\left. + 2\left[1 + \frac{1}{2}\left(\frac{1}{R_1} + \frac{1}{R_2}\right)z\right]z\tau\right\}\frac{dz}{1+z/R_1},$$

$$T_{21} = \frac{E}{2(1+\mu)}\int_{-\delta/2}^{+\delta/2}\left\{\left(1 - \frac{z^2}{R_1 R_2}\right)\omega + \right.$$
$$\left. + 2\left[1 + \frac{1}{2}\left(\frac{1}{R_1} + \frac{1}{R_2}\right)z\right]z\tau\right\}\frac{dz}{1+z/R_2}, \qquad (10.14)$$

$$M_1 = \frac{E}{1-\mu^2}\int_{-\delta/2}^{+\delta/2}\left\{\frac{1+z/R_2}{1+z/R_1}(\varepsilon_1 + z\kappa_1) + \mu(\varepsilon_2 + z\kappa_2)\right\} z\,dz,$$

$$M_2 = \frac{E}{1-\mu^2}\int_{-\delta/2}^{+\delta/2}\left\{\frac{1+z/R_1}{1+z/R_2}(\varepsilon_2 + z\kappa_2) + \mu(\varepsilon_1 + z\kappa_1)\right\} z\,dz,$$

$$M_{12} = \frac{E}{2(1+\mu)}\int_{-\delta/2}^{+\delta/2}\left\{\left(1 - \frac{z^2}{R_1 R_2}\right)\omega + \right.$$
$$\left. + 2\left[1 + \frac{1}{2}\left(\frac{1}{R_1} + \frac{1}{R_2}\right)z\right]z\tau\right\}\frac{z\,dz}{1+z/R_1},$$

$$M_{21} = \frac{E}{2(1+\mu)} \int_{-\delta/2}^{+\delta/2} \left\{ \left(1 - \frac{z^2}{R_1 R_2}\right) \omega + \right.$$
$$\left. + 2\left[1 + \frac{1}{2}\left(\frac{1}{R_1} + \frac{1}{R_2}\right) z\right] z\tau \right\} \frac{z\,dz}{1 + z/R_2}.$$

All the integrals in (10.14) can be evaluated. However, if no simplifications are made, one obtains very cumbersome expressions involving a number of small terms the magnitude of which is such that they lie beyond the limits of accuracy of the initial assumptions of the theory of shells.

Taking into consideration the accuracy of these assumptions, one has to omit in (10.14) quantities of order δ/R compared with unity. Hence, at a first glance, one must put there

$$1 + z/R_1 \approx 1 + z/R_2 \approx 1 - z^2/R_1 R_2 \approx 1,$$
since
$$-\delta/2 \leqslant z \leqslant \delta/2. \tag{10.15}$$

As a result of this simplification, the following relations between the forces, moments and the deformations of the middle surface are obtained:

$$T_1 = \frac{E\delta}{1-\mu^2}(\varepsilon_1 + \mu\varepsilon_2), \quad T_2 = \frac{E\delta}{1-\mu^2}(\varepsilon_2 + \mu\varepsilon_1),$$
$$T_{12} = T_{21} = \frac{E\delta}{2(1+\mu)}\omega,$$
$$M_1 = \frac{E\delta^3}{12(1-\mu^2)}(\kappa_1 + \mu\kappa_2), \quad M_2 = \frac{E\delta^3}{12(1-\mu^2)}(\kappa_2 + \mu\kappa_1), \tag{10.16}$$
$$M_{12} = M_{21} = \frac{E\delta^3}{12(1+\mu)}\tau.$$

These are very similar to the formulae (10.10) in that they differ from them only by the absence of terms in the expressions for T_{12} and T_{21}. The formulae (10.16) are used more often in the theory of shells. Love recommended them as first approximations, but he made the mistake of stating that these expressions correspond to the definition of the potential energy of shells by the expression (9.12), (cf. Love [49] p. 557). The merit of the formulae (10.16) lies in their simplicity and their complete analogy with the corresponding formulae of the theory of plates.

However, the adoption of these formulae introduces several contradictions into the theory of shells. For example, substituting these expressions into the equilibrium equations for an element of the shell, one obtains differential equations in the displacement components for which it is impossible to select stress functions. In other words, a shell, the elastic properties of which are expressed by (10.16), is such that not all deformations of the elastic body are expressible as potential energy of deformation. A further consequence of this discrepancy is that the reciprocal theorem does not hold. Further, substituting (10.16) into the six equilibrium equations (7.9) for an element of the shell, it may be verified that they are not identically satisfied. Meanwhile, as has already been shown in § 7, the fact that these equations are satisfied identically secures the symmetry of the stress tensor from which follows that (10.16) contradict these fundamental properties of the stresses. These contradictions, introduced into the theory of shells by the formulae (10.16), were first pointed out by V. Z. Vlasov. The author of this book does not agree with the excessive exaggeration of the harmful role played by these contradictions in the theory of shells, since no examples are known where they would introduce mathematical inaccuracies exceeding those due to the initial assumptions underlying the theory of thin shells. However, at the same time, he cannot, of course, deny that the presence of such contradictions is a defect of a theory which admits them. All these contradictions may be removed, if the laws relating the forces, moments and deformations of the middle surface are selected in the form (10.8), (10.10) rather than in the form (10.16), as it will be recommended here.

It should be pointed out that until quite recently differences in opinion have existed regarding the best way of writing the relations between the forces, moments and deformations, and a considerable number of variations of these formulae have been proposed which were the result of neglecting (sometimes in a quite arbitrary manner) one or the other term in (10.14). Only one of these variations will be mentioned here which is more logical and was proposed by Lur'e [17].

Since Lur'e did not have at his disposal a criterion for discarding small terms, he proceeded, in order to avoid errors, by expanding the integrands in (10.14) in power series of z, omitting all terms of order larger than z^2. As a result he obtained the expressions

$$
\begin{aligned}
T_1 &= \frac{E\delta}{1-\mu^2}\left[\varepsilon_1 + \mu\varepsilon_2 - \frac{\delta^2}{12}\left(\frac{1}{R_1} - \frac{1}{R_2}\right)(\kappa_1 - \varepsilon_1/R_1)\right],\\
T_2 &= \frac{E\delta}{1-\mu^2}\left[\varepsilon_2 + \mu\varepsilon_1 + \frac{\delta^2}{12}\left(\frac{1}{R_1} - \frac{1}{R_2}\right)(\kappa_2 - \varepsilon_2/R_2)\right], \qquad (10.17)\\
T_{12} &= \frac{E\delta}{2(1+\mu)}\left[\omega - \frac{\delta^2}{12}\left(\frac{1}{R_1} - \frac{1}{R_2}\right)\left(\tau - \frac{\omega}{R_1}\right)\right],
\end{aligned}
$$

$$
\begin{aligned}
T_{21} &= \frac{E\delta}{2(1+\mu)}\left[\omega + \frac{\delta^2}{12}\left(\frac{1}{R_1}-\frac{1}{R_2}\right)\left(\tau - \frac{\omega}{R_2}\right)\right],\\
M_1 &= \frac{E\delta^3}{12(1-\mu^2)}\left[\kappa_1 + \mu\kappa_2 - \left(\frac{1}{R_1}-\frac{1}{R_2}\right)\varepsilon_1\right],\\
M_2 &= \frac{E\delta^3}{12(1-\mu^2)}\left[\kappa_2 + \mu\kappa_1 + \left(\frac{1}{R_1}-\frac{1}{R_2}\right)\varepsilon_2\right],\\
M_{12} &= \frac{E\delta^3}{12(1+\mu)}\left(\tau - \frac{\omega}{2R_1}\right),\quad M_{21} = \frac{E\delta^3}{12(1+\mu)}\left(\tau - \frac{\omega}{2R_2}\right).
\end{aligned} \tag{10.17}
$$

Comparing these formulae with (10.10) and (10.16), it is seen that they are considerably more cumbersome and it is clear from the above discussion that these additional terms introduce into the theory corrections which do not exceed the accuracy of the initial assumptions. Hence the formulae (10.17) have come to be regarded as inconsistent and, for shells of arbitrary form, they introduce into the equations quite unnecessary complications. However, for the particular case of spherical shells, the equations are essentially simplified thanks to the presence of the factor $(1/R_1 - 1/R_2)$ and it turns out that for such shells, which are of great practical importance, simpler equations are obtained by use of (10.17) than by (10.10). Therefore, when considering spherical shells, it will be expedient to employ the formulae (10.17).

In conclusion, it will be noted that the formulae (10.10) were first obtained in 1944 by the author of this book [32] and, almost simultaneously, by L. I. Balabukh. Balabukh sought the simplest form of relation between the forces, moments and deformations of the middle surface which would satisfy the reciprocal law and the six equilibrium equations of an element of a shell and arrived at his result by trial.

The present author sought that form of the same relation which, apart from being simplest, would simultaneously ensure solution of any problem of the theory of shells to an accuracy, not exceeding that of the initial assumptions of the given theory. It turned out that the results of these two different approaches coincided.

§ 11. The boundary conditions for the problem of stresses in a shell

The solutions of the differential equations (8.2), (10.8) naturally do not yet determine completely the state of stress in a shell as long as they are not subject to boundary conditions, i.e., until a certain number of relations between the forces, moments, displacements or functions of these quantities at the supporting edge or edges of the shell are specified.

It will now be shown how many boundary conditions must be imposed in any given case; in addition, the accepted formulation of these boundary conditions will be presented.

Consider first the case when the shell has no boundary, i.e., when it is completely closed or complete. It is evident that in this case the concept of the boundary condition loses its meaning. However, it does not follow from this that any solution satisfying (8.2), (10.8) and (4.23) is in this case a solution of the stress problem of the shell. The coordinate lines α_1, α_2 on the middle surface of a complete shell will themselves be closed curves. Letting $\alpha_2 =$ const. and allowing α_1 to vary (or vice versa), one will periodically return to the same point of the middle surface. Since the displacements of this point will be completely determined quantities, one must impose (in this case) on the solutions of the system (8.2), (10.8) the conditions that the solutions are to be periodic functions of α_1 and α_2 (with such a period that the above mentioned single-valuedness of the displacements will be insured). Thus, in the case of complete shells, the boundary conditions are replaced by the conditions of periodicity.

It may at times happen that a shell is complete with respect to the one coordinate and incomplete with regard to the other, as is, for example, the case for cylindrical tubes. In this case, a condition of periodicity will apply only for one of the coordinates.

Proceeding now to shells with boundaries, consideration will be limited to the simplest case where the boundaries coincide with lines of principal curvature of the middle surface, since this case is encountered most frequently in practice. The formulations of the boundary conditions for the general case may be found, for example, in the book by Love [49].

Thus, let one of the coordinate lines, for example, the line $\alpha_2 = \alpha_2^0 =$ $=$ const. be the boundary of a shell which, as has been stated in § 1, will be assumed to be generated by the motion of the normal to the middle surface along this line.

In the present theory, the stresses acting on the boundary will be integrated over the thickness and replaced by three forces per unit length (in this case T_{21}, T_2 and N_2) and by two moments per unit length (here M_2 and M_{21}). Hence one would think that the number of boundary conditions, completely determining the solution, must be equal to five (the number of quantities, determining the state of stress).

However, the problem of the stresses in a plate (as in a shell, since the analogous reasoning may also be applied in this case) is not completely defined by five, but only by four boundary conditions. In fact, the twisting moment on the boundary of the shell can be replaced by corresponding, distributed, tangential and transverse forces. This statement will now be proved.

For this purpose, consider a segment of the boundary of the middle surface near some point m_1 (Fig. 11 which shows the plane, containing the tangent to the edge at the point m_1 and the normal to the middle surface at this point). The boundary curve will now be approximated by a polygon, formed by a number of very small segments of equal length. The length of each segment, with high order accuracy, may be set equal to the length of the arc spanning it

$$m_1m_2 = mm_1 \approx ds_1.$$

Next, place half way between the points m, m_1 and m_1, m_2 the points n_1 and n_2. Then it is evident that

$$n_1n_2 \approx ds_1.$$

The value of the twisting moment at the point n_1 will be M_{21}, while at the point n_2 it will be given by

$$M_{21} + \frac{\partial M_{21}}{\partial s_1} ds_1.$$

Further, replace the twisting moment acting over the boundary mm_1 by two forces, equal in magnitude and opposite in direction,

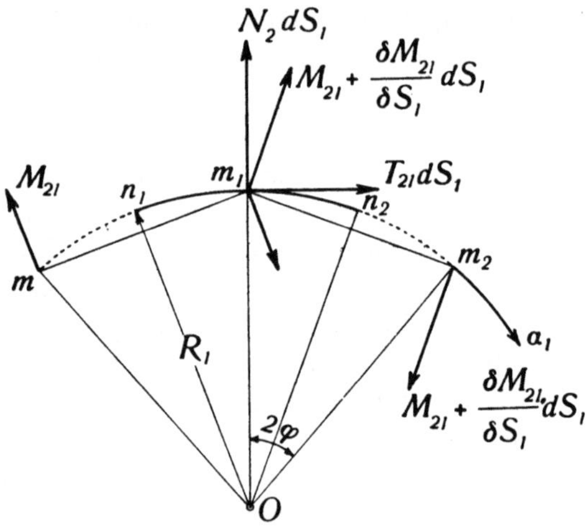

Fig. 11.

which are placed at the ends of the arc mm_1. For this purpose, the force placed at the point m_1 will be directed along the normal to the middle surface at the point n_1 (i.e., parallel to On_1 in Fig. 11) and the force at the point m in the opposite direction. In a similar manner, the twisting moment in the adjacent section m_1m_2 will be replaced by two forces, parallel to the normal at the point n_2, etc.

Then, at each of the corners of the above polygon, there will apply two forces. One will be directed along the normal at the point n_1 and it will be equal to the twisting moment, belonging to the arc ds_1 and divided by that moment arm. The second will be directed along the normal at the point n_2 a distance ds_1 from n_1 (and pointing in the opposite direction to the first force). Its magnitude will be

$$M_{21} + \frac{\partial M_{21}}{ds_1} ds_1.$$

Projecting both these forces, acting at the point m_1, on the direction of the normal to the middle surface (the direction of N_2) and on the direction of the tangent to the middle surface (the direction of T_{21}), one obtains:

a) in the first case

$$\frac{\partial M_{21}}{ds_1} ds_1 + M_{21} - M_{21} = \frac{\partial M_{21}}{ds_1} ds_1;$$

b) in the second case

$$\left(2M_{21} + \frac{\partial M_{21}}{\partial s_1} ds_1\right) \varphi,$$

where 2φ is the angle between the normals to the middle surface at the points n_1 and n_2, i.e.,

$$2\varphi = \frac{ds_1}{R_1}.$$

Substituting this value for φ into the expression b) and retaining only first order quantities, one finds

$$\frac{M_{21}}{R_1} ds_1.$$

Thus, along the edge of the shell, the twisting moment can be replaced by distributed normal shear forces

$$\frac{\partial M_{21}}{\partial s_1}$$

and distributed tangential forces

$$\frac{M_{21}}{R_1}.$$

By replacing the twisting moments by statically equivalent forces, one admits a certain redistribution of the stresses over the thickness of the shell along its edges. By the principle of St. Venant, the effect of this approximation cannot be significant; it can only be noticeable within the immediate neighbourhood of the edges (at distances of the order of the thickness of the shell).

It follows from the above that, in the given case when the boundary is a line $\alpha_2 = \alpha_2^0 =$ constant, the four quantities

$$T_2,\ T_{21} + \frac{M_{21}}{R_1},\ N_2 + \frac{1}{A_1}\frac{\partial M_{21}}{\partial \alpha_1},\ M_2$$

completely determine the state of stress at the edge of the shell; therefore, the number of boundary conditions at each of its edges must be equal to four, and not five.

Notice further that, by (8.1) and the second of the equations (7.8),

$$T_{21} + \frac{M_{21}}{R_1} = T_{21} - \frac{M_{12}}{R_1} + \frac{1}{R_1}(M_{21} + M_{12}) = S + 2\frac{H}{R_1}, \qquad (11.1)$$

$$N_2 + \frac{1}{A_1}\frac{\partial M_{21}}{\partial \alpha_1} = \frac{1}{A_1 A_2}\left(2\frac{\partial A_2 H}{\partial \alpha_1} + \frac{\partial A_1 M_2}{\partial \alpha_2} - \frac{\partial A_1}{\partial \alpha_2} M_1\right). \qquad (11.2)$$

Hence it is clear that the forces T_{12}, T_{21} and the moments M_{12}, M_{21} do not enter separately into the boundary conditions, but only in the combinations denoted earlier by S and H.

Using the above results, it is not difficult to deduce by analogy the boundary conditions for the case when the edge of the shell is a line

$$\alpha_1 = \alpha_1^0 = \text{const}.$$

Of course, it is not always the case that the boundary conditions will be expressed just in terms of the four quantities above. The former may also be given indirectly, for example, by prescribing displacements or angles of rotation at the edge. This leads to a number of possible variations of the boundary conditions:

a) Free edge

$$T_2 = 0,\ T_{21} + \frac{M_{21}}{R_1} = 0,\ N_2 + \frac{1}{A_1}\frac{\partial M_{21}}{\partial \alpha_1} = 0,\ M_2 = 0;$$

b) Hinged edge with fixed support

$$M_2 = 0,\ u = 0,\ v = 0,\ w = 0;$$

c) Hinged edge with support free to move in the normal direction

$$M_2 = 0,\ N_2 + \frac{1}{A_1}\frac{\partial M_{21}}{\partial \alpha_1} = 0,\ u = 0,\ v = 0;$$

d) Clamped edge

$$u = 0,\ v = 0,\ w = 0,\ \psi = -\frac{1}{A_2}\frac{\partial w}{\partial \alpha_2} + \frac{v}{R_2} = 0$$

(where the last of these equalities expresses the fact that the angle of rotation of the edge of the shell is equal to zero — formula (3.16)).

The above alternatives of the boundary conditions do not exhaust all possibilities. Depending on the character of the supports on which the shell rests, its boundary conditions may assume a great variety of forms. Although the number of boundary conditions for plates as well as for shells equals four, the problem of the stresses in a plate[1] breaks up into two independent problems (the plane problem and the problem of bending) and therefore the boundary conditions likewise comprise two independent groups. Therefore, the bending of plates is determined by two conditions at each of their boundaries. For shells, excluding the simplest problems, one is obliged, as a rule, to deal with all four boundary conditions. Hence, the boundary conditions in the theory of shells are more diversified than in the theory of beams and plates. In particular, the concepts of "support" and "clamping" are quite determined in the theory of beams and plates, while in the theory of shells there is no such degree of definiteness, since a shell, completely unmovable with regard to one displacement, may be free or freely supported with regard to the other displacement.

§ 11a. On the deformation of an element of the edge surface of a shell

As in the preceding section, consider the case when the edge of the shell is a coordinate line, for example, the line α_1. In this case the shell edge will be a ruled surface, generated by motion of the perpendicular to the mid-surface along this coordinate line.

As parameters determining the position of points on the edge surface, one must use z, the distance of the point from the mid-surface, and α_1, the coordinate of the middle surface corresponding to the perpendicular drawn from the point to the middle surface.

The coordinate lines of this system are, firstly, the family of straight lines, normal to the middle surface along its edge, the lines z, and, secondly, the family of curves each of which lies at equal distance from the edge of the middle surface, the lines α_1. Construct now at an arbitrary point N of the edge surface the triad of unit vectors $\mathbf{k}_1$, $\mathbf{k}_z$, $\mathbf{k}_n$, where $\mathbf{k}_1$, $\mathbf{k}_z$ will be tangential to the lines α_1 and z through the selected point and $\mathbf{k}_n$ is normal to the edge surface.

Further, construct also at the point M of the middle surface, which is the projection of the point N, the triad of directions $\mathbf{e}_1$, $\mathbf{e}_2$, $\mathbf{e}_n$. It follows from Fig. 11a, which shows a section of the edge of the shell

[1] Having in mind the linear theory of plates.

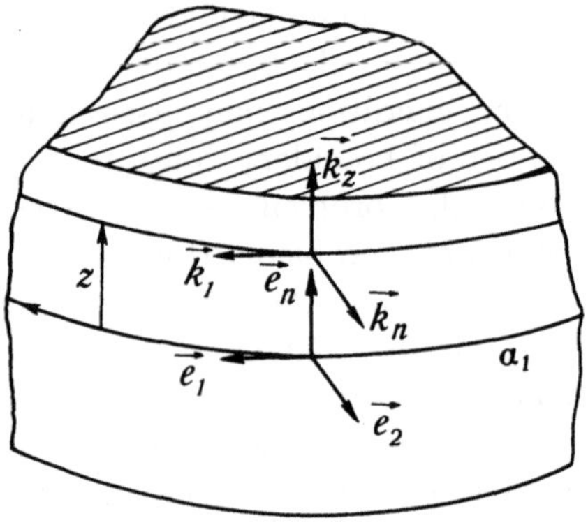

Fig. 11a.

with an adjoining section of the middle surface, that

$$\mathbf{k}_1 = \mathbf{e}_1, \quad \mathbf{k}_2 = \mathbf{e}_n, \quad \mathbf{k}_n = \mathbf{e}_2, \tag{11.1a}$$

that the element of arc along the coordinate line z of the edge is equal to

$$ds_2' = dz$$

and the element of arc of the coordinate line α_1 of the edge is equal to

$$ds_1' = \left(1 + \frac{z}{R_1}\right) ds_1 = A_1\left(1 + \frac{z}{R_1}\right) d\alpha_1,$$

where ds_1 is the element of arc of the line α_1 on the middle surface.

Hence the Lamé parameters of the curvilinear coordinate system α_1, z are given by

$$A_1' = A_1\left(1 + \frac{z}{R_1}\right), \quad A_2' = 1. \tag{11.2a}$$

It is easy now to determine also the radii of curvature of normal sections of the boundary surface, tangential to α_1 and z, respectively. One has

$$\frac{1}{R_1'} = -\frac{\partial \mathbf{k}_1}{\partial s_1'}\mathbf{k}_n = -\frac{1}{A_1(1 + z/R_1)}\frac{\partial \mathbf{e}_1}{\partial \alpha_1}\mathbf{e}_2 = \frac{1}{1 + z/R_1}\frac{1}{A_1A_2}\frac{\partial A_1}{\partial \alpha_2},$$

$$\frac{1}{R_2'} = -\frac{\partial \mathbf{k}_z}{\partial s_2'}\mathbf{k}_n = -\frac{\partial \mathbf{e}_n}{\partial z}\mathbf{e}_2 = 0. \tag{11.3a}$$

Note that the above coordinate system on the edge is not only orthogonal, but also principal. The last follows from the fact that for the displacement of the triad $\mathbf{k}_1$, $\mathbf{k}_z$, $\mathbf{k}_n$ along the line z the unit vector $\mathbf{k}_n$ remains always in the same plane; as has been mentioned in § 2, this is the property of the lines of principal curvature. Thus, for the purpose of a study of the edge, one can use the formulae (4.23), derived earlier for the middle surface, since they apply obviously to any surface, defined in terms of its principal coordinates.

In view of the correspondence between the triads defined on the middle surface and on the edge surface, expressed by (11.1a), one obtains the following formulae for the components of the displacement of an arbitrary point of the boundary surface:

$$u' = u_z = u + z\vartheta, \quad w' = v_z = v + z\psi. \qquad v' = w_z = w, \tag{11.4a}$$

Substituting (11.4a) into (4.23) and taking into consideration the Gauss-Codazzi relations (2.28) and the fact that for the edge $\alpha_2 = z$, one obtains

$$\varepsilon_2' = \omega' = \kappa_2' = 0, \tag{11.5a}$$

where ε_2', κ_2' are the direct strain and change of curvature of the edge surface in the direction z, and ω' is the shear of the edge surface, i.e., the change of angle between the lines z and α_1, caused by the deformation.

The fact that these three strain parameters of the edge surface turn out to be equal to zero is equivalent to the introduction of the Kirchhoff hypothesis that the perpendiculars to the middle surface remain perpendicular during the deformation and that the length of the corresponding fibres of the shell is invariant.

The remaining three strain parameters of the edge surface then are

$$\begin{aligned}
\varepsilon_1' &= \frac{1}{1 + z/R_1}(\varepsilon_1 + z\kappa_1),\\
\kappa_1' &= \frac{1}{1 + z/R_1}\left\{\frac{1}{R_1}\psi + \frac{1}{A_1}\frac{\partial\omega_1}{\partial\alpha_1} - \frac{z}{A_1}\frac{\partial}{\partial\alpha_1}\left[\frac{\tau_1 - \omega_1/R_1}{1 + z/R_1}\right]\right\},\\
\tau' &= -\frac{\tau_1 - \omega_1/R_1}{(1 + z/R_1)^2}.
\end{aligned} \tag{11.6a}$$

Thus, it is observed that the deformation of the edge is entirely determined when on the corresponding edge of the middle surface the following four parameters are given as functions of the displacement of the middle surface:

$$\varepsilon_1, \kappa_1, \tau_1 - \omega_1/R_1, \psi/R_1 + (1/A_1)(\partial\alpha_1/\partial\omega_1). \tag{11.7a}$$

In fact, if these parameters are given on the edge of the middle surface $\alpha = \alpha_2^0$, i.e., as functions of α_1, then, using (11.6a), one can find all non-zero parameters of the deformation of the edge at any point. In this context, the parameter ε_1 corresponds to the relative elongation of the edge line of the middle surface and the parameter κ_1 to its distortion in the plane normal to the middle surface.

As can be seen from (11.6a), $\psi/R_1 + (1/A_1)(\partial\omega_1/\partial\alpha_1)$ and $\tau_1 - \omega_1/R_1$ measure the distortion of the edge of the shell in the plane, perpendicular to the edge surface, and hence tangential to the middle surface, and the twist of the edge of the middle surface, respectively.

Thus, if the Kirchhoff hypothesis is accepted, the deformation of the edge surface of the shell is completely determined by the four parameters, characterizing the detormation of the corresponding edge of the middle surface.

Reasoning in an analogous manner, one can arrive at the conclusion that the deformation of the edge surface $\alpha_1 = \alpha_1{}^0 =$ const. is completely determined by specification on the corresponding edge of the middle surface of the four parameters

$$\varepsilon_2,\ \kappa_2,\ \tau_2 - \omega_2/R_2,\ (\vartheta/R_2) + (1/A_2)(\partial\omega_2/\partial\alpha_2) \tag{11.8a}$$

An important advantage which these parameters have over the customary geometrical quantities is that they can be formulated as strain components. This follows from the fact that

$$\tau_2 - \omega_2/R_2 = \tau - \omega/R_2,$$

$$\frac{\vartheta}{R_2} + \left(\frac{1}{A_2}\right)\left(\frac{\partial\omega_1}{\partial\alpha_2}\right) = -\left(\frac{1}{A_1A_2}\right)\left(\frac{\partial A_2\varepsilon_2}{\partial\alpha_1} - \frac{1}{A_1}\frac{\partial A_1\omega}{\partial\alpha_2} - \frac{\partial A_2}{\partial\alpha_1}\varepsilon_1\right).$$

The above discussion makes it possible to modify somewhat the formulation of the boundary conditions in comparison with those stated in the last section. Thus, if the edge of a shell $\alpha_1 = \alpha_1{}^0$ is clamped, then, as has already been indicated, one must equate to zero all three components of the displacement and the corresponding angle of rotation, i.e., one will have

$$u = v = w = \psi = 0. \tag{11.9a}$$

On the basis of the above discussion, the same boundary condition can be formulated in a different manner, namely:

$$\varepsilon_2 = \kappa_2 = \tau - \frac{\omega}{R_2} = \frac{1}{A_1A_2}\left(\frac{\partial A_2\varepsilon_2}{\partial\alpha_1} - \frac{1}{A_1}\frac{\partial A_1\omega}{\partial\alpha_2} - \frac{\partial A_2}{\partial\alpha_1}\varepsilon_1\right), \tag{11.10a}$$

since these equalities express absence of any deformation of the edge surface of the shell, which is equivalent to its being rigidly clamped.

In the stress analysis of shells, one encounters often problems posed by the junction of two shells. In such a case, it is required to analyse a shell consisting of several shells of the same (or different) thickness, the edges of which are rigidly welded to each other. Problems of this type will be considered in Chapters III and IV. If in this context along the joint between two shells the middle surfaces have a common tangent plane, then the condition for joining the shells reduces to the requirement that on the common edge of the two shells one must ensure continuity of the displacements, of the angle of rotation, as well as of the forces and moments acting in the edge section (cf. last section).

Thus, if the common edge of two shells is the line $\alpha_1 = \alpha_1^0 = \text{const.}$, then for transition through this boundary the following eight quantities must be continuous:

$$u,\ v,\ w,\ \vartheta,\ T_1,\ T_{12} + \frac{M_{12}}{R_2},\ M_1,\ N_1 + \frac{1}{A_2}\frac{\partial M_{12}}{\partial \alpha_2}. \qquad (11.11a)$$

However, in accordance with the results of this and the last sections, the conditions at the junction of two shells can also be formulated in a different manner, namely in the form of a requirement of the continuity on the common edge of the shells $\alpha_1 = \alpha_1^0 = \text{const.}$ of the eight quantities

$$\begin{gathered} T_1,\ S + 2\frac{H}{R_2},\ M_1,\ \frac{1}{A_1 A_2}\left(\frac{\partial A_2 M_1}{\partial \alpha_1} + 2\frac{\partial A_1 H}{\partial \alpha_2} - \frac{\partial A_2}{\partial \alpha_1} M_2\right), \\ \varepsilon_2,\ \kappa_2,\ \tau - \frac{\omega}{R_2},\ \frac{1}{A_1 A_2}\left(\frac{\partial A_2 \varepsilon_2}{\partial \alpha_1} - \frac{1}{A_1}\frac{\partial A_1 \omega}{\partial \alpha_2} - \frac{\partial A_2}{\partial \alpha_1}\varepsilon_1\right). \end{gathered} \qquad (11.12a)$$

The equalities (11.12a) express that on the common edge of the shells one must have, first of all, the same deformation of their edge surfaces and, secondly, the same corresponding forces and moments.

In what follows, this possibility for modifying the formulation of the boundary conditions of certain problems of the theory of shells will be found to be very convenient.

§ 12. Formulation of the compatibility conditions in terms of forces and moments

On the basis of the formulae (10.8), the strain of the middle surface is expressed in terms of the forces and moments by the following relations:

$$\varepsilon_1 = \frac{1}{E\delta}(T_1 - \mu T_2),\quad \varepsilon_2 = \frac{1}{E\delta}(T_2 - \mu T_1),$$

$$\omega = \frac{2(1+\mu)}{E\delta} S = \frac{2(1+\mu)}{E\delta}\left(T_{12} - \frac{H}{R_2}\right) = \frac{2(1+\mu)}{E\delta}\left(T_{21} - \frac{H}{R_1}\right),$$

$$\kappa_1 = \frac{12}{E\delta^3}(M_1 - \mu M_2),\quad \kappa_2 = \frac{12}{E\delta^3}(M_2 - \mu M_1), \tag{12.1}$$

$$\tau = \frac{12(1+\mu)}{E\delta^3} H.$$

Substituting these expressions into the conditions for the compatibility of strain in the middle surface, i.e., in (5.1), one finds

$$\frac{\partial A_2(M_2 - \mu M_1)}{\partial\alpha_1} - (1+\mu)\left(\frac{\partial A_1 H}{\partial\alpha_2} + \frac{\partial A_1}{\partial\alpha_2} H\right) - \frac{\partial A_2}{\partial\alpha_1}(M_1 - \mu M_2) -$$

$$-\frac{\delta^2}{12R_1}\left[\frac{\partial A_2(T_2 - \mu T_1)}{\partial\alpha_1} - \frac{\partial A_2}{\partial\alpha_1}(T_1 - \mu T_2) - 2(1+\mu)\frac{\partial A_1 S}{\partial\alpha_2} -\right.$$

$$\left. - 2(1+\mu)\frac{R_1}{R_2}\frac{\partial A_1}{\partial\alpha_2} S\right] = 0,$$

$$\frac{\partial A_1(M_1 - \mu M_2)}{\partial\alpha_2} - (1+\mu)\left(\frac{\partial A_2 H}{\partial\alpha_1} + \frac{\partial A_2}{\partial\alpha_1} H\right) - \frac{\partial A_1}{\partial\alpha_2}(M_2 - \mu M_1) -$$

$$-\frac{\delta^2}{12R_2}\left[\frac{\partial A_1(T_1 - \mu T_2)}{\partial\alpha_2} - \frac{\partial A_1}{\partial\alpha_2}(T_2 - \mu T_1) - 2(1+\mu)\frac{\partial A_2 S}{\partial\alpha_1} -\right.$$

$$\left. - 2(1+\mu)\frac{R_2}{R_1}\frac{\partial A_2}{\partial\alpha_1} S\right] = 0, \tag{12.2}$$

$$\frac{M_2 - \mu M_1}{R_1} + \frac{M_1 - \mu M_2}{R_2} + \frac{\delta^2}{12}\frac{1}{A_1A_2}\left\{\frac{\partial}{\partial\alpha_1}\frac{1}{A_1}\left[\frac{\partial A_2(T_2 - \mu T_1)}{\partial\alpha_1} -\right.\right.$$

$$\left. - (1+\mu)\left(\frac{\partial A_1 S}{\partial\alpha_2} + \frac{\partial A_1}{\partial\alpha_2} S\right) - \frac{\partial A_2}{\partial\alpha_1}(T_1 - \mu T_2)\right] +$$

$$+ \frac{\partial}{\partial\alpha_2}\frac{1}{A_2}\left[\frac{\partial A_1(T_1 - \mu T_2)}{\partial\alpha_2} - (1+\mu)\left(\frac{\partial A_2 S}{\partial\alpha_1} + \frac{\partial A_2}{\partial\alpha_1} S\right) -\right.$$

$$\left.\left. - \frac{\partial A_1}{\partial\alpha_2}(T_2 - \mu T_1)\right]\right\} = 0.$$

These three differential relations are the compatibility conditions in terms of the forces and moments. Their name is based on the following considerations. It will be assumed that the forces and moments are known. Substituting their values in (12.1) and replacing in these formulae the strains of the middle surface by their expressions in terms of the displacement components (using 4.23), six partial differential equations for the determination of the three displacements u, v, w will be obtained. It is evident that the given system may turn out to be inconsistent, i.e., it may be found that there do not exist functions u, v, w, satisfying all the six equations.

The necessary and sufficient conditions for such functions to exist are that the forces and moments must satisfy (12.2). Thus the differential relations between the forces and moments (12.2) guarantee the compatibility of the formulae (12.1), which can be regarded as a system of differential equations for the determination of the displacements.

It will be recalled that in the general theory of elasticity the conditions of St. Venant which, when written in terms of stresses are the so-called Beltrami-Michell equations, likewise ensure the possibility of integrating the formulae connecting the displacements and the stresses. Hence the formulae (12.2) are the analogue to the equations of Beltrami-Michell in the theory of shells. In view of the existence of such an analogue, there will be no need to prove the above statements.

It should still be mentioned that the compatibility conditions of the middle surface (12.2), after transformations employing the equilibrium equations for shells, can be reduced to the form

$$(1+\mu)N_1 = \frac{1}{A_1}\frac{\partial M}{\partial \alpha_1} - \frac{\delta^2}{12m_1}\left\{\frac{1}{R_1A_1}\frac{\partial T}{\partial \alpha_1} - \frac{(1+\mu)}{A_2}\left(\frac{1}{R_1}\frac{\partial S}{\partial \alpha_2} + \frac{2S}{R_2A_1}\frac{\partial A_1}{\partial \alpha_2}\right) + \right.$$

$$\left. + \frac{1}{R_1}\left[\frac{1}{R_1A_1}\frac{\partial M}{\partial \alpha_1} + \frac{1+\mu}{A_2}\left(\frac{1}{R_1}\frac{\partial H}{\partial \alpha_2} + \frac{2H}{R_2A_1}\frac{\partial A_1}{\partial \alpha_2}\right)\right] + \frac{1+\mu}{R_1}q_1\right\},$$

$$(1+\mu)N_2 = \frac{1}{A_2}\frac{\partial M}{\partial \alpha_2} - \frac{\delta^2}{12m_2}\left\{\frac{1}{R_2A_2}\frac{\partial T}{\partial \alpha_2} - \frac{1+\mu}{A_1}\left(\frac{1}{R_2}\frac{\partial S}{\partial \alpha_1} + \frac{2S}{R_1A_2}\frac{\partial A_2}{\partial \alpha_1}\right) + \right.$$

$$\left. + \frac{1}{R_2}\left[\frac{1}{R_2A_2}\frac{\partial M}{\partial \alpha_2} + \frac{1+\mu}{A_1}\left(\frac{1}{R_2}\frac{\partial H}{\partial \alpha_1} + \frac{2H}{R_1A_2}\frac{\partial A_2}{\partial \alpha_1}\right)\right] + \frac{1+\mu}{R_2}q_2\right\}, \quad (12.3)$$

$$\frac{M_1-\mu M_2}{R_2} + \frac{M_2-\mu M_1}{R_1} + \frac{\delta^2}{12}\left\{\Delta(T) + \frac{1+\mu}{A_1A_2}\left[\frac{\partial}{\partial \alpha_1}\left(\frac{A_2}{R_1}N_1\right) + \frac{\partial}{\partial \alpha_2}\left(\frac{A_1}{R_2}N_2\right)\right] + \right.$$

$$+ \frac{1+\mu}{A_1A_2}\left[\frac{\partial}{\partial \alpha_1}\left(\frac{1}{R_1}\frac{\partial H}{\partial \alpha_2} + \frac{2H}{R_2A_1}\frac{\partial A_1}{\partial \alpha_2}\right) + \frac{\partial}{\partial \alpha_2}\left(\frac{1}{R_2}\frac{\partial H}{\partial \alpha_1} + \frac{2H}{R_1A_2}\frac{\partial A_2}{\partial \alpha_1}\right)\right] +$$

$$\left. + \frac{1+\mu}{A_1A_2}\left[\frac{\partial}{\partial \alpha_1}(A_2q_1) + \frac{\partial}{\partial \alpha_2}(A_1q_2)\right]\right\} = 0,$$

where

$$T = T_1 + T_2,$$
$$M = M_1 + M_2, \tag{12.4}$$

$$m_1 = 1 + \frac{\delta^2}{12R_1^2}, \quad m_2 = 1 + \frac{\delta^2}{12R_2^2}, \tag{12.5}$$

and the symbol $\Delta\,(\ldots)$ denotes the Laplace operator in the curvilinear coordinates α_1, α_2, i.e.,

$$\Delta(\ldots) = \frac{1}{A_1 A_2}\left\{\frac{\partial}{\partial \alpha_1}\left[\frac{A_2}{A_1}\frac{\partial(\ldots)}{\partial \alpha_1}\right] + \frac{\partial}{\partial \alpha_2}\left[\frac{A_1}{A_2}\frac{\partial(\ldots)}{\partial \alpha_2}\right]\right\}. \tag{12.6}$$

§ 13. The two methods of solution of problems in the theory of shells

In the general theory of elasticity, there exist two methods for solving problems. The first involves replacement in the equilibrium equations of the element of the elastic body of the stresses by their expressions in terms of the displacements. In this way, one obtains for the displacements in a homogeneous, isotropic body a system of three equations, the so-called "equations of Lamé". The second method replaces in the compatibility conditions of St. Venant the strains by their expressions in terms of the stresses. In this way, one obtains the six equations of Beltrami-Michell which, in conjunction with the equilibrium equations, permit the solution of the problem directly in terms of the stresses.

In a similar manner, problems in the theory of shells may be solved by two methods — in terms of the displacements of the middle surface or in terms of the forces and moments. The first method replaces in the equilibrium equations (8.2) the forces and moments by their expressions in terms of the strains of the middle surface. If for this purpose the forces and moments are determined by use of (10.8), one obtains for the displacements of the middle surface a system of three partial differential equations. This system will be of eighth order. Here and in the future discussion, the statement that a system of partial differential equations is of eighth (or, in general, of n-th) order will be understood to mean that the system is equivalent to some single eighth (or n-th) order partial differential equation.

The second method consists in supplementing the equilibrium equations (8.2) by the compatibility relations in terms of the forces and moments which, if use is made of (10.8), will have the form (12.2) or (12.3). Then one obtains for the determination of the unknowns T_1, T_2, S, M_1, M_2, H a system of six partial differential equations (which again are of eighth order).

Because of the hypothesis that the displacements are to be small in comparison with the thickness of the shell, all the equations will be linear in both cases, a fact which, of course, greatly simplifies their study. However, not withstanding this simplification, the problems associated with shell calculations remain very difficult, in view of the complexity and high order of the governing equations. Even for the simplest circular cylindrical shells for which the calculations reduce to the study of a system of equations with constant coefficients, the computations remain involved and by far not all problems of practical interest for such shells have been solved. This statement applies to an even greater extent to other shells for which the computations require the study of differential equations with variable coefficients. In view of this situation, instruction in shell theory has lately been directed towards the search for possibilities of simplifying the problems. Several such possibilities will be considered later on.

§ 14. Complex transformation of the differential equations in the theory of thin shells when Poisson's ratio is equal to zero

Poisson's ratio will now be assumed to be equal to zero and for this particular case the equilibrium equations (8.2) will be written in such a way that they alternate with the continuity equations (12.2). Then, the following system of six equations in six unknowns is obtained:

$$\text{(I)}\quad \frac{\partial A_2T_1}{\partial\alpha_1}+\frac{\partial A_1S}{\partial\alpha_2}+\frac{\partial A_1}{\partial\alpha_2}S-\frac{\partial A_2}{\partial\alpha_1}T_2+\frac{1}{R_1}\left[\frac{\partial A_2M_1}{\partial\alpha_1}-\frac{\partial A_2}{\partial\alpha_1}M_2+\right.$$
$$\left.+2\frac{\partial A_1H}{\partial\alpha_2}+2\frac{R_1}{R_2}\frac{\partial A_1}{\partial\alpha_2}H\right]=-A_1A_2q_1,$$

$$\text{(II)}\quad \frac{\partial A_2M_2}{\partial\alpha_1}-\frac{\partial A_1H}{\partial\alpha_2}-\frac{\partial A_1}{\partial\alpha_2}H-\frac{\partial A_2}{\partial\alpha_1}M_1-\frac{\delta^2}{12R_1}\left[\frac{\partial A_2T_2}{\partial\alpha_1}-\right.$$
$$\left.-\frac{\partial A_2}{\partial\alpha_1}T_1-2\frac{\partial A_1S}{\partial\alpha_2}-2\frac{R_1}{R_2}\frac{\partial A_1}{\partial\alpha_2}S\right]=0,$$

$$\text{(III)}\quad \frac{\partial A_2S}{\partial\alpha_1}+\frac{\partial A_1T_2}{\partial\alpha_2}+\frac{\partial A_2}{\partial\alpha_1}S-\frac{\partial A_1}{\partial\alpha_2}T_1+\frac{1}{R_2}\left[\frac{\partial A_1M_2}{\partial\alpha_2}-\frac{\partial A_1}{\partial\alpha_2}M_1+\right.$$
$$\left.+2\frac{\partial A_2H}{\partial\alpha_1}+2\frac{R_2}{R_1}\frac{\partial A_2}{\partial\alpha_1}H\right]=-A_1A_2q_2,\qquad(14.1)$$

$$\text{(IV)}\quad -\frac{\partial A_2 H}{\partial \alpha_1}+\frac{\partial A_1 M_1}{\partial \alpha_2}-\frac{\partial A_2}{\partial \alpha_1}H-\frac{\partial A_1}{\partial \alpha_2}M_2-\frac{\delta^2}{12R_2}\left[\frac{\partial A_1 T_1}{\partial \alpha_2}-\frac{\partial A_1}{\partial \alpha_2}T_2-\right.$$

$$\left.-2\frac{\partial A_2 S}{\partial \alpha_1}-2\frac{R_2}{R_1}\frac{\partial A_2}{\partial \alpha_1}S\right]=0,$$

$$\text{(V)}\quad \frac{T_1}{R_1}+\frac{T_2}{R_2}-\frac{1}{A_1A_2}\left\{\frac{\partial}{\partial \alpha_1}\frac{1}{A_1}\left[\frac{\partial A_2 M_1}{\partial \alpha_1}+\frac{\partial A_1 H}{\partial \alpha_2}+\frac{\partial A_1}{\partial \alpha_2}H-\frac{\partial A_2}{\partial \alpha_1}M_2\right]+\right.$$

$$\left.+\frac{\partial}{\partial \alpha_2}\frac{1}{A_2}\left[\frac{\partial A_2 H}{\partial \alpha_1}+\frac{\partial A_1 M_2}{\partial \alpha_2}+\frac{\partial A_2}{\partial \alpha_1}H-\frac{\partial A_1}{\partial \alpha_2}M_1\right]\right\}=q_n,$$

$$\text{(VI)}\quad \frac{M_2}{R_1}+\frac{M_1}{R_2}+\frac{\delta^2}{12}\frac{1}{A_1A_2}\left\{\frac{\partial}{\partial \alpha_1}\frac{1}{A_1}\left[\frac{\partial A_2 T_2}{\partial \alpha_1}-\frac{\partial A_1 S}{\partial \alpha_2}-\frac{\partial A_1}{\partial \alpha_2}S-\frac{\partial A_2}{\partial \alpha_1}T_1\right]+\right.$$

$$\left.+\frac{\partial}{\partial \alpha_2}\frac{1}{A_2}\left[\frac{\partial A_1 T_1}{\partial \alpha_2}-\frac{\partial A_2 S}{\partial \alpha_1}-\frac{\partial A_2}{\partial \alpha_1}S-\frac{\partial A_1}{\partial \alpha_2}T_2\right]\right\}=0. \quad (14.1)$$

A study of this system shows that the equations possess the property of symmetry; in fact, the unknowns T_1, T_2, S, M_1, M_2, H enter into equations (I), (III), (IV) in the same way as M_2, M_1, $-H$, $-T_2$, $-T_1$, $-S$ enter into equations (II), (IV), (VI) (overlooking the constant multiplier $\delta^2/12$). This property of symmetry of the continuity equations and the equilibrium equations has already been exploited earlier, namely in § 8, when finding functions which identically satisfy the homogeneous equations of equilibrium of an element of the shell. In this way, the solution of the system (14.1) could be reduced to the solution of a system of altogether three equations of half the order of the system (14.1).

Introduce the three complex auxiliary functions

$$\tilde{T}_1 = T_1 - \frac{i}{c_0}M_2,$$

$$\tilde{T}_2 = T_2 - \frac{i}{c_0}M_1, \quad (14.2)$$

$$\tilde{S} = S + \frac{i}{c_0}H,$$

where

$$c_0 = \frac{\delta}{\sqrt{12}}.$$

Further, express T_1, T_2, S in terms of M_1, M_2, H, $\tilde{T}_1$, $\tilde{T}_2$, $\tilde{S}$ and substitute these expressions in (14.1). As a result, one finds

$$\text{(I)}\quad \frac{i}{c_0}\left[\frac{\partial A_2M_2}{\partial\alpha_1}-\frac{\partial A_1H}{\partial\alpha_2}-\frac{\partial A_1}{\partial\alpha_2}H-\frac{\partial A_2}{\partial\alpha_1}M_1\right]+\frac{\partial A_2\tilde{T}_1}{\partial\alpha_1}+\frac{\partial A_1\tilde{S}}{\partial\alpha_2}+$$

$$+\frac{\partial A_1}{\partial\alpha_2}\tilde{S}-\frac{\partial A_2}{\partial\alpha_1}\tilde{T}_2+\frac{1}{R_1}\left[\frac{\partial A_2M_1}{\partial\alpha_1}-\frac{\partial A_2}{\partial\alpha_1}M_2+2\frac{\partial A_1H}{\partial\alpha_2}+\right.$$

$$\left.+2\frac{R_1}{R_2}\frac{\partial A_1}{\partial\alpha_2}H\right]=-A_1A_2q_1, \tag{14.3}$$

$$\text{(II)}\quad -\frac{c_0^2}{R_1}\left[\frac{\partial A_2\tilde{T}_2}{\partial\alpha_1}-\frac{\partial A_2}{\partial\alpha_1}\tilde{T}_1-2\frac{\partial A_1\tilde{S}}{\partial\alpha_2}-2\frac{R_1}{R_2}\frac{\partial A_1}{\partial\alpha_2}\tilde{S}\right]+\frac{\partial A_2M_2}{\partial\alpha_1}-$$

$$-\frac{\partial A_1H}{\partial\alpha_2}-\frac{\partial A_1}{\partial\alpha_2}H-\frac{\partial A_2}{\partial\alpha_1}M_1-i\frac{c_0}{R_1}\left[\frac{\partial A_2M_1}{\partial\alpha_1}-\frac{\partial A_2}{\partial\alpha_1}M_2+\right.$$

$$\left.+2\frac{\partial A_1H}{\partial\alpha_2}+2\frac{R_1}{R_2}\frac{\partial A_1}{\partial\alpha_2}H\right]=0,$$

$$\text{(III)}\quad \frac{i}{c_0}\left[\frac{\partial A_1M_1}{\partial\alpha_2}-\frac{\partial A_2H}{\partial\alpha_1}-\frac{\partial A_2}{\partial\alpha_1}H-\frac{\partial A_1}{\partial\alpha_2}M_2\right]+\frac{\partial A_2\tilde{S}}{\partial\alpha_1}+\frac{\partial A_1\tilde{T}_2}{\partial\alpha_2}+$$

$$+\frac{\partial A_2}{\partial\alpha_1}\tilde{S}-\frac{\partial A_1}{\partial\alpha_2}\tilde{T}_1+\frac{1}{R_2}\left[\frac{\partial A_1M_2}{\partial\alpha_2}-\frac{\partial A_1}{\partial\alpha_2}M_1+2\frac{\partial A_2H}{\partial\alpha_1}+\right.$$

$$\left.+2\frac{R_2}{R_1}\frac{\partial A_2}{\partial\alpha_1}H\right]=-A_1A_2q_2,$$

$$\text{(IV)}\quad -\frac{c_0^2}{R_2}\left[\frac{\partial A_1\tilde{T}_1}{\partial\alpha_2}-\frac{\partial A_1}{\partial\alpha_2}\tilde{T}_2-2\frac{\partial A_2\tilde{S}}{\partial\alpha_1}-2\frac{R_2}{R_1}\frac{\partial A_2}{\partial\alpha_1}\tilde{S}\right]+\frac{\partial A_1M_1}{\partial\alpha_2}-$$

$$-\frac{\partial A_2H}{\partial\alpha_1}-\frac{\partial A_2}{\partial\alpha_1}H-\frac{\partial A_1}{\partial\alpha_2}M_2-i\frac{c_0}{R_2}\left[\frac{\partial A_1M_2}{\partial\alpha_2}-\frac{\partial A_1}{\partial\alpha_2}M_1+\right.$$

$$\left.+2\frac{\partial A_2H}{\partial\alpha_1}+2\frac{R_2}{R_1}\frac{\partial A_2}{\partial\alpha_1}H\right]=0, \tag{14.3}$$

$$\text{(V)}\quad \frac{\tilde{T}_1}{R_1}+\frac{\tilde{T}_2}{R_2}+\frac{i}{c_0}\left(\frac{M_2}{R_1}+\frac{M_1}{R_2}\right)-\frac{1}{A_1A_2}\left\{\frac{\partial}{\partial\alpha_1}\frac{1}{A_1}\times\right.$$

$$\times\left[\frac{\partial A_2M_1}{\partial\alpha_1}+\frac{\partial A_1H}{\partial\alpha_2}+\frac{\partial A_1}{\partial\alpha_2}H-\frac{\partial A_2}{\partial\alpha_1}M_2\right]+$$

$$\left.+\frac{\partial}{\partial\alpha_2}\frac{1}{A_2}\left[\frac{\partial A_2H}{\partial\alpha_1}+\frac{\partial A_1M_2}{\partial\alpha_2}+\frac{\partial A_2}{\partial\alpha_1}H-\frac{\partial A_1}{\partial\alpha_2}M_1\right]\right\}=q_n,$$

$$
\text{(VI)}\ \frac{M_2}{R_1}+\frac{M_1}{R_2}+\frac{c_0^2}{A_1A_2}\left\{\frac{\partial}{\partial\alpha_1}\frac{1}{A_1}\left[\frac{\partial A_2\tilde{T}_2}{\partial\alpha_1}-\frac{\partial A_1\tilde{S}}{\partial\alpha_2}-\frac{\partial A_1}{\partial\alpha_2}\tilde{S}-\frac{\partial A_2}{\partial\alpha_1}\tilde{T}_1\right]+\right.
$$

$$
+\left.\frac{\partial}{\partial\alpha_2}\frac{1}{A_2}\left[-\frac{\partial A_2\tilde{S}}{\partial\alpha_1}+\frac{\partial A_1\tilde{T}_1}{\partial\alpha_2}-\frac{\partial A_2}{\partial\alpha_1}\tilde{S}-\frac{\partial A_1}{\partial\alpha_2}\tilde{T}_2\right]\right\}+
$$

$$
+i\frac{c_0}{A_1A_2}\left\{\frac{\partial}{\partial\alpha_1}\frac{1}{A_1}\left[\frac{\partial A_2M_1}{\partial\alpha_1}+\frac{\partial A_1H}{\partial\alpha_2}+\frac{\partial A_1}{\partial\alpha_2}H-\frac{\partial A_2}{\partial\alpha_1}M_2\right]+\right.
$$

$$
+\left.\frac{\partial}{\partial\alpha_2}\frac{1}{A_2}\left[\frac{\partial A_2H}{\partial\alpha_1}+\frac{\partial A_1M_2}{\partial\alpha_2}+\frac{\partial A_2}{\partial\alpha_1}H-\frac{\partial A_1}{\partial\alpha_2}M_1\right]\right\}=0.
$$

Now multiply equations (II), (IV), (VI) by $\dfrac{i}{c_0}$ and subtract them from equations (I), (III), (V) respectively. This operation eliminates the terms containing moments and one arrives at the following system of three partial differential equations in three unknowns — the auxiliary functions $\tilde{T}_1$, $\tilde{T}_2$, $\tilde{S}$:

$$
\frac{\partial A_2\tilde{T}_1}{\partial\alpha_1}+\frac{\partial A_1\tilde{S}}{\partial\alpha_2}+\frac{\partial A_1}{\partial\alpha_2}\tilde{S}-\frac{\partial A_2}{\partial\alpha_1}\tilde{T}_2+
$$

$$
+i\frac{c_0}{R_1}\left[\frac{\partial A_2\tilde{T}_2}{\partial\alpha_1}-\frac{\partial A_2}{\partial\alpha_1}\tilde{T}_1-2\frac{\partial A_1\tilde{S}}{\partial\alpha_2}-2\frac{R_1}{R_2}\frac{\partial A_1}{\partial\alpha_2}\tilde{S}\right]=-A_1A_2q_1,
$$

$$
\frac{\partial A_2\tilde{S}}{\partial\alpha_1}+\frac{\partial A_1\tilde{T}_2}{\partial\alpha_2}+\frac{\partial A_2}{\partial\alpha_1}\tilde{S}-\frac{\partial A_1}{\partial\alpha_2}\tilde{T}_1+ \qquad (14.4)
$$

$$
+i\frac{c_0}{R_2}\left[\frac{\partial A_1\tilde{T}_1}{\partial\alpha_2}-\frac{\partial A_1}{\partial\alpha_2}\tilde{T}_2-2\frac{\partial A_2\tilde{S}}{\partial\alpha_1}-2\frac{R_2}{R_1}\frac{\partial A_2}{\partial\alpha_1}\tilde{S}\right]=-A_1A_2q_2,
$$

$$
\frac{\tilde{T}_1}{R_1}+\frac{\tilde{T}_2}{R_2}-i\frac{c_0}{A_1A_2}\left\{\frac{\partial}{\partial\alpha_1}\frac{1}{A_1}\left[\frac{\partial A_2\tilde{T}_2}{\partial\alpha_1}-\frac{\partial A_1\tilde{S}}{\partial\alpha_2}-\frac{\partial A_1}{\partial\alpha_2}\tilde{S}-\frac{\partial A_2}{\partial\alpha_1}\tilde{T}_1\right]+\right.
$$

$$
+\left.\frac{\partial}{\partial\alpha_2}\frac{1}{A_2}\left[\frac{\partial A_1\tilde{T}_1}{\partial\alpha_2}-\frac{\partial A_2\tilde{S}}{\partial\alpha_1}-\frac{\partial A_2}{\partial\alpha_1}\tilde{S}-\frac{\partial A_1}{\partial\alpha_2}\tilde{T}_2\right]\right\}=q_n.
$$

The solution of this system will be equivalent to the solution of the system (14.1), since, once the auxiliary functions $\tilde{T}_1$, $\tilde{T}_2$, $\tilde{S}$ have been found, the quantities T_1, T_2, S, M_1, M_2, H may be determined from (14.2) by separation of the real and imaginary parts.

The system (14.4) includes the equilibrium equations of the shell element and the equations of compatibility for the strains of the middle surface. It is of fourth order, i.e., half the order of the system (14.1), and it contains half the number of unknowns and equations. In view of this fact, this reduction of the system (14.1) to the system (14.4) must be considered an essential simplification of the problem.

It will be expedient to modify somewhat the equations (14.4). Taking into consideration the identities

$$\frac{\partial A_2\tilde{T}_2}{\partial\alpha_1}-\frac{\partial A_1\tilde{S}}{\partial\alpha_2}-\frac{\partial A_1}{\partial\alpha_2}\tilde{S}-\frac{\partial A_2}{\partial\alpha_1}\tilde{T}_1=$$
$$=A_2\frac{\partial\tilde{T}}{\partial\alpha_1}-\frac{\partial A_2\tilde{T}_1}{\partial\alpha_1}-\frac{\partial A_1\tilde{S}}{\partial\alpha_2}-\frac{\partial A_1}{\partial\alpha_2}\tilde{S}+\frac{\partial A_2}{\partial\alpha_1}\tilde{T}_2, \quad (14.5)$$

the first two of the equations (14.4) may be reduced to the form

$$\frac{1}{A_1A_2}\left[\frac{\partial A_2\tilde{T}_1}{\partial\alpha_1}+\frac{\partial A_1\tilde{S}}{\partial\alpha_2}+\frac{\partial A_1}{\partial\alpha_2}\tilde{S}-\frac{\partial A_2}{\partial\alpha_1}\tilde{T}_2\right]+$$
$$+i\frac{c_0}{1-i\dfrac{c_0}{R_1}}\left[\frac{1}{R_1A_1}\frac{\partial\tilde{T}}{\partial\alpha_1}-\frac{1}{R_1A_2}\frac{\partial\tilde{S}}{\partial\alpha_2}-\frac{2\tilde{S}}{R_2}\frac{1}{A_1A_2}\frac{\partial A_1}{\partial\alpha_2}\right]=-\frac{q_1}{1-i\dfrac{c_0}{R_1}}, \quad (14.6)$$

$$\frac{1}{A_1A_2}\left[\frac{\partial A_2\tilde{S}}{\partial\alpha_1}+\frac{\partial A_1\tilde{T}_2}{\partial\alpha_2}+\frac{\partial A_2}{\partial\alpha_1}\tilde{S}-\frac{\partial A_1}{\partial\alpha_2}\tilde{T}_1\right]+$$
$$+i\frac{c_0}{1-i\dfrac{c_0}{R_2}}\left[\frac{1}{R_2A_2}\frac{\partial\tilde{T}}{\partial\alpha_2}-\frac{1}{R_2A_1}\frac{\partial\tilde{S}}{\partial\alpha_1}-\frac{2\tilde{S}}{R_1}\frac{1}{A_1A_2}\frac{\partial A_2}{\partial\alpha_1}\right]=-\frac{q_2}{1-i\dfrac{c_0}{R_2}}, \quad (14.7)$$

where $\tilde{T}=\tilde{T}_1+\tilde{T}_2$.

By use of the same identities, the third of the equations (14.4) can be written in the form

$$\frac{\tilde{T}_1}{R_1}+\frac{\tilde{T}_2}{R_2}-ic_0\Delta(\tilde{T})+i\frac{c_0}{A_1A_2}\left\{\frac{\partial}{\partial\alpha_1}\frac{1}{A_1}\left[\frac{\partial A_2\tilde{T}_1}{\partial\alpha_1}+\frac{\partial A_1\tilde{S}}{\partial\alpha_2}+\frac{\partial A_1}{\partial\alpha_2}\tilde{S}-\frac{\partial A_2}{\partial\alpha_1}\tilde{T}_2\right]+\right.$$
$$\left.+\frac{\partial}{\partial\alpha_2}\frac{1}{A_2}\left[\frac{\partial A_2\tilde{S}}{\partial\alpha_1}+\frac{\partial A_1\tilde{T}_2}{\partial\alpha_2}+\frac{\partial A_2}{\partial\alpha_1}\tilde{S}-\frac{\partial A_1}{\partial\alpha_2}\tilde{T}_1\right]\right\}=q_n, \quad (14.8)$$

where, as before,

$$\Delta(\ldots)=\frac{1}{A_1A_2}\left\{\frac{\partial}{\partial\alpha_1}\left[\frac{A_2}{A_1}\frac{\partial(\ )}{\partial\alpha_1}\right]+\frac{\partial}{\partial\alpha_2}\left[\frac{A_1}{A_2}\frac{\partial(\ )}{\partial\alpha_2}\right]\right\}. \quad (14.9)$$

Further, substituting for the expressions in the square brackets in (14.8) their values from (14.6) and (14.7), one finds

$$\frac{\tilde{T}_1}{R_1}+\frac{\tilde{T}_2}{R_2}-ic_0\Delta(\tilde{T})+\frac{c_0^2}{A_1A_2}\left\{\frac{\partial}{\partial\alpha_1}\frac{1}{1-ic_0/R_1}\left[\frac{1}{R_1}\frac{A_2}{A_1}\frac{\partial\tilde{T}}{\partial\alpha_1}-\frac{1}{R_1}\frac{\partial\tilde{S}}{\partial\alpha_2}-\right.\right.$$

$$\left.\left.-\frac{2\tilde{S}}{R_2}\frac{1}{A_1}\frac{\partial A_1}{\partial\alpha_2}\right]+\frac{\partial}{\partial\alpha_2}\frac{1}{1-ic_0/R_2}\left[\frac{1}{R_2}\frac{A_1}{A_2}\frac{\partial\tilde{T}}{\partial\alpha_2}-\frac{1}{R_2}\frac{\partial\tilde{S}}{\partial\alpha_1}-\frac{2\tilde{S}}{R_1}\frac{1}{A_2}\frac{\partial A_2}{\partial\alpha_1}\right]\right\}=$$

$$=q_n+i\frac{c_0}{A_1A_2}\left[\frac{\partial}{\partial\alpha_1}\left(\frac{A_2q_1}{1-ic_0/R_1}\right)+\frac{\partial}{\partial\alpha_2}\left(\frac{A_1q_2}{1-ic_0/R_2}\right)\right]. \tag{14.10}$$

Hence the system (14.4) may now be written in the following final form:

$$\frac{1}{A_1A_2}\left[\frac{\partial A_2\tilde{T}_1}{\partial\alpha_1}+\frac{\partial A_1\tilde{S}}{\partial\alpha_2}+\frac{\partial A_1}{\partial\alpha_2}\tilde{S}-\frac{\partial A_2}{\partial\alpha_1}\tilde{T}_2\right]+$$

$$+i\frac{c_0}{1-i\dfrac{c_0}{R_1}}\left[\frac{1}{R_1A_1}\frac{\partial\tilde{T}}{\partial\alpha_1}-\frac{1}{R_1A_2}\frac{\partial\tilde{S}}{\partial\alpha_2}-\frac{2\tilde{S}}{R_2}\frac{1}{A_1A_2}\frac{\partial A_1}{\partial\alpha_2}\right]=-\tilde{q}_1(\alpha_1,\alpha_2),$$

$$\frac{1}{A_1A_2}\left[\frac{\partial A_2\tilde{S}}{\partial\alpha_1}+\frac{\partial A_1\tilde{T}_2}{\partial\alpha_2}+\frac{\partial A_2}{\partial\alpha_1}\tilde{S}-\frac{\partial A_1}{\partial\alpha_2}\tilde{T}_1\right]+$$

$$+i\frac{c_0}{1-i\dfrac{c_0}{R_2}}\left[\frac{1}{R_2A_2}\frac{\partial\tilde{T}}{\partial\alpha_2}-\frac{1}{R_2A_1}\frac{\partial\tilde{S}}{\partial\alpha_1}-\frac{2\tilde{S}}{R_1}\frac{1}{A_1A_2}\frac{\partial A_2}{\partial\alpha_1}\right]=-\tilde{q}_2(\alpha_1,\alpha_2),$$

$$\frac{\tilde{T}_1}{R_1}+\frac{\tilde{T}_2}{R_2}-ic_0\Delta(\tilde{T})+ \tag{14.11}$$

$$+\frac{c_0^2}{A_1A_2}\left\{\frac{\partial}{\partial\alpha_1}\frac{1}{1-i\dfrac{c_0}{R_1}}\left[\frac{1}{R_1}\frac{A_2}{A_1}\frac{\partial\tilde{T}}{\partial\alpha_1}-\frac{1}{R_1}\frac{\partial\tilde{S}}{\partial\alpha_2}-\frac{2\tilde{S}}{R_2}\frac{1}{A_1}\frac{\partial A_1}{\partial\alpha_2}\right]+\right.$$

$$\left.+\frac{\partial}{\partial\alpha_2}\frac{1}{1-i\dfrac{c_0}{R_2}}\left[\frac{1}{R_2}\frac{A_1}{A_2}\frac{\partial\tilde{T}}{\partial\alpha_2}-\frac{1}{R_2}\frac{\partial\tilde{S}}{\partial\alpha_1}-\frac{2\tilde{S}}{R_1}\frac{1}{A_2}\frac{\partial A_2}{\partial\alpha_1}\right]\right\}=\tilde{q}_n(\alpha_1,\alpha_2),$$

where

$$\tilde{q}_1(\alpha_1,\alpha_2)=\frac{1}{1-i\dfrac{c_0}{R_1}}q_1,$$

$$\tilde{q}_2(\alpha_1, \alpha_2) = \frac{1}{1 - i\dfrac{c_0}{R_2}} q_2, \tag{14.12}$$

$$\tilde{q}_n(\alpha_1, \alpha_2) = q_n + i\frac{c_0}{A_1 A_2}\left[\frac{\partial}{\partial\alpha_1}\left(\frac{A_2 q_1}{1 - i\dfrac{c_0}{R_1}}\right) + \frac{\partial}{\partial\alpha_2}\left(\frac{A_1 q_2}{1 - i\dfrac{c_0}{R_2}}\right)\right].$$

The auxiliary functions $\tilde{T}_1$, $\tilde{T}_2$, $\tilde{S}$ will in future be called complex stress resultants or stresses. The name is justified by the fact that knowledge of these functions immediately determines all forces and moments.

§ 15. Other forms of the complex transformation of the equations of the theory of thin shells (for $\mu = 0$)

Continuing with the particular case when Poisson's ratio is equal to zero, introduce into the consideration three further complex auxiliary functions, denoted by $\tilde{u}$, $\tilde{v}$, $\tilde{w}$ and related to the complex stresses by the six differential relations

$$\tilde{\varepsilon}_1 = \frac{1}{E\delta}\left(\tilde{T}_1 + \frac{i}{c_0}M_2^*\right), \qquad \tilde{\kappa}_1 = \frac{i}{c_0}\frac{1}{E\delta}(\tilde{T}_2 - T_2^*),$$

$$\tilde{\varepsilon}_2 = \frac{1}{E\delta}\left(\tilde{T}_2 + \frac{i}{c_0}M_1^*\right), \quad (15.1) \qquad \tilde{\kappa}_2 = \frac{i}{c_0}\frac{1}{E\delta}(\tilde{T}_1 - T_1^*), \quad (15.2)$$

$$\tilde{\omega} = \frac{2}{E\delta}\left(\tilde{S} - \frac{i}{c_0}H^*\right), \qquad \tilde{\tau} = -\frac{i}{c_0}\frac{1}{E\delta}(\tilde{S} - \tilde{S}^*).$$

In these relations $\tilde{\varepsilon}_1$, $\tilde{\varepsilon}_2$, $\tilde{\omega}$, $\tilde{\kappa}_1$, $\tilde{\kappa}_2$, $\tilde{\tau}$ will be understood to be the expressions (4.23) after replacing in them the displacements u, v, w by $\tilde{u}$, $\tilde{v}$, $\tilde{w}$, respectively. As regards T_1^*, T_2^*, S^*, M_1^*, M_2^*, H^*, these functions must be selected in such a manner that the system of six equations in the three unknowns $\tilde{u}$, $\tilde{v}$, $\tilde{w}$ will be compatible.

From a mathematical point of view, the problem of the determination of $\tilde{u}$, $\tilde{v}$, $\tilde{w}$ for different right-hand sides of the equations (15.1), (15.2) is identical to the problem of the determination of the displacements of the middle surface for a given deformation, the only inessential difference being that the last are assumed to be complex.

Therefore the condition of compatibility of the equations (15.1), (15.2) requires that their right-hand sides must be subjected to the three differential equations obtained by substitution of $\tilde{\varepsilon}_1$, $\tilde{\varepsilon}_2$, $\tilde{\omega}$, $\tilde{\kappa}_1$, $\tilde{\kappa}_2$, $\tilde{\tau}$ in the compatibility conditions for the strains of the middle surface (5.6) which are simultaneously the compatibility conditions for a system of the form (15.1), (15.2).

After this substitution, one obtains

$$\frac{\partial A_2\tilde{T}_1}{\partial\alpha_1}+\frac{\partial A_1\tilde{S}}{\partial\alpha_2}+\frac{\partial A_1}{\partial\alpha_2}\tilde{S}-\frac{\partial A_2}{\partial\alpha_1}\tilde{T}_2+$$
$$+i\frac{c_0}{R_1}\left[\frac{\partial A_2\tilde{T}_2}{\partial\alpha_1}-\frac{\partial A_2}{\partial\alpha_1}\tilde{T}_1-2\frac{\partial A_1\tilde{S}}{\partial\alpha_2}-2\frac{R_1}{R_2}\frac{\partial A_1}{\partial\alpha_2}\tilde{S}\right]=$$
$$=\frac{\partial A_2T_1^*}{\partial\alpha_1}+\frac{\partial A_1S^*}{\partial\alpha_2}+\frac{\partial A_1}{\partial\alpha_2}S^*-\frac{\partial A_2}{\partial\alpha_1}T_2^*+$$
$$+\frac{1}{R_1}\left[\frac{\partial A_2M_1^*}{\partial\alpha_1}-\frac{\partial A_2}{\partial\alpha_1}M_2^*+2\frac{\partial A_1H^*}{\partial\alpha_2}+2\frac{R_1}{R_2}\frac{\partial A_1}{\partial\alpha_2}H^*\right],$$

$$\frac{\partial A_2\tilde{S}}{\partial\alpha_1}+\frac{\partial A_1\tilde{T}_2}{\partial\alpha_2}+\frac{\partial A_2}{\partial\alpha_1}\tilde{S}-\frac{\partial A_1}{\partial\alpha_2}\tilde{T}_1+$$
$$+i\frac{c_0}{R_2}\left[\frac{\partial A_1\tilde{T}_1}{\partial\alpha_2}-\frac{\partial A_1}{\partial\alpha_2}\tilde{T}_2-2\frac{\partial A_2\tilde{S}}{\partial\alpha_1}-2\frac{R_2}{R_1}\frac{\partial A_2}{\partial\alpha_1}\tilde{S}\right]=$$
$$=\frac{\partial A_2S^*}{\partial\alpha_1}+\frac{\partial A_1T_2^*}{\partial\alpha_2}+\frac{\partial A_2}{\partial\alpha_1}S^*-\frac{\partial A_1}{\partial\alpha_2}T_1^*+$$
$$+\frac{1}{R_2}\left[\frac{\partial A_1M_2^*}{\partial\alpha_2}-\frac{\partial A_1}{\partial\alpha_2}M_1^*+2\frac{\partial A_2H^*}{\partial\alpha_1}+2\frac{R_2}{R_1}\frac{\partial A_2}{\partial\alpha_1}H^*\right],\quad(15.3)$$

$$\frac{\tilde{T}_1}{R_1}+\frac{\tilde{T}_2}{R_2}-i\frac{c_0}{A_1A_2}\left\{\frac{\partial}{\partial\alpha_1}\frac{1}{A_1}\left[\frac{\partial A_2\tilde{T}_2}{\partial\alpha_1}-\frac{\partial A_1\tilde{S}}{\partial\alpha_2}-\frac{\partial A_1}{\partial\alpha_2}\tilde{S}-\frac{\partial A_2}{\partial\alpha_1}\tilde{T}_1\right]+\right.$$
$$\left.+\frac{\partial}{\partial\alpha_2}\frac{1}{A_2}\left[\frac{\partial A_1\tilde{T}_1}{\partial\alpha_2}-\frac{\partial A_2\tilde{S}}{\partial\alpha_1}-\frac{\partial A_2}{\partial\alpha_1}\tilde{S}-\frac{\partial A_1}{\partial\alpha_2}\tilde{T}_2\right]\right\}=\frac{T_1^*}{R_1}+\frac{T_2^*}{R_2}.$$
$$-\frac{1}{A_1A_2}\left\{\frac{\partial}{\partial\alpha_1}\frac{1}{A_1}\left[\frac{\partial A_2M_1^*}{\partial\alpha_1}+\frac{\partial A_1H^*}{\partial\alpha_2}+\frac{\partial A_1}{\partial\alpha_2}H^*-\frac{\partial A_2}{\partial\alpha_1}M_2^*\right]+\right.$$
$$\left.+\frac{\partial}{\partial\alpha_2}\frac{1}{A_2}\left[\frac{\partial A_2H^*}{\partial\alpha_1}+\frac{\partial A_1M_2^*}{\partial\alpha_2}+\frac{\partial A_2}{\partial\alpha_1}H^*-\frac{\partial A_1}{\partial\alpha_2}M_1^*\right]\right\}.$$

But the complex stresses $\tilde{T}_1$, $\tilde{T}_2$, $\tilde{S}$ satisfy the differential equations (14.4), the left hand sides of which are identical with the left hand sides of the relations (15.3). Hence it follows that the right hand sides of the equations (14.4) must be equal to the right hand sides of the relations (15.3) so that one may write

$$\frac{\partial A_2T_1^*}{\partial \alpha_1} + \frac{\partial A_1S^*}{\partial \alpha_2} + \frac{\partial A_1}{\partial \alpha_2} S^* - \frac{\partial A_2}{\partial \alpha_1} T_2^* + \frac{1}{R_1}\left[\frac{\partial A_2M_1^*}{\partial \alpha_1} - \frac{\partial A_2}{\partial \alpha_1} M_2^* + 2\frac{\partial A_1H^*}{\partial \alpha_2} + 2\frac{R_1}{R_2}\frac{\partial A_1}{\partial \alpha_2} H^*\right] = - A_1A_2q_1,$$

$$\frac{\partial A_2S^*}{\partial \alpha_1} + \frac{\partial A_1T_2^*}{\partial \alpha_2} + \frac{\partial A_2}{\partial \alpha_1} S^* - \frac{\partial A_1}{\partial \alpha_2} T_1^* + \frac{1}{R_2}\left[\frac{\partial A_1M_2^*}{\partial \alpha_2} - \frac{\partial A_1}{\partial \alpha_2} M_1^* + 2\frac{\partial A_2H^*}{\partial \alpha_1} + 2\frac{R_2}{R_1}\frac{\partial A_2}{\partial \alpha_1} H^*\right] = - A_1A_2q_2, \quad (15.4)$$

$$\frac{T_1^*}{R_1} + \frac{T_2^*}{R_2} - \frac{1}{A_1A_2}\left\{\frac{\partial}{\partial \alpha_1}\frac{1}{A_1}\left[\frac{\partial A_2M_1^*}{\partial \alpha_1} + \frac{\partial A_1H^*}{\partial \alpha_2} + \frac{\partial A_1}{\partial \alpha_2} H^* - \frac{\partial A_2}{\partial \alpha_1} M_2^*\right] + \frac{\partial}{\partial \alpha_2}\frac{1}{A_2}\left[\frac{\partial A_2H^*}{\partial \alpha_1} + \frac{\partial A_1M_2^*}{\partial \alpha_2} + \frac{\partial A_2}{\partial \alpha_1} H^* - \frac{\partial A_1}{\partial \alpha_2} M_1^*\right]\right\} = q_n.$$

However, the equations (15.4) are the same as the equilibrium equations (8.2); hence it is necessary for the compatibility of the equations (15.1), (15.2) that the quantities T_1^*, T_2^*, S^*, M_1^*, M_2^*, H^* in these equations should represent some particular solution of the equilibrium equations of a shell in terms of forces and moments. Since the system (15.4) contains six unknowns and there are altogether three equations, one has great freedom in the choice of T_1^*, T_2^*, S^*, M_1^*, M_2^*, H^*. For example, one may select $M_1^* = M_2^* = H^* = 0$. Then the remaining quantities T_1^*, T_2^*, S^* will be subject to the system of three equations in three unknowns

$$\frac{1}{A_1A_2}\left[\frac{\partial A_2T_1^*}{\partial \alpha_1} + \frac{\partial A_1S^*}{\partial \alpha_2} + \frac{\partial A_1}{\partial \alpha_2} S^* - \frac{\partial A_2}{\partial \alpha_1} T_2^*\right] + q_1 = 0,$$

$$\frac{1}{A_1A_2}\left[\frac{\partial A_2S^*}{\partial \alpha_1} + \frac{\partial A_1T_2^*}{\partial \alpha_2} + \frac{\partial A_2}{\partial \alpha_1} S^* - \frac{\partial A_1}{\partial \alpha_2} T_1^*\right] + q_2 = 0, \quad (15.5)$$

$$\frac{T_1^*}{R_1} + \frac{T_2^*}{R_2} = q_n.$$

which are well known as the equations of the membrane theory of shells. The following chapter will deal with methods of solution of this system.

One may use for the determination of $T_1^*, \ldots, H^*$ any other simplified version of the equations (15.4), the only restriction being that the supplementary limitations imposed on $T_1^*, \ldots, H^*$ must not violate the system (15.4).

Revert now to the equations (15.1), (15.2) and eliminate from them the complex forces $\tilde{T}_1$, $\tilde{T}_2$, $\tilde{S}$. Then one arrives at the three equalities

$$\begin{aligned} \tilde{\varepsilon}_1 + ic_0\tilde{\kappa}_2 &= \frac{1}{E\delta}\left(T_1^* + \frac{i}{c_0}\right)M_2^*, \\ \tilde{\varepsilon}_2 + ic_0\tilde{\kappa}_1 &= \frac{1}{E\delta}\left(T_2^* + \frac{i}{c_0}M_1^*\right), \\ \tfrac{1}{2}\tilde{\omega} - ic_0\tilde{\tau} &= \frac{1}{E\delta}\left(S^* - \frac{i}{c_0}H^*\right), \end{aligned} \tag{15.6}$$

which, after replacing $\tilde{\varepsilon}_1$, $\tilde{\varepsilon}_2$, $\tilde{\omega}$, $\tilde{\kappa}_1$, $\tilde{\kappa}_2$, $\tilde{\tau}$ in terms of the new unknowns $\tilde{u}$, $\tilde{v}$, $\tilde{w}$, become three linear partial differential equations for these three unknowns.

This system which is similar to the system (14.11) for complex forces will be of fourth order.

In conclusion, the meaning of the complex unknowns $\tilde{u}$, $\tilde{v}$, $\tilde{w}$ will be explained. Let

$$\begin{aligned} \tilde{u} &= u' + iu'', \\ \tilde{v} &= v' + iv'', \\ \tilde{w} &= w' + iw'', \end{aligned} \tag{15.7}$$

Substituting (15.7) and the expressions for the complex forces (14.2) into (15.1), (15.2), one obtains the twelve relations

$$\begin{aligned} \varepsilon_1' &= \frac{1}{E\delta}T_1, \\ \varepsilon_2' &= \frac{1}{E\delta}T_2, \\ \omega' &= \frac{2}{E\delta}S, \\ \kappa_1' &= \frac{12}{E\delta^3}M_1, \end{aligned} \tag{15.8}$$

$$\begin{aligned} \varepsilon_1'' &= \frac{\sqrt{12}}{E\delta^2}(-M_2 + M_2^*), \\ \varepsilon_2'' &= \frac{\sqrt{12}}{E\delta^2}(-M_1 + M_1^*), \\ \omega'' &= \frac{2\sqrt{12}}{E\delta^2}(H - H^*), \\ \kappa_1'' &= \frac{\sqrt{12}}{E\delta^2}(T_2 - T_2^*), \end{aligned} \tag{15.9}$$

$$\kappa_2' = \frac{12}{E\delta^3} M_2, \qquad \tau' = \frac{12}{E\delta^3} H, \tag{15.8}$$

$$\kappa_2'' = \frac{\sqrt{12}}{E\delta^2}(T_1 - T_1^*), \qquad \tau'' = \frac{\sqrt{12}}{E\delta^2}(-S + S^*), \tag{15.9}$$

where $\varepsilon_1', \ldots, \tau'$ are the expressions (4.23) after replacing there u, v, w by u', v', w' and $\varepsilon_1'', \ldots, \tau''$ are the same expressions after replacement of u, v, w by u'', v'', w'', respectively.

Comparing now (15.8) with (12.1) (letting beforehand $\mu = 0$) it is readily verified that u', v', w' are related to the forces and moments by the same relations as the displacements of the middle surface u, v, w. Thus, the real parts of the complex quantities $\tilde{u}, \tilde{v}, \tilde{w}$ are equal (apart from rigid body motion of the shell) to the displacements u, v, w, respectively.

As regards interpretation of the imaginary parts of these unknowns, it will be convenient to introduce new quantities which are proportional to them, namely

$$u''' = \frac{E\delta^2}{\sqrt{12}} u'', \quad v''' = \frac{E\delta^2}{\sqrt{12}} v'', \quad w''' = \frac{E\delta^2}{\sqrt{12}} w''.$$

Then one has instead of (15.9)

$$\begin{aligned} T_1 - T_1^* &= \kappa_2(u''', v''', w''') = \kappa_2''', \\ T_2 - T_2^* &= \kappa_1''', \\ S - S^* &= \tau''', \\ M_1 - M_1^* &= -\varepsilon_2''', \\ M_2 - M_2^* &= -\varepsilon_1''', \\ H - H^* &= \tfrac{1}{2}\omega'''. \end{aligned} \tag{15.10}$$

It follows from these formulae that, selecting from the infinity of particular solutions of the equilibrium equations any completely defined combination $T_1^*, \ldots, H^*$, one can speak of u''', v''', w''' (and, consequently, also of u'', v'', w'') as completely defined quantities (for given forces and moments $T_1, \ldots, H$). Thus, the imaginary parts of the functions are not as well defined as the real parts.

In the particular case when a shell is free from surface loading

($q_1 = q_2 = q_n = 0$), the equilibrium equations turn out to be homogeneous and one can use as statically possible forces and moments

$$T_1^* = T_2^* = S^* = M_1^* = M_2^* = H^* = 0. \tag{15.11}$$

Introducing (15.11) into (15.10) and comparing the resulting formulae with (8.3), it is seen that for the homogeneous problem the imaginary parts of $\tilde{u}$, $\tilde{v}$, $\tilde{w}$ can be identified with the Lure'-Gol'denweizer functions (allowing for an additional constant factor). Thus

$$u'' = \frac{\sqrt{12}}{E\delta^2}\, a, \quad v'' = \frac{\sqrt{12}}{E\delta^2}\, b, \quad w'' = \frac{\sqrt{12}}{E\delta^2}\, c.$$

It follows from (15.11) that for the homogeneous problem the equations (15.6) assume the form

$$\begin{aligned} &\tilde{\varepsilon}_1 + ic_0\tilde{\kappa}_2 = 0, \\ &\tilde{\varepsilon}_2 + ic_0\tilde{\kappa}_1 = 0, \\ &\frac{\tilde{\omega}}{2} - ic_0\tilde{\tau} = 0. \end{aligned} \tag{15.12}$$

In what follows the complex variables $\tilde{u}$, $\tilde{v}$, $\tilde{w}$, introduced in this section, will be called complex displacements (on the basis that their real parts are the displacements of the middle surface u, v, w). Correspondingly, the system of equations (15.6) will be referred to as the equations of the theory of thin shells in terms of complex displacements.

§ 15a. On the possibility of a Complex Formulation of Certain Boundary Conditions

Following K. F. Chernykh [27], consider the four complex expressions

$$\tilde{T}_1, \quad \tilde{T}_{12} + \frac{\tilde{M}_{12}}{R_2} = \left(1 - i\frac{2c_0}{R_2}\right)\tilde{S}, \quad \tilde{M}_1 = ic_0\tilde{T}_2,$$

$$\tilde{N}_1 + \frac{1}{A_2}\frac{\partial \tilde{M}_{12}}{\partial \alpha_2} = \frac{ic_0}{A_1A_2}\left(\frac{\partial A_2\tilde{T}_2}{\partial \alpha_1} - 2\frac{\partial A_1\tilde{S}}{\partial \alpha_2} - \frac{\partial A_2}{\partial \alpha_1}\tilde{T}_1\right). \tag{15.1a}$$

Separating real and imaginary parts in (15.1a) and using (14.2) and (10.8) for $\mu = 0$, one obtains

$$Re\,(\tilde{T}_1) = T_1, \quad Re\left(\tilde{T}_{12} + \frac{\tilde{M}_{12}}{R_2}\right) = S + \frac{2H}{R_2}, \quad Re\,(\tilde{M}_1) = M_1,$$

$$Re\left(\tilde{N}_1 + \frac{1}{A_2}\frac{\partial \tilde{M}_{12}}{\partial \alpha_2}\right) = \frac{1}{A_1A_2}\left(2\frac{\partial A_1H}{\partial \alpha_2} + \frac{\partial A_2M_1}{\partial \alpha_1} - \frac{\partial A_2}{\partial \alpha_1}M_2\right), \quad (15.2a)$$

$$Im\,(\tilde{T}_1) = -\frac{E\delta^2}{\sqrt{12}}\kappa_2, \quad Im\left(\tilde{T}_{12} + \frac{\tilde{M}_{21}}{R_2}\right) = \frac{E\delta^2}{\sqrt{12}}\left(\tau - \frac{\omega}{R_2}\right),$$

$$Im\,(\tilde{M}_1) = \frac{E\delta^2}{\sqrt{12}}\varepsilon_2,$$

$$Im\left(\tilde{N}_1 + \frac{1}{A_2}\frac{\partial \tilde{M}_{12}}{\partial \alpha_2}\right) = \frac{c_0}{A_1A_2}\left(\frac{\partial A_2T_2}{\partial \alpha_1} - 2\frac{\partial A_1S}{\partial \alpha_2} - \frac{\partial A_2}{\partial \alpha_1}T_1\right) =$$

$$= \frac{E\delta^2}{\sqrt{12}}\frac{1}{A_1A_2}\left(\frac{\partial A_2\varepsilon_2}{\partial \alpha_1} - \frac{\partial A_1\omega}{\partial \alpha_2} - \frac{\partial A_2}{\partial \alpha_1}\varepsilon_1\right).$$

It is evident that the expressions (15.1a) have been constructed in such a manner that their real parts are

$$T_1, \quad T_{12} + \frac{M_{12}}{R_2}, \quad M_1, \quad N_1 + \frac{1}{A_2}\frac{\partial M_{12}}{\partial \alpha_2},$$

i.e.,such combinations of forces and moments that, on being specified on the edge of the shell $\alpha_1 = \text{const.}$, they will determine completely the state of stress in the shell (§ 11).

Substituting for ε_1, ε_2, ω in the last formula (15.2a) their expressions in terms of the displacements u, v, w and rearranging, this formula can be reduced to the form

$$Im\left(\tilde{N}_1 + \frac{1}{A_2}\frac{\partial \tilde{M}_{12}}{\partial \alpha_2}\right) = -\frac{E\delta^2}{\sqrt{12}}\left(\frac{\vartheta}{R_2} + \frac{1}{A_2}\frac{\partial \omega_2}{\partial \alpha_2}\right).$$

Hence the imaginary parts of the quantities (15.1a) are proportional to parameters (depending on the displacements u, v, w and their derivatives) such that, if they are given on the boundary of the shell $\alpha_1 = \alpha_1{}^0 = \text{const.}$, they determine completely the deformation of the corresponding edge surface (cf. § 11a).

If the edge $\alpha_1 = \alpha_1{}^0$ is free, then

$$Re\,(\tilde{T}_1) = Re\left(\tilde{T}_{12} + \frac{\tilde{M}_{12}}{R_2}\right) = Re\,(\tilde{M}_1) = Re\left(\tilde{N}_1 + \frac{1}{A_2}\frac{\partial \tilde{M}_{12}}{\partial \alpha_2}\right) = 0.$$

On the other hand, if the edge $\alpha_1 = \alpha_1{}^0$ is clamped, then

$$Im\,(\tilde{T}_1) = Im\left(\tilde{T}_{12} + \frac{\tilde{M}_{12}}{R_2}\right) = Im\,(\tilde{M}_1) = Im\left(\tilde{N}_1 + \frac{1}{A_2}\frac{\partial \tilde{M}_{12}}{\partial \alpha_2}\right) = 0.$$

From the point of view of the use of the complex equations of shell theory, derived in the last two sections, special interest attaches to those variants of the boundary conditions which have been formulated directly in terms of complex forces (or displacements) without the necessity of separation of real and imaginary parts. Problems of this type are, in particular, those of the stress analysis of closed shells (for example, closed spherical or ellipsoidal shells), since in these cases one has to fulfill instead of boundary conditions corresponding conditions of periodicity.

A frequently used boundary condition is that which can be called a freely hinged edge (also referred to as Navier boundary condition). This condition, in terms of the complex auxiliary functions, has the mathematical formulation

$$\tilde{T}_1 = \tilde{M}_1 = 0 \text{ for } \alpha_1 = \alpha_1^0.$$

Separating real and imaginary parts, one obtains

$$T_1 = M_1 = \kappa_2 = \varepsilon_2 = 0.$$

Thus, this particular boundary condition corresponds to the case, where the edge of the shell is not subject to a force normal to the edge and a bending moment. In addition, it is assumed that the edge of the shell does not extend and cannot twist in a plane, normal to the middle surface.

Special attention must be given to the important problem of the junction of two shells. As has been shown in § 11a, in that case one must ensure on the common edge of the two shells $\alpha_1 = \alpha_1^0$ the continuity of the eight quantities (11.12a). But the first four of these quantities are the real parts of the complex quantities (15.1a), while the remaining four (apart from a multiplier of the form $E\delta^2/\sqrt{12}$) are the imaginary parts of the same complex quantities. Thus, if the two adjoining shells have equal thickness, the continuity of the eight real quantities (11.12a) is equivalent to the continuity of the four complex quantities (15.1a).

As has been shown by K. F. Chernykh [27], the complex formulation of the problem of the junction of two shells of equal thickness can be generalized also to the case where along the common edge the shells form an angle. However, no attention will be given to this problem and the reader will be referred to the paper [29].

§ 16. Generalization of the concepts of "complex forces" and "complex displacements" to the case of arbitrary Poisson's ratio

The method which has been studied in the preceding two sections and which led to a halving of the order of the differential equations of the theory of shells has only been deduced for the particular case when Poisson's ratio is zero.

In this case, there was no need to neglect any terms in the equilibrium equations of the element of the shell or in the compatibility equations for its middle surface, so that the above theory conforms completely to these two groups of equations. The only point at which errors enter into this theory is represented by the relations between stress and moment resultants and the strains of the middle surface, i.e., by the formulae (10.8). However, since during the deduction of these last relations it has been shown that they involved errors not exceeding those inherent in the initial assumptions of the theory of thin shells, it may be stated that this is also true for the errors in the complex equations (14.11) and (15.5).

Since the complex transformation of the equations of the theory of shells simplifies the solution of problems, it will be desirable to generalize this method to the case of arbitrary Poisson's ratio. Although it is impossible to retain the same accuracy as in the above development, nevertheless, after neglecting some small terms in the original equations, the required results can be attained at the price of additional errors which, for the majority of problems, are not large.

A beginning will be made with the discussion of the possible simplifications in the system (14.11). From this point of view, one may first of all put

$$1 - ic_0/R_1 \approx 1 - ic_0/R_2 \approx 1, \tag{16.1}$$

since this leads to a probable error of order δ/R in comparison with unity.

Then the system (14.11) can be written in the following way:

$$\begin{gathered}
\frac{1}{A_1A_2}\left[\frac{\partial A_2\tilde{T}_1}{\partial\alpha_1} + \frac{\partial A_1\tilde{S}}{\partial\alpha_2} + \frac{\partial A_1}{\partial\alpha_2}\tilde{S} - \frac{\partial A_2}{\partial\alpha_1}\tilde{T}_2\right] + \\
+ i\frac{c_0}{R_1}\underline{\left(\frac{1}{A_1}\frac{\partial\tilde{T}}{\partial\alpha_1} - \frac{1}{A_2}\frac{\partial\tilde{S}}{\partial\alpha_2} - 2\frac{R_1}{R_2}\frac{\tilde{S}}{A_1A_2}\frac{\partial A_1}{\partial\alpha_2}\right)} = -q_1, \\
\frac{1}{A_1A_2}\left[\frac{\partial A_2\tilde{S}}{\partial\alpha_1} + \frac{\partial A_1\tilde{T}_2}{\partial\alpha_2} + \frac{\partial A_2}{\partial\alpha_1}\tilde{S} - \frac{\partial A_1}{\partial\alpha_2}\tilde{T}_1\right] + \\
+ i\frac{c_0}{R_2}\underline{\left[\frac{1}{A_2}\frac{\partial\tilde{T}}{\partial\alpha_2} - \frac{1}{A_1}\frac{\partial\tilde{S}}{\partial\alpha_1} - 2\frac{R_2}{R_1}\frac{\tilde{S}}{A_1A_2}\frac{\partial A_2}{\partial\alpha_1}\right]} = -q_2, \\
\frac{\tilde{T}_1}{R_1} + \frac{\tilde{T}_2}{R_2} - ic_0\Delta(\tilde{T}) + \\
+ \frac{c_0^2}{A_1A_2}\left\{\frac{\partial}{\partial\alpha_1}\underline{\left(\frac{1}{R_1}\frac{A_2}{A_1}\frac{\partial\tilde{T}}{\partial\alpha_1} - \frac{1}{R_1}\frac{\partial\tilde{S}}{\partial\alpha_2} - \frac{2\tilde{S}}{R_2}\frac{1}{A_1}\frac{\partial A_1}{\partial\alpha_2}\right)} + \right. \\
\left. + \frac{\partial}{\partial\alpha_2}\underline{\left(\frac{1}{R_2}\frac{A_1}{A_2}\frac{\partial\tilde{T}}{\partial\alpha_2} - \frac{1}{R_2}\frac{\partial\tilde{S}}{\partial\alpha_1} - \frac{2\tilde{S}}{R_1}\frac{1}{A_2}\frac{\partial A_2}{\partial\alpha_1}\right)}\right\} = \tilde{q}_n \approx
\end{gathered} \tag{16.2}$$

$$\approx q_n + i\frac{c_0}{A_1A_2}\left(\frac{\partial A_2 q_1}{\partial \alpha_1} + \frac{\partial A_1 q_2}{\partial \alpha_2}\right). \tag{16.2}$$

Further, it will be noted that the underlined terms in the first two equations occur in the same equations without the factor $i\frac{c_0}{R}$, the modulus of which is small. As regards the underlined terms in the third equation, they involve the factor $\frac{c_0^2}{A_1A_2}$. On this basis, all the underlined terms in (16.2) can be neglected and one arrives at the equations

$$\frac{1}{A_1A_2}\left[\frac{\partial A_2\tilde{T}_1}{\partial\alpha_1} + \frac{\partial A_1\tilde{S}}{\partial\alpha_2} + \frac{\partial A_1}{\partial\alpha_2}\tilde{S} - \frac{\partial A_2}{\partial\alpha_1}\tilde{T}_2\right] + i\frac{c_0}{R_1}\frac{1}{A_1}\frac{\partial\tilde{T}}{\partial\alpha_1} + q_1 = 0,$$

$$\frac{1}{A_1A_2}\left[\frac{\partial A_2\tilde{S}}{\partial\alpha_1} + \frac{\partial A_1\tilde{T}_2}{\partial\alpha_2} + \frac{\partial A_2}{\partial\alpha_1}\tilde{S} - \frac{\partial A_1}{\partial\alpha_2}\tilde{T}_1\right] + i\frac{c_0}{R_2}\frac{1}{A_2}\frac{\partial\tilde{T}}{\partial\alpha_2} + q_2 = 0, \tag{16.3}$$

$$\frac{\tilde{T}_1}{R_1} + \frac{\tilde{T}_2}{R_2} - ic_0\Delta(\tilde{T}) = \tilde{q}_n.$$

It is clear from the character of these simplifications that the error incurred by replacing the system of equations (14.11) by the system (16.3) will, as a rule, be a quantity of the order δ/R in comparison with unity. Those special cases for which the above statement is violated will be indicated below.

Further, consider the form in which the initial shell equations (for $\mu = 0$) must be written, so that they will correspond to the complex equations (16.3). After some manipulations it may be verified that in order to achieve this object one must put in the equilibrium equations for an element of the shell (7.4) and (7.8)

$$T_{12} \approx T_{21} \approx S. \tag{16.4}$$

In addition, one must neglect in the conditions of compatibility (12.3) a number of terms by giving these relations the following simplified form:

$$(1+\mu)N_1 = \frac{1}{A_1}\frac{\partial M}{\partial\alpha_1} - \frac{\delta^2}{12}\underline{\frac{1}{R_1A_1}\frac{\partial T}{\partial\alpha_1}},$$

$$(1+\mu)N_2 = \frac{1}{A_2}\frac{\partial M}{\partial\alpha_2} - \frac{\delta^2}{12}\underline{\frac{1}{R_2A_2}\frac{\partial T}{\partial\alpha_2}}, \tag{16.5}$$

$$\frac{M_1-\mu M_2}{R_2} + \frac{M_2-\mu M_1}{R_1} + \frac{\delta^2}{12}\Delta(T) = -\frac{\delta^2}{12}\frac{(1+\mu)}{A_1A_2}\left[\frac{\partial A_2 q_1}{\partial\alpha_1} + \frac{\partial A_1 q_2}{\partial\alpha_2}\right].$$

The terms underlined in these formulae are likewise of negligible magnitude, as has been indicated first in the papers [28] and [31]. Thus, for a wide range of problems of the theory of shells, the first two of the conditions of compatibility of the middle surface can be written in the following simple manner:

$$(1+\mu)N_1 \approx \frac{1}{A_1}\frac{\partial M}{\partial \alpha_1},$$
$$(1+\mu)N_2 \approx \frac{1}{A_2}\frac{\partial M}{\partial \alpha_2}. \tag{16.6}$$

Eliminate from the first three equilibrium equations (7.4), in conformity with (16.4), the normal shearing forces by use of (16.6) and from the remaining equations (7.8) the shearing forces by use of (16.5). In this way, one obtains

$$\text{(I)}\quad \frac{1}{A_1A_2}\left[\frac{\partial A_2T_1}{\partial\alpha_1}+\frac{\partial A_1S}{\partial\alpha_2}+\frac{\partial A_1}{\partial\alpha_2}S-\frac{\partial A_2}{\partial\alpha_1}T_2\right]+\frac{1}{1+\mu}\frac{1}{R_1A_1}\frac{\partial M}{\partial\alpha_1}+q_1=0.$$

$$\text{(II)}\quad \frac{1}{A_1A_2}\left[-\frac{\partial A_2(M_2-\mu M_1)}{\partial\alpha_1}+(1+\mu)\left(\frac{\partial A_1H}{\partial\alpha_2}+\frac{\partial A_1}{\partial\alpha_2}H\right)+\right.$$
$$\left.+\frac{\partial A_2}{\partial\alpha_1}(M_1-\mu M_2)\right]+\frac{\delta^2}{12R_1A_1}\frac{\partial T}{\partial\alpha_1}=0, \tag{16.7}$$

$$\text{(III)}\quad \frac{1}{A_1A_2}\left[\frac{\partial A_2S}{\partial\alpha_1}+\frac{\partial A_1T_2}{\partial\alpha_2}+\frac{\partial A_2}{\partial\alpha_1}S-\frac{\partial A_1}{\partial\alpha_2}T_1\right]+\frac{1}{1+\mu}\frac{1}{R_2A_2}\frac{\partial M}{\partial\alpha_2}+q_2=0,$$

$$\text{(IV)}\quad \frac{1}{A_1A_2}\left[-\frac{\partial A_1(M_1-\mu M_2)}{\partial\alpha_2}+(1+\mu)\left(\frac{\partial A_2H}{\partial\alpha_1}+\frac{\partial A_2}{\partial\alpha_1}H\right)+\right.$$
$$\left.+\frac{\partial A_1}{\partial\alpha_2}(M_2-\mu M_1)\right]+\frac{\delta^2}{12R_2A_2}\frac{\partial T}{\partial\alpha_2}=0,$$

$$\text{(V)}\quad \frac{T_1}{R_1}+\frac{T_2}{R_2}-\frac{1}{1+\mu}\Delta(M)-q_n=0,$$

$$\text{(VI)}\quad \frac{M_2-\mu M_1}{R_1}+\frac{M_1-\mu M_2}{R_2}+\frac{\delta^2}{12}\Delta(T)=-$$
$$-\frac{\delta^2}{12}\frac{1+\mu}{A_1A_2}\left[\frac{\partial}{\partial\alpha_2}(A_2q_1)+\frac{\partial}{\partial\alpha_1}(A_1q_2)\right]. \tag{16.7}$$

Letting in these equations $\mu = 0$, introducing the complex stresses $\tilde{T}_1$, $\tilde{T}_2$, $\tilde{S}$ defined in (14.2) and performing the computations in the same manner as in § 14, one arrives at the system of differential equations (16.3).

Thus it is seen that the equations for the complex stresses, simplified to the form (16.3), are equivalent to writing the initial equations in the approximate form (16.7). However, as regards this last system of equations, the complex transformation may be applied without neglecting Poisson's ratio.

In fact, introduce now the auxiliary functions

$$\tilde{T}_1 = T_1 - \frac{i}{c}\,\frac{M_2 - \mu M_1}{1-\mu^2},$$

$$\tilde{T}_2 = T_2 - \frac{i}{c}\,\frac{M_1 - \mu M_2}{1-\mu^2}, \tag{16.8}$$

$$\tilde{S} = S + \frac{i}{c}\,\frac{H}{1-\mu},$$

where

$$c = \frac{\delta}{\sqrt{12(1-\mu^2)}}, \tag{16.9}$$

which are the generalizations of the complex stresses to the case $\mu \neq 0$. Replace in (16.7) the stresses T_1, T_2, S by their expressions in terms of M_1, M_2, H, $\tilde{T}_1$, $\tilde{T}_2$, $\tilde{S}$ as indicated by (16.8). In this way one arrives at a system of equations from which one may eliminate the unknowns M_1, M_2, H in the same manner as was done in § 14. This process leads to the following system of partial differential equations for the complex stresses:

$$\frac{1}{A_1A_2}\left[\frac{\partial A_2\tilde{T}_1}{\partial\alpha_1} + \frac{\partial A_1\tilde{S}}{\partial\alpha_2} + \frac{\partial A_1}{\partial\alpha_2}\tilde{S} - \frac{\partial A_2}{\partial\alpha_1}\tilde{T}_2\right] + i\,\frac{c}{R_1A_1}\,\frac{\partial\tilde{T}}{\partial\alpha_1} + q_1 = 0,$$

$$\frac{1}{A_1A_2}\left[\frac{\partial A_2\tilde{S}}{\partial\alpha_1} + \frac{\partial A_1\tilde{T}_2}{\partial\alpha_2} + \frac{\partial A_2}{\partial\alpha_1}\tilde{S} - \frac{\partial A_1}{\partial\alpha_2}\tilde{T}_1\right] + i\,\frac{c}{R_2A_2}\,\frac{\partial\tilde{T}}{\partial\alpha_2} + q_2 = 0,$$

$$\frac{\tilde{T}_1}{R_1} + \frac{\tilde{T}_2}{R_2} - ic\Delta(T) = \tilde{q}_n, \tag{16.10}$$

where

$$\tilde{q}_n = q_n + i\,\frac{(1+\mu)c}{A_1A_2}\left[\frac{\partial A_2q_1}{\partial\alpha_1} + \frac{\partial A_1q_2}{o\alpha_2}\right] \approx q_n.$$

This system is the generalization of the differential equations (14.11) to the case when Poisson's ratio is not zero. If this system had been deduced from the simplified equations (16.7), one would have obtained simpler equations than those of the system (14.11), the deduction of which did not require omission of terms either from the equilibrium equations or from the compatibility conditions.

Next, the concept of complex displacements will be generalized to the case $\mu \neq 0$ by defining them as the auxiliary functions $\tilde{u}$, $\tilde{v}$, $\tilde{w}$, related to the complex stresses by means of the six equations

$$\tilde{\varepsilon}_1 = \frac{1}{E\delta}(\tilde{T}_1 - \mu \tilde{T}_2),$$

$$\tilde{\varepsilon}_2 = \frac{1}{E\delta}(\tilde{T}_2 - \mu \tilde{T}_1), \qquad (16.11)$$

$$\tilde{\omega} = \frac{2(1+\mu)}{E\delta}\tilde{S};$$

$$\tilde{\kappa}_1 = \frac{i}{c}\,\frac{1}{E\delta}(\tilde{T}_2 - T_2^*),$$

$$\tilde{\kappa}_2 = \frac{i}{c}\,\frac{1}{E\delta}(\tilde{T}_1 - T_1^*), \qquad (16.12)$$

$$\tilde{\tau} = -\frac{i}{c}\,\frac{1}{E\delta}(\tilde{S} - S^*).$$

As before, the quantities $\tilde{\varepsilon}_1, \tilde{\varepsilon}_2, \ldots, \tilde{\tau}$ are the values of the strains of the middle surface after replacing u, v, w by $\tilde{u}$, $\tilde{v}$, $\tilde{w}$ respectively and T_1^*, T_2^*, S^* are particular (or the general) solutions of the equations of the membrane theory, that is, of the system (15.4).

Substituting $\tilde{\varepsilon}_1, \tilde{\varepsilon}_2, \ldots, \tilde{\tau}$ in (5.1) and proceeding in a manner analogous to that in §§ 14, 15, one obtains the following three differential equations for the quantities $\tilde{T}_1$, $\tilde{T}_2$, $\tilde{S}$ which will be the conditions of compatibility for the formulae (16.11) and (16.12), considered as differential equations for $\tilde{u}$, $\tilde{v}$, $\tilde{w}$:

$$\frac{1}{A_1A_2}\left[\frac{\partial A_2\tilde{T}_1}{\partial\alpha_1} + \frac{\partial A_1\tilde{S}}{\partial\alpha_2} + \frac{\partial A_1}{\partial\alpha_2}\tilde{S} - \frac{\partial A_2}{\partial\alpha_1}\tilde{T}_2\right] +$$

$$+ i\frac{c}{1-i(1+\mu)c/R_1}\left[\frac{1}{R_1A_1}\frac{\partial\tilde{T}}{\partial\alpha_1} - (1+\mu)\left(\frac{1}{R_1A_2}\frac{\partial\tilde{S}}{\partial\alpha_2} +\right.\right.$$

$$\left.\left. + \frac{2\tilde{S}}{R_2}\frac{1}{A_1A_2}\frac{\partial A_1}{\partial\alpha_2}\right)\right] = \frac{q_1}{1-i(1+\mu)c/R_1};$$

$$\frac{1}{A_1A_2}\left[\frac{\partial A_2\tilde{S}}{\partial\alpha_1} + \frac{\partial A_1\tilde{T}_2}{\partial\alpha_2} + \frac{\partial A_2}{\partial\alpha_1}\tilde{S} - \frac{\partial A_1}{\partial\alpha_2}\tilde{T}_1\right] +$$

$$+ i\frac{c}{1-i(1+\mu)c/R_2}\left[\frac{1}{R_2A_2}\frac{\partial\tilde{T}}{\partial\alpha_2} - (1+\mu)\left(\frac{1}{R_2A_1}\frac{\partial\tilde{S}}{\partial\alpha_1} +\right.\right.$$

$$\left.\left. + \frac{2\tilde{S}}{R_1}\frac{1}{A_1A_2}\frac{\partial A_2}{\partial\alpha_1}\right)\right] = \frac{q_2}{1-i(1+\mu)c/R_2}, \qquad (16.13)$$

$$\frac{\tilde{T}_1}{R_1} + \frac{\tilde{T}_2}{R_2} - ic\Delta(\tilde{T}) +$$

$$+\frac{(1+\mu)c^2}{A_1A_2}\left\{\frac{\partial}{\partial\alpha_1}\frac{1}{1-i(1+\mu)c/R_1}\left[\frac{1}{R_1}\frac{A_2}{A_1}\frac{\partial\tilde{T}}{\partial\alpha_1}-\right.\right.$$

$$\left.-(1+\mu)\left(\frac{1}{R_1}\frac{\partial\tilde{S}}{\partial\alpha_2}+\frac{2\tilde{S}}{R_2}\frac{1}{A_1}\frac{\partial A_1}{\partial\alpha_2}\right)\right]+$$

$$+\frac{\partial}{\partial\alpha_2}\frac{1}{1-i(1+\mu)c/R_2}\left[\frac{1}{R_2}\frac{A_1}{A_2}\frac{\partial\tilde{T}}{\partial\alpha_2}-\right.$$

$$\left.\left.-(1+\mu)\left(\frac{1}{R_2}\frac{\partial\tilde{S}}{\partial\alpha_1}+\frac{2\tilde{S}}{R_1}\frac{1}{A_2}\frac{\partial A_2}{\partial\alpha_1}\right)\right]\right\}=q_n+$$

$$+i\frac{(1+\mu)c}{A_1A_2}\left[\frac{\partial}{\partial\alpha_1}\left(\frac{A_2q_1}{1-i(1+\mu)c/R_1}\right)+\frac{\partial}{\partial\alpha_2}\left(\frac{A_1q_2}{1-i(1+\mu)c/R_2}\right)\right].$$

But the above complex stresses $\tilde{T}_1$, $\tilde{T}_2$, $\tilde{S}$ were solutions of the system of differential equations (16.10) which differs from (16.13) and, therefore, the equations (16.11) and (16.12) remain incompatible.

However, comparing (16.10) with (16.13), it is seen that these two systems differ by terms which, as has been established above, are in the majority of problems of order δ/R [cf. the reasoning for the transition from the system (14.4) to the system (16.3)].

Hence it is concluded that the compatibility of the equations (16.11), (16.12) will be violated only by terms of corresponding order which may not be taken into consideration, so that these are to be eliminated during the computations as their identity is established. Thus, the equations (16.11) and (16.12) will be assumed to be compatible in the above stated approximate sense.

Eliminating from these equations the complex stresses, one obtains the following system of differential equations for $\tilde{u}$, $\tilde{v}$, and $\tilde{w}$:

$$\frac{\tilde{\varepsilon}_1+\mu\tilde{\varepsilon}_2}{1-\mu^2}+ic\tilde{\kappa}_2=\frac{1}{E\delta}\left[T_1^*+\frac{i}{c}\frac{M_2^*+\mu M_1^*}{1-\mu^2}\right],$$

$$\frac{\tilde{\varepsilon}_2+\mu\tilde{\varepsilon}_1}{1-\mu^2}+ic\tilde{\kappa}_1=\frac{1}{E\delta}\left[T_2^*+\frac{i}{c}\frac{M_1^*+\mu M_2^*}{1-\mu^2}\right],\qquad(16.14)$$

$$\frac{\tilde{\omega}}{2(1+\mu)}-ic\tilde{\tau}=\frac{1}{E\delta}\left[S^*-\frac{i}{c}H^*\right],$$

This is the generalized system (15.6) for the case $\mu\neq 0$. Consequently, the concept of complex forces as well as that of complex displacements has now been extended to arbitrary values of Poisson's ratio. However, this has been done at the price of certain simplifications in the equilibrium equations for the shell element and the compatibility conditions for the middle surface which did not have to be considered when dealing with the particular case $\mu = 0$.

A mathematical consequence of these simplifications is that the solutions obtained from (16.10) and those obtained from (16.14) will not agree in all respects. However, in the majority of problems, this discrepancy will not be significant as it does not exceed the limits of accuracy of the initial assumptions of shell theory.

However, it turns out to be difficult to give an exhaustive estimate of the accuracy of the above approximate complex theory and to classify the particular cases in which it may surpass the accuracy of the Kirchhoff assumption. This is the reason why so much attention has been given earlier to the complex transformation of shell theory for $\mu = 0$. Since for this value of Poisson's ratio it has been possible to establish an exact correspondence between the real and imaginary equations of shell theory, the results of §§ 14, 15, 15a are in agreement with those of the Kirchhoff theory.

In solving any concrete problem on the basis of the approximate complex equations of this section, by letting in the equation $\mu = 0$ one can determine to what degree it contradicts the exact complex equations derived for any particular value of Poisson's ratio.

As a consequence, the presence of a value of μ for which the complex transformation of shell theory is applicable with any error (in addition to that inherent in the Kirchhoff hypothesis) makes it possible to bring under control the results obtained by use of (16.10) and (16.14).

The possibility of complex transformation of the equations of shell theory for the particular case of axi-symmetric deformation of shells of revolution (§ 54) has been established by E. Meissner already in 1913 [155]. The Author has generalized this method to all problems of the linear theory of isotropic shells (Dissertation of 1945, [106, 134]). The role and importance of such a transformation in shell theory is determined, in the Author's opinion, by the following circumstances.

1. Introducing the complex auxiliary functions (i.e., complex forces and complex displacements), one can lower the order of the differential equations of shell theory and significantly reduce the number of terms. As a result, the equations become less complicated and more lucid which makes it possible to display possibilities of their transformation and simplification. Every transformation of a general character, and likewise every exhibition of the general properties of solutions of shell theory, is undoubtedly much more conveniently performed on the basis of the above equations in complex form. This is especially evident in the case of a study of solutions of the theory of shells of revolution (Chap. IV), where

this method has already been used with success for forty years.

A. L. Gol'denweizer's general theory of asymptotic integration of the equations of shell theory [5a] would undoubtedly have been improved (in regard to a reduction of manipulations as well as more compact formulae), if he had made use of the complex transformation.

2. In those cases when the solution of problems of the theory of shells reduces to the solution of an ordinary differential equation (for example, if the solution can be sought in the form of a simple trigonometric series, or if deformations of a particular form are being studied which do not depend or depend in a definite manner on one of the curvilinear coordinates) the complex transformation eases the derivation of a general solution of a problem, since, for this purpose, one has to integrate equations of half the order of the real equations.

3. In order to subject solutions to the boundary conditions, one has, as a rule, to reduce the complex solutions to real form, since the boundary conditions are formulated in terms of the real and imaginary parts of the complex auxiliary functions. This has usually been viewed as a basic disadvantage of the complex transformation which reduced its effectiveness as a method of solution of problems of shell theory. However, it should be noted that there are also problems for which the boundary conditions have been formulated in complex form (§ 15a). For example, such a problem is the case of the hinged edge and the problem of the junction of two shells of equal thickness. For all such problems the facilities of the complex transformation can be exploited more fully. However, even if the boundary conditions of a problem are such that they require separation of real and imaginary parts of the complex auxiliary functions, application of the complex transformation may still turn out to be useful in view of the advantages indicated under 1. and 2. above. In Chapters III and IV, when dealing with examples of cylindrical shells and shells of revolution, the possibilities of the method will be demonstrated with regard to the study and transformation of the equations of various actual problems of shell theory as well as with regard to their solution.

The solution of a number of problems of practical importance, illustrating the effectiveness of the complex transformation in shell theory, and also a further theoretical development of this method are given in the monograph by K. F. Chernykh [29a].

§ 17. Simplification of the theory of shells by the method of Mushtari-Donnell-Vlazov

As has already been mentioned, the first attempt at the construction of a theory of shells on the basis of the hypothesis that normals remain normals was made by Aron (§ 1) who, by the way, disregarded for this purpose terms depending on the tangential displacements u and v in the formulae for change of curvature and twist, i.e., he defined these parameters by the expressions

$$\begin{aligned}\kappa_1 &= -\frac{1}{A_1}\frac{\partial}{\partial\alpha_1}\left(\frac{1}{A_1}\frac{\partial w}{\partial\alpha_1}\right) - \frac{1}{A_1A_2^2}\frac{\partial w}{\partial\alpha_2},\\ \kappa_2 &= -\frac{1}{A_2}\frac{\partial}{\partial\alpha_2}\left(\frac{1}{A_2}\frac{\partial w}{\partial\alpha_2}\right) - \frac{1}{A_1^2A_2}\frac{\partial w}{\partial\alpha_1},\\ \tau &= -\frac{1}{A_1A_2}\left(\frac{\partial^2 w}{\partial\alpha_1\partial\alpha_2} - \frac{1}{A_1}\frac{\partial A_1}{\partial\alpha_2}\frac{\partial w}{\partial\alpha_1} - \frac{1}{A_2}\frac{\partial A_2}{\partial\alpha_1}\frac{\partial w}{\partial\alpha_2}\right).\end{aligned} \tag{17.1}$$

In his paper [48], Love corrected a number of Aron's inaccuracies and gave the formulae for κ_1, κ_2, τ, used earlier. However, even without a special study, it is obvious on purely intuitive grounds that changes in curvature and twist caused by displacement components tangential to the middle surface must, as a rule, be unimportant. In the bending of shells a comparatively big role must be played by the displacement w of points of the middle surface along its normal. Hence, Aron's errors mentioned above will often be admissible as deliberate simplifications.

A well founded application of this method was introduced almost simultaneously by Kh. M. Mushtari [20, 21] and L. Donnell [41, 42] who applied it to a number of problems of the stability of shells.

These authors established that the neglecting of the displacements u, v in the formulae for the changes of curvature and twist will be the more justified the smaller are the stresses due to the moments in comparison with the stresses due to the forces.

Only in problems where flexure and twist are the predominant deformations (in the sense that the maximal stresses in the shells are determined largely by them) one may expect that neglect of the displacements u and v in the formulae for κ_1, κ_2, τ can lead to essential errors.

Neglecting the displacements u and v in κ_1, κ_2, τ one must, in order to be consistent, introduce in the equations of shell theory further simplifications. This can be demonstrated by substituting the simplified expressions for κ_1, κ_2, τ into the expressions for the potential energy of the shells (9.12) and deriving then from it, by use of the principle of minimum total energy, the equilibrium equations for an element of the middle surface in terms of the displacements. Writing these equations in terms of forces and moments and comparing them with (7.4) and (7.8), it may be shown that the neglecting of the displacements u, v in the formulae for κ_1, κ_2, τ leads to a loss of terms, which depend on the forces N_1, N_2 in the first two equations (7.4) and, besides this, to the following equalities, expressing these forces in terms of the displacement w:

$$N_1 = -\frac{E\delta^3}{12(1-\mu^2)}\frac{1}{A_1}\frac{\partial \Delta w}{\partial \alpha_1},$$
$$N_2 - = \frac{E\delta^3}{12(1-\mu_2)}\frac{1}{A_2}\frac{\partial \Delta w}{\partial \alpha_2}, \tag{17.2}$$

where $\Delta(\ )$ is the Laplace operator (14.9).

Using this result the equation (8.2) can be rewritten in the form

$$\frac{1}{A_1A_2}\left[\frac{\partial A_2T_1}{\partial \alpha_1}+\frac{\partial A_1S}{\partial \alpha_2}+\frac{\partial A_1}{\partial \alpha_2}S-\frac{\partial A_2}{\partial \alpha_1}T_2\right]+q_1=0,$$
$$\frac{1}{A_1A_2}\left[\frac{\partial A_2S}{\partial \alpha_1}+\frac{\partial A_1T_2}{\partial \alpha_2}+\frac{\partial A_2}{\partial \alpha_1}S-\frac{\partial A_1}{\partial \alpha_2}T_1\right]+q_2=0, \tag{17.3}$$
$$\frac{E\delta^3}{12(1-\mu^2)}\Delta\Delta w+\frac{T_1}{R_1}+\frac{T_2}{R_2}=q_n.$$

This is the form of the equations of shell theory employed in the work of Kh. M. Mushtari.

Further transformations of the equations (17.3) have been given by S. Feinberg [36] and V. Z. Vlasov [3, 4]. Attention will be given here only to the study by V. Z. Vlasov. This author introduced into the considerations an auxiliary function Φ, related to the forces by means of the equations

$$T_1 = -\frac{1}{A_2}\frac{\partial}{\partial \alpha_2}\left(\frac{1}{A_2}\frac{\partial \Phi}{\partial \alpha_2}\right)-\frac{1}{A_1A_2}\frac{\partial A_2}{\partial \alpha_1}\frac{1}{A_1}\frac{\partial \Phi}{\partial \alpha_1},$$
$$T_2 = -\frac{1}{A_1}\frac{\partial}{\partial \alpha_1}\left(\frac{1}{A_1}\frac{\partial \Phi}{\partial \alpha_1}\right)-\frac{1}{A_1A_2}\frac{\partial A_1}{\partial \alpha_2}\frac{1}{A_2}\frac{\partial \Phi}{\partial \alpha_2}, \tag{17.4}$$
$$S = \frac{1}{A_1A_2}\left(\frac{\partial^2 \Phi}{\partial \alpha_1 \partial \alpha_2}-\frac{1}{A_1}\frac{\partial A_1}{\partial \alpha_2}\frac{\partial \Phi}{\partial \alpha_1}-\frac{1}{A_2}\frac{\partial A_2}{\partial \alpha_1}\frac{\partial \Phi}{\partial \alpha_2}\right),$$

Substitution of (17.4) into the first two equations (17.3) and use of the Gauss-Codazzi relations leads then to

$$\frac{1}{A_1A_2}\left[\frac{\partial A_2T_1}{\partial \alpha_1}+\frac{\partial A_1S}{\partial \alpha_2}+\frac{\partial A_1}{\partial \alpha_2}S-\frac{\partial A_2}{\partial \alpha_1}T_2\right]=\frac{1}{R_1R_2}\frac{1}{A_1}\frac{\partial \Phi}{\partial \alpha_1}, \tag{17.5}$$

$$\frac{1}{A_1A_2}\left[\frac{\partial A_2 S}{\partial \alpha_1}+\frac{\partial A_1 T_2}{\partial \alpha_2}+\frac{\partial A_2}{\partial \alpha_1}S-\frac{\partial A_1}{\partial \alpha_2}T_1\right]=\frac{1}{R_1R_2}\frac{1}{A_2}\frac{\partial \Phi}{\partial \alpha_2}.$$

If the Gaussian curvature of the middle surface is equal to zero (as is, for example, the case for cylindrical or conical shells), then the right-hand sides of (17.5) vanish. Thus, the expressions (17.4) turn out to satisfy the first two equations (17.3) identically (for $q_1 = q_2 = 0$). The third of the equations (17.3) reduces, after substituting in it (17.4) and transforming by use of the Gauss-Codazzi relations, to the following equation in the two unknowns w and Φ:

$$\frac{E\delta^3}{12(1-\mu^2)}\Delta\Delta(w) - D(\Phi) = q_n, \tag{17.6}$$

where

$$D(\ldots) = \frac{1}{A_1A_2}\left\{\frac{\partial}{\partial \alpha_1}\left[\frac{1}{R_2}\frac{A_2}{A_1}\frac{\partial(\ldots)}{\partial \alpha_1}\right]+\frac{\partial}{\partial \alpha_2}\left[\frac{1}{R_1}\frac{A_1}{A_2}\frac{\partial(\ldots)}{\partial \alpha_2}\right]\right\}. \tag{17.7}$$

Strictly speaking, this equation is true only for shells of zero Gaussian curvature; however, as an approximation, it can be used also in certain other cases. For example, if the stresses in the shell are expressible in terms of sufficiently rapidly varying functions of the arc of the coordinate lines, then the right-hand sides of this formula can be assumed to vanish approximately, even if the Gaussian curvature of the shell is not zero. This statement is based on the fact that on substituting (17.4) on the left-hand side of (17.3) there occur side by side with the first partial derivatives of Φ second and third order derivatives which, by virtue of the assumption of rapidly varying stresses (and hence, of rapidly varying Φ) will essentially exceed the first order derivatives. Hence it may be said that for a sufficiently rapidly varying state of stress the expressions (17.4) satisfy the first two equations (17.3) (for $q_1 = q_2 = 0$) exactly apart from neglecting the first order derivatives of Φ compared with its second and third order derivatives.

Another example, when the employment of the equations (17.6) is possible even if the Gaussian curvature of the middle surface is non-zero, is the case when the shell is sufficiently shallow. In this context, shells will be said to be shallow when their middle surface at all points is sufficiently close to some plane and at the same time sufficiently smooth.

For shallow shells, the Gaussian curvature $1/R_1R_2$ at any point of the middle surface is small compared with $1/l^2$, where l is the width of the plane outline of the shallow shell. However, if this is so, then one may neglect for shallow shells the terms on the right-hand sides of (17.5) on the basis that they contain as factors the Gaussian curvature, a small parameter which is not contained in the remaining terms occurring as a result of the substitution of (17.4) into the first two equations (17.3).

The second differential equation, relating the two functions Φ and w, was obtained by V. Z. Vlasov from the third condition of compatibility of deformations (5.6). He introduced instead of κ_1, κ_2, τ their expressions in terms of w from the formulae

$$\varepsilon_1 = \frac{1}{E\delta}(T_1 - \mu T_2), \quad \varepsilon_2 = \frac{1}{E\delta}(T_2 - \mu T_1), \quad \omega = \frac{2(1+\mu)}{E\delta} S.$$

and (17.4).

The final result of this substitution (after some rearrangement using the Gauss-Codazzi relations) is the equation

$$E\delta D(w) + \Delta[\Delta(\Phi)] + \\ + \underline{\frac{1+\mu}{A_1 A_2}\left[\frac{\partial}{\partial\alpha_1}\left(\frac{1}{R_1R_2}\frac{A_2}{A_1}\frac{\partial\Phi}{\partial\alpha_1}\right) + \frac{\partial}{\partial\alpha_2}\left(\frac{1}{R_1R_2}\frac{A_1}{A_2}\frac{\partial\Phi}{\partial\alpha_2}\right)\right]} = 0, \quad (17.8)$$

where the underlined term will vanish for all shells of zero Gaussian curvature.

Even if the Gaussian curvature is non-zero, the underlined term can be neglected, if the state of stress under consideration varies sufficiently rapidly in the direction of one of the coordinate lines, or if the shell is sufficiently shallow. The basis for this statement is the same as that earlier for the derivation of (17.6).

Neglecting in (17.8) the underlined terms and adding to it (17.6), one obtains the following system of two linear partial differential equations in the two unknowns, the normal deflection w and the auxiliary function Φ:

$$\begin{gathered} \frac{1}{E\delta}\Delta\Delta(\Phi) + D(w) = 0, \\ \frac{E\delta^3}{12(1-\mu^2)}\Delta\Delta(w) - D(\Phi) = q_n. \end{gathered} \quad (17.9)$$

The equations of shell theory in this simplified form are very popular and form the basis of many investigations devoted to the solution of specific problems.

As is clear from the above, apart from the errors due to the basic assumptions of shell theory (Kirchhoff hypothesis), these equations involve additional errors, caused, firstly, by the neglecting of the tangential displacements u, v in the formulae for κ_1, κ_2, τ and, secondly, by neglecting in the equilibrium equations and the compatibility equations for the strains terms involving lower derivatives of the unknown functions and having the Gaussian curvature of the middle surface as a multiplier.

In particular, the equations (17.9) are used often in the analysis of shallow shells and hence they are frequently called the equations of the theory of shallow shells. However, it must be remembered that the range of the equations (17.9) is not restricted to shallow shells. They have been applied with success in the analysis of shells of zero Gaussian curvature as well as of moment edge effects (cf. below), since in the last case the displacements and stresses in the shell are expressible in terms of rapidly varying functions of the coordinates of the middle surface.

Finally, the equations (17.9) can be readily obtained from the complex equations (16.10), if one neglects there certain terms. For this purpose let in (16.10)

$$q_1 = q_2 = 0,$$

neglect in the first two equations the imaginary terms

$$i\frac{c}{R_1}\frac{1}{A_1}\frac{\partial \tilde{T}}{\partial \alpha_1} \text{ and } i\frac{c}{R_2}\frac{1}{A_2}\frac{\partial \tilde{T}}{\partial \alpha_2},$$

and introduce the auxiliary complex function $\tilde{\Psi}$ related to the complex stresses by means of the equalities

$$\begin{aligned}
\tilde{T}_1 &= -\frac{1}{A_2}\frac{\partial}{\partial \alpha_2}\left(\frac{1}{A_2}\frac{\partial \tilde{\Psi}}{\partial \alpha_2}\right) - \frac{1}{A_1A_2}\frac{\partial A_2}{\partial \alpha_1}\frac{1}{A_1}\frac{\partial \tilde{\Psi}}{\partial \alpha_1},\\
\tilde{T}_2 &= -\frac{1}{A_1}\frac{\partial}{\partial \alpha_1}\left(\frac{1}{A_1}\frac{\partial \tilde{\Psi}}{\partial \alpha_1}\right) - \frac{1}{A_1A_2}\frac{\partial A_1}{\partial \alpha_2}\frac{1}{A_2}\frac{\partial \tilde{\Psi}}{\partial \alpha_2},\\
\tilde{S} &= \frac{1}{A_1A_2}\left(\frac{\partial^2 \tilde{\Psi}}{\partial \alpha_1 \partial \alpha_2} - \frac{1}{A_1}\frac{\partial A_1}{\partial \alpha_2}\frac{\partial \tilde{\Psi}}{\partial \alpha_1} - \frac{1}{A_2}\frac{\partial A_2}{\partial \alpha_1}\frac{\partial \tilde{\Psi}}{\partial \alpha_2}\right).
\end{aligned} \tag{17.10}$$

Under these conditions, if $\tilde{\Psi}$ is a rapidly varying function (even if only of one of the curvilinear coordinates α_1, α_2) or if the shell has sufficient slope, it may be assumed that the expressions (17.10) satisfy the first two of the equations (16.10) approximately in the above stated simplified manner. The third of the equations (16.10) then becomes the following equation with the auxiliary function $\tilde{\Psi}$ as single variable,

$$-D(\Psi) + ic\Delta\Delta(\tilde{\Psi}) = q_n, \tag{17.11}$$

where $D(\ldots)$ is the operator (17.7).

Letting in this equation

$$\tilde{\Psi} = \Phi - i\frac{E\delta^2}{\sqrt{12(1-\mu^2)}}w \tag{17.12}$$

and separating real and imaginary parts, one arrives at the system of equations (17.9). Substituting now (17.12) into (17.10) and again separating real and imaginary parts, one obtains the formulae (17.4).

In conclusion, it should be noted that, although the equations (17.9) in their general form (for shells of arbitrary shape) were only deduced in 1944 [3], they were used much earlier for the particular case of circular cylindrical shells [113, 27].

§ 17a. The theory of shallow shells

A shell can be called shallow, if at every point of the middle surface

$$\left(\frac{\partial z}{\partial x}\right)^2 \ll 1, \qquad \left(\frac{\partial z}{\partial y}\right)^2 \ll 1, \tag{17.1a}$$

where

$$z = z(x, y) \tag{17.2a}$$

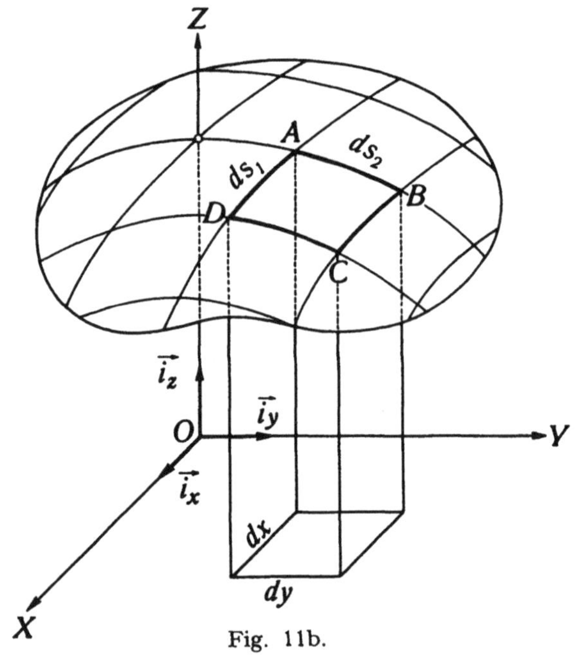

Fig. 11b.

is the equation of the middle surface in the Cartesian system of coordinates XYZ (Fig. 11b).

The theory of shallow shells is appropriately based on the following simplifications:

1. One must neglect squares of the derivatives $\partial z/\partial \alpha$, $\partial z/\partial y$ and their

products in comparison with unity, as follows from the above definition of a shallow shell.

2. One must neglect the tangential displacements u, v in the formulae for κ_1, κ_2, τ as well as the shearing forces N_1, N_2 in the first two equilibrium equations for the shell element. As has already been noted in the preceding section, these two approximations are interrelated.

For plates, the quantities κ_1, κ_2, τ do not depend at all on u, v which is the basis for neglecting the corresponding terms in a theory of slightly curved (shallow) shells.

It should be recalled that in the derivation of (17.9) use has been made only of the second group of simplifications, i.e., in the given system the possible simplifications arising from the assumption of a shallow shell have not been completely exhausted. This is the reason why the equations (17.9) have a wider range of applications. Thus, if one is only concerned with the problem of the analysis of shallow shells, additional simplifications become possible. A corresponding theory has been given by A. A. Nazarov [18a] and developed further by S. G. Mikhlin [12a]. In this theory, the middle surface of the shell is represented in the form (17.2a), where points of the middle surface are determined by the Cartesian coordinates x, y of their projections on the plane xOy. In this context, the parameters x, y serve as curvilinear coordinates of the middle surface.

The lines x of this curvilinear system of coordinates are plane curves, which are generated by intersection of the middle surface with planes parallel to the plane xOz. The lines y are plane curves, formed by intersection of the middle surface with planes parallel to the plane yOz. This system of coordinates is not orthogonal, except in the particular case when the middle surface is part of a cylindrical surface and one of the coordinate axes, for example x, is parallel to its axis.

Consider an element of the middle surface ABCD, projected on the plane xOy in the form of a rectangle with sides dx, dy (Fig. 11b). The sides of this element are

$$ds_1 = \sqrt{1 + (\partial z/\partial x)^2}\,dx, \quad ds_2 = \sqrt{1 + (\partial z/\partial y)^2}\,dy. \tag{17.3a}$$

However, since the shell is shallow, one can write

$$ds_1 = dx, \qquad ds_2 = dy, \tag{17.4a}$$

i.e., one can identify the increments of the arcs of the coordinate lines with the increments of the corresponding coordinates. The cosine of the angle between the lines x, y will be

$$\cos (\widehat{ds_1, ds_2}) = \frac{(\partial z/\partial x)(\partial z/\partial y)}{\sqrt{1 + (\partial z/\partial x)^2}\,\sqrt{1 + (\partial z/\partial y)^2}}, \tag{17.5a}$$

i.e., one can justify the assumption that the curvilinear system x, y is approximately orthogonal. and assume that

$$\cos (\widehat{ds_1, ds_2}) = 0 \tag{17.6a}$$

The possibility of using (17.4a), (17.6a) as well as the other modifications noted above considerably eases the task of the derivation of the equations of the theory of shallow shells and simplifies their final form.

Referring the reader for details to the original publications [18a, 12a], only the final results of these manipulations will be stated in the form of the system of two equations in two unknowns

$$\Delta\Delta\Phi + E\delta\left(\frac{\partial^2 z}{\partial x^2}\frac{\partial^2 w}{\partial y^2} - 2\frac{\partial^2 z}{\partial x \partial y}\frac{\partial^2 w}{\partial x \partial y} + \frac{\partial^2 z}{\partial y^2}\frac{\partial^2 w}{\partial x^2}\right) =$$

$$= (1 + \mu)\left(\frac{\partial^2 Q_x}{\partial y^2} + \frac{\partial^2 Q_y}{\partial x^2}\right) - \mu\Delta(Q_x + Q_y) = Q, \tag{17.7a}$$

$$\frac{E\delta^3}{12(1 - \mu^2)}\Delta\Delta w - \left(\frac{\partial^2 z}{\partial x^2}\frac{\partial^2 \Phi}{\partial y^2} - 2\frac{\partial^2 z}{\partial x \partial y}\frac{\partial^2 \Phi}{\partial x \partial y} + \frac{\partial^2 z}{\partial y^2}\frac{\partial^2 \Phi}{\partial x^2}\right) = q_z,$$

where

$$\Delta(\ ) = \frac{\partial^2(\ )}{\partial x^2} + \frac{\partial^2(\ )}{\partial y^2},$$

w is the projection of the displacements of the middle surface of the shell on the z-axis,

$$Q_x = \int_b^x q_x dx, \qquad Q_y = \int_c^x q_y dy, \tag{17.8a}$$

q_x, q_y, q_z are the projections of the loading of the middle surface onto the coordinate axes.

The forces in the shell are linked to the stress function by the equalities

$$T_1 = T_{xx} = \frac{\partial^2 \Phi}{\partial y^2}, \qquad S = T_x \approx T_{yx} = -\frac{\partial^2 \Phi}{\partial x \partial y}, \tag{17.9a}$$

$$T_2 = T_{yy} = \frac{\partial^2 \Phi}{\partial x^2}.$$

The direct strains in the directions of the lines x and y, the shear strain and also the change of curvature and twist of the middle surface can be expressed in terms of the displacements of the middle surface by the formulae

$$\varepsilon_1 = \frac{\partial u}{\partial x} - \frac{\partial^2 z}{\partial x^2} w, \qquad \varepsilon_2 = \frac{\partial v}{\partial y} - \frac{\partial^2 z}{\partial y^2} w,$$

$$\omega = \frac{\partial u}{\partial y} + \frac{\partial v}{\partial x} - 2\frac{\partial^2 z}{\partial x \partial y} w, \tag{17.10a}$$

$$\kappa_1 = -\frac{\partial^2 w}{\partial x^2}, \qquad \kappa_2 = -\frac{\partial^2 w}{\partial y^2}, \qquad \tau = -\frac{\partial^2 w}{\partial x \partial y},$$

where u, v are the components of the displacements of the middle surface in the directions of the respective coordinate lines. In accordance with the accepted system of simplifications, one must omit in κ_1, κ_2, τ terms which depend on these components.

The forces and moments are connected with ε_1, ε_2, ω, κ_1, κ_2, τ by the usual relations (10.8).

This set of formulae forms together with the boundary conditions a complete system of equations of a shell theory. By their structure, the equations (17.7a) are analogous to (17.8); however, in contrast to (17.8), these are not connected with the coordinate lines of the shell surface. This facilitates the approach to the solution of those problems in which the edge of the shell does not coincide with a line of its curvature; such problems are encountered, for example, in the analysis of roofing shells.

Introducing the complex auxiliary function

$$\tilde{\Phi} = \Phi - i \frac{E\delta^2}{\sqrt{12(1-\mu^2)}} w, \tag{17.11a}$$

one can obtain from (17.7a), in the same manner as this has been done earlier, the equation

$$ic\Delta\Delta\tilde{\Phi} - \frac{\partial^2 z}{\partial y^2}\frac{\partial^2 \tilde{\Phi}}{\partial x^2} + 2\frac{\partial^2 z}{\partial x \partial y}\frac{\partial^2 \tilde{\Phi}}{\partial x \partial y} - \frac{\partial^2 z}{\partial x^2}\frac{\partial^2 \tilde{\Phi}}{\partial y^2} = \tilde{q}_z, \tag{17.12a}$$

where

$$\tilde{q}_z = q_z + icQ. \tag{17.13a}$$

The system (17.7a) becomes especially simple for shells the middle surfaces of which are sections of elliptical or hyperbolical paraboloids or circular cylinders. In this case, one has

$$z = H - c_1 x - c_2 y - k_{11} x^2 - k_{12} xy - k_{22} y^2, \tag{17.14a}$$

where H, c_1, c_2, k_{11}, k_{12}, k_{22} are constants.

Substituting (17.14a) into (17.7a), one obtains the following system of two linear differential equations with constant coefficients

$$\Delta\Delta\Phi - 2E\delta\left[k_{22}\frac{\partial^2 w}{\partial x^2} - k_{12}\frac{\partial^2 w}{\partial x \partial y} + k_{11}\frac{\partial^2 w}{\partial y^2}\right] = Q,$$

$$\frac{E\delta^3}{24(1-\mu^2)}\Delta\Delta w + k_{22}\frac{\partial^2 \Phi}{\partial x^2} - k_{12}\frac{\partial^2 \Phi}{\partial x \partial y} + k_{11}\frac{\partial^2 \Phi}{\partial y^2} = \frac{q_z}{2}. \qquad (17.15a)$$

Solution of this system for certain boundary conditions can be achieved by use of the method of separation of variables.

It is customary to assume that each sufficiently shallow shell, in a first approximation, can be represented by the equation (17.14a), i.e., it can be roughly considered as an elliptic or hyperbolic paraboloid or circular cylinder. Thus, the equations (17.15a) acquire a general significance and can be considered as a mathematical apparatus suitable for the analysis of any shallow shells. The errors inherent in the equations (17.7a) as well as those in the equation (17.15a) approximating them have not yet been completely clarified and hence the limits of their applicability are not definite. It is accepted practice to assume, on the basis of experiment, that application is possible in the case of shallow shells with a central rise H which is not more than one fifth of the width of the roof span. Apparently, this estimate is in agreement with actual conditions.

Finally, the role of shallow shells in structural mechanics will be discussed. It is a fact that even very shallow shells with a central rise of an order of only a few times their thickness differ radically in character from plates. Under otherwise equal conditions, a shell turns out to be considerably more rigid than a plate and reacts not only in bending, but also (in a principal manner) through stresses distributed uniformly over its thickness. The fact that shallow shells with relatively large central rise have these properties is responsible for their practical value. In particular, shallow shells find wide application in roof structures.

However, it has to be noted that there must be a definite relationship between the degree of shallowness of a roof and the roof span. The greater is the span, the larger will be, under otherwise equal conditions, the shell stresses and the smaller the range of stability.

It is essentially impossible to counteract this effect by an increase in thickness, since one would simultaneously increase the loading. The only proper and effective measure of improving the strength and stability of a roof, without introduction of additional ribs or stringers, is an increase of its shallowness. Even a small increase of the central rise of a roof changes essentially the curvature of the shell with a consequent salutary effect on stresses and stability. Therefore, in recognising the use of shallow roofs in present day practice, it must be noted as a general rule that the choice of degree of shallowness of a shell must be correlated with the dimensions of the required roof structure.

If the central rise of a roof exceeds one fifth of its span, then the analysis of the shell by use of (17.7a) can turn out to be too inaccurate. This will be due to the error inherent in the simplifications 1. The error arising from the simplifications 2. will be less important and they can be used for a larger range of shallowness, since they apply not only to shallow shells. This fact forms the basis of the recommendation that for the analysis of roofs with a ratio of central rise to span exceeding one fifth the equations (17.7a) should be replaced by the equations (17.9a), since only the second set of simplifications was employed for their derivation.

§ 18. On the basic properties of the solutions of the equations of shell theory

The radii of curvature of the middle surface have the dimension of length. The Lamé parameters A_1, A_2 can be made to be dimensional or not, depending on the dimensions of the curvilinear coordinates α_1, α_2. If the Lamé parameters are assumed to have the dimension of length, the corresponding coordinates α_1, α_2 will be dimensionless.

Introduce the notation

$$\frac{1}{r_1} = \frac{R_0}{R_1}, \quad \frac{1}{r_2} = \frac{R_0}{R_2}, \quad a_1 = \frac{A_1}{R_0}, \quad a_2 = \frac{A_2}{R_0}, \tag{18.1}$$

where R_0 is a constant which is characteristic for the middle surface and has the dimension of length. For example, one may choose in the capacity of R_0 the average value of one of the radii of curvature. The dimensionless quantities r_1, r_2, a_1, a_2 will be functions of α_1, α_2.

In the notation (18.1), the equations of shell theory in the complex form (16.10) assume the form

$$\begin{gathered}
\frac{1}{a_1 a_2}\left[\frac{\partial a_2 \tilde{T}_1}{\partial \alpha_1} + \frac{\partial a_1 \tilde{S}}{\partial \alpha_2} + \frac{\partial a_1}{\partial \alpha_2}\tilde{S} - \frac{\partial a_2}{\partial \alpha_1}\tilde{T}_2\right] + i\frac{\lambda^2}{r_1 a_1}\frac{\partial \tilde{T}}{\partial \alpha_1} = -R_0 q_1, \\
\frac{1}{a_1 a_2}\left[\frac{\partial a_2 \tilde{S}}{\partial \alpha_1} + \frac{\partial a_1 \tilde{T}_2}{\partial \alpha_2} + \frac{\partial a_2}{\partial \alpha_1}\tilde{S} - \frac{\partial a_1}{\partial \alpha_2}\tilde{T}_1\right] + i\frac{\lambda^2}{r_2 a_2}\frac{\partial \tilde{T}}{\partial \alpha_2} = -R_0 q_2, \\
\frac{\tilde{T}_1}{r_1} + \frac{\tilde{T}_2}{r_2} - i\lambda^2 \Delta'(\tilde{T}) = R_0 \tilde{q}_n,
\end{gathered} \tag{18.2}$$

where

$$\Delta'(\ ) = R_0^2\Delta(\ ) = \frac{1}{a_1a_2} \times \left\{\frac{\partial}{\partial\alpha_1}\left[\frac{a_2}{a_1}\frac{\partial(\ )}{\partial\alpha_1}\right] + \frac{\partial}{\partial\alpha_2}\left[\frac{a_1}{a_2}\frac{\partial(\ )}{\partial\alpha_2}\right]\right\}, \quad (18.4)$$

$$\lambda^2 = \frac{c}{R_0} = \frac{1}{\sqrt{12(1-\nu^2)}}\frac{\delta}{R_0} > 0.$$

For thin shells $\lambda \ll 1$, and hence the system (18.1) contains terms multiplied by a small parameter which presents the possibility of constructing approximate solutions.

The first idea arising in this connection is an attempt to neglect all these terms throughout. Then one obtains instead of (18.2) the system

$$\frac{1}{a_1a_2}\left[\frac{\partial a_2\tilde{T}_1}{\partial\alpha_1} + \frac{\partial a_1\tilde{S}}{\partial\alpha_2} + \frac{\partial a_1}{\partial\alpha_2}\tilde{S} - \frac{\partial a_2}{\partial\alpha_1}\tilde{T}_2\right] = -R_0q_1,$$

$$\frac{1}{a_1a_2}\left[\frac{\partial a_2\tilde{S}}{\partial\alpha_1} + \frac{\partial a_1\tilde{T}_2}{\partial\alpha_2} + \frac{\partial a_2}{\partial\alpha_1}\tilde{S} - \frac{\partial a_1}{\partial\alpha_1}\tilde{T}_1\right] = -R_0q_2, \quad (18.5)$$

$$\frac{\tilde{T}_1}{r_1} + \frac{\tilde{T}_2}{r_2} = R_0q_n,$$

which has already been encountered in § 15. This system, in terms of the notation (18.1), can be obtained from the equilibrium equations of a shell element (7.4), (7.8) by letting

$$N_1 = N_2 = M_1 = M_2 = M_{12} = M_{21} = 0,$$

i.e., by neglecting all the moments and shearing forces.

Correspondingly, the system (18.5) is referred to as the equations of the membrane theory of shells. Attention should be drawn to the fact that the system (18.5) has half the order of the original system (18.2), i.e., it is of second instead of fourth order. Hence, clearly, its general solution has half the number of arbitrary constants which are required for satisfying four boundary conditions. As has been shown in § 11, four such conditions must be given on the edge of the shell, and therefore one can replace only in particular cases, for a special choice of boundary conditions, a solution of (18.2) by one of (18.5).

In addition, it should be noted that, even if the solution of (18.5) could be subjected to all the boundary conditions of a particular problem, it would still be impossible to assert that it is an approximation to the solution of (18.2) for these boundary conditions. In order to verify this, one has to substitute the solution into (18.2) and find out whether the terms which were neglected in the transition from (18.2) to (18.5) turn out to be comparable in magnitude with the general terms of both systems.

This may happen, if the solution contains functions which vary rapidly with one or both coordinates α_1, α_2, since then the influence of the small parameter λ can be compensated, so that it is multiplied by terms of the equations with higher derivatives. Hence it is seen that the replacement of the general equations of shell theory by the membrane equations as a method of shell analysis has important limitations. A condition under which this method may nevertheless be used will be considered in detail in the next chapter. It follows from this that also constructions of solutions to the system (18.2) in the form of power series in λ^2 is not applicable. Indeed, this method is, in essence, identical with the method of successive approximations, where one has taken as first approximation the solution of (18.5); however, this system is not sufficiently arbitrary to permit subjection of its solution to four boundary conditions. Hence it is also impossible to apply boundary conditions of a general form to solutions of (18.5) in the form of power series in λ^2.

It has been noted above that the terms rejected in the transition from (18.2) to (18.5) can turn out to be important, if the solution varies rapidly, i.e., if it contains functions the absolute values of which essentially increase for every differentiation with respect to α_1 or α_2 (or with respect to one or the other coordinate), except at points belonging to the neighbourhood of zeros of the derivatives, if such exist. Examples of such functions are

$$e^{k_1\alpha_1+k_2\alpha_2}, \quad \sin(k_1\alpha_1 + k_2\alpha_2), \quad \cos(k_1\alpha_1 + k_2\alpha_2),$$

if one of the constants k_1, k_2 happens to be of order $1/\lambda \gg 1$.

Hence one may conclude that the parts of the general solution lost during the transition from (18.2) to (18.5) must change rapidly with respect to one of the curvilinear coordinates. This is actually the case; however, it must be noted that the role of the terms of the system (18.2) with a small multiplier λ^2 can also sometimes be essential for sufficiently smooth solutions, for example, if the stresses due to bending of the shell exceed significantly the stresses due to extension. In order to reveal and approximately determine quickly varying solutions of the system (18.2), one can use the method of asymptotic integration essential details of which will now be demonstrated by means of a simple example.

Consider the second order ordinary differential equation

$$\lambda^2[y'' + p(x)y'] - q^2(x)y = 0, \tag{18.6}$$

where $p(x)$, $q(x)$ are sufficiently smooth functions the absolute values of which are of order unity and λ is a small positive parameter. Seek a solution of (18.6) in the form

$$y = \varphi(x)\, e^{(1/\lambda)\psi(x)}, \tag{18.7}$$

where $\varphi(x)$, $\psi(x)$ will be determined from the condition that the expression (18.7) should satisfy the equation (18.6) exactly apart from terms of order λ, inclusively.

Substituting (18.7) into (18.6), one obtains

$$\left\{ (\psi')^2 - q^2(x) + \lambda\left[\frac{2\varphi'}{\varphi}\,\psi' + \psi'' + p\psi'\right] + \lambda^2 \frac{[\varphi'' + p\varphi']}{\varphi} \right\} e^{(1/\lambda)\psi} = 0. \tag{18.8}$$

In order to achieve the stated accuracy, one must take

$$(\psi')^2 = q^2, \qquad \frac{2\varphi'}{\varphi}\psi' + \psi'' + p\psi' = 0. \tag{18.9}$$

The first equation yields

$$\psi_1 = \int_b^x q\,dx, \qquad \psi_2 = -\int_c^x q\,dx, \tag{18.10}$$

where the lower integration limits b and c can be chosen arbitrarily.

Substituting (18.10) into (18.9), one finds

$$\varphi_1 = C_1 \frac{1}{\sqrt{q}} e^{-\frac{1}{2}\int_a^x p\,dx}, \qquad \varphi_2 = C_2 \frac{1}{\sqrt{q}} e^{-\frac{1}{2}\int_a^x p\,dx}, \tag{18.11}$$

where C_1, C_2 are constants of integration and the lower limit of integration a can again be chosen arbitrarily. In the end, one obtains an approximate solution of (18.6) in the form

$$y = \frac{1}{\sqrt{q}} e^{-\frac{1}{2}\int_a^x p\,dx}\left[C_1 e^{(1/\lambda)\int_b^x q\,dx} + C_2 e^{(1/-\lambda)\int_c^x q\,dx}\right]. \tag{18.12}$$

It is now seen from (18.8) that the preceding reasoning is justified only if

$$\frac{\varphi'' + p\varphi'}{\varphi} = f' + pf + f^2, \tag{18.13}$$

where

$$f = -\frac{1}{2}\left(\frac{q'}{q} + p\right) \tag{18.14}$$

is a quantity the order of which does not exceed unity.

Otherwise the last term in the braces of (18.8) may turn out to be comparable with the preceding terms in the braces or it may even exceed them. Then (18.12) ceases, of course, to be an approximate solution of (18.6). It follows from (18.13) that, if the above is true, the coefficient q^2 in the equation (18.6) is not only subject to the limitations imposed formerly on it, but it also must not vanish (or approach sufficiently close to zero) in the interval of x which is of interest for the problem under consideration.

However, sometimes one may succeed in constructing an asymptotic solution even in cases when the above reasoning turns out to be unjustified. In that case one may replace the equation (18.6) by another simpler equation the solution of which is known and approximates the solution of (18.6) in the range of x which is of interest. This method will be applied in Chapter IV to the important problems of the stress analysis of toroidal shells.

The reasoning studied above by means of a very simple example can be extended to cover more complicated problems such as equations of higher order, systems of equations and partial differential equations. In the last case, the possibility of using this method will depend not only on the properties of the equations, but also on the type of contour on which the boundary conditions must be imposed.

In shell theory, the method of asymptotic integration has already been applied for a long time. It forms the basis of the development of the effective methods of analysis of the symmetrical deformation of shells of revolution [140, 150]. Further, it has been applied to shells of revolution bounded by one or two parallel circles subject to deformations of a general form [49, 151]. The first attempt to apply the method to shells of arbitrary form was made by S. Feinberg [36]. A detailed development of the corresponding theory has been given by A. L. Gol'denweizer [9, 5a, 7a] and others who consider the method of asymptotic integration to be a universal device which permits, on the one hand, construction of approximate solutions of problems in shell theory. On the other hand, it makes it possible to classify these problems with common qualitative features displaying in this way the possibility of simplifications of the general equations of shell theory which are admissible for this or that group of particular cases.

However, a disadvantage of this theory is that, while it is cumbersome, it is also insufficiently general. This is explained by the fact that the possibilities of the asymptotic method are limited, that it depends in an essential manner on the properties of the coefficients of the differential equations and for partial differential equations also on the properties of the boundaries, on which the boundary conditions must be fulfilled. One may add also that the concept of a rapidly varying part of a solution in exponential form, as introduced by A. L. Gol'denweizer, does not exhaust all the possibilities of the asymptotic method.

It is sometimes possible to succeed in constructing an asymptotic solution on the basis of different rapidly varying functions; for example, in the analysis of toroidal shells and the solution of certain problems for spherical shells, Bessel functions have been employed with success.

In this context, it should be noted that the problems of shell theory for which the theory of [5a] makes it possible to suggest means for simplifying the equations and constructing approximate solutions, in the majority of cases, have been considered earlier with some success. Further, for difficult problems which have not yet been fully investigated, such as the analysis of shells subject to concentrated forces, stress analysis in regions where the coefficients of the differential equations of shell theory have singularities, the analysis of stress concentrations near openings in a shell, these methods of solution cannot be used.

Summing up the above it may be emphasised that the asymptotic method is based on a very simple idea. It is sufficient to consider several examples of its application, in order to learn how to use it in particular problems and to distinguish those cases where application is possible. Therefore the practical aspects of the method in obtaining approximate solutions of problems of shell theory have been presented, since there seemed to be no sense in studying the method in its most general form. The possibilities of the method will be demonstrated below by examples.

Chapter II

THE MEMBRANE THEORY OF SHELLS

§ 19. Preliminary remarks. History

Historically, the general theory of shells studied in the first chapter was preceded by the so called membrane theory which is considerably simpler and, at the same time, gives in certain cases a completely correct notion of the shell mechanism. In the study of the equilibrium of the shell element, this theory neglects all moments.

It follows from the relations

$$M_1 = \frac{E\delta^3}{12(1-\mu^2)}(\kappa_1 + \mu\kappa_2), \qquad M_2 = \frac{E\delta^3}{12(1-\mu^2)}(\kappa_2 + \mu\kappa_1),$$
$$H = \frac{E\delta^3}{12(1+\mu)}\tau, \tag{19.1}$$

that the neglecting of the moments will be well justified either when the shell has very small bending stiffness or when the changes of curvature and twist of the middle surface are very small.

In the first case, one is concerned with the equilibrium of absolutely flexible shells (membranes); in the second case, with the momentless state of stress of shells which have finite bending stiffness. Although both these problems are described by one and the same theory, nevertheless a distinction must be made between them since they exhibit special features. Thus, an absolutely flexible shell (for example, made from cloth) is not able to sustain compression forces, since any compression, however small, will cause loss of stability of its shape, i.e., the formation of wrinkles. Therefore, the calculations for such a shell will only correspond to a real situation in the case where tensile forces are obtained at all sections. For example, this condition is the basic requirement which must be satisfied by the body of a flexible (or semi-rigid) airship in order to ensure its longitudinal stability.

On the other hand, shells with finite bending stiffness, in contrast to absolutely flexible shells, may sustain in the momentless state of stress compressive as well as tensile forces. They will loose stability only when the compressive forces exceed a certain critical value. For absolutely flexible shells, since they do not possess resistance to bending, only the momentless state of stress is possible; for shells with finite stiffness such a state of stress is only one of the possible stress conditions and for its existence a number of supplementary conditions must be fulfilled which refer to the shape of the shell, the type of the external loads and the support of its edges.

It is not out of place here to compare the properties of shells with those of curved beams. As is well known, an arch of arbitrary form functions, as a rule, not only as a compression member, but also in bending. However, it is possible to specify its shape and the manner of its loading in such a way that it does not experience bending, so that, in agreement with the above terminology, it is in a momentless state of stress. Thus, for example, for a parabolic arch bending will not be induced by a vertical load uniformly distributed over its chord, while for a catenary arch this is true, if it is loaded by its own weight. The ability of arches to support certain transverse loads without bending is the reason for their technical advantage over straight beams.

Shells possess an analogous advantage over plates with, however, the essential difference that, while an arch of a given form will support only one completely determined load without bending, a shell of given shape has, as a rule, the same property for a wide range of loads which satisfy only very general requirements, provided its edges are suitably supported. In fact, this property of shells to react, under certain conditions, without or with insignificant bending is the reason for the extended practical application which they have found lately in different fields of industry. It should be emphasized as an important point in connection with the above that the concept of the momentless state of stress is not at all necessarily linked to infinitely large flexibility of shells (or, what is the same thing, to their infinitely small thickness). Even a thick shell, under suitable conditions, may exhibit momentless behaviour (in the sense that its bending stresses will be of order δ/R compared with the direct stresses).

The conditions for the existence of the momentless state of stress will be elucidated in the following section. However, these conditions may not always be fulfilled in a design and then a mixed stress field will be superimposed on the momentless state of stress (as has already been stated at the end of the last chapter, the pure bending condition has no advantage and should be avoided). Thus, the momentless stress condition occupies a place of honor in the design of shells and it is an ideal (sometimes, unfortunately, unobtainable) at which the designer should aim.

Brief mention will now be made of the stages of the development of the momentless or membrane theory. Its origin goes back to the work of G. Lamé and B. P. E. Clapeyron [70] who considered the symmetrical loading of shells of revolution. The general form of the equations of the membrane theory were established by E. Beltrami [65] and L. Lecornu [72].

The greatest successes of the momentless theory are connected with work of Russian origin. Among the contributions to the general momentless theory, the paper by V. V. Sokolovskii [64] should be noted since he reduced the equations of the problem to cannonical form and revealed a number of their characteristic properties.

The study of Iu. N. Rabotnov [61] is likewise of basic importance for the general membrane theory. In this paper, the problem of the momentless theory was linked to the problem of bending of a surface. The determination of the forces and displacements in a momentless shell was thus reduced to the solution of two independent partial differential equations.

The differential equations of the membrane theory are easily integrated for shells of zero Gaussian curvature (in particular, for cylindrical shells). It is likewise not difficult to obtain solutions for the practically important cases of symmetrical and wind type loadings on shells of revolution. Shells of revolution, loaded in an arbitrary manner, were investigated by V. Z. Vlasov [53, 4], Sokolovskii [63] and other authors. The momentless theory of shells (the middle surface being a second order surface) has been worked out by Vlasov [54, 55] who applied to this problem the theory of functions of a complex variable.

§ 20. The fundamental equations of the membrane theory

The equations of the membrane theory can be obtained directly from the equations of the general theory. For this purpose, it will be assumed that, although in principle the shell may be resistant to bending, in view of the smallness of the changes of curvature and twist, the moment terms in the equations of equilibrium for the shell element are unimportant. Rejecting these terms in the equations (7.8), one will have

$$\begin{aligned} T_{12} &= T_{21} = S^*, \\ N_1 &= N_2 = 0. \end{aligned} \tag{20.1}$$

Thus, the neglect of the moments leads to the neglect of the normal shearing forces.

Introducing these results into the equations (7.4), one arrives at the system of differential equations

$$\frac{1}{A_1A_2}\left[\frac{\partial A_2T_1^*}{\partial\alpha_1}+\frac{\partial A_1S^*}{\partial\alpha_2}+\frac{\partial A_1}{\partial\alpha_2}S^*-\frac{\partial A_2}{\partial\alpha_1}T_2^*\right]+q_1=0,$$
$$\frac{1}{A_1A_2}\left[\frac{\partial A_2S^*}{\partial\alpha_1}+\frac{\partial A_1T_2^*}{\partial\alpha_2}+\frac{\partial A_2}{\partial\alpha_1}S^*-\frac{\partial A_1}{\partial\alpha_2}T_1^*\right]+q_2=0, \quad (20.2)$$
$$\frac{T_1^*}{R_1}+\frac{T_2^*}{R_2}=q_n,$$

which, in conjunction with (20.1), are the conditions of equilibrium for the shell element in the momentless or membrane state of stress.

Here and in what follows, the stress resultants corresponding to the membrane stress condition will be distinguished by means of an asterisk.

In the system (20.2), the number of unknowns is equal to the number of equations, so that the problem of the theory of shells in the above stated formulation is statically determinate (in relation to the equilibrium of an infinitely small element of the shell, but not always in relation to the equilibrium of the shell as a whole). The reader is reminded that an analogous example is the problem of the bending of a beam in its technical formulation, where likewise the number of unknowns in the equilibrium equations corresponds to the number of equations and where nevertheless the determination of the reactions at the supports requires sometimes first the determination of the displacements. It will be assumed that the system (20.2) may be solved (the question of the boundary conditions which must be supplied for this purpose will be overlooked for the time being). Then, for the determination of the displacements in a shell which is in a membrane state of stress, one may write down the following system of three partial differential equations

$$\varepsilon_1^*=\frac{1}{A_1}\frac{\partial u}{\partial\alpha_1}+\frac{1}{A_1A_2}\frac{\partial A_1}{\partial\alpha_2}v+\frac{w}{R_1}=\frac{1}{E\delta}(T_1^*-\mu T_2^*),$$
$$\varepsilon_2^*=\frac{1}{A_2}\frac{\partial v}{\partial\alpha_2}+\frac{1}{A_1A_2}\frac{\partial A_2}{\partial\alpha_1}u+\frac{w}{R_2}=\frac{1}{E\delta}(T_2^*-\mu T_1^*), \quad (20.3)$$
$$\omega^*=\frac{A_2}{A_1}\frac{\partial}{\partial\alpha_1}\left(\frac{v}{A_2}\right)+\frac{A_1}{A_2}\frac{\partial}{\partial\alpha_2}\left(\frac{u}{R_1}\right)=\frac{2(1+\mu)}{E\delta}S^*,$$

the right hand sides of which are known functions of α_1 and α_2. The general solutions of this system can be presented in the form

$$u=u^*+u^{**},$$
$$v=v^*+v^{**}, \quad (20.4)$$
$$w=w^*+w^{**},$$

where u^*, v^*, w^* are understood to be any particular integrals of the system (20.3) and u^{**}, v^{**}, w^{**} are the general solutions of the homogeneous system of equations

$$\frac{1}{A_1}\frac{\partial u^{**}}{\partial \alpha_1} + \frac{1}{A_1 A_2}\frac{\partial A_1}{\partial \alpha_2} v^{**} + \frac{w^{**}}{R_1} = 0,$$

$$\frac{1}{A_2}\frac{\partial v^{**}}{\partial \alpha_2} + \frac{1}{A_1 A_2}\frac{\partial A_2}{\partial \alpha_1} u^{**} + \frac{w^{**}}{R_2} = 0, \qquad (20.5)$$

$$\frac{A_2}{A_1}\frac{\partial}{\partial \alpha_1}\left(\frac{v^{**}}{A_2}\right) + \frac{A_1}{A_2}\frac{\partial}{\partial \alpha_2}\left(\frac{u^{**}}{A_1}\right) = 0.$$

It is now seen that the displacements u^{**}, v^{**}, w^{**} give zero values for the direct and shear strains of the middle surface. Hence it is clear that the corresponding displacements may only be either displacements due to pure bending of the shell or displacements of the shell as a rigid body.

Thus, in general, the solutions of the equations of the membrane theory present pure bending displacements on an equal basis with displacements of the shell as a rigid body. Physically, this means that a freely flexible shell admits the appearance of such displacements without offering any resistance.

For the statement of the problems of the membrane theory, the pure bending displacements must either be eliminated completely or, at least, bounded in a proper manner. This remark applies not only to freely flexible shells (the shape of which otherwise will be undetermined), but also to shells with bending stiffness which are in such a state of stress that the bending stresses may be neglected in comparison with the membrane stresses. Actually, for such shells, the presence of essentially large bending displacements will be incompatible with the hypothesis of small bending stresses.

The principal possibility of the indeterminacy of the magnitudes of the displacements in the membrane theory and the consequent necessity for imposing bounds on the displacements, which is one of the conditions for the existence of the membrane state of stress, has an effect on the formulation of the boundary conditions in the membrane theory, as will become clear in the next section.

§ 21. The boundary conditions

The system of differential equations (20.2) for determining the stresses in the membrane theory is of the second order. Accordingly, also the system (20.3) for the displacements is of second order. However, the forces entering into the right hand sides of the equations of the last system are themselves solutions of a second order system.

Hence, the displacements in the membrane theory satisfy a fourth order system of equations which can be obtained by substituting into the equations (20.2) for the forces T_1^*, T_2^*, S^* their expressions in terms of the strains. However, this will not be done, because it is always more convenient to obtain the solutions in two successive steps — determination of the stresses from the system (20.2) and of the displacements from the system (20.3). Thus, in the membrane theory, the differential equations for the displacements have half the order of those in the general (moment) theory of shells; therefore the number of edge conditions which must be satisfied in the first theory will be half the number of edge conditions in the second theory.

In fact, in the membrane theory, only two boundary conditions may be specified on each edge of the shell.

For the sake of definiteness, consider the case when the edge of the shell is the line $\alpha_1 =$ constant. In the general theory of shells, there must be given on this edge the quantities

$$T_1,\ T_{12} + \frac{M_{12}}{R_2},\ N_1 + \frac{1}{A_2}\,\frac{\partial M_{12}}{\partial \alpha_2},\ M_1,$$

(when the boundary conditions have been formulated in terms of the stresses). In the membrane theory, one has

$$N_1 = N_2 = M_1 = M_2 = M_{12} = M_{21} = 0$$

at all points of the middle surface, and hence also on the boundary of the shell.

Therefore the edges of a shell in the membrane state of stress must be free from external edge loading in the form of normal shearing stresses and bending moments. Membranes can only support tangential edge loadings, as a result of which only the forces T_1 and $S = T_{12}$ may act on the edge under consideration. Hence the boundary conditions must be formulated in terms of these quantities.

Consider further the case where the boundary conditions are given in terms of the displacements. Then, in order to make the solutions of the equations of the general theory of shells unique, one must specify (on the boundary $\alpha_1 =$ constant) the quantities

$$u,\ v,\ w,\ \vartheta = -\frac{1}{A_1}\,\frac{\partial w}{\partial \alpha_1} + \frac{u}{R_1}.$$

In the membrane theory, it is impossible to specify the edge displacement w and the angle of twist ϑ, since this would immediately affect the values of the corresponding generalized forces N_1 and M_1. For example, letting $w = \vartheta = 0$ on the edge of the shell (i.e., "clamping" the edge in relation to radial displacement and angle of rotation), it is, of course, impossible to assume that on the same edge $N_1 = 0$, $M_1 = 0$, since the last directly contradicts the first condition. It follows from this that on the edge of the membrane one may only specify the tangential displacement components of the middle surface, i.e., u and v. Therefore, the boundary conditions of the membrane theory, if they are to be given in terms of displacements, must be formulated in terms of these quantities. Further, it must be taken into consideration that the differential equations of the membrane theory in terms of forces and displacements are of different order, i.e., of second and fourth order respectively. As a consequence of this, the edge conditions for membranes may not be given in terms of forces only, but half of these conditions must be given in terms of the displacements. This necessity for giving half the edge conditions in terms of the displacements has the following physical meaning: as has been shown in the previous section, a shell without bending stiffness is not a rigid body, but a mechanism which is free to admit displacements corresponding to pure bending. As a rule, displacements corresponding to an appropriate support of the edges may be eliminated, i.e., the shell can be converted into a rigid system. For this purpose, use must be made of the "compulsory" boundary conditions referred to above.

The remaining edge conditions can be given either in terms of forces or displacements. In the case when all these remaining boundary conditions are given in terms of forces, the problem of the membrane theory will be statically determinate (the stresses in the shell may be found independently of the displacements). However, if one of these boundary conditions is given in terms of displacements, the problem of the membrane theory is statically indeterminate.

In one particular case, all the boundary conditions for the membrane can be formulated in terms of the displacements. The rigid clamping of the edges of such a shell corresponds to the conditions

$$u = v = 0$$

which must be satisfied on every edge.

In conclusion, it must be noted that the possibility of retaining arbitrariness in the general solution of the equations of the membrane theory depends on the shape of the shell. This is explained by the fact that the partial differential equations of second order, which must be solved to determine the forces and displacements in membrane shells, belong to different classes for shells of positive, negative and zero Gaussian curvature; in fact, they are elliptic for the first, hyperbolic for the second and parabolic for the third type of shells. This difference becomes apparent in the formulation of the boundary conditions, as will be shown by examples below.

§ 22. The conditions for the existence of the membrane state of stress

It has been established in the previous section that one of the sources of contradiction in the membrane theory stems from the fact that the solutions of this last theory may not be subjected to the general boundary conditions of § 11.

The membrane may not be loaded along its edges by transverse forces or moments. The angles of twist and the normal displacements at its edges may not be constrained. Finally, it is necessary to give on its edges boundary conditions (in terms of displacements) which will ensure appropriate bounds on the values of the pure bending displacements (which sometimes may prove to be impossible, cf. § 37). Violation of any of these requirements is equivalent to disturbing the membrane state of stress to a greater or lesser degree. However, while these requirements are necessary, they are not sufficient; in addition to these conditions, there are supplementary requirements without which the membrane state of stress may not exist.

Another source of contradiction in the membrane theory is the fact that its equations determine the forces in the shell without dependence on the compatibility relations for the middle surface [equations (5.1)]. This also leads to violations of a greater or smaller magnitude. If the form of the shell and its surface loading are smooth, so that the derivatives of R_1, R_2, δ, q_1, q_2, q_n with respect to α_1 and α_2 are continuous, then it is sufficient for the purpose of satisfying the compatibility conditions to assume the presence of small bending moments and normal shear forces which are such that they may be neglected in the equilibrium equations of the shell element. This will be different if the curvature of the shell, its thickness or its loading exhibit at some sections sudden variations. Under these circumstances the displacements in the membrane theory will behave in the same manner, thus essentially violating the continuity of the middle surface. For this purpose, in order to satisfy the conditions of continuity, one must assume the presence of large normal shear stresses and moments in the regions of the stated sections which is equivalent to a rejection of the use of the membrane theory (in these particular regions of the shell). It may be expected that this conclusion remains in force in those cases where the radii of curvature, the thickness or the loading on the shell, while not being discontinuous, change sufficiently abruptly at certain sections.

In particular, it follows from this that a membrane may not be loaded by concentrated forces, perpendicular to its middle surface. The state of stress in a shell loaded by such forces will involve moments (at least in the neighbourhoods of the points of application of these forces).

Next, consider a situation which may be explained by the following example. A long circular cylindrical shell of constant thickness with a hole of arbitrary shape is stretched longitudinally by forces T_1, distributed uniformly over its ends (Fig. 11c). In this problem, all the above conditions are fulfilled, i.e.,

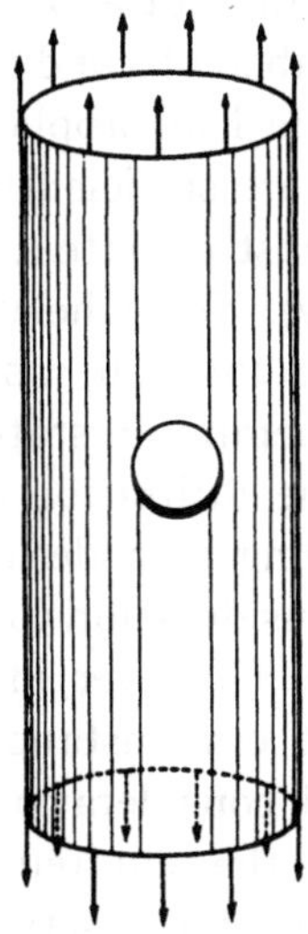

Fig. 11c.

a) The boundaries of the shell are free from shear forces and moments,
b) the normal displacements and rotations at the edges of the shell are unconstrained,
c) the geometrical parameters of the shell are smooth functions of the coordinates,
d) the components of the surface and edge loadings are sufficiently smooth.

Neverthless, the problem cannot be treated as a membrane problem, since on the edge of the hole both force components (normal and tangential) vanish, and the general solution of the equations of the membrane theory does not contain in this case sufficient arbitrariness to satisfy the two boundary conditions stated above. This fact will be readily confirmed by application of the formulae of §§ 17a, 18.

Hence it is seen that the class of problems in which the boundary conditions are formulated tangentially, i.e., in terms of displacements and forces, tangential to the middle surface, is essentially wider than the class of problems with tangential boundary conditions which may be satisfied within the framework of membrane theory. Thus, it does not follow from the fact that a shell is loaded along its edges and is only supported tangentially that it is in a membrane state.

In several sections of this chapter attention will be given in a brief manner to the question of the tangential boundary conditions under which a membrane state of stress can be realised.

Summarizing the above results, it should again be emphasized that the existence of the membrane state of stress is connected with the necessity of satisfying a number of conditions relating to the shape of the shell, the character of the load applied and the attachment of its edges. Since these conditions are sufficiently numerous one would think that the membrane theory must deviate a great deal in particular cases (from the general shell theory). Nevertheless, the membrane theory is of great practical importance. Above all, its value is determined by the technical advantage of the membrane state of stress (from the point of view of the uniformity of the utilization of the material of the shell).

In view of this, the planning of a design consisting of shells should always give a great deal of consideration to the fulfillment of the requirements of the membrane theory. However, in this direction the possibilities are limited by a number of circumstances — in particular, by constructional and technological considerations. Thus, for example, (from the point of view of satisfying the requirements of the membrane theory) the most advantageous shapes for reservoirs subject to large external or internal pressures are closed surfaces of sufficiently smooth form (sphere, ellipsoid with small eccentricity). However, it would seem unlikely that one could design a fire tube steam-boiler in one of the above stated shapes, since they are from all other points of view (except from the consideration of strength) unsuitable for this type of structure.

As another example, it may be said that (from the point of view of satisfying the requirements of the membrane theory) the most convenient method of production of metallic shells is casting, since in this way the designer has at his disposal many possibilities of joining shells with all accompanying details. Casting likewise permits variation of the thickness of the shell according to any law, increasing it where large stresses are expected and decreasing it where the stresses are smaller. However, the casting of very thin structures which have large dimensions is very difficult and in certain cases presents an insoluble technological problem. In view of this situation, casting is relatively rarely used for manufacturing shells. Quite often use is made of manufacturing processes such as pressing, forging and rolling, although these methods offer fewer possibilities than casting with regard to the choice of the form and the smoothness of the transitions between the components of the shell.

However, all this does not alter the fact that the struggle with bending stresses is one of the most important problems facing the designer of shells. If he does not succeed in eliminating these stresses completely, he must aim at localizing them and limiting their magnitude. He must also be in a position to correctly evaluate the forces in those regions of the shell where bending occurs. Thus, the solution given by the membrane theory must in many cases be supplemented by a solution of the equations of the bending theory in those parts of the shell where bending proves to be important. Such a combination of the bending and membrane theories is one of the important ideas on the basis of which the majority of the problems of shell theory are being solved at the present time.

Hence, practical interest will even attach to a study based on the membrane theory of such problems for which not all the requirements of that theory can be satisfied (for example, the analysis of pressure vessel heads, of thin cylindrical roofs, of shells subject to concentrated forces, etc.).

In conclusion, it should be noted that the role played by the membrane theory in the general theory of shells may be compared with that of the theory of the velocity potential of an ideal fluid in hydrodynamics. Potential flow is not realized exactly in practice, but, starting from this concept, there have been deduced many important practical results of hydrodynamics. The combination of the theories of ideal and viscous fluids (by use of the concept of the boundary layer) has some similarity (as a method of approach) with the above combination of the membrane and bending theories.

§ 23. The membrane theory of shells of revolution of arbitrary shape

Consider a shell the middle surface of which is a surface of revolution. For such a shell the lines of principal curvature are its meridians and parallel circles (the lines of intersection of the surface with the meridian planes and the planes perpendicular to the axis). Accordingly, one may take as the principal curvilinear co-ordinates of the middle surface the angle θ (between the normal to the middle surface and the axis of the shell) and the angle φ determining the position of a point on the corresponding parallel circle (Fig. 12).

Let R_1 be the radius of curvature of the meridian. The second radius of curvature R_2 will always be the length of the intercept of the normal to the middle surface between this surface and the axis of the shell, since, considering two adjacent points on the same parallel circle, the normals from these points intersect on the axis of the shell.

The element of arc of a meridian will be given by

$$ds_1 = R_1 d\theta,$$

whence follows

$$A_1 = R_1. \tag{23.1}$$

Correspondingly, the element of arc of a parallel circle is given by

$$ds_2 = R_2 \sin\theta\, d\varphi,$$

and therefore

$$A_2 = R_2 \sin\theta. \tag{23.2}$$

Thus, the middle surface of a shell of revolution is completely determined by knowledge of its two principal radii of curvature R_1, R_2 which

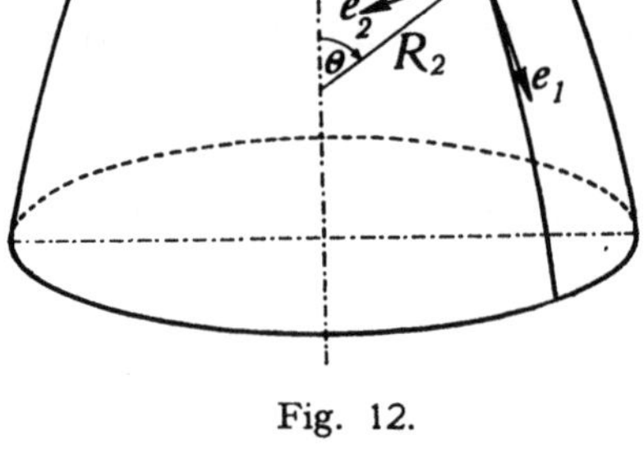

Fig. 12.

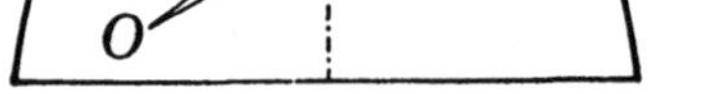

$$EB = R_2;\ DB = R_2 \sin\theta$$
$$AC - DB = \Delta(R_2 \sin\theta)$$
$$(AC - DB) : \cos\theta = AB;\ AB = R_1 \Delta\theta$$
$$\frac{\Delta(R_2 \sin\theta)}{\Delta\theta} = R_1 \cos\theta$$

Fig. 13.

will obviously be functions of only one of the curvilinear coordinates, namely θ.

The first two of the relations of Gauss-Codazzi (§ 2) are in this case satisfied identically, while the third gives

$$\frac{dR_2 \sin\theta}{d\theta} = R_1 \cos\theta. \tag{23.3}$$

The necessity for the existence of this differential relationship between the radii of curvature of shells of revolution is clear from the simple calculations reproduced underneath Fig. 13.

Substituting in the general differential equations of the membrane theory (20.2) for A_1, A_2 their values from (23.1) and (23.2) and taking into consideration the differential relation (23.3), one will have

$$\begin{gathered}
\frac{1}{R_1}\frac{\partial T_1^*}{\partial\theta} + \frac{\cot\theta}{R_2}(T_1^* - T_2^*) + \frac{1}{R_2\sin\theta}\frac{\partial S^*}{\partial\varphi} + q_1 = 0, \\
\frac{1}{R_1}\frac{\partial S^*}{\partial\theta} + 2\frac{\cot\theta}{R_2}S^* + \frac{1}{R_2\sin\theta}\frac{\partial T_2^*}{\partial\varphi} + q_2 = 0, \\
\frac{T_1^*}{R_1} + \frac{T_2^*}{R_2} = q_n,
\end{gathered} \tag{23.4}$$

where the q_1, q_2 and q_n are the projections of the surface loading on the tangent to the meridian, on the tangent to the parallel circle and on the normal to the middle surface (Fig. 12).

Solving the third of the above equations for T_2^* and substituting the value so found into the remaining two equations, one finds

$$\begin{gathered}
\frac{1}{R_1}\frac{\partial T_1^*}{\partial\theta} + \left(\frac{1}{R_1} + \frac{1}{R_2}\right)\cot\theta\, T_1^* + \frac{1}{R_2\sin\theta}\frac{\partial S^*}{\partial\varphi} = q_n\cot\theta - q_1, \\
\frac{1}{R_1}\frac{\partial S^*}{\partial\theta} + 2\frac{\cot\theta}{R_2}S^* - \frac{1}{R_1\sin\theta}\frac{\partial T_1^*}{\partial\varphi} = -q_2 - \frac{1}{\sin\theta}\frac{\partial q_n}{\partial\varphi}.
\end{gathered} \tag{23.5}$$

Introducing instead of T_1^* and S^* the new unknowns

$$\begin{aligned}
U^* &= T_1^* R_2 \sin^2\theta, \\
V^* &= S^* R_2^2 \sin^2\theta,
\end{aligned} \tag{23.6}$$

and taking into consideration (23.3) for purposes of differentiation, the system (23.5) reduces to the form

$$\begin{aligned}
&\frac{R_2^2 \sin\theta}{R_1} \frac{\partial U^*}{\partial \theta} + \frac{\partial V^*}{\partial \varphi} = (q_n \cos\theta - q_1 \sin\theta) R_2^3 \sin^2\theta, \\
&\frac{\partial V^*}{\partial \theta} - \frac{R_2}{\sin\theta} \frac{\partial U^*}{\partial \varphi} = -\left(q_2 \sin\theta + \frac{\partial q_n}{\partial \varphi}\right) R_1 R_2^2 \sin\theta.
\end{aligned} \tag{23.7}$$

Further, eliminating the unknown V^*, one arrives at the following partial differential equation of second order in the single unknown U^*:

$$\frac{1}{R_1 R_2 \sin\theta} \frac{\partial}{\partial \theta} \left[\frac{R_2^2 \sin\theta}{R_1} \frac{\partial U^*}{\partial \theta}\right] + \frac{1}{R_1 \sin^2\theta} \frac{\partial^2 U^*}{\partial \varphi^2} = F(\theta, \varphi), \tag{23.8}$$

where

$$F(\theta, \varphi) = \frac{1}{R_1 R_2 \sin\theta} \frac{\partial}{\partial \theta} [(q_n \cos\theta - q_1 \sin\theta) R_2^3 \sin^2\theta] + \\ + R_2 \left(\frac{\partial^2 q_n}{\partial \varphi^2} + \frac{\partial q_2}{\partial \varphi} \sin\theta\right). \tag{23.9}$$

Thus, the determination of the stresses in a shell of revolution by means of the membrane theory is reduced to the solution of the differential equation (23.8).

It will now be shown that the determination of the displacements leads in this case to the study of an analogous equation. For shells of revolution, the system (20.3) may be written

$$\begin{aligned}
&R_1 \varepsilon_1 = \frac{\partial u}{\partial \theta} + w = \frac{R_1}{E\delta} (T_1^* - \mu T_2^*), \\
&R_2 \varepsilon_2 = \frac{1}{\sin\theta} \frac{\partial v}{\partial \varphi} + u \operatorname{cotan}\theta + w = \frac{R_2}{E\delta} (T_2^* - \mu T_1^*), \\
&R_2 \omega = \frac{R_2}{R_1} \frac{\partial v}{\partial \theta} - v \operatorname{cotan}\theta + \frac{1}{\sin\theta} \frac{\partial u}{\partial \varphi} = \frac{2R_2(1+\mu)}{E\delta} S^*.
\end{aligned} \tag{23.10}$$

Eliminating from these equations w and taking into consideration the third of the equations (23.4), one has

$$\begin{aligned}
&\frac{\partial u}{\partial \theta} - u \operatorname{cotan}\theta - \frac{1}{\sin\theta} \frac{\partial v}{\partial \varphi} = \frac{R_1^2 + R_2^2 + 2\mu R_1 R_2}{R_1} \frac{T_1^*}{E\delta} - \frac{R_2(R_2 + \mu R_1)}{E\delta} q_n, \\
&\frac{R_2}{R_1} \frac{\partial v}{\partial \theta} - v \operatorname{cotan}\theta + \frac{1}{\sin\theta} \frac{\partial u}{\partial \varphi} = \frac{2(1+\mu)}{E\delta} R_2 S^*.
\end{aligned} \tag{23.11}$$

Introduce now instead of u and v the new variables

$$\xi = \frac{u}{\sin\theta}, \qquad \zeta = \frac{v}{R_2 \sin\theta}. \tag{23.12}$$

Then the system (23.11) reduces to the form

$$\frac{\partial \xi}{\partial \theta} - \frac{R_2}{\sin\theta}\frac{\partial \zeta}{\partial \varphi} = \frac{R_1{}^2+R_2{}^2+2\mu R_1R_2}{R_1\sin\theta}\frac{T_1{}^*}{E\delta} - \frac{R_2(R_2+\mu R_1)}{E\delta\sin\theta}q_n,$$
$$\frac{R_2{}^2\sin\theta}{R_1}\frac{\partial \zeta}{\partial \theta} + \frac{\partial \xi}{\partial \varphi} = \frac{2(1+\mu)R_2}{E\delta}S^*. \tag{23.13}$$

Finally, eliminating ξ, one finds

$$\frac{1}{R_1R_2\sin\theta}\frac{\partial}{\partial\theta}\left[\frac{R_2{}^2\sin\theta}{R_1}\frac{\partial\zeta}{\partial\theta}\right] + \frac{1}{R_1\sin^2\theta}\frac{\partial^2\zeta}{\partial\varphi^2} = f(\theta, \varphi), \tag{23.14}$$

where

$$f(\theta, \varphi) = \frac{1}{R_1R_2\sin\theta}\left\{2(1+\mu)\frac{\partial R_2S^*}{\partial\theta} - \frac{R_1{}^2+R_2{}^2+2\mu R_1R_2}{R_1\sin\theta}\frac{\partial T_1{}^*}{\partial\varphi} + \right.$$
$$\left. + \frac{R_2(R_2+\mu R_1)}{\sin\theta}\frac{\partial q_n}{\partial\varphi}\right\}\frac{1}{E\delta}. \tag{23.15}$$

Denoting now by $B(\ldots)$ the differential operator

$$B(\ldots) = \frac{1}{R_1R_2\sin\theta}\frac{\partial}{\partial\theta}\left[\frac{R_2{}^2\sin\theta}{R_1}\frac{\partial(\ldots)}{\partial\theta}\right] + \frac{1}{R_1\sin^2\theta}\frac{\partial^2(\ldots)}{\partial\varphi^2}, \tag{23.16}$$

the equations (23.8) and (23.14) can be written in the form

$$B(U^*) = F(\theta, \varphi),$$
$$B(\zeta) = f(\theta, \varphi). \tag{23.17}$$

Thus, the determination of the stresses as well as of the displacements in the membrane theory of shells of revolution leads to the study of essentially one and the same differential equation, so that, having solved the first part of the problem, it is always possible to solve its second part.

§ 24. Method of solution of the differential equations of the membrane theory of shells of revolution

Consider a shell of revolution bounded by one or two parallel circles (Fig. 14). In this case, the components of the surface loading as well as those of the edge loading will be periodic functions of the angle φ (with period 2π) as a result of the closed character of the shell with respect to the stated curvilinear coordinate. Hence, the components of the arbitrary surface loading can be represented in the form of the series

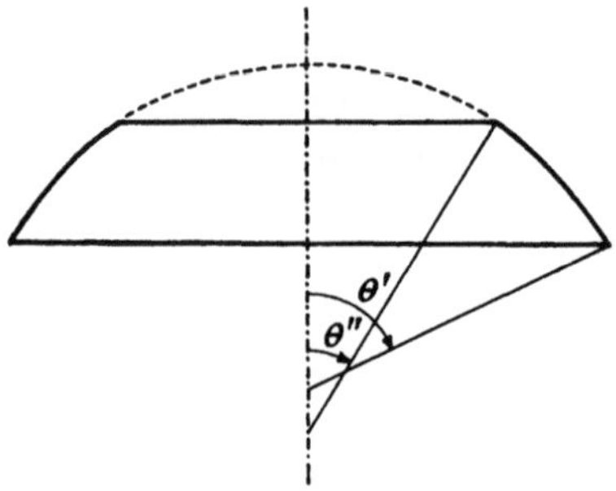

Fig. 14.

$$q_1 = \sum_{k=0}^{\infty} (q_{1,k} \cos k\varphi + q'_{1,k} \sin k\varphi),$$

$$q_2 = \sum_{k=0}^{\infty} (q'_{2,k} \cos k\varphi + q_{2,k} \sin k\varphi), \quad (24.1)$$

$$q_n = \sum_{k=0}^{\infty} (q_{n,k} \cos k\varphi + q'_{n,k} \sin k\varphi),$$

the coefficients of which (being functions only of the angle θ) are determined in any particular case by the known formulae of the theory of Fourier Series. Therefore, in order to analyze shells of revolution loaded in an arbitrary manner, it is sufficient to consider loadings of the form

$$q_1 = q_{1,k} \cos k\varphi,$$
$$q_2 = q_{2,k} \sin k\varphi, \quad (24.2)$$
$$q_n = q_{n,k} \cos k\varphi,$$

since knowing the solution for this particular case, one may always by use of the series (24.1) obtain the solution for any other distributed loading.

Substitute (24.2) into (23.9) and seek the solution of the differential equation (23.8) in the form

$$U^* = U_k^* \cos k\varphi, \quad (24.3)$$

where U_k^* is a function of θ only. Then, dividing the equation (23.8) by the trigonometric term $\cos k\varphi$, one obtains for the determination of U_k^* the ordinary, second order differential equation

$$\frac{1}{R_1 R_2 \sin\theta} \frac{d}{d\theta}\left[\frac{R_2^2 \sin\theta}{R_1} \frac{dU_k^*}{d\theta}\right] - \frac{k_2}{R_1 \sin^2\theta} U_k^* = F_k(\theta), \quad (24.4)$$

where

$$F_k(\theta) = \frac{1}{R_1 R_2 \sin\theta} \frac{d}{d\theta} [(q_{n,k} \cos\theta - q_{1,k} \sin\theta) R_2^3 \sin^2\theta] -$$
$$- k(kq_{n,k} - q_{2,k} \sin\theta) R_2. \quad (24.5)$$

Integrating this equation and then using the formulae (23.6), (23.7) and (23.4), one arrives at the following expressions for the stresses:

$$T_1^* = T_{1,k}^* \cos k\varphi, \; T_2^* = T_{2,k}^* \cos k\varphi, \; S^* = S_k^* \sin k\varphi, \quad (24.6)$$

in which $T_{1,k}^*$, $T_{2,k}^*$, S_k^* are known functions of θ [including two constants of integration, entering into the general solution of the equation (24.4)]. Substituting these expressions and also those for q_n as given by (24.2) into (23.15), one obtains the function $f(\theta, \varphi)$ in the form

$$f(\theta, \varphi) = f_k(\theta) \sin k\varphi, \tag{24.7}$$

where

$$f_k(\theta) = \frac{1}{E\delta} \frac{1}{R_1 R_2 \sin\theta} \left\{ 2(1+\mu) \frac{dR_2 S_k^*}{d\theta} + \right.$$
$$\left. + k \frac{R_1^2 + R_2^2 + 2\mu R_1 R_2}{R_1 \sin\theta} T_{1,k}^* - k \frac{R_2(R_2 + \mu R_1)}{\sin\theta} q_{n,k} \right\}. \tag{24.8}$$

Introducing (24.7) on the right hand side of the second of the equations (23.17), its solution will be sought in the form

$$\zeta = \zeta_k(\theta) \sin k\varphi. \tag{24.9}$$

Then the following second order, ordinary differential equation is obtained for the determination of $\zeta_k(\theta)$:

$$\frac{1}{R_1 R_2 \sin\theta} \frac{d}{d\theta} \left[\frac{R_2^2 \sin\theta}{R_1} \frac{d\zeta_k}{d\theta} \right] - \frac{k^2}{R_1 \sin^2\theta} \zeta_k = f_k(\theta). \tag{24.10}$$

This last equation is identical with the equation (24.4) from which it differs only by the free term on the right hand side.

After finding $\zeta_k(\theta)$ and then using the formulae (23.12) and (23.10), it is possible to obtain the following expressions for the displacements:

$$u = u_k(\theta) \cos k\varphi, \quad v = v_k(\theta) \sin k\varphi, \quad w = w_k(\theta) \cos k\varphi, \tag{24.11}$$

in which u_k, v_k, w_k are known functions of θ, including four constants of integration: two coming from the general solution of the equation (24.4) entering into the right hand side of the equation (24.10), two from the integration of the equation (24.10).

Hence, in order that the solution will be unique, there must be given on each of the two parallel circles bounding the shell two boundary conditions. Out of the total number (four) of conditions, at least two conditions must be given in terms of displacements, in order to eliminate the possibility of pure bending of the shell, and also its displacement as a rigid body. The remaining two conditions may be given in terms of stresses or displacements.

As will become clear later on, in the particular case when the shell is closed at the top and when, therefore, there is only one edge (dome), half the arbitrary constants disappear from the general solutions of the equations (24.4) and (24.10) (since the functions multiplied by these constants will become infinite at the top of the shell).

For this purpose, one has to retain in the formulae for the stresses (24.6) altogether one arbitrary constant and in the formulae for the displacements (24.11) only two arbitrary constants. In correspondence with this, on the only edge of a shell which is closed at the top there must be prescribed two conditions, at least one of which must be formulated in terms of displacements.

§ 25. Symmetrical deformation of shells of revolution

Consider the particular case when the edge and surface loadings do not depend on the angle φ. Then also the deformation of the shell is independent of this angle, i.e., it will be symmetrical about the axis of the shell. This case is obtained from the series (24.1) by omitting there all terms but the first, corresponding to $k = 0$. In this way, one will have

$$q_1 = q_{1,0}(\theta), \quad q_2 = q_{2,0}(\theta), \quad q_n = q_{n,0}(\theta). \tag{25.1}$$

In order to avoid in what follows the multiple subscripts, the left notation in (25.1) will be used in preference to that on the right, where q_1, q_2, q_n will be understood to be certain functions of the angle θ only.

The partial differential equations (23.7) for this particular case become the two independent ordinary differential equations

$$\begin{aligned} \frac{dU^*}{d\theta} &= (q_n \cos\theta - q_1 \sin\theta) R_1 R_2 \sin\theta, \\ \frac{dV^*}{d\theta} &= -\, q_2 R_1 R_2{}^2 \sin^2\theta, \end{aligned} \tag{25.2}$$

which may be integrated to give

$$\begin{aligned} U^* &= C_1 + \int\limits_{\theta''}^{\theta} (q_n \cos\theta - q_1 \sin\theta) R_1 R_2 \sin\theta \, d\theta, \\ V^* &= C_2 - \int\limits_{\theta''}^{\theta} q_2 R_1 R_2{}^2 \sin^2\theta \, d\theta, \end{aligned} \tag{25.3}$$

where C_1 and C_2 are constants of integration and the lower integration limit θ'' can be chosen arbitrarily, to suit computational convenience.

This angle usually will be identified with the angle corresponding to one of the edges of the shell (Fig. 14).

Having found the auxiliary functions U^*, V^*, it is possible to write down at once, on the basis of the formulae (23.6), expressions for the stresses S^*, T_1^*

$$S^* = \frac{C_2}{R_2^2 \sin^2 \theta} - \frac{1}{R_2^2 \sin^2 \theta} \int_{\theta''}^{\theta} q_2 R_1 R_2^2 \sin^2 \theta \, d\theta,$$

$$T_1^* = \frac{C_1}{R_2 \sin^2 \theta} + \frac{1}{R_2 \sin^2 \theta} \int_{\theta''}^{\theta} (q_n \cos \theta - q_1 \sin \theta) R_1 R_2 \sin \theta \, d\theta, \tag{25.4}$$

and also, on the basis of the third equation (23.4), an expression for the stress T_2^*

$$T_2^* = q_n R_2 - \frac{C_1}{R_1 \sin^2 \theta} - \frac{1}{R_1 \sin^2 \theta} \int_{\theta''}^{\theta} (q_n \cos \theta - q_1 \sin \theta) R_1 R_2 \sin \theta \, d\theta. \tag{25.5}$$

It follows from these formulae that the components of symmetrical surface loading q_n, q_1 cause only direct stresses T_1^*, T_2^* in the shell, while the component q_2 causes only shear stresses.

It is readily understood that the first of the formulae (25.4) corresponds to the torsion of the shell about its axis. If for this purpose q_2 is equal to zero, the shell will be twisted by edge forces only. Let the shear force S'' act on the edge $\theta = \theta''$. Then the first of the formulae (25.4) with $q_2 = 0$ gives the following law of variation of the shear stress in the shell:

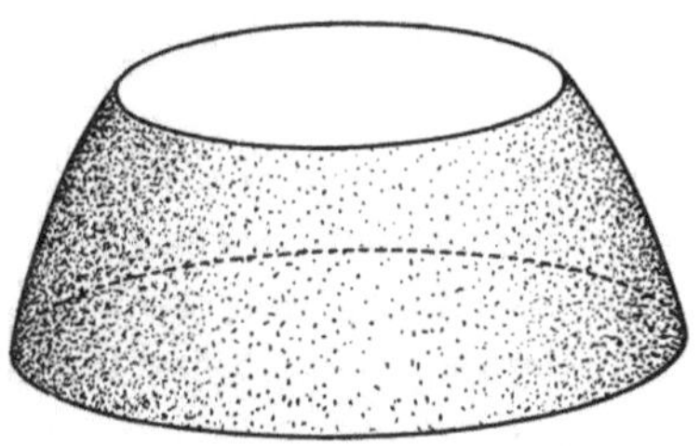

Fig. 15.

$$S^* = \frac{(R_2'' \sin \theta'')^2}{R_2^2 \sin^2 \theta} S''. \tag{25.6}$$

It will be noted that

$$2\pi (R_2'' \sin \theta'')^2 S'' = \mathfrak{M}, \tag{25.7}$$

where $\mathfrak{M}$ is the moment, twisting the shell about its axis.

Hence, in a shell of revolution, the shear stress due to torsion by moments, applying to its edges, is given by

$$\sigma_t = \frac{S^*}{\delta} = \frac{\mathfrak{M}}{2\pi\delta R_2^2 \sin^2 \theta} = \frac{\mathfrak{M}}{2\delta\Omega}, \tag{25.8}$$

where Ω is the area of the parallel circle along which the stress is being considered.

Great practical interest attaches to another variation of the symmetrical deformation of shells of revolution, in fact, to the case when the shell is deformed by symmetrical loadings q_1, q_n causing in it only direct stresses. This second case must be employed in the analysis of domes and reservoirs, and in future, when using the term "symmetrical deformation of shells of revolution", it will in fact refer to this case.

Given the stress $T_1^* = T_1''$ on the edge $\theta = \theta''$, one can find the constant C_1 in the second of the formulae (25.4)

$$C_1 = T_1'' R_2'' \sin^2 \theta''. \tag{25.9}$$

Hence the second of the formulae (25.4) and the formula (25.5) may be written

$$\begin{aligned} T_1^* &= \frac{R_2'' \sin^2 \theta''}{R_2 \sin^2 \theta} T_1'' + \frac{1}{R_2 \sin^2 \theta} \int_{\theta''}^{\theta} (q_n \cos \theta - q_1 \sin \theta) R_1 R_2 \sin \theta \, d\theta, \\ T_2^* &= q_n R_2 - \frac{R_2'' \sin^2 \theta''}{R_1 \sin^2 \theta} T_1'' - \frac{1}{R_1 \sin^2 \theta} \int_{\theta''}^{\theta} (q_n \cos \theta - q_1 \sin \theta) R_1 R_2 \sin \theta \, d\theta. \end{aligned} \tag{25.10}$$

These last formulae completely determine the state of stress in symmetrically deformed shells of revolution (on the basis of the membrane theory). It will be noted that the first of these formulae may be obtained if the shell shown in Fig. 15 and subjected to surface loads $q_1(\theta)$, $q_n(\theta)$ and forces T_1'' along the upper edge is cut by an arbitrary plane parallel to the upper edge and if one sets equal to zero the sum of the projections on the axis of the shell of all forces acting on its severed part. Thus, this formula is the condition for the equilibrium of an element of shell which has finite dimensions. The requirement for satisfying this condition uniquely determines in the case under consideration all forces in the shell down to its lower edge, as soon as the loading on the upper edge is given.

In conclusion, it will be observed that in the particular case when the shell has only one edge, it being closed at the top $\theta'' = 0$ (dome without opening, Fig. 12), the formulae (25.10) take the form

$$\begin{aligned} T_1^* &= \frac{1}{R_2 \sin^2 \theta} \int_0^{\theta} (q_n \cos \theta - q_1 \sin \theta) R_1 R_2 \sin \theta \, d\theta, \\ T_2^* &= q_n R_2 - \frac{1}{R_1 \sin^2 \theta} \int_0^{\theta} (q_n \cos \theta - q_1 \sin \theta) R_1 R_2 \sin \theta \, d\theta. \end{aligned} \tag{25.11}$$

§ 26. The membrane theory of domes

As an example of the application of the formulae derived in the previous section, several different forms of domes will now be compared from the point of view of the membrane theory. For these calculations, it will be assumed that the domes are loaded by their own weight. The components of this loading for a dome of arbitrary form are given by (Fig. 16)

$$q_1 = q \sin \theta, \; q_n = - q \cos \theta, \tag{26.1}$$

where q is the weight of the dome per unit area of its surface.

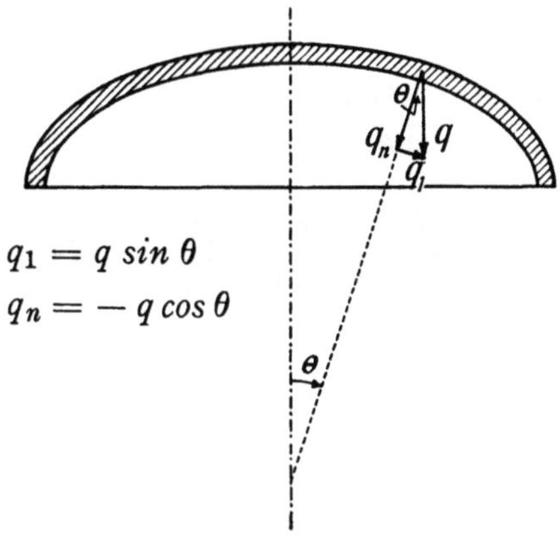

Fig. 16.

Substituting these expressions for q_1 and q_n into the formulae (25.11), one finds

$$T_1^* = - \frac{q}{R_2 \sin^2 \theta} \int_0^\theta R_1 R_2 \sin \theta \, d\theta,$$

$$T_2^* = - q R_2 \cos \theta + \tag{26.2}$$

$$+ \frac{q}{R_1 \sin^2 \theta} \int_0^\theta R_1 R_2 \sin \theta \, d\theta.$$

These results will now be applied to shells the radii of curvature of which are given by the formulae

$$R_1 = \frac{R_0}{(1 + \gamma \sin^2 \theta)^{3/2}}, \qquad R_2 = \frac{R_0}{(1 + \gamma \sin^2 \theta)^{1/2}}. \tag{26.3}$$

These radii of curvature correspond to surfaces, generated by rotation of second order curves about their axis of symmetry; thus, one obtains

for $\gamma = 0$, spheres,
for $\gamma = -1$, paraboloids,
for $\gamma > -1$, ellipsoids,
for $\gamma < -1$, hyperboloids.

The parameter R_0 is equal to the value of the radius of curvature for $\theta = 0$, i.e., for the vertex of the dome. Substituting (26.3) into (26.2) and carrying out the integration, the following expressions for the forces T_1^*, T_2^* are obtained:

for the spherical dome

$$T_1^* = -\frac{qR_0}{1+\cos\theta}, \quad T_2^* = -qR_0\left(\cos\theta - \frac{1}{1+\cos\theta}\right); \tag{26.4}$$

for the parabolic dome

$$\begin{aligned} T_1^* &= -\frac{qR_0}{3}\left(\frac{1}{\cos^2\theta} + \frac{1}{1+\cos\theta}\right), \\ T_2^* &= -qR_0\left[1 - \frac{1}{3}\left(1 + \frac{\cos^2\theta}{1+\cos\theta}\right)\right]; \end{aligned} \tag{26.5}$$

for the elliptic dome

$$\begin{aligned} T_1^* = -\frac{qa}{2}\,\frac{\sqrt{1-\varepsilon^2\cos^2\theta}}{\sin^2\theta}\Bigg\{&\frac{(1-\cos\theta)(1+\varepsilon^2\cos\theta)}{1-\varepsilon^2\cos^2\theta} + \\ &+ \frac{1-\varepsilon^2}{\varepsilon}\ln\left[\sqrt{\frac{1+\varepsilon}{1-\varepsilon}}\sqrt{\frac{1-\varepsilon\cos\theta}{1+\varepsilon\cos\theta}}\right]\Bigg\}, \end{aligned} \tag{26.6}$$

$$T_2^* = -\frac{qa\cos\theta}{\sqrt{1-\varepsilon^2\cos^2\theta}} - \frac{1-\varepsilon^2\cos^2\theta}{1-\varepsilon^2}T_1^*.$$

In the last two formulae, 'a' is the horizontal semi-axis of the ellipse (the radius of the base of the dome) and ε its eccentricity, so that

$$\varepsilon^2 = 1 - \frac{b^2}{a^2} = \frac{\gamma}{1+\gamma}.$$

These results will now be used to compare the four different domes, shown in Fig. 17:

Dome I is spherical ($R_0 = R = 1.25r$);

Dome II is parabolic ($R_0 = r$);

Dome III is elliptic ($R_0 = 2r$, $\varepsilon^2 = 0.75$);

Dome IV is also elliptic, but differs from Dome III in that it is not formed by half an ellipsoid, but by a smaller part of it, since the tangent to the meridian of the dome along its edge is not vertical ($R_0 = 1.625r$, $\varepsilon^2 = 0.6$).

All four domes have the same height ($h = \frac{1}{2}r$) and cover the same area (radius r). The weight q likewise will be assumed to be the same for the four shells.

In Fig. 17, the variations of the forces T_1^*, T_2^* with the height of the domes are shown. They were calculated from the formulae (26.4),

(26.5), (26.6), and have been graphed to the same scale. It follows from these curves that, from the point of view of the magnitude of the stresses, the greatest advantage is offered by that dome which has the smallest radius of curvature at the apex, i.e., the parabolic dome. From the same point of view, the least advantageous dome is the elliptic dome III. However, these are not the only factors which are essential for the choice of the form of a dome. The dome must be such that the bending in the shell will be unimportant. In order to remove bending, it is necessary that the line of support of the dome must assimilate the load acting on it and for this purpose its deformation must coincide with the circumferential lengthening of the shell in accordance with the membrane theory. Only under these conditions will the membrane state of stress prevail and the calculations correspond to reality (§ 20).

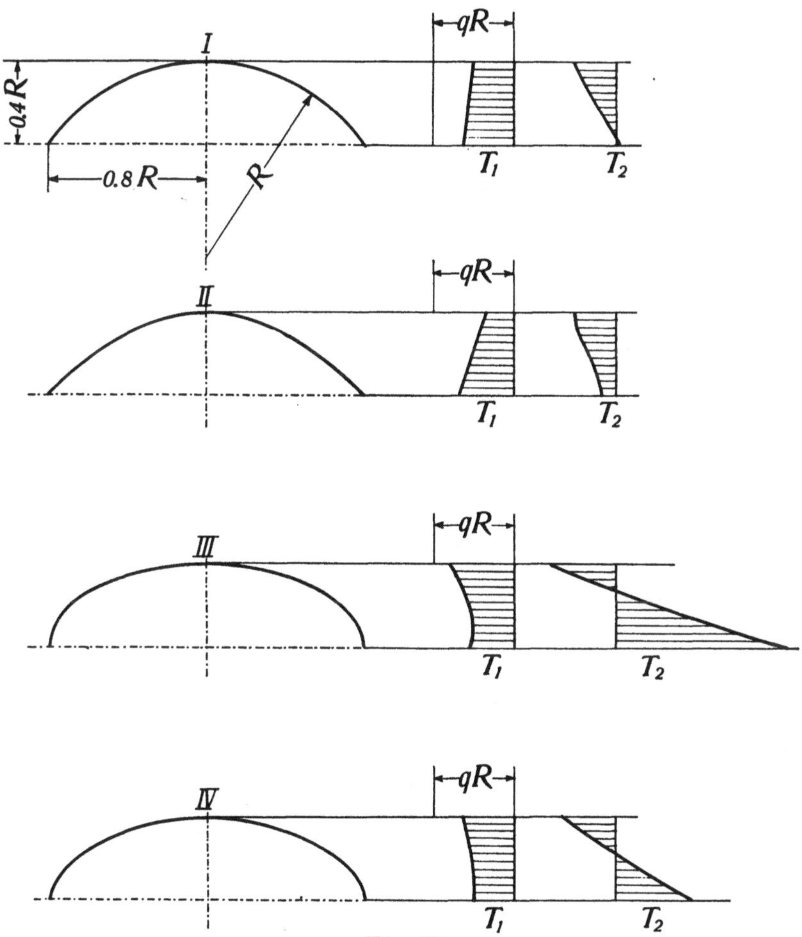

Fig. 17.

Let the forces acting along the edge of the dome be resolved into two components, acting vertically and horizontally respectively. The vertical component Q_z will be transmitted to the wall on which the dome rests, while the assimilation of the horizontal component Q_x requires a special structural element, called a thrust ring (Fig. 18).

If one compares the domes under consideration from the point of view of the required thrusts, one arrives at the following results:

$$Q_{xI} = 0.375qR,$$
$$Q_{xII} = 0.487qR,$$
$$Q_{xIII} = 0.$$
$$Q_{xIV} = 0.190qR,$$

where R is the radius of the spherical dome.

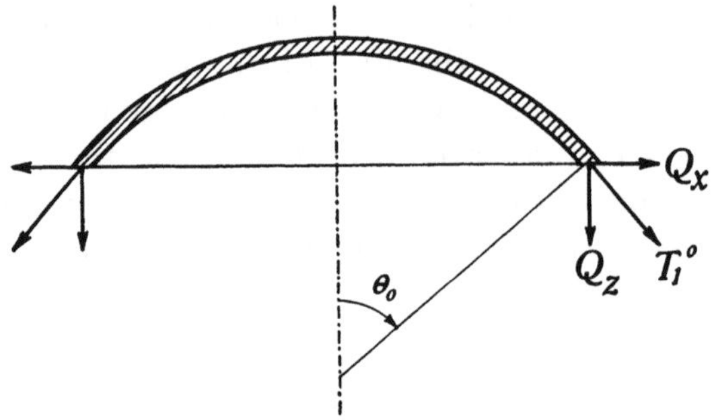

Fig. 18.

Thus, the thrust ring of the largest area would be required for dome II which, from this point of view, will be the least advantageous design. Dome III does not require a ring at all. However, its shape has the essential defect that it leads to considerable hoop stresses near the edge of the shell. Among the materials used for thin domes is reinforced concrete which is much weaker in tension than in compression. In view of this, if the shell of shape III is selected, it has to be constructed with variable thickness, the thickness increasing toward the edge. Further, attention should be drawn to the fact that in shell II the hoop force $T_2{}^*$ does not change its sign along the whole of the meridian, it being compressive everywhere. But since the thrust ring absorbing the loading Q_x will be stretched, it is evident in this case that the deformation of the support ring cannot be arranged to be the same as that of the edge of the shell (since they are of opposite sign). In the domes I, III and IV, the hoop forces change sign from compression at the apex of the dome to tension near the edge. The horizontal circles on which the change of sign of the circumferential force occurs will be called the 'line of transition'. The presence of a line of transition in a shell is favourable in the sense that for similar domes the sign of the circumferential stress will coincide with that of the stress in the support ring. Hence the area of the section of the support ring may be selected in such a way that the stretching of the edge of the shell will be equal to that of the support ring.

The stress in the support ring may be calculated by use of the formula

$$\sigma_k = \frac{Q_x r}{\Omega_k} = -\frac{T_1{}^0 r}{\Omega_k} \cos \theta_0, \tag{26.7}$$

where

$T_1{}^0$ is the value of the force $T_1{}^*$ at the edge of the shell,

r is the radius of the support ring (approximately equal to the radius of the edge of the shell),

Ω_k is the sectional area of the support ring,

θ_0 is the angle corresponding to the edge of the shell.

Hence the hoop stress at the edge of the dome will be equal to

$$\sigma = \frac{T_2{}^0}{\delta}. \tag{26.8}$$

Further, let Young's modulus of the material of the support ring be equal to E_k and that of the material of the shell equal to E. As a rule, Poisson's ratio for reinforced concrete is assumed to be equal to zero. Then, comparison of the extension of the edge of the shell and of the support ring leads to the formula

$$-\frac{T_1{}^0 r}{E_k \Omega_k} \cos \theta_0 = \frac{T_2{}^0}{E\delta}, \tag{26.9}$$

whence one obtains for the required area of the cross-section of the support ring

$$\Omega_k = -\frac{E}{E_k} \frac{T_1{}^0}{T_2{}^0} r\delta \cos \theta_0. \tag{26.10}$$

It follows from this formula that the area of the cross section of the support ring, necessary for the dome to be in a membrane state of stress, is proportional to $\frac{T_1{}^0}{T_2{}^0}$, i.e., to the ratio of the meridional and the hoop forces at the edge of the shell.

As is seen from the graphs in Fig. 17, this ratio is very large for shell I and hence a very rigid support ring is required. In this sense, the more advantageous shape is exhibited by the dome IV, since it has a sufficiently steep edge ($\cos \theta_0$ is relatively small) and the quantity $-\frac{T_1{}^0}{T_2{}^0}$ is less than unity.

With respect to the domes, shown in Fig. 17, the following final assessments may be given:

Dome I is uneconomical, since it requires a very stiff support ring;

Dome II is unacceptable, since there is no transition line and therefore the hoop force at its edge will have the same sign as at the apex. Formula (26.10) gives a negative value for Ω_k which physically corresponds to it being impossible to find a dome of this shape which is stressed in agreement with the membrane theory;

Dome III does not require a support ring, but considerable tensile hoop stresses act in the neighbourhood of its support, so that the thickness of the shell must be increased there;

Dome IV is suitable, since it may be placed in a membrane state of stress by use of a support ring of relatively small stiffness.

Thus, dome IV is the most acceptable design of the four types under consideration. Other equally or even more acceptable shapes are possible and it is clear from the above discussion what criteria must be used to compare the shapes of domes.

Sometimes domes are spherical and in order to obtain a sufficiently high transition line and a sufficiently steep tangent at the edge of the shell one has to resort near the support to a transition arc by gradually diminishing the radius of curvature of the meridian. For details regarding this problem the reader is referred to the book [66].

§ 27. The membrane theory of pressure vessels

As a second example, consider the application of the membrane theory to the stress analysis of pressure vessels and their caps.

In this case, the loading will be uniform normal pressure

$$q_n = p = \text{constant},$$

the other components of the loading being equal to zero.

Substituting this loading into the formulae (25.10), one finds

$$T_1^* = T_1'' \frac{R_2'' \sin^2 \theta''}{R_2 \sin^2 \theta} + \frac{p}{R_2 \sin^2 \theta} \int_{\theta''}^{\theta} R_1 R_2 \sin \theta \cos \theta \, d\theta =$$

$$= T_1'' \frac{R_2'' \sin^2 \theta''}{R_2 \sin^2 \theta} + \frac{p}{R_2 \sin^2 \theta} \int_{\theta''}^{\theta} R_2 \sin \theta \, d(R_2 \sin \theta) =$$

$$= \frac{pR_2}{2} + \frac{C}{R_2 \sin^2 \theta}, \tag{27.1}$$

$$T_2^* = -\frac{R_2}{R_1} T_1^* + pR_2 = \frac{pR_2}{2}\left(2 - \frac{R_2}{R_1}\right) - \frac{C}{R_1 \sin^2 \theta},$$

where

$$C = (T_1'' - \tfrac{1}{2}pR_2'')R_2'' \sin^2 \theta''. \tag{27.2}$$

Consider first a vessel having the form of a complete sphere. Then $R_1 = R_2 = R =$ constant and the formulae (27.1) give

$$T_1^* = \frac{pR}{2} + \frac{C}{R \sin^2 \theta}, \quad T_2^* = \frac{pR}{2} - \frac{C}{R \sin^2 \theta}. \tag{27.3}$$

The constant C in this case must necessarily be set equal to zero, since otherwise the solution would become infinite at the points $\theta = 0$, $\theta = \pi$ (which belong to the complete sphere).

Hence one obtains

$$T_1^* = T_2^* = \frac{pR}{2} = \text{const.} \tag{27.4}$$

Thus the complete thin-walled sphere, loaded by uniform internal pressure, is found to be in a state of uniform tension. (For external pressure, the stresses would change sign and instead of uniform tension one would obtain uniform compression). Hence it appears that the complete sphere is the ideal form for a shell loaded by uniform normal pressure (in the sense of uniform utilization of the material). However, reservoirs of this form are relatively rarely used, mainly due to the difficulty experienced in the manufacture of complete spherical shells. (Since no section of a sphere can be developed on to the plane, the complete sphere must be composed of a large number of segments, each of which must be bent beforehand in a suitable manner).

A much more convenient shape from the point of view of production as well as from other considerations (for example, dimensional) is exhibited by cylindrical reservoirs, closed by lids of one shape or other. The larger part of the surface of such vessels is developable and, as a consequence, it can be bent from plane sheets. The ends which represent a relatively small part of the surface of the reservoir and which as a rule are very curved, can without difficulty be manufactured by forging or rolling using a matrix.

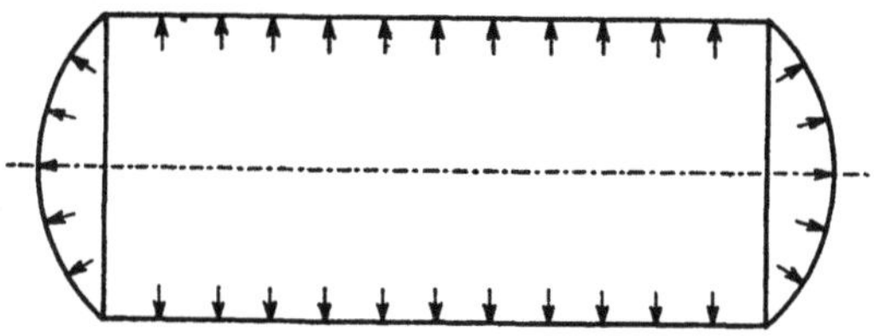

Fig. 19.

The membrane theory will now be applied to composite shells consisting of a cylinder and two ends. For this purpose, the cylinder and the cap will be treated separately and in this way the possibility of joining given shells without violation of the requirements of the membrane theory will be explained. Applying the formulae (27.1) to the cylindrical part of the vessel, i.e., letting there

$$R_1 = \infty,\ R_2 = r = \text{const},\ \theta = \theta'' = \frac{\pi}{2},$$

one obtains

$$T_1{}^* = T_1'',\ T_2{}^* = pr. \tag{27.5}$$

The force T_1'' is determined from the condition that the axial forces stretching the reservoir must be equal to the resultant force due to the pressure acting on the ends of the vessel. Therefore

$$T_1{}^* = T_1'' = p\,\frac{\pi r^2}{2\pi r} = \frac{pr}{2}, \tag{27.6}$$

$$T_2{}^* = pr.$$

Thus, by the membrane theory, the cylindrical part of the pressure vessel is stretched by constant forces:

in the longitudinal direction

$$T_1{}^* = \frac{pr}{2}, \tag{27.7}$$

in the circumferential direction

$$T_2{}^* = pr.$$

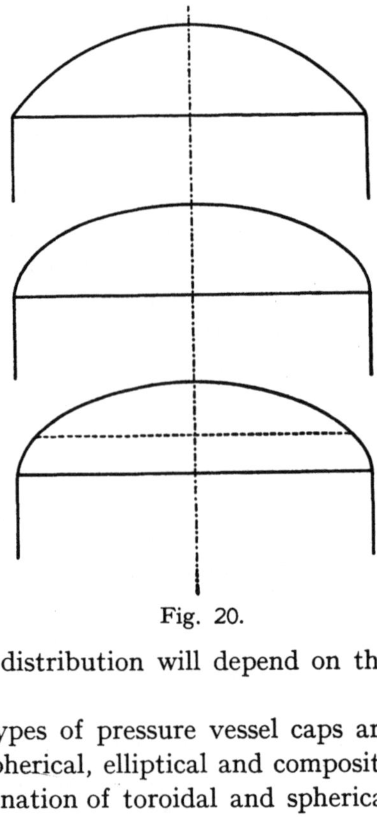

Fig. 20.

As regards the forces arising in the ends of the vessel, their magnitude and distribution will depend on the shape of the ends.

Three of the most commonly used types of pressure vessel caps are shown in Fig. 20, namely, caps of the spherical, elliptical and composite types. The last one is formed by a combination of toroidal and spherical shells.

Applying the formulae (27.1) to the spherical cap, one obtains

$$T_1{}^* = T_2{}^* = \frac{pR}{2}. \tag{27.8}$$

In other words, by the membrane theory, the spherical end is subject to uniform tensile forces $T^* = \frac{pR}{2}$.

Consider now the cylindrical shell to which the spherical end is attached. It is seen from Fig. 20 that in joining these two shells one of the basic conditions for the existence of the membrane state of stress has been violated; namely, the force

$$Q_x = T_1^* \cos\theta_0 = \frac{pR}{2}\cos\theta_0, \tag{27.9}$$

may be transferred to the cylindrical shell only by shearing forces $\left(\text{cf. below for the case } \theta_0 = \frac{\pi}{2}\right)$. Hence it is clear that the stresses actually arising in vessels of the above type will differ from the stresses which follow from the membrane theory, since it is impossible to neglect the bending in the neighbourhood of the joints between spherical and cylindrical shells.

Next, the formulae (27.1) will be applied to the elliptical end for which (cf. the previous section) the radii of curvature are given by the expressions

$$R_1 = \frac{r(1+\gamma)^{1/2}}{(1+\gamma\sin^2\theta)^{3/2}}, \quad R_2 = \frac{r(1+\gamma)^{1/2}}{(1+\gamma\sin^2\theta)^{1/2}}. \tag{27.10}$$

Hence

$$\begin{aligned} T_1^* &= \frac{pR_2}{2} = \frac{pr}{2}\,\frac{(1+\gamma)^{1/2}}{(1+\gamma\sin^2\theta)^{1/2}}, \\ T_2^* &= \frac{pR_2}{2}\left(2 - \frac{R_2}{R_1}\right) = \frac{pr}{2}\,\frac{1-\gamma\sin^2\theta}{(1+\gamma\sin^2\theta)^{1/2}}(1+\gamma)^{1/2}, \end{aligned} \tag{27.11}$$

where $\gamma = \frac{r^2}{h^2} - 1$, h being the height of the cap and r the radius of the reservoir.

Studying now the junction between the cylindrical and elliptical shell, it is seen that in this case there is no radial component at the edge of the cap $\left(\text{i.e., at } \theta = \frac{\pi}{2}\right)$ and therefore the cap of elliptical form does not demand the presence of shearing forces at the edge of the cylindrical shell. Thus, in this case, one of the conditions for the existence of the membrane state of stress is fulfilled.

However, in addition, the hoop forces in the cylindrical and elliptical shells must be equal along their junction (in order to ensure the continuity of the circumferential deformations).

Comparing the force at the edge of the cylinder

$$T_2^* = pr$$

with the force on the edge of the ellipsoidal shell

$$T_2^* = \frac{pr}{2}(1 - \gamma),$$

it appears that they will only be equal for $\gamma = -1$, i.e., for $\frac{h}{r} \to \infty$, a case which is of no practical interest. Thus, although the ellipsoidal cap violates the requirements of the membrane theory to a lesser degree than the spherical cap, in this case also not all the conditions for the existence of the membrane state of stress can be fulfilled.

For most elliptical caps the height is half the radius of the vessel. In this case $\gamma = 3$ and the hoop forces at the edge of the cap are $T_2^* = -pr$, i.e., equal in value and opposite in sign to the hoop force at the edge of the cylindrical part of the reservoir. Hence, for this choice of dimensions of the cap, bending moments will occur;

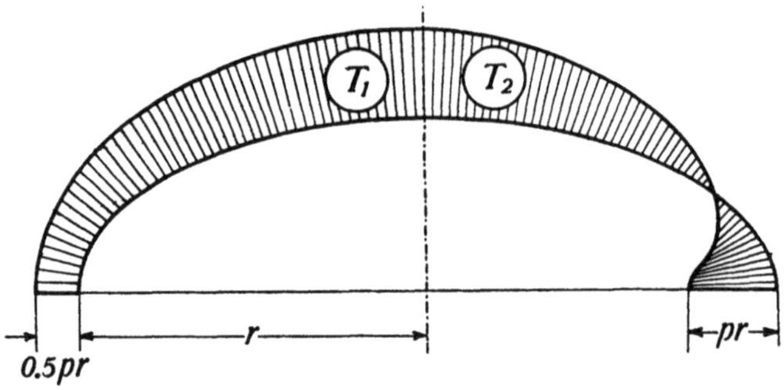

Fig. 21.

Table 1

θ_0	0	5	10	15	20	25	30	35	40	45	50
$T_1^* : pr$	1.000	0.988	0.960	0.912	0.860	0.810	0.756	0.709	0.669	0.634	0.603
$T_2^* : pr$	1.000	0.968	0.875	0.732	0.559	0.381	0.165	0.087	−0.161	−0.317	−0.459

θ_0	55	60	65	70	75	80	85
$T_1^* : pr$	0.576	0.555	0.539	0.524	0.514	0.506	0.502
$T_2^* : pr$	−0.581	−0.693	−0.787	−0.863	−0.925	−0.966	−0.977

however, they will not be large (cf. § 59), so that the sum of the stresses due to direct forces and moments in the cap and the cylindrical part of the reservoir will only slightly differ from that given by the membrane theory, i.e., it will be close to

$$\sigma = \frac{pr}{\delta}.$$

Table 1 gives the values for T_1^*, T_2^* in the elliptical cap as calculated from (27.11), using $\gamma = 3$. These results are shown graphically in Fig. 21.

The composite cap is characterized by the radius r of its base, the transverse radius (radius of the meridian of the torus) r_0, the radius R and the marginal angle θ_0 of the sphere. Since the joint between the sphere and the torus is to be smooth, the following relation must hold between these quantities (Fig. 22):

$$r - r_0 = (R - r_0) \sin \theta_0. \tag{27.12}$$

Denoting by ν the ratio of the height of the cap to the radius r of the cylinder, one finds

$$\nu = \frac{R - (R - r_0) \cos \theta_0}{r},$$

whence, using the previous formula, one can write

$$\frac{r_0}{R} = \frac{1 - \cos \theta_0 - \nu \sin \theta_0}{\nu(1 - \sin \theta_0) - \cos \theta_0}. \tag{27.13}$$

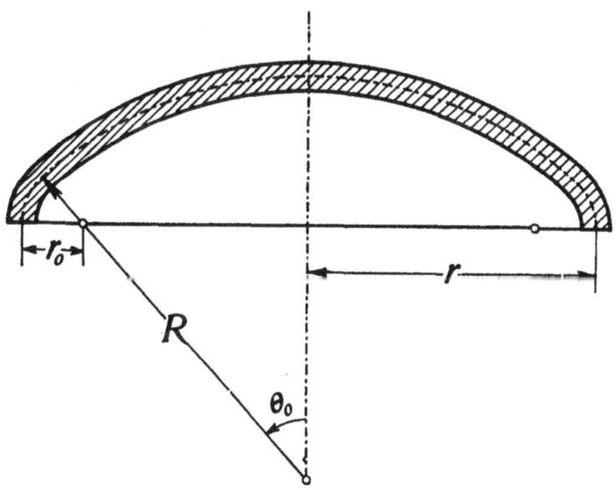

Fig. 22.

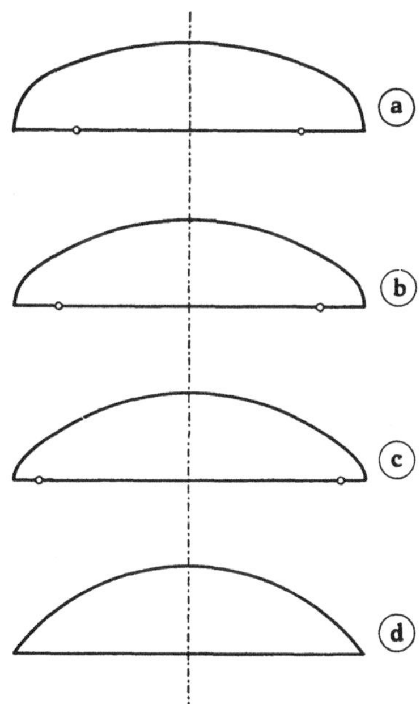

Fig. 23.

This last formula gives the relationship between the radius of the torus and the radius of the sphere as a function of the marginal angle θ_0 and the relative height of the end.

The most frequently used case is $\nu = \frac{1}{2}$ when

$$\frac{r_0}{R} = \frac{2 - \sin\theta_0 - 2\cos\theta_0}{1 - \sin\theta_0 - 2\cos\theta_0}. \tag{27.14}$$

Fig. 23 shows four different composite caps having the same height ($\nu = \frac{1}{2}$) and different ratios $\frac{r_0}{R}$. In fact, the dimensions of the caps in Fig. 23 are

a) $R = 2r, \quad r_0 = \frac{3}{8}r, \quad \sin\theta_0 = \frac{5}{13}$;

b) $R = 1.5r, \quad r_0 = 0.25r, \quad \sin\theta_0 = 0.6$;

c) $R = 1.353r, \quad r_0 = 0.1466r, \quad \theta_0 = 45°$;

d) $R = 1.25r, \quad r_0 = 0, \quad \sin\theta_0 = 0.8$.

Thus, the spherical end is a special case of the composite end (for $r_0 = 0$).

Cap a) of Fig. 23 has the same radius of curvature of the meridian at the apex, cap b) the same radius of curvature of the meridian at the edge as the elliptical end with the same value of ν which, as has been stated above, corresponds to the optimal case.

Applying (27.1) to the torus and taking into consideration that in this case

$$R_1 = r_0, \; R_2 = \frac{r - r_0(1 - \sin\theta)}{\sin\theta} \tag{27.15}$$

and that the force $T_1{}^*$ on the upper edge of the torus must (for equilibrium) be equal to the corresponding force at the edge of the sphere $\left(\text{i.e., } T_1 = \frac{pR}{2}\right)$, one will have

$$T_1{}^* = \frac{pR_2}{2} = \frac{pr}{2}\,\frac{1 - \frac{r_0}{r}(1 - \sin\theta)}{\sin\theta}, \tag{27.16}$$

$$T_2{}^* = \frac{pR_2}{2}\left(2 - \frac{R_2}{R_1}\right) = \frac{pr_0}{2} - \frac{p(r - r_0)^2}{2r_0\sin^2\theta}.$$

As regards the forces in the spherical part of the cap, they have been shown earlier to be equal to

$$T_1{}^* = T_2{}^* = \frac{pR}{2}. \tag{27.17}$$

Fig. 24 shows a typical distribution of the forces along the meridian of the composite end (according to the membrane theory). It may be verified from the above formulae that the conditions for the existence of the membrane state of stress may not be completely satisfied either at the joint between the torus and the cylinder or at the joint between the torus and the sphere. In fact, it turns out to be impossible to ensure continuity of the circumferential deformations along the above mentioned joints between the shells for any relations between the dimensions of the composite cap. Thus composite ends must be analysed by use of a theory which takes bending into consideration.

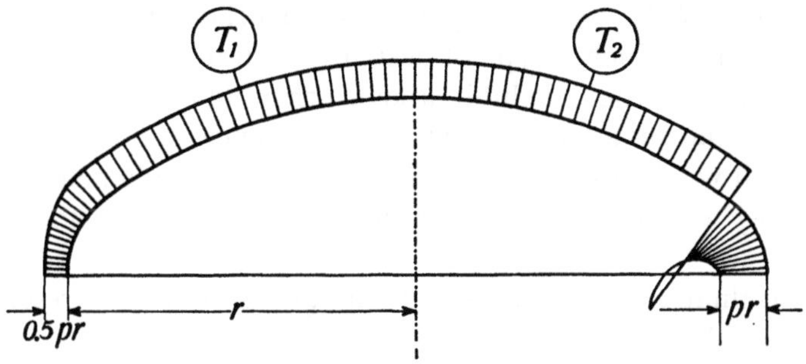

Fig. 24.

Summing up the above results, it may be established that all the caps of the types discussed here fail to satisfy the requirements of the membrane theory to a greater or lesser degree. This result has general importance and it may be stated that it is impossible to design cylindrical pressure vessels which are in a membrane state of stress. (It will be noted that in the previous section, when dealing with domes, the opposite conclusion was reached).

Therefore the stress analysis of such reservoirs must take into account bending; however, as will be shown later, this fact only partly invalidates the above results, since these enter into the refined theory of pressure vessels as an essential part of the solution.

In conclusion, it will be noted that the above results may be considered as examples illustrating the truth of the statement in § 22 that the membrane theory cannot yield correct results, if the radii of curvature of the middle surface of the shell have discontinuities. In fact, the cylindrical vessel of Fig. 19, closed by lids, can be considered as a single closed shell of revolution in which there occur at two parallel circles (the joints between the cylinder and the lids) discontinuities of one (elliptic lid) or both (spherical lid) radii of curvature. For box lids, the radius of curvature of the meridian has, in addition, discontinuities on parallel circles corresponding to the transition from the ring shaped inset to a sphere. It follows from the above that at all these parallel circles the membrane theory leads to discontinuities in the hoop forces and, hence, to a violation of the continuity of the deformations.

§ 28. Shells of uniform strength

As can be seen from the formulae (25.10), the forces in a shell depend on its shape as well as on its external loading. There arises the question whether it is impossible to select the shape of a shell in such a manner that the forces T_1^*, T_2^* are equal to each other and constant at all points of the shell. Such a type of design (which may be called a shell of uniform strength) would exhibit above other advantages the uniform utilization of the material, and as a consequence, all other conditions being equal, it would give the smallest weight. Two fundamental problems of the theory of shells of uniform strength will be considered below, namely:

1. Determination of the most advantageous shapes of reservoirs for the storage of liquids;
2. Determination of the most advantageous shapes of domes.

1. Examine first the case of a tank for the storage of oil or other fluids, starting from the principle of uniform strength. The loading in this case will be the hydrostatic pressure

$$q_n = p_0 + z\rho, \tag{28.1}$$

where p_0 is the pressure at the top of the reservoir and ρ is the specific gravity of the fluid. The z axis will be directed downwards along the axis of the reservoir with the origin at the top of the reservoir.

As a result of the symmetry of the loading, the unknown shape of the vessel will undoubtedly be that of a body of revolution. Letting in the equations (23.4)

$$q_1 = q_2 = S^* = 0, \quad T_1^* = T_2^* = T_0^* = \text{constant},$$

it is then found that the first two of these equations are satisfied identically, while the third assumes the form

$$\left(\frac{1}{R_1} + \frac{1}{R_2}\right) T_0 = p_0 + z\rho. \tag{28.2}$$

Using (23.3), one may write

$$\frac{1}{\cos\theta}\,\frac{d(R_2 \sin\theta)}{d\theta} = \frac{d(R_2 \sin\theta)}{d(\sin\theta)} = R_1, \tag{28.3}$$

or, introducing the new notation

$$x = R_2 \sin\theta, \ \xi = \sin\theta, \tag{28.4}$$

one arrives at

$$R_1 = \frac{dx}{d\xi}. \tag{28.5}$$

Substituting in (28.2) the values of R_1 and R_2 in terms of the new variables, one finds

$$T_0\left(\frac{d\xi}{dx} + \frac{\xi}{x}\right) = p_0 + zp. \tag{28.6}$$

There remains to deduce the dependence between z, x and ξ. The tangent of the angle between the normal to the shell and the z axis (i.e., the angle θ) will be equal to

$$\tan\theta = \frac{dz}{dx}.$$

Expressing $\tan\theta$ in terms of $\sin\theta$ and introducing ξ, one can write

$$\frac{dz}{dx} = \frac{\xi}{\sqrt{1-\xi^2}} \quad \text{for the interval } 0 \leqslant \theta \leqslant \frac{\pi}{2},$$

$$\frac{dz}{dx} = -\frac{\xi}{\sqrt{1-\xi^2}} \quad \text{for the interval } \frac{\pi}{2} \leqslant \theta \leqslant \pi.$$

Replacing z in these formulae by the expression (28.6), one obtains the two differential equations

$$\begin{aligned} \frac{d^2\xi}{dx^2} + \frac{1}{x}\frac{d\xi}{dx} - \frac{\xi}{x^2} - \frac{p}{T_0}\frac{\xi}{\sqrt{1-\xi^2}} &= 0, \\ \frac{d^2\xi}{dx^2} + \frac{1}{x}\frac{d\xi}{dx} - \frac{\xi}{x^2} + \frac{p}{T_0}\frac{\xi}{\sqrt{1-\xi^2}} &= 0, \end{aligned} \tag{28.7}$$

the first of which determines the shape of the meridian of the shell of uniform strength in the interval $0 \leqslant \theta \leqslant \frac{\pi}{2}$ (i.e., the shape of the upper part of the reservoir), the second the shape of the meridian in the interval $\frac{\pi}{2} \leqslant \theta \leqslant \pi$ (i.e., the shape of the lower part of the reservoir).

The equations (28.7) are non-linear and may only be integrated approximately. Replacing the absolute coordinates of the meridian of the shell x and z by the non-dimensional quantities

$$x_1 = \sqrt{\frac{\rho}{T_0}}\, x, \quad z_1 = \sqrt{\frac{\rho}{T_0}}\, z, \tag{28.8}$$

the formulae (28.6) and (28.7) are reduced to the form

$$\frac{d\xi}{dx_1} + \frac{\xi}{x_1} = \frac{p_0}{\sqrt{\rho T_0}} + z_1 = \lambda + z_1, \tag{28.9}$$

$$\begin{aligned} \frac{d^2\xi}{dx_1^2} + \frac{1}{x_1}\frac{d\xi}{dx_1} - \left(\frac{1}{x_1^2} + \frac{1}{\sqrt{1-\xi^2}}\right)\xi &= 0, \\ \frac{d^2\xi}{dx_1^2} + \frac{1}{x_1}\frac{d\xi}{dx_1} - \left(\frac{1}{x_1^2} - \frac{1}{\sqrt{1-\xi^2}}\right)\xi &= 0. \end{aligned} \tag{28.10}$$

It is now clear that the unknown shape of the shell is completely determined by the single non-dimensional parameter

$$\lambda = \frac{p_0}{\sqrt{T_0 \rho}}. \tag{28.11}$$

The purpose of the second parameter $\frac{\rho}{T_0}$ is to determine the dimensions of the shell, since its knowledge permits not only the calculation of the non-dimensional coordinates x_1, z_1, but also of the absolute coordinates x, z.

Fig. 25 shows a typical shape of the meridian of a shell of uniform strength. As is clear from this sketch, the reservoir has the shape of a drop in that it has a plane base on which it rests and which smoothly fits into the meridian.

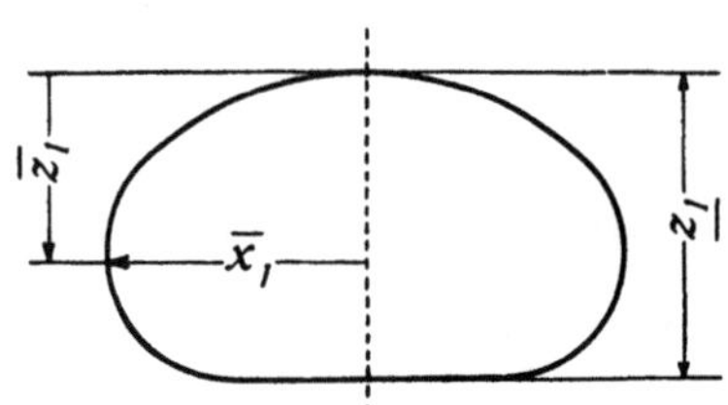

Fig. 25.

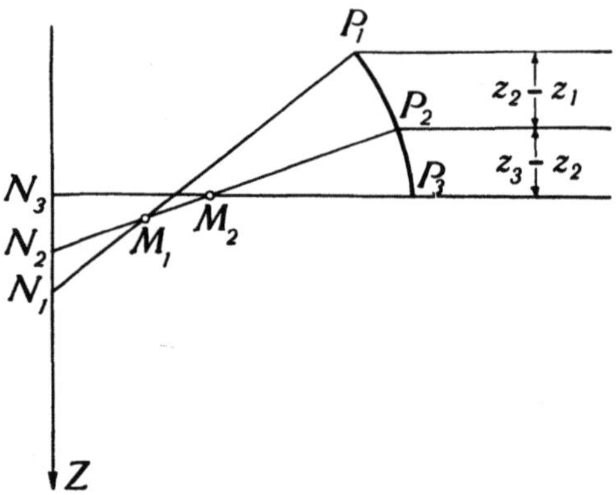

Fig. 26.

That the unknown shape of the reservoir must exactly coincide with the form of a supported drop may be verified by taking into consideration that the surface tension of a liquid is constant, while the pressure inside a drop varies according to a hydrostatic law. Hence the equations deduced above are essentially the equations of the theory of capillarity, where they have been sufficiently well studied. In particular, equation (28.2) was obtained by Laplace and it is, as a rule, associated with his name.

No time will be devoted here to methods of analytic solution of the equations (28.9), (28.10). However, a simple graphical method due to Kelvin for determining the form of drops (or, what is the same thing, the shape of reservoirs of uniform strength) will be described.

In this method, the meridian of the drop is made up of a series of small arcs of circles, the radius of each of these circles being given by the following scheme. Let (Fig. 26) $R_1^1 = P_1M_1$ and $R_2^1 = P_1N_1$ be the magnitudes of the radii of curvature of the surface of the drop at the point P_1. Their knowledge permits construction of the arc of the circle P_1P_2 which will be very close to the corresponding element of the arc of the meridian of the drop. Next, the equation (28.2) will be written for the points P_1 and P_2,

$$\begin{aligned} \frac{1}{R_1'} + \frac{1}{R_2'} &= \frac{p_0}{T_0} + \frac{\rho}{T_0} z_1, \\ \frac{1}{R_1''} + \frac{1}{R_2''} &= \frac{p_0}{T_0} + \frac{\rho}{T_0} z_2, \end{aligned} \tag{28.12}$$

Here z_1, z_2 are the values of z at the points P_1, P_2 and R_1'', R_2'' are the values of the radii of curvature at the point P_2. On the basis of P_1P_2, one may now determine from the drawing the difference of the heights of the points P_1 and P_2 and likewise the radius $R_2'' = P_2N_2$.

Subtracting the first equation (28.12) from the second, one finds

$$R_1'' = \frac{1}{\dfrac{\rho}{T_0}(z_2 - z_1) + \dfrac{1}{R_1'} + \dfrac{1}{R_2'} - \dfrac{1}{R_2''}}. \tag{28.13}$$

By calculating the radius of curvature of the meridian at the point P_2 by means of this formula, one may construct the arc of the circle P_2P_3 (with centre at M_2 and $M_2P_2 = R_1''$), etc.

Denoting the distance $z_{i+1} - z_i$ by Δz and beginning the process of construction at the top of the reservoir, where

$$R_1 = R_2 = 2\frac{T_0}{p_0}, \tag{28.14}$$

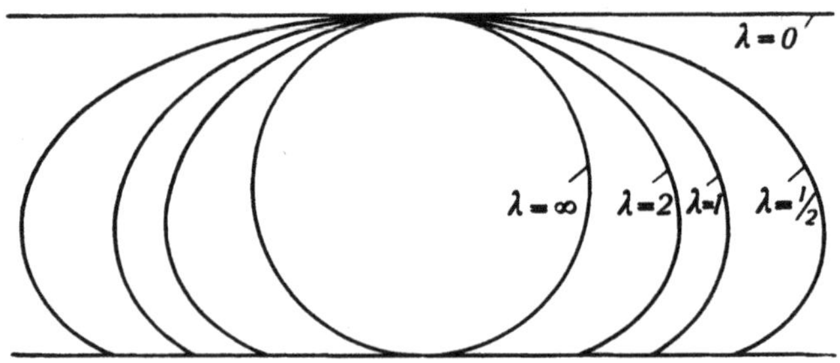

Fig. 28.

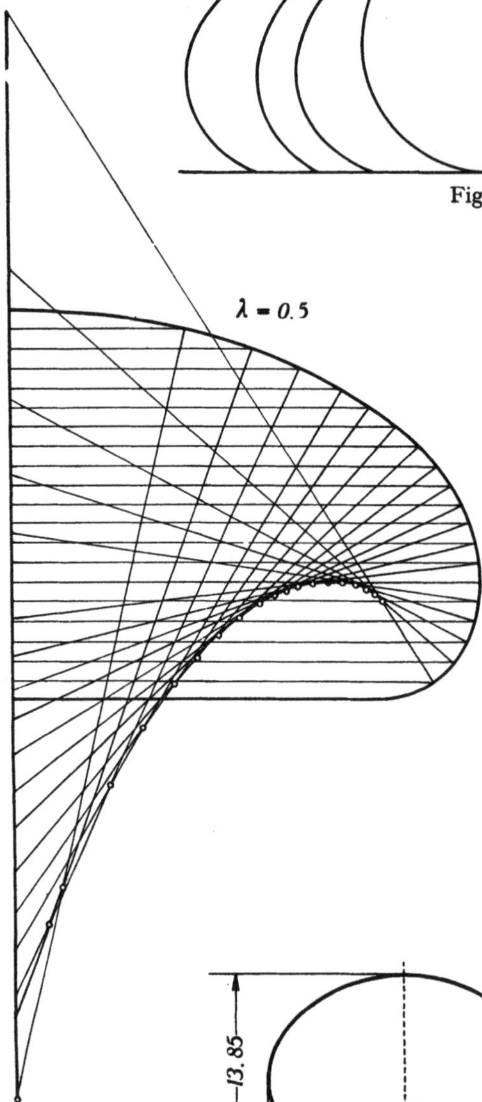

Fig. 27.

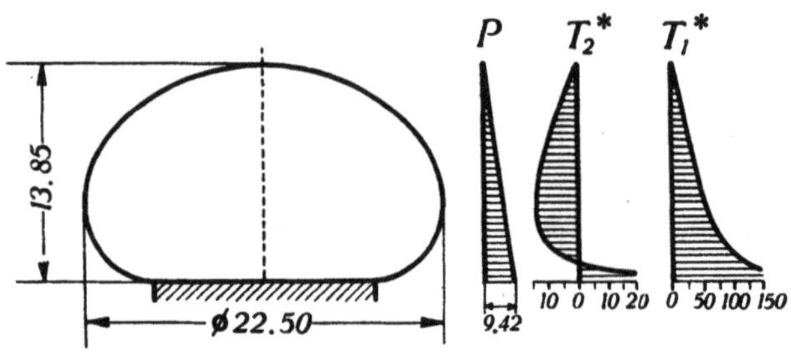

Fig. 29.

one may thus draw step by step the entire meridian of the drop (Fig. 27). Fig. 28 shows the meridians of shells of uniform strength obtained for $\lambda = 0.5$, $\lambda = 1.0$, $\lambda = 2.0$. It is seen that a decrease in the value of the pressure at the top of the reservoir leads to a decrease of the height over width ratio. Two extreme cases will be noted: when $\lambda \to \infty$, the form of the reservoir approaches the form of a sphere, while for $\lambda \to 0$, the transverse dimension of the reservoir becomes infinitely large, so that in the limit the fluid is contained between two parallel walls.

The pressure at the top of the vessel of a considered shape must be maintained constant and equal to the calculated value, since otherwise the shell will not be stressed uniformly and, besides, it may develop wrinkles. Fig. 29 shows a reservoir of uniform strength (constructed starting from the value $\lambda \approx 1.1$) with zero pressure at the top, in order to study the effect of the pressure. As is seen from the graphs in the same figure, the stress distribution in that shell is not at all uniform and near the plane base the forces tend to infinity which indicates that in this region there occurs appreciable bending. This is, of course, not taken into consideration by the membrane theory. Reservoirs having the shape of drops are sometimes used in practice.

2. As has already been mentioned, the basic loading for domes is ordinarily considered to be due to their weight. An attempt will now be made to select the shape of a dome, starting from the requirement that the stresses throughout the shell are to be constant.

The components of loading in this case will be

$$q_1 = q \sin \theta, \quad q_n = - q \cos \theta. \tag{28.15}$$

Putting in the equation (23.4)

$$T_1^* = T_2^* = T^*,$$

assuming besides that the deformation of the shell does not depend on the angle φ and that

$$S^* = q_2 = 0,$$

one will have

$$\begin{aligned} \frac{dT^*}{d\theta} &= - qR_1 \sin \theta, \\ \left(\frac{1}{R_1} + \frac{1}{R_2}\right) T^* &= - q \cos \theta. \end{aligned} \tag{28.16}$$

The essential difference between this and the preceding problem is that it is impossible to assume T^* to be constant, since then the first of the equations (28.16) cannot be satisfied.

Solving the second of these equations for R_1, one finds

$$R_1 = -\frac{R_2}{1 + \dfrac{q}{T^*} R_2 \cos\theta}. \tag{28.17}$$

At the same time, the relation (23.3) can be used to obtain a different expression for R_1 in terms of R_2, in fact,

$$R_1 = \frac{1}{\cos\theta} \frac{d(R_2 \sin\theta)}{d\theta}. \tag{28.18}$$

Comparing the equations (28.17) and (28.18), one obtains the differential equation

$$\frac{d(R_2 \sin\theta)}{d\theta} = -\frac{R_2 \cos\theta}{1 + \dfrac{q}{T^*} R_2 \cos\theta}, \tag{28.19}$$

which may be simplified by introduction of the new variable

$$x = R_2 \sin\theta \tag{28.20}$$

and by use of the formulae

$$q = \rho\delta, \quad T^* = \sigma\delta, \tag{28.21}$$

where ρ is the specific gravity of the material of the shell and σ is the stress in the shell under the action of the force T^*. Then one will have

$$\frac{dx}{d\theta} = -\frac{x}{\tan\theta + \dfrac{\rho}{\sigma} x}. \tag{28.22}$$

Since the stresses at all sections of the shell are equal, the ratio $\dfrac{\rho}{\sigma}$ will be constant. For this purpose, one has to consider the thickness δ of the shell to be variable in agreement with the second of the formulae (28.21) and the first of the equations (28.16).

Thus, in the present problem, a uniform stress field throughout the shell may not be achieved at the expense of the shape of the dome only; it is also necessary to select a suitable law of variation of the thickness of the shell along the meridian. This law can be found, if one introduces into the first of the equations (28.16) the values for q and T^* from (28.21).

Then one will have

$$\frac{d\delta}{d\theta} = -\frac{\rho}{\sigma}\,\delta R_1 \sin\theta, \tag{28.23}$$

whence

$$\ln\left(\frac{\delta}{\delta_0}\right) = -\frac{\rho}{\sigma}\int_0^\theta R_1 \sin\theta\, d\theta, \tag{28.24}$$

where δ_0 is the thickness of the shell at $\theta = 0$, i.e., at the apex of the dome. The integral on the right-hand side of the formula (28.24) has a simple geometrical meaning, since the quantity

$$R_1 \sin\theta\, d\theta$$

is the projection of an element of arc of the meridian on the axis of the shell. Therefore

$$\int_0^\theta R_1 \sin\theta\, d\theta = z, \tag{28.25}$$

where z is the vertical coordinate of the point under consideration (the origin of z being at the apex of the dome). Thus the thickness of the shell must change according to the law

$$\delta = \delta_0 e^{-\frac{\varrho}{\sigma} z}. \tag{28.26}$$

The functional dependence of z on θ (determination of which is equivalent to finding the shape of the meridian of the dome) can be established by integration of the equation (28.22). Introducing for this purpose instead of the actual coordinates x and z of the meridian the non-dimensional coordinates

$$x_1 = -\frac{\rho}{\delta}\,x, \quad z_1 = -\frac{\rho}{\delta}\,z, \tag{28.27}$$

the equation (28.22) may be written in the form

$$\frac{dx_1}{d\theta} = -\frac{x_1}{\tan\theta - x_1}. \tag{28.28}$$

Integrating this equation, using the initial condition $\theta = 0$, $x_1 = 0$, one finds $x_1 = f(\theta)$; the second coordinate of the meridian of the dome can now be calculated from the formula

$$z_1 = -\frac{\rho}{\sigma}\int_0^\theta R_1 \sin\theta\, d\theta = \int_0^\theta \frac{dx_1}{d\theta}\tan\theta\, d\theta, \tag{28.29}$$

which is obtained by taking into consideration the expressions (23.3). (28.20), (28.25), and (28.27).

Thus, the determination of the non-dimensional coordinates of the meridian of the shell is reduced to the solution of the differential equation (28.28). Apparently this last integration can only be carried out approximately, as this has been done by G. Megareus [73] from whose paper the following table has been taken:

Table 2

x_1	z_1	e^{z_1}	x_1	z_1	e^{z_1}
0.2	0.0100	1.0101	2.2	1.4138	4.1116
0.4	0.0402	1.0410	2.4	1.6320	5.6519
0.6	0.0910	1.0953	2.6	2.096	8.139
0.8	0.1632	1.1773	2.8	2.508	12.28
1.0	0.2580	1.2943	3.0	2.972	19.52
1.2	0.3768	1.4576	3.2	3.496	32.98
1.4	0.5216	1.6847	3.4	4.100	60.34
1.6	0.6946	2.0029	3.6	4.800	121.51
1.8	0.8988	2.4567	3.8	5.76	317.53
2.0	1.1374	3.1187	4.0	7.22	1365.5

It is clear from the above study that this table exhausts the problem under consideration, since the thickness of the shell δ is determined by (28.26) as a function of e^{z_1}. The form of the dome of uniform strength, constructed on the basis of Table 2, is shown in Fig. 30.

It follows from this sketch, and also directly from Table 2, that the thickness of the dome must be changed considerably along the meridian. For example, if the height of the dome is taken to be equal to half the radius of the base, the thickness at the edge of the shell must be approximately 2.5 times larger than the thickness at the apex. But if the height of the dome is made equal to the radius of the base, this ratio is raised to 20. Thus it is impossible to make the height of the dome too large if it is intended to avoid the necessity of a considerable change of the thickness of the shell. The dome then turns out to be too shallow and it leads to an appreciable thrust. Further, since a shell of the shape under consideration is subject to uniform compression at all points, it has, of course, no line of transition and therefore, as has been established in § 26, it is impossible to create for it the conditions necessary for the existence of the membrane state of stress. In view of this situation, the above shape of dome may be called "most advantageous" in a limited sense only.

Fig. 30.

The shape is convenient from the point of view of the stress distribution at some distance from the edge, but it is quite useless from the point of view of the possibility of ensuring the proper boundary conditions. But the last are very important and must always be taken into consideration in a dome design. Therefore a dome of the above type has, to the author's knowledge, not found practical application.

§ 29. "Wind" type loading (The case k = 1)

In the stress analysis of shells of revolution, the second place in practical importance (after symmetrical loading) belongs to external loads the components of which are given by the formulae

$$q_1 = q_{1,1} \cos \varphi, \; q_2 = q_{2,1} \sin \varphi, \; q_n = q_{n,1} \cos \varphi, \tag{29.1}$$

obtained from (24.2) for $k = 1$.

This loading is ordinarily said to be of the "wind type" since the wind pressure on a dome may, to a first approximation, be represented by components of the form (29.1). It will be shown that in the case of such loading (as for the symmetrical type) the equations of the membrane theory can be integrated by quadrature for shells of revolution of arbitrary shape.

The basic differential equation (24.4) takes in this case the form

$$\frac{1}{R_1 R_2 \sin\theta} \frac{d}{d\theta}\left[\frac{R_2{}^2 \sin\theta}{R_1} \frac{dU_1{}^*}{d\theta}\right] - \frac{1}{R_1 \sin^2\theta} U_1{}^* = F_1(\theta), \tag{29.2}$$

where

$$F_1(\theta) = \frac{1}{R_1 R_2 \sin\theta} \frac{d}{d\theta} [(q_{n,1} \cos\theta - q_{1,1} \sin\theta) R_2{}^3 \sin^2\theta] - \\ - R_2(q_{n,1} - q_{2,1} \sin\theta). \tag{29.3}$$

In this equation, the new auxiliary function

$$Y = U_1{}^* R_2 \sin\theta. \tag{29.4}$$

will now be introduced instead of $U_1{}^*$. Taking into consideration the relation (23.3), the equation (29.2) can then be rewritten

$$\frac{d}{d\theta}\left(\frac{1}{R_1 \sin\theta}\,\frac{dY}{d\theta}\right) = R_1 F(\theta), \tag{29.5}$$

so that Y can be determined by two successive integrations.

The first integration gives

$$\frac{1}{R_1\sin\theta}\,\frac{dY}{d\theta} = \int R_1 F_1(\theta)\,d\theta =$$

$$= \int \frac{1}{R_2 \sin\theta}\,\frac{d}{d\theta}\,[(q_{n,1}\cos\theta - q_{1,1}\sin\theta)R_2^3 \sin^2\theta]\,d\theta -$$

$$- \int (q_{n,1} - q_{2,1}\sin\theta) R_1 R_2\, d\theta. \tag{29.6}$$

Further, integrating the right-hand side by parts, one finds

$$\frac{dY}{d\theta} = C_1 R_1 \sin\theta + \Phi R_1 \sin\theta, \tag{29.7}$$

where C_1 is an arbitrary constant,

$$\Phi = (q_{n,1}\cos\theta - q_{1,1}\sin\theta)R_2^2\sin\theta -$$

$$- \int_{\theta''}^{\theta} (q_{n,1}\sin\theta + q_{1,1}\cos\theta - q_{2,1}) R_1 R_2 \sin\theta\, d\theta, \tag{29.8}$$

and the lower limit of integration θ'' can be fixed arbitrarily. In the further discussion, θ'' will be identified with the angle determining the upper edge of the shell (which in the general case will be bounded by two parallel circles — Fig. 14).

The second integration now leads to

$$Y = C_2 + C_1 \int_{\theta''}^{\theta} R_1 \sin\theta\, d\theta + \int_{\theta''}^{\theta} \Phi R_1 \sin\theta\, d\theta, \tag{29.9}$$

where C_2 is another arbitrary constant.

Reverting now from Y to $U_1{}^*$ and then, with the help of the formulae (24.3) and (23.6), from $U_1{}^*$ to $T_1{}^*$, one finds

$$T_1{}^* = \frac{1}{R_2^2 \sin^3\theta}\left\{C_2 + C_1 \int_{\theta''}^{\theta} R_1 \sin\theta\, d\theta + \int_{\theta''}^{\theta} \Phi R_1 \sin\theta\, d\theta\right\}\cos\varphi = T^*{}_{1,1}\cos\varphi. \tag{29.10}$$

Further, the third equation (23.4) gives

$$T_2{}^* = \left\{q_{n,1}R_2 - \frac{1}{R_1 R_2 \sin^3\theta}\left[C_2 + C_1\int_{\theta''}^{\theta} R_1 \sin\theta\, d\theta +\right.\right.$$

$$+\int_{\theta''}^{\theta} \Phi R_1 \sin\theta\, d\theta\Big]\Big\}\cos\varphi = T^*_{2,1}\cos\varphi. \tag{29.11}$$

Finally, the shear force S^* may be determined from the first equation (23.7)

$$\frac{\partial V^*}{d\varphi} = R_2^2\sin^2\theta\,\frac{\partial S^*}{\partial\varphi} = -\frac{R_2^2\sin\theta}{R_1}\frac{\partial U^*}{\partial\theta} + (q_n\cos\theta - q_1\sin\vartheta)R_2^3\sin^2\theta =$$

$$= \Big[C_2\,\mathrm{cotan}\,\theta - C_1\Big(R_2\sin\theta - \mathrm{cotan}\,\theta\int_{\theta''}^{\theta} R_1\sin\theta\, d\theta\Big) - \Phi R_2\sin\theta +$$

$$+ (q_{n,1}\cos\theta - q_{1,1}\sin\theta)\,R_2^3\sin^2\theta + \mathrm{cotan}\,\theta\int_{\theta''}^{\theta}\Phi R_1\sin\theta\, d\theta\Big]\cos\varphi. \tag{29.12}$$

Substituting here

$$S^* = S_1^*\sin\varphi,$$

one will have

$$S^* = \frac{\cos\theta}{R_2^2\sin^3\theta}\Big\{C_2 - C_1\Big(R_2\frac{\sin^2\theta}{\cos\theta} - \int_{\theta''}^{\theta} R_1\sin\theta\, d\theta\Big) - \Phi\frac{R_2\sin^2\theta}{\cos\theta} +$$

$$+ (q_{n,1}\cos\theta - q_{1,1}\sin\theta)\frac{R_2^3\sin^3\theta}{\cos\theta} + \int_{\theta''}^{\theta}\Phi R_1\sin\theta\, d\theta\Big\}\sin\varphi =$$

$$= \Big[T^*_{1,1}\cos\theta - \frac{C_1}{R_2\sin\theta} + \chi(\theta)\Big]\sin\varphi = S_1^*\sin\varphi, \tag{29.13}$$

where

$$\chi(\theta) = \frac{1}{R_2\sin\theta}\int_{\theta''}^{\theta}(q_{n,1}\sin\theta + q_{1,1}\cos\theta - q_{2,1})R_1R_2\sin\theta\, d\theta. \tag{29.14}$$

The two constants C_1 and C_2 in the obtained solution must be determined from given values of the forces T_1^* and S^* at the upper edge of the shell. Letting in the formulae (29.10) and (29.13) $\theta = \theta''$, one finds

$$\begin{aligned} C_2 &= (R_2''\sin\theta'')^3 T^*_{1,1}, \\ C_1 &= T_{1,1}''R_2''\sin\theta''\cos\theta'' - S_1''R_2''\sin\theta''. \end{aligned} \tag{29.15}$$

It follows from these formulae that for $\theta'' = 0$, i.e., if the dome does not have an opening at the apex,

$$C_1 = C_2 = 0.$$

By the way, all the formulae obtained in this section may likewise be deduced by cutting the shell along an arbitrary parallel circle and by considering the equilibrium of the severed part of the shell under the influence of all forces acting on it.

As an example for the above theory, the forces arising in a spherical dome subject to wind pressure will be determined. The dome will be assumed to be a hemisphere and the wind pressure will be represented by the components

$$q_1 = q_2 = 0, \qquad q_n = -p \sin\theta \cos\varphi. \tag{29.16}$$

In other words, it will be assumed that the wind exerts only normal pressure which varies in such a manner that it vanishes at the apex of the dome and along the meridian determined by the plane perpendicular to the wind direction. The pressure attains its maximum on the windward side and its minimum on the leeward side (Fig. 31).

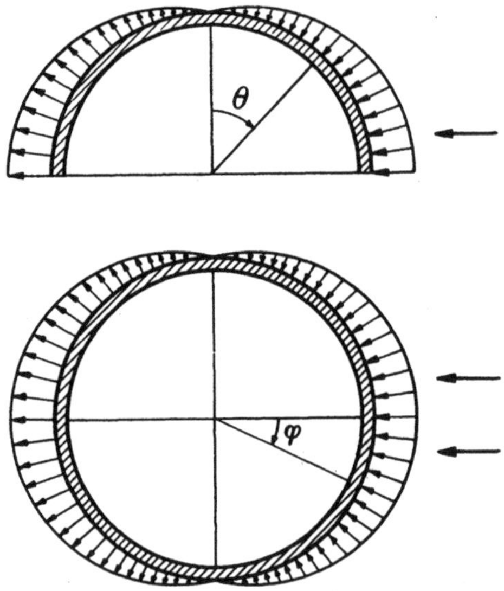

Fig. 31.

The formulae (29.16) may, of course, only be regarded as giving first approximations to the stresses due to wind loading; however, they are often used in practical calculations. For the determination of T_1^*, T_2^*, S^*, the formulae (29.10), (29.11) and (29.13) with C_1 and C_2 equal to zero may be used directly, since the dome is closed at the apex.

Letting in these formulae

$$R_1 = R_2 = R,\ q_{1,1} = q_{2,1} = 0,\ q_{n,1} = -p \sin\theta,\ \theta'' = 0,$$

one will have

$$T^*_{1,1} = -\frac{pR}{\sin^3\theta}\int_0^\theta \left[\cos\theta \sin\theta - \frac{(2+\cos\theta)(1-\cos\theta)^2}{3\sin\theta}\right] d\theta =$$

$$= -\frac{pR}{3}\,\frac{\cos\theta}{\sin^3\theta}\,(2+\cos\theta)(1-\cos\theta)^2, \tag{29.17}$$

$$S_1^* = -\frac{pR}{3}\,\frac{(2+\cos\theta)(1-\cos\theta)^2}{\sin^3\theta},$$

$$T^*_{2,1} = -pR\left[\sin\theta - \frac{\cos\theta}{3\sin^3\theta}(2+\cos\theta)(1-\cos\theta)^2\right].$$

Fig. 32 shows the dependence of the coefficients $T^*_{1,1}$, $T^*_{2,1}$, S_1^* on the height of the dome.

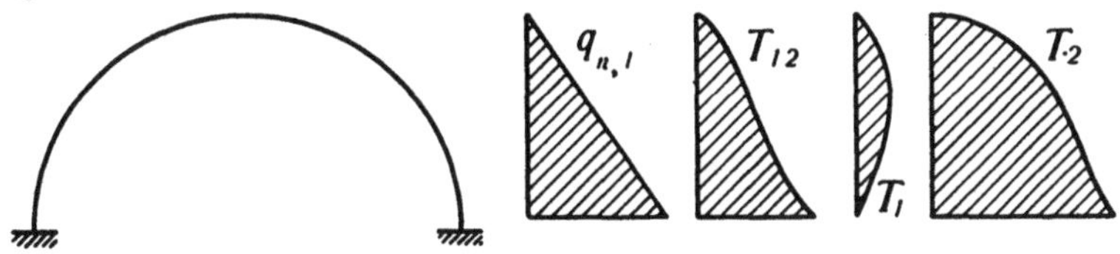

Fig. 32.

§ 30. Deformation of shells of revolution which do not stretch and shear the middle surface

Consider the determination of displacements in shells of revolution which deform without extension and shear [i.e., the solutions of the system (23.5)]. As in previous examples attention will be restricted to the case when the boundaries of the shell are two of its parallel circles.

As has been seen in § 23, this problem reduces to the solution of the homogeneous auxiliary system of equations

$$\frac{\partial \xi}{\partial \theta} - \frac{R_2}{\sin\theta}\frac{\partial \zeta}{\partial \varphi} = 0, \qquad \frac{\partial \zeta}{\partial \theta} + \frac{R_1}{R_2^2\sin\theta}\frac{\partial \xi}{\partial \varphi} = 0, \tag{30.1}$$

where

$$\xi = \frac{u}{\sin\theta}, \qquad \zeta = \frac{v}{R_2\sin\theta}. \tag{30.2}$$

Due to the type of shell under consideration, the solution must be periodic in the coordinate φ with period 2π. Hence it can be sought in the form

$$\xi = \sum_{k=0}^{\infty}(\xi_k'\cos k\varphi + \xi_k''\sin k\varphi), \quad \zeta = \sum_{k=0}^{\infty}(\zeta_k'\sin k\varphi + \zeta_k''\cos k\varphi), \tag{30.3}$$

where ξ_k', ξ_k'', ζ_k', ζ_k'' are functions of θ. Substituting (30.3) into (30.1), one finds for the ξ_k', ζ_k' the system of two ordinary differential equations

$$\frac{d\xi_k'}{d\theta} - k\frac{R_2}{\sin\theta}\zeta_k' = 0, \qquad \frac{d\zeta_k'}{d\theta} - k\frac{R_1}{R_2^2\sin\theta}\xi_k' = 0. \tag{30.4}$$

Identical systems are derived for the coefficients ξ_k'', ζ_k'', so that the search for the k-th mode of pure bending reduces to the solution of equations (30.4).

It will now be shown that for $k = 0$, $k = 1$, the equations (30.4) can be integrated for arbitrary shells of revolution.

A. $k = 0$

In this case the solution of (30.1) does not depend on φ, since it assumes the form

$$\frac{d\xi_0}{d\theta} = 0, \qquad \frac{d\zeta_0}{d\theta} = 0, \tag{30.5}$$

whence

$$\xi_0 = A_1, \qquad \zeta_0 = A_2, \tag{30.6}$$

where A_1, A_2 are arbitrary constants.

Thus,

$$u_0 = A_1 \sin\theta, \qquad v_0 = A_2 R_2 \sin\theta. \tag{30.7}$$

Further, using the equation

$$R_1 \varepsilon_1 = \frac{\partial u}{\partial \theta} + w = 0, \tag{30.8}$$

one can find also the normal deflection

$$w_0 = -\frac{du_0}{d\theta} = -A_1 \cos\theta. \tag{30.9}$$

Substitution in the corresponding formulae verifies readily that the displacements u_0, v_0, w_0 vanish not only for extension and shearing of the middle surface, but also for changes of curvature and torsion. Hence it is clear that u_0, v_0, w are displacements of the shell as a rigid body; since they do not depend on φ, they are displacements of the shell which are symmetric about its axis.

Noting that the displacement component of points of the middle surface in the direction of the axis of the shell is

$$\Delta_z = w\cos\theta - u\sin\theta \tag{30.10}$$

and substituting here for u, w from (30.7), one finds

$$\Delta_z^0 = -A_1. \tag{30.11}$$

Hence, the displacements u_0, w_0 correspond to a translation of the shell along its axis by the distance $-A_1$.

The meaning of the other constant of integration A_2 becomes clear, if one realizes that

$$\Omega = \frac{v}{R_2 \sin \theta} \tag{30.12}$$

is the angle by which an arbitrary point of the middle surface rotates about the shell axis. Substituting for v_0, one obtains

$$\Omega^0 = A_2, \tag{30.13}$$

i.e., the displacement v_0 corresponds to a rotation of all points of the shell about its axis by the same angle A_2.

B. $k = 1$

The equations (30.4) assume the form

$$\frac{d\xi_1'}{d\theta} - \frac{R_2}{\sin \theta}\zeta_1' = 0, \qquad \frac{d\zeta_1'}{d\theta} - \frac{R_1}{R_2^2 \sin \theta}\xi_1' = 0. \tag{30.14}$$

Eliminating ξ_1', one finds

$$\frac{d}{d\theta}\left[\frac{R_2^2 \sin \theta}{R_1}\frac{d\zeta_1'}{d\theta}\right] - \frac{R_2}{\sin \theta}\zeta_1' = 0. \tag{30.15}$$

This equation is the same as (30.2) for $F_1 = 0$. Introducing the new unknown

$$z = \zeta_1' R_2 \sin \theta, \tag{30.16}$$

the equation (30.15) may be reduced to the form

$$\frac{d}{d\theta}\left(\frac{1}{R_1 \sin \theta}\frac{dz}{d\theta}\right) = 0. \tag{30.17}$$

Hence

$$z = B_1 + B_2 \int_{\theta_0}^{\theta} R_1 \sin \theta \, d\theta, \tag{30.18}$$

where the lower limit of integration can be chosen arbitrarily.

Thus,

$$\zeta_1' = \frac{B_1}{R_2 \sin \theta} + \frac{B_2}{R_2 \sin \theta}\int_{\theta_0}^{\theta} R_1 \sin \theta \, d\theta \tag{30.19}$$

and ξ_1' can be found from

$$\xi_1' = \frac{R_2^2 \sin \theta}{R_1}\frac{d\zeta_1'}{d\theta}. \tag{30.20}$$

Substituting from (30.19) and taking for the differentiation the Gauss-Codazzi relation (23.3) into account, one finds

$$\xi_1' = -B_1 \operatorname{cotan}\theta + B_2[R_2 \sin\theta - \operatorname{cotan}\theta \int_{\theta_0}^{\theta} R_1 \sin\theta \, d\theta]. \quad (30.21)$$

Changing over from the auxiliary functions ξ_1' and ζ_1' to the displacements u, v yields

$$u = u_1(\theta)\cos\varphi = \{-B_1\cos\theta + B_2[R_2\sin^2\theta - \cos\theta\int_{\theta_0}^{\theta} R_1\sin\theta\, d\theta]\}\cos\varphi,$$

$$v = v_1(\theta)\sin\varphi = \{B_1 + B_2\int_{\theta_0}^{\theta} R_1\sin\theta\, d\theta\}\sin\varphi. \quad (30.22)$$

Finally, the equation (30.8) renders the normal displacement

$$w = w_1(\theta)\cos\varphi = -\{B_1 + B_2[R_2\cos\theta + \int_{\theta_0}^{\theta} R_1\sin\theta\; d\theta]\}\sin\theta\cos\varphi. \quad (30.23)$$

Substitution of (30.22), (30.23) into the formulae for the changes of curvature and twist of the middle surface shows that these displacements make all deformations of the middle surface vanish. Hence, they correspond to motion of the shell as a rigid body, the displacements having the form

$$\begin{aligned} u &= -B_1\cos\theta\cos\varphi, \\ v &= B_1\sin\varphi, \\ w &= -B_1\sin\theta\cos\varphi. \end{aligned} \quad (30.24)$$

These are a translation of the shell, perpendicular to its axis in the plane which contains the origin of the angle φ.

In fact, the projection of the displacement of an arbitrary point

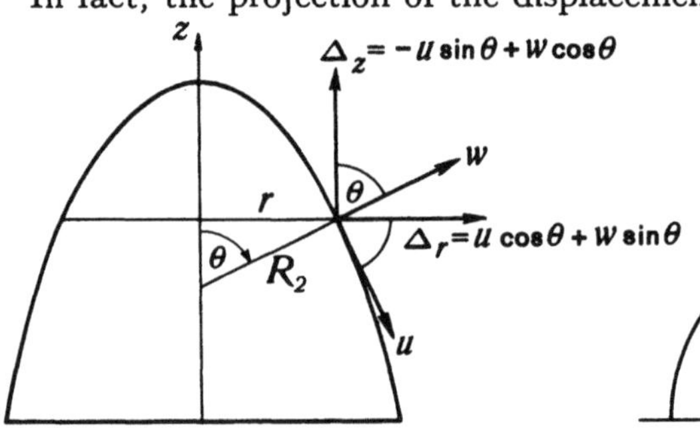

Fig. 32a.

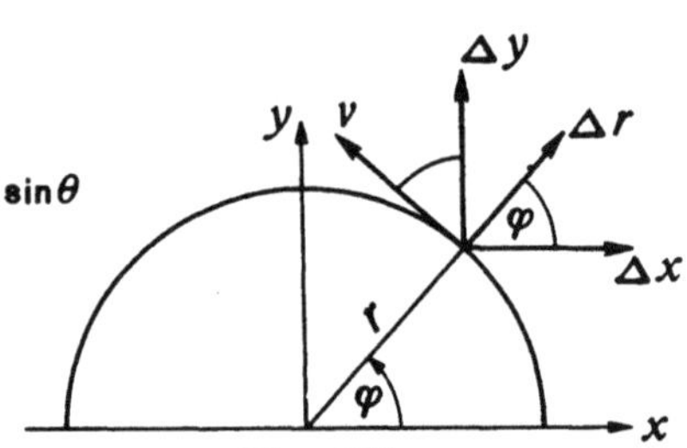

Fig. 32b.

of the middle surface on this direction can be expressed in terms of the displacements u, v, w by the formula [Fig. 32a]

$$\Delta_x = u \cos\theta \cos\varphi + w \sin\theta \cos\varphi - v \sin\varphi. \tag{30.25}$$

Substituting from (30.24), one finds

$$\Delta_x = -B_1. \tag{30.26}$$

Hence, in the case of the displacements (30.24), all points of the middle surface undergo the same translation in the direction of the x-axis [Fig. 32b].

However, the projections of the displacements of an arbitrary point of the middle surface on the other two axes, i.e.,

$$\begin{aligned} \Delta_y &= u \cos\theta \sin\varphi + w \sin\theta \sin\varphi + v \cos\varphi, \\ \Delta_z &= -u \sin\theta + w \cos\theta \end{aligned} \tag{30.27}$$

vanish in this case. Thus, (30.24) yields a translation $-B_1$ of the shell in the direction of the x-axis.

The displacements

$$\begin{aligned} u &= B_2[R_2 \sin^2\theta - \cos\theta \int_{\theta_0}^{\theta} R_1 \sin\theta \, d\theta] \cos\varphi, \\ v &= B_2(\int_{\theta_0}^{\theta} R_1 \sin\theta \, d\theta) \sin\varphi, \\ w &= -B_2[R_2 \cos\theta + \int_{\theta_0}^{\theta} R_1 \sin\theta \, d\theta] \sin\theta \cos\varphi \end{aligned} \tag{30.28}$$

can be shown to correspond to a rotation of the shell as a rigid body [cf. Fig. 32a], but this will not be done here.

It is now readily verified that a study of the coefficients ξ_k'', ζ_k'' in the series (30.3) leads to displacements u_1'', v_1'', w_1'' which also correspond to displacements of the shell as a rigid body, namely, a translation in the direction of the y-axis and a rotation about the x-axis.

It follows from these considerations that one can fix the position of the shell in space by fixing the values of the six arbitrary constants in the series (30.3). Since the terms in these series which correspond to $k = 0$, $k = 1$ take care of all degrees of freedom of the shell as a rigid body, it is obvious that the remaining terms will not lead to vanishing of the curvatures κ_1, κ_2 and the twist τ (since otherwise one would have more arbitrary constants than are required to take account of an arbitrary rigid body motion). Thus, the terms in (30.3) which correspond to $k \geqslant 2$ will yield deformations of the shell without extension and shearing of the middle surface, i.e., pure bending deformations.

In what follows, the displacements u_k^{**}, v_k^{**}, w_k^{**} corresponding to the k-th term in the series (30.3) will be referred to as the k-th mode of pure bending deformation. Already for $k = 2$, the equations (30.4) cannot be integrated until the deformation of the middle surface is given.

In order to explain the basic properties of pure bending deformation of shells, the following two particular cases will be studied for which the system (30.4) can be solved.

a. *Shells of positive Gaussian curvature formed by revolution of second order curves.*

As has been stated in § 26, for such shells the radii of curvature of the middle surface may be given in the form

$$R_1 = \frac{R_0}{(1 + \gamma \sin^2 \theta)^{3/2}}, \quad R_2 = \frac{R_0}{(1 + \gamma \sin^2 \theta)^{1/2}}, \tag{30.29}$$

whence

$$\frac{R_2{}^3}{R_1} = R_0{}^2 = \text{constant}. \tag{30.30}$$

Using (30.30), the system (30.4) assumes the form

$$\frac{\sin \theta}{R_2} \frac{\partial \xi_k'}{d\theta} - k\zeta_k' = 0, \quad \frac{\sin \theta}{R_2} \frac{d\zeta_k}{d\theta} - \frac{k}{R_0{}^2} \xi_k' = 0. \tag{30.31}$$

Replace now θ by the new non-dimensional variable

$$\psi = \frac{1}{R_0} \int_{\pi/2}^{\theta} \frac{R_2}{\sin \theta} \, d\theta = \int_{\pi/2}^{\theta} \frac{d\theta}{\sin \theta \sqrt{1 + \gamma \sin^2 \theta}} = \log \left(\frac{\sqrt{1 + \gamma} \sin \theta}{\sqrt{1 + \gamma \sin^2 \theta} + \cos \theta} \right). \tag{30.32}$$

Then the trigonometric functions of θ can be expressed in terms of ψ by the formulae

$$\begin{aligned} \sin \theta &= -\frac{1}{\cosh \psi} \frac{1}{\sqrt{1 + \gamma \tanh^2 \psi}}, \\ \operatorname{cotan} \theta &= -\sqrt{1 + \gamma} \sinh \psi, \\ \cos \theta &= \sqrt{1 + \gamma} \frac{\tanh \psi}{\sqrt{1 + \gamma \tanh^2 \psi}}, \end{aligned} \tag{30.33}$$

and the system (30.31) becomes

$$\frac{d\xi_k'}{d\psi} - kR_0\zeta_k' = 0, \quad \frac{d\zeta_k'}{d\psi} - \frac{k}{R_0}\xi_k' = 0, \tag{30.34}$$

and hence

$$\frac{d^2\xi_k'}{d\psi^2} - k^2\xi_k' = 0. \tag{30.35}$$

Consequently,

$$\xi_k' = A_1e^{k\psi} + A_2e^{-k\psi}, \tag{30.36}$$

or

$$\zeta_k' = \frac{1}{R_0}(A_1e^{k\psi} - A_2e^{-k\psi}). \tag{30.37}$$

Further, recalling the expressions linking ξ and ζ to the displacements, one finds

$$\begin{aligned} u_k^{**} &= \xi_k' \sin\theta = -\frac{1}{\cosh\psi}\frac{1}{\sqrt{1+\gamma\tanh^2\psi}}(A_1e^{k\psi} + A_2e^{-k\psi}), \\ v_k^{**} &= \zeta_k' R_2 \sin\theta = -\frac{1}{\sqrt{1+\gamma}}\frac{1}{\cosh\psi}(A_1e^{k\psi} - A_2e^{-k\psi}). \end{aligned} \tag{30.38}$$

Replacing in (30.18) the variable θ by ψ, one obtains

$$w = -\frac{\cosh\psi(1+\gamma\tanh^2\psi)}{(1+\gamma)^{\frac{1}{2}}}\frac{du}{d\psi}. \tag{30.39}$$

Substituting here for u_k^{**} from (30.38), one arrives at

$$\begin{aligned} w_k^{**} = &-\frac{k}{(1+\gamma)^{\frac{1}{2}}}\sqrt{1+\gamma\tanh^2\psi}\,(A_1e^{k\psi} - A_2e^{-k\psi}) + \\ &+ \frac{(1+\gamma)^{\frac{1}{2}}\tanh\psi}{\sqrt{1+\gamma\tanh^2\psi}}(A_1e^{k\psi} + A_2e^{-k\psi}). \end{aligned} \tag{30.40}$$

For $k = 1$, the formulae (30.38), (30.40) yield a rigid body displacement of the shell. This and the analogous case $k = 0$ have been considered above for arbitrary shells of revolution. For $k \geqslant 2$, these expressions give different modes of pure bending deformation of the type of shells under consideration.

As can be seen from (30.32), the poles of these shells, i.e., the points $\theta = 0$, $\theta = \pi$, correspond to $\psi = -\infty$, $\psi = +\infty$, respectively. If one of these points belongs to the middle surface of the shell, one must set in (30.38), (30.40) that constant equal to zero, for which the displacement remains finite at the pole.

However, if both these points belong to the middle surface, one must set $A_1 = A_2 = 0$ in (30.38) and (30.40), to obtain

$$u_k^{**} = v_k^{**} = w_k^{**} = 0.$$

The physical meaning of this case is that a closed shell of positive Gaussian curvature, formed by rotation of a second order curve, cannot bend without stretching and shearing of its middle surface, i.e., pure bending of such shells is impossible. This result has general significance and can be extended to all shells of positive Gaussian curvature.

If a shell is bounded by one or two parallel circles, it follows from (30.38) and (30.40) that there exists an infinite number of modes of pure bending. In this context, there arises the question regarding the type of support of the edge (or edges) of a shell which will make pure bending impossible.

Let the two parallel circles bounding the shell correspond to $\psi = \psi_1$, $\psi = \psi_2$, and let at these edges

$$u = v = 0. \tag{30.41}$$

Subjecting (30.38), (30.40) to these boundary conditions, one finds the following system of four, linear, homogeneous, algebraic equations for A_1, A_2:

$$\begin{aligned} A_1 e^{k\psi} + A_2 e^{-k\psi_1} &= 0, \quad A_1 e^{k\psi_1} - A_2 e^{-k\psi_1} = 0, \\ A_1 e^{k\psi_2} + A_2 e^{-k\psi_2} &= 0, \quad A_1 e^{k\psi_2} - A_2 e^{-k\psi_2} = 0; \end{aligned} \tag{30.42}$$

the only solution of this system is

$$A_1 = A_2 = 0 \tag{30.43}$$

from which follows that complete tangential fixing of both edges of a shell of the type under consideration removes the possibility of its deformation in any mode of pure bending. However, one might say that the conditions (30.41) are more than necessary to ensure the rigidity of the shell in bending.

In order to eliminate displacements in pure bending, one need only remove on each of the two edges of a shell one tangential displacement component u or v. Likewise, one can achieve bending rigidity of a shell by complete tangential clamping (30.41) of only one of the edges, leaving the other edge free. All these conclusions are readily obtained by testing any two combinations of the equations (30.42) and verifying that they will always lead to the unique solution (30.43). Similarly, it is not difficult to establish that for shells closed at one end and bounded by a parallel circle, the elimination of all modes of pure bending is achieved by the boundary condition $u = 0$ (or, alternating, $v = 0$).

b. *Shells of negative Gaussian curvature formed by rotation of curves of second order.*

Consider a shell the middle surface of which is generated by rotation of the hyperbola

$$\frac{z^2}{c^2} - \frac{x^2}{a^2} = 1 \tag{30.44}$$

about the z-axis. The principal radii of curvature of the middle surface will then be

$$R_2 = \frac{a}{\sqrt{1 - (1 + \alpha^2)\cos^2\theta}},$$
$$R_1 = -\frac{c^2}{a[1 - (1 + \alpha^2)\cos^2\theta]^{3/2}}, \qquad \alpha = \frac{c}{a}. \tag{30.45}$$

For such shells

$$\frac{R_2{}^3}{R_1} = -\frac{a^4}{c^2} = -r_0{}^2 = \text{const.} < 0 \tag{30.46}$$

and the equations (30.4) reduce to the form

$$\frac{\sin\theta}{R_2}\frac{d\xi_k{}'}{d\theta} - k\zeta_k{}' = 0, \qquad r_0{}^2\frac{\sin\theta}{R_2}\frac{d\zeta_k}{d\theta} + k\xi_k{}' = 0. \tag{30.47}$$

Introducing the new independent variable

$$\eta = -\frac{1}{r_0}\int\limits_{\pi/2}^{\theta}\frac{R_2}{\sin\theta}\,d\theta = -\frac{a}{r_0}\int\limits_{\pi/2}^{\theta}\frac{d\theta}{\sin\theta\sqrt{(1 - (1 + \alpha^2)\cos^2\theta}} =$$
$$= \frac{1}{2}\left\{\arcsin\left[\frac{1 + (1 + \alpha^2)\cos\theta}{\sqrt{1 + \alpha^2}(1 + \cos\theta)}\right] - \arcsin\left[\frac{1 - (1 + \alpha^2)\cos\theta}{\sqrt{1 + \alpha^2}(1 - \cos\theta)}\right]\right\}$$
$$-\frac{\pi}{2} \leqslant \eta \leqslant \frac{\pi}{2}, \tag{30.48}$$

one finds

$$\frac{1}{r_0}\frac{d\xi_k{}'}{d\eta} + k\zeta_k{}' = 0, \qquad r_0\frac{d\zeta_k{}'}{d\eta} - k\xi_k{}' = 0, \tag{30.49}$$

whence

$$\frac{d^2\xi_k{}'}{d\eta^2} + k^2\xi_k{}' = 0. \tag{30.50}$$

Thus

$$\xi_k{}' = A_1\sin k\eta + A_2\cos k\eta. \tag{30.51}$$

Substitution of (30.51) in the first equation (30.49) now yields

$$\zeta_k' = -\frac{1}{r_0}[A_1 \cos k\eta - A_2 \sin k\eta]. \tag{30.52}$$

Let the boundaries of the shell under consideration be the two parallel circles $\eta = \eta_1$, $\eta = \eta_2$ and let the shell be subject to complete tangential clamping at these edges

$$u = v = 0. \tag{30.53}$$

It will be shown that this type of support removes the possibility of deformation of the shell in any form of pure bending. Obviously, it will be sufficient to show that this is true for any particular shape.

Noting that for (30.53) $u_k^{**} = v_k^{**} = \xi_k' = \zeta_k' = 0$, one arrives at the following linear algebraic system of four equations in two unknowns:

$$\begin{aligned} &A_1 \sin k\eta_1 + A_2 \cos k\eta_1 = 0, \quad A_1 \cos k\eta_1 - A_2 \sin k\eta_1 = 0, \\ &A_1 \sin k\eta_2 + A_2 \cos k\eta_2 = 0, \quad A_1 \cos k\eta_2 - A_2 \sin k\eta_2 = 0 \end{aligned} \tag{30.54}$$

with the unique solution

$$A_1 = A_2 = 0, \tag{30.55}$$

which proves the above statement. However, there arises the question whether the conditions (30.53) must be imposed on both edges and whether they can be relaxed.

For example, consider the case when (30.53) is only fulfilled on one of the two edges of a shell, say, on $\eta = \eta_1$. One then finds

$$A_1 \sin k\eta_1 + A_2 \cos k\eta_1 = 0, \quad A_1 \cos k\eta_1 - A_2 \sin k\eta_1 = 0; \tag{30.56}$$

the determinant of this algebraic system

$$D = \begin{vmatrix} \sin k\eta_1 & \cos k\eta_1 \\ \cos k\eta_1 & -\sin k\eta_1 \end{vmatrix} = -1 \tag{30.57}$$

will always be non-zero, and hence its unique solution is the trivial solution (30.55).

Next, consider the case when on each of the two edges of a shell only one of the two components of tangential displacement are removed, for example,

$$u = 0 \text{ for } \eta = \eta_1, \ \eta = \eta_2. \tag{30.58}$$

One then arrives at the system of equations

$$\begin{aligned} A_1 \sin k\eta_1 + A_2 \cos k\eta_1 = 0, \\ A_1 \sin k\eta_2 + A_2 \cos k\eta_2 = 0, \end{aligned} \tag{30.59}$$

for which the determinant of the coefficients is

$$D = \sin k\eta_1 \cos k\eta_2 - \sin k\eta_2 \cos k\eta_2 = \sin k(\eta_1 - \eta_2). \quad (30.60)$$

This determinant vanishes if

$$k(\eta_1 - \eta_2) = m\pi, \quad (30.61)$$

where m is an integer. If (30.61) is fulfilled, the system (30.59) has non-trivial solutions (in fact, an infinity of such solutions). It follows from this that the variant of the boundary conditions (30.58), in general, does not eliminate deformation of a shell in the k-th mode of pure bending, since the constants A_1, A_2 may have non-zero values. Next, consider the boundary conditions

$$\begin{aligned} u &= 0 \text{ for } \eta = \eta_1, \\ v &= 0 \text{ for } \eta = \eta_2; \end{aligned} \quad (30.62)$$

it is readily verified that again it is not sufficient to eliminate in all cases bending of a shell.

So far, detailed consideration has been given to the simplest modes of tangential clamping of two parallel circles of shells of revolution when along these edges either $u = 0$ or $v = 0$ or both these conditions apply. It was the purpose of the present study to explain the circumstances under which bending rigidity of a shell can be achieved.

One could still consider from the same point of view more complicated bending conditions, for example, the case when one is given the linear combinations of tangential displacements

$$\alpha_{11}u + \alpha_{12}v = 0$$

on one parallel circle and

$$\alpha_{21}u + \alpha_{22}v = 0$$

on the other parallel circle bounding the shell, where the α_{ij} are periodic functions in φ with period 2π. These conditions correspond to the case when at each point of a boundary of a shell one tangential component of the displacement, forming with the edge of the shell a certain angle, is removed.

For shells of positive Gaussian curvature, this type of condition has been studied in detail by A. L. Gol'denweizer [5a, 6a]. It can be shown that they ensure only partially the bending rigidity of such shells which retain their freedom of deformation in certain modes of pure bending. As has already been mentioned earlier, the problem of the conditions under which pure bending deformation of a shell becomes impossible is closely linked to the problem of the realization of the membrane state of stress in shells.

Let it be assumed that the boundary conditions are such that they ensure the freedom of deformation of a shell in the k-th mode of pure bending. If the edge and surface loadings applied to the shell perform work over the displacements corresponding to this form of bending, a membrane state of stress turns out to be impossible, since the shell is unable to resist the action of these forces. However, if external forces, applied to a shell, do not perform work over these displacements, the membrane problem can be solved and, correspondingly, a membrane state of stress established.

However, from a practical point of view, it is unreasonable to assume freedom of pure bending in any of the modes whether the shell belongs to either of these classes or not. The fact is that a shell which is not clamped with respect to one mode of pure bending will lose stability in this mode for very small critical loads and will be very sensitive to small accidental bending loads. Therefore complete removal (or at least sufficient limitation) of deformation of shells in all modes of pure bending becomes a law for the designer of shell structures.

Neglect of this rule can only occur in cases when either the loading of the shell is insignificant or when, by the conditions of its employment, the shell has to exhibit less strength but considerable flexibility; for example, this will be the case in scientific and musical instruments.

It follows from the above that shells of revolution of positive and negative Gaussian curvature have typical peculiarities with regard to the conditions which will ensure their rigidity in bending; certain of these conditions, which will be sufficient to ensure bending rigidity of the middle surface of shells of positive Gaussian curvature, will sometimes, for unfavorable ratios of the shell dimensions, turn out to be insufficient to ensure bending rigidity of shells of negative Gaussian curvature.

In this context, the opinion has been expressed [4] that in a number of cases shells of negative Gaussian curvature cannot, in general, fail to work in a membrane state of stress. Strictly speaking, this is not true, because, as has been shown above by means of examples, a shell of negative Gaussian curvature bounded by two parallel circles can always be clamped tangentially in such a manner that all its modes of pure bending are excluded. For this purpose, at least one of its edges must be clamped with respect to both tangential displacement components. Nevertheless, from a practical engineering point of view, it can be said that shells of positive Gaussian curvature have certain advantages compared with those of negative Gaussian curvature.

It is a fact that a restriction of the tangential displacement component, tangential to the edge of a shell, is more readily achieved in a construction than elimination of the tangential displacement component normal to the edge. In order to restrain the first component, one needs only have along the edge a structural component which is able to resist extension and compression. Theoretically speaking, this can be an inextensible, flexible string.

As regards the second component, one requires for its elimination along the edge a structural element which has sufficient bending rigidity in a plane tangential to the middle surface. Such an element inevitably must have a larger size and weight than an element designed for the elimination of the tangential component of the tangential edge displacement which only has to be rigid in tension or compression. Since shells of positive Gaussian curvature can be made inflexible simply by elimination or restraint along their edges of the tangential component of the tangential displacement, while shells of negative Gaussian curvature in some cases require elimination on one of the edges of both components of the tangential displacement, the first type of shells, purely from a point of view of structural engineering, can be said to exhibit certain advantages in the sense that such shells may be subjected to a membrane state of stress.

§ 31. The membrane theory of shells of positive Gaussian curvature, formed by rotation of curves of second order

So far the membrane theory has only been applied to shells of revolution bounded by one or two parallel circles. It has been shown that this type of problem lends itself to solution by Fourier series, the coefficients of which are determined by integration of the linear ordinary differential equations (24.10).

The problem of the membrane theory of shells of revolution turns out to be much more difficult, if the edges of a shell are arbitrary. A method of solution can only be stated for shells of specific shape and, in particular, for shells of positive and negative Gaussian curvature, formed by revolution of second order curves. In this and the next section, shells of positive and negative curvatures, respectively, will be considered.

For shells of positive Gaussian curvature, generated by rotation of a second order curve, the expressions (30.29) apply. Hence the new variable ψ, defined in (30.32), may be introduced; the system of equations (23.7) then assumes the form

$$\frac{\partial U}{\partial \psi} + \frac{\partial V}{\partial \varphi} = f_1, \quad \frac{\partial V}{\partial \psi} - \frac{\partial U}{\partial \varphi} = f_2, \tag{31.1}$$

where, for the sake of simplicity, the constant multiplier $1/R_0$ has been included in the functions V, i.e., the auxiliary function in (31.1) differs by this factor from the function V, defined by (23.6); f_1 and f_2 are known functions of the coordinates φ, ψ, since they, the components of the external surface loading, are assumed to be known. For $q_1 = q_2 = q_n = 0$, one has also $f_1 = f_2 = 0$.

Consider first this particular case, namely, when the only forces acting on the shell are applied to its edges. Then the equations (31.1) are homogeneous

$$\frac{\partial V}{\partial \psi} = \frac{\partial U}{\partial \varphi}, \qquad \frac{\partial V}{\partial \varphi} = -\frac{\partial U}{\partial \psi}, \tag{31.2}$$

and they are found to be identical with the Cauchy-Riemann relations. Thus, in this particular case, U and V are conjugate harmonic functions and, correspondingly, they may be considered to be real and imaginary parts of some function $\Phi(z)$ of the complex variable

$$z = \psi + i\varphi. \tag{31.3}$$

Hence, for this class of shells, the membrane problem for shells loaded only along their edges reduces to the search for pairs of conjugate harmonic functions which assume on the boundary of the shell

$$f(\varphi, \psi) = 0$$

values, corresponding to the applied tangential external stresses.

However, this problem has no solution for all possible given boundary loadings, since it is impossible to specify arbitrarily the values of both conjugate harmonic functions on the boundary of a region. Only such versions of the tangential boundary conditions will be admissible for which these functions exist and are fully determined. In particular, one may be given on the boundary one of two conjugate harmonic functions, i.e., one of the two forces T_1 or S. Its determination inside the region reduces to the solution of a Dirichlet problem. Further, one can find from (31.2) also the second harmonic function apart from an arbitrary constant, the value of which must be given separately.

Thus, the law of change of one of the two components of the tangential loading, i.e., T_1 or S, specified on the whole boundary of a shell, fixes also the law of change of the other component. Although the above arguments relate only to the homogeneous problem of membrane theory, they have nevertheless general validity. The point here is that every non-homogeneous problem, described by the linear equations (31.1), can be reduced to a homogeneous problem, if a particular solution of this system is known. But in a given case, a particular solution is readily constructed for any $f_1(\varphi, \psi)$, $f_2(\varphi, \psi)$. Therefore the above approach to the membrane analysis of shells of positive Gaussian curvature, formed by rotation of second order curves, which reduces this problem to a search for a pair of conjugate harmonic functions prescribed in the region, bounded by the shell contour, can even be used in the presence of surface loading. More details relating to this type of problems are given in [4, 5a].

§ 32. The membrane theory of shells of negative Gaussian curvature generated by rotation of second order curves

If the middle surface of the shell has been generated by rotation of the hyperbola (30.44) about the z-axis, then the principal radii of curvature of its middle surface are given by (30.45). Hence one may introduce for the angle θ the new variable η of (30.48) and reduce the system of equations (23.7) to the form

$$\frac{\partial U}{\partial \eta} - \frac{\partial V}{\partial \varphi} = f_1(\eta, \varphi), \qquad \frac{\partial V}{\partial \eta} - \frac{\partial U}{\partial \varphi} = f_2(\eta, \varphi), \tag{32.1}$$

where, for the sake of simplicity, the constant factor $1/r_0$ has been absorbed in the function V and f_1, f_2 are known functions of η, φ which vanish if

$$q_1 = q_2 = q_n = 0.$$

Eliminating one of the unknown functions, for example U, from (32.1), one obtains for the other function the equation

$$\frac{\partial^2 V}{\partial \varphi^2} - \frac{\partial^2 V}{\partial \eta^2} = F(\eta, \varphi) \tag{32.2}$$

which is identical to the equation for the problem of forced vibrations of a string subject to normal loading $F(\eta, \varphi)$. The solution of this problem is completely determined for given boundary and initial conditions. In terms of the above notation, these conditions may be formulated as

$$V = V_1(\eta) \text{ for } \varphi = \varphi_1, \quad V = V_2(\eta) \text{ for } \varphi = \varphi_2, \tag{32.3}$$

$$V = F_1(\varphi), \quad \frac{\partial V}{\partial \eta} = F_2(\varphi) \text{ for } \eta = \eta_1. \tag{32.4}$$

These conditions, when applied to problems of the membrane theory of shells, can be interpreted in the following manner.

Consider a part of a shell of revolution, bounded by two parallel circles $\eta = \eta_1$, $\eta = \eta_2$ and two meridians $\varphi = \varphi_1$, $\varphi = \varphi_2$. The solution of a problem of membrane theory for such a shell is fully determined, if on the bounding meridians one is given the value of the function V (i.e., the force S), and on one of the bounding parallels this stress as well as its derivative along the normal to the boundary.

By (32.1), specification of the normal drivative of V for $\eta = \eta_1$ is equivalent to specification on the line $\eta = \eta_1$ of the function U (i.e., the stress T_1) apart from an arbitrary constant.

One cannot eliminate from (32.1) the function U, but the function V. In this way one finds for U an equation analogous to (32.2). Applying to this equation the above reasoning, based on the analogy with the problem of string vibrations, one arrives at the conclusion that it is possible to specify the force T_1 (or what is the same thing, T_2), on the edges $\varphi = \varphi_1$, and $\varphi = \varphi_2$, since these two forces are interrelated by the third (algebraic) equation (23.4), and on the edge $\eta = \eta_2$ the force T_1 as well as the force S (apart from an arbitrary constant).

In (32.2), the coordinate η may be interpreted as time and φ as the space coordinate of points on the string. In (32.3) and (32.4), the variables φ and η interchange roles and one arrives at conditions $V = V_1(\varphi)$,

$V = V_2(\varphi)$ for $\eta = \eta_1$ and $\eta = \eta_2$, respectively,

$$V = F_1(\eta), \quad \frac{\partial V}{\partial \varphi} = F_2(\eta) \text{ for } \varphi = \varphi_1.$$

It is essential to note that in all the above variants of the boundary conditions on one of the four edges of the shell the unknown functions are not subject to any conditions, while on the opposite side conditions are imposed on both functions. Such a formulation of the boundary conditions is characteristic for equations of the hyperbolic type (in particular, for the equation for string vibrations).

There arises now the question what will happen if one of the forces, say S, is given on all four sides of a shell of the type under consideration.

As has been shown in the last section, such a formulation of the boundary conditions is typical for shells of positive Gaussian curvature. However, it is not always admissible for shells of negative Gaussian curvature. In order to demonstrate this fact, construct the solution of such a problem for the boundary conditions

$$V = 0 \text{ for } \eta = \eta_1, \ \eta = \eta_2, \ \varphi = \varphi_1, \ \varphi = \varphi_2. \tag{32.5}$$

Assume the function $F(\eta, \varphi)$ to be continuous and finite in the region of values of η and φ under consideration. Then one can represent this function in this domain in the form of the Fourier series

$$F(\eta, \varphi) = -\sum_{m=1}^{\infty} F_m(\eta) \sin \lambda_m(\varphi - \varphi_1), \tag{32.6}$$

where

$$\lambda_m = \frac{m\pi}{\varphi_2 - \varphi_1}. \tag{32.7}$$

Substituting (32.6) into (32.2), the function V will be sought in the form of a similar series

$$V = \sum_{m=1}^{\infty} V_m(\eta) \sin \lambda_m(\varphi - \varphi_1), \tag{32.8}$$

satisfying the boundary conditions $V = 0$ for $\varphi = \varphi_1$ and $\varphi = \varphi_2$. Then one finds for the coefficients $V_m(\eta)$ the ordinary differential equations

$$\frac{d^2 V_m}{d\eta^2} + \lambda_m{}^2 V_m = F_m(\eta) \tag{32.9}$$

with general solutions in the form

$$V_m = A_m' \sin \lambda_m(\eta - \eta_1) + A_m'' \cos \lambda_m(\eta - \eta_1) + \\ + \frac{1}{\lambda_m} \int_{\eta_1}^{\eta} F_m(t) \sin \lambda_m(\eta - t)\, dt. \tag{32.10}$$

The constants of integration in (32.10) will be determined by subjecting the solution to the boundary conditions $V_m = 0$ for $\eta = \eta_1$, $\eta = \eta_2$. Then

$$A_m'' = 0,$$

$$A_m' \sin \lambda_m(\eta_2 - \eta_1) + \frac{1}{\lambda_m} \int_{\eta_1}^{\eta_2} F_m(t) \sin \lambda_m(\eta_2 - t)\, dt = 0. \tag{32.11}$$

Now let

$$\sin \lambda_m(\eta_2 - \eta_1) = 0, \tag{32.12}$$

which will be true for

$$\frac{\eta_2 - \eta_1}{\varphi_2 - \varphi_1} = \frac{k}{m}, \tag{32.13}$$

where k is an integer.

If now

$$\int_{\eta_1}^{\eta_2} F_m(t) \sin \lambda_m(\eta_2 - t)\, dt \neq 0, \tag{32.14}$$

then the constants A_m' turn out to be infinite.

However, if, instead of equality, one has inequality, the constants

A_m' turn out to be indeterminate. The first indicates the absence of a solution to the problem under consideration, the second the indeterminateness of its solution for the boundary conditions (32.5).

It follows from the above example that specification on the entire boundary of a shell of negative Gaussian curvature of one of the tangential forces may determine a problem of membrane theory completely; at times, it may not determine the solution and at other times it may lead to nonexistence of a solution.

It has been shown in the last section that for shells of positive Gaussian curvature such a formulation of the boundary conditions is always possible. From a mathematical point of view, this difference can be explained by the fact that the membrane equations for shells of positive Gaussian curvature are of the elliptic type, while those for shells of negative Gaussian curvature are of the hyperbolic type.

From the physical point of view, this difference is explained by the difference in the conditions of clamping for which shells of positive and negative Gaussian curvature lose their freedom in pure bending.

§ 33. Affine transformations in the membrane theory

In the preceding sections, the membrane theory has been applied to shells of revolution and a general method of solution has been established, based on the expansion of the external loading and of all the forces in the shell in trigonometric series in terms of the angle φ. As was noted by F. Dischinger [68], this method may be used for the analysis of more complicated shells (for example, domes and caps which have oval planform).

Consider a shell of arbitrary form in a membrane state of stress under the influence of surface loadings q_1, q_2, q_n and the corresponding boundary forces. Let the Cartesian coordinates of its middle surface be x, y, z. Further, let there be obtained from this shell by means of a linear transformation another shell with the following coordinates of its middle surface:

$$\bar{x} = c_1 x, \quad \bar{y} = c_2 y, \quad \bar{z} = c_3 z, \tag{33.1}$$

where c_1, c_2, c_3 are constant coefficients.

It turns out that, if this second shell is loaded in a completely definite manner (its load distribution being related to that of the first shell), all the forces in the second shell may be determined by a very simple calculation in terms of the forces in the first shell (which will be assumed known) instead of by solving again differential equations.

The relation between the loadings of the two shells, which must be satisfied so that the above procedure will be possible, will now be established.

For this purpose, consider a small right-angled element of the first shell, the sides of the element being lines of principal curvature of the surface. The sides of this element will then be given by

$$ds_1 = A_1 d\alpha_1, \; ds_2 = A_2 d\alpha_2. \tag{33.2}$$

Denote the direction cosines of the angles between ds_1 and the axes x, y, z by l_1, l_2, l_3. Similarly denote by m_1, m_2, m_3 the direction cosines of the angles between ds_2 and the same axes.

To this element of the first shell there will correspond an element of the second shell with the sides

$$\begin{aligned} d\bar{s}_1 &= \sqrt{c_1^2 l_1^2 + c_2^2 l_2^2 + c_3^2 l_3^2}\, ds_1, \\ d\bar{s}_2 &= \sqrt{c_1^2 m_1^2 + c_2^2 m_2^2 + c_3^2 m_3^2}\, ds_2, \end{aligned} \tag{33.3}$$

which (in the general case) will no longer be right-angled. The cosine of the angle between the directions $d\bar{s}_1$, $d\bar{s}_2$ may be calculated from the formula

$$\cos(d\bar{s}_1, d\bar{s}_2) = \frac{c_1^2 l_1 m_1 + c_2^2 l_2 m_2 + c_3^2 l_3 m_3}{\sqrt{c_1^2 l_1^2 + c_2^2 l_2^2 + c_3^2 l_3^2} \cdot \sqrt{c_1^2 m_1^2 + c_2^2 m_2^2 + c_3^2 m_3^2}}. \tag{33.4}$$

All the internal and external forces acting on the element of the first shell will now be resolved in the directions x, y, z. Then the conditions for equilibrium of this element can be written as follows:

$$\begin{gathered} \Sigma X_j = 0, \; \Sigma Y_j = 0, \; \Sigma Z_j = 0, \\ \Sigma (y_j Z_j - z_j Y_j) = 0, \; \Sigma (z_j X_j - x_j Z_j) = 0, \\ \Sigma (x_j Y_j - y_j X_j) = 0, \end{gathered} \tag{33.5}$$

where X_j, Y_j, Z_j denote the projections of a definite (internal or external) force acting on the element and x_j, y_j, z_j are the coordinates of its point of application. The summation extends over all internal and external forces.

Let the forces X_j, Y_j, Z_j of the first element correspond to forces $\bar{X}_j$, $\bar{Y}_j$, $\bar{Z}_j$ of the second element, applied to the points

$$\bar{x}_j = c_1 x_j, \; \bar{y}_j = c_2 y_j, \; \bar{z}_j = c_3 z_j. \tag{33.6}$$

Then the equilibrium conditions for the second element are

$$\begin{gathered} \Sigma \bar{X}_j = 0, \; \Sigma \bar{Y}_j = 0, \; \Sigma \bar{Z}_j = 0, \\ \Sigma (c_2 y_j \bar{Z}_j - c_3 z_j \bar{Y}_j) = 0, \; \Sigma (c_3 z_j \bar{X}_j - c_1 x_j \bar{Z}_j) = 0, \\ \Sigma (c_1 x_j \bar{Y}_j - c_2 y_j \bar{X}_j) = 0, \end{gathered} \tag{33.7}$$

from which it is clear that, if one lets

$$\bar{X}_j = X_j, \ \bar{Y}_j = Y_j, \ \bar{Z}_j = Z_j,$$

i.e., if one simply transfers the forces acting on the first element to the corresponding points of the second element without changing their magnitudes, then the last three equations (33.7) will remain unsatisfied.

But if one takes

$$\bar{X}_j = c_1 X_j, \ \bar{Y}_j = c_2 Y_j, \ \bar{Z}_j = c_3 Z_j, \tag{33.8}$$

then fulfillment of the equilibrium conditions of the first element (33.5) will automatically ensure satisfaction of all the equilibrium conditions of the second element.

However, it must still be shown that the internal forces which will act in the second shell in accordance with (33.8) will be tangential to the middle surface (since the appearance of normal shearing forces would be equivalent to a violation of the membrane state of stress.)

Consider now the internal force $T_1{}^* ds_2$, acting on an element of the first shell in the direction ds_1. In transferring the force to the element of the second shell in agreement with (33.8), its components will be changed in the same manner as the components of ds_1. Therefore the internal force $\bar{T}_1{}^* d\bar{s}_2$ of the second element, corresponding to the force

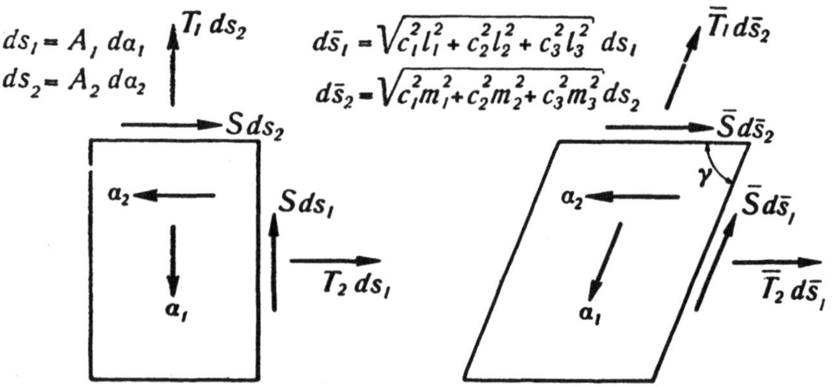

Fig. 37.

$T_1{}^* ds_2$, will have the same direction as $d\bar{s}_1$, i.e., it will act tangentially to the middle of the second surface shell. This same reasoning may be repeated for all the forces and in this way it may be shown that the second shell for the above law of calculation of all external and internal forces will be in a membrane state of stress, provided this state of stress applies to the first shell.

There remains now to establish formulae giving the forces and the surface loading for the transition from the first to the second shell (or vice versa).

Fig. 37 shows an element of the first shell with the forces acting on it and the corresponding rhombic element of the second shell. The internal forces acting on this second element have components in the directions $d\bar{s}_1$, $d\bar{s}_2$. In this way, by the law of transition (33.8), one finds

$$\bar{T}_1^* d\bar{s}_2 = \sqrt{c_1^2 l_1^2 + c_2^2 l_2^2 + c_3^2 l_3^2}\, T_1^* ds_2. \tag{33.9}$$

Substituting here for ds_2 its value from the second of the formulae (33.3), one will have

$$\bar{T}_1^* = \frac{\sqrt{c_1^2 l_1^2 + c_2^2 l_2^2 + c_3^2 l_3^2}}{\sqrt{c_1^2 m_1^2 + c_2^2 m_2^2 + c_3^2 m_3^2}}\, T_1^*. \tag{33.10}$$

In a similar manner it can be shown that

$$\bar{T}_2^* = \frac{\sqrt{c_1^2 m_1^2 + c_2^2 m_2^2 + c_3^2 m_3^2}}{\sqrt{c_1^2 l_1^2 + c_2^2 l_2^2 + c_3^2 l_3^2}}\, T_2^*, \quad \bar{S}^* = S^*. \tag{33.11}$$

Further, denote by X, Y, Z the components in the directions x, y, z of the surface loading on the first shell and by $\bar{X}$, $\bar{Y}$, $\bar{Z}$ those on the second shell.

By (33.8), one will have the following formulae

$$\begin{aligned} \bar{X}\, d\bar{\Omega} &= c_1 X\, d\Omega, \\ \bar{Y}\, d\bar{\Omega} &= c_2 Y\, d\Omega, \\ \bar{Z}\, d\bar{\Omega} &= c_3 Z\, d\Omega, \end{aligned} \tag{33.12}$$

where $d\Omega$ is the area of an element of the middle surface of the first shell and $d\bar{\Omega}$ is the area of the corresponding element of the second shell.

Denote by n_1, n_2, n_3 the cosines of the angles between the normal to the middle surface of the first shell and the axes x, y, z. Then the quantity $n_1 d\Omega$ will be the projection of the area $d\Omega$ on the plane yz. Under the affine coordinate transformation, this area becomes $c_2 c_3 n_1 d\Omega$. In a similar manner, the other two projections of $d\Omega$ may be transformed, as a result of which one obtains for $d\bar{\Omega}$ the expression

$$d\bar{\Omega} = \sqrt{c_2^2 c_3^2 n_1^2 + c_3^2 c_1^2 n_2^2 + c_1^2 c_2^2 n_3^2}\, d\Omega. \tag{33.13}$$

Taking this result into consideration, one finds the following final formulae for the calculation of the surface loading

$$\bar{X} = \frac{c_1}{\sqrt{c_2^2 c_3^2 n_1^2 + c_3^2 c_1^2 n_2^2 + c_1^2 c_2^2 n_3^2}} X,$$
$$\bar{Y} = \frac{c_2}{\sqrt{c_2^2 c_3^2 n_1^2 + c_3^2 c_1^2 n_2^2 + c_1^2 c_2^2 n_3^2}} Y, \qquad (33.14)$$
$$\bar{Z} = \frac{c}{\sqrt{c_2^2 c_3^2 n_1^2 + c_3^2 c_1^2 n_2^2 + c_1^2 c_2^2 n_3^2}} Z.$$

The above results may be used (for membrane analysis) to replace a shell of one shape by one of another shape which permits the deduction of a number of interesting results by very simple means.

This information will now be applied to the problem of the stresses in an oval-shaped cap, loaded by uniform normal pressure p. For this purpose, it will be assumed that the cap is half an ellipsoid with semi-axes a, b and h (h being the height of the cap, $2a$ and $2b$ its length and width).

The analytic expressions for the middle surface of such a cap are

$$\bar{x} = a \sin \theta \cos \varphi,$$
$$\bar{y} = b \sin \theta \sin \varphi, \qquad (33.15)$$
$$\bar{z} = h \cos \theta.$$

Further, consider the hemisphere given by

$$x = h \sin \theta \cos \varphi,$$
$$y = h \sin \theta \sin \varphi, \qquad (33.16)$$
$$z = h \cos \theta.$$

Comparing the formulae (33.15) and (33.16), it is clear that the equations for the ellipsoidal cap may be obtained from those for the hemisphere by means of the linear coordinate transformation

$$c_1 = \frac{a}{h}, \quad c_2 = \frac{b}{h}, \quad c_3 = 1. \qquad (33.17)$$

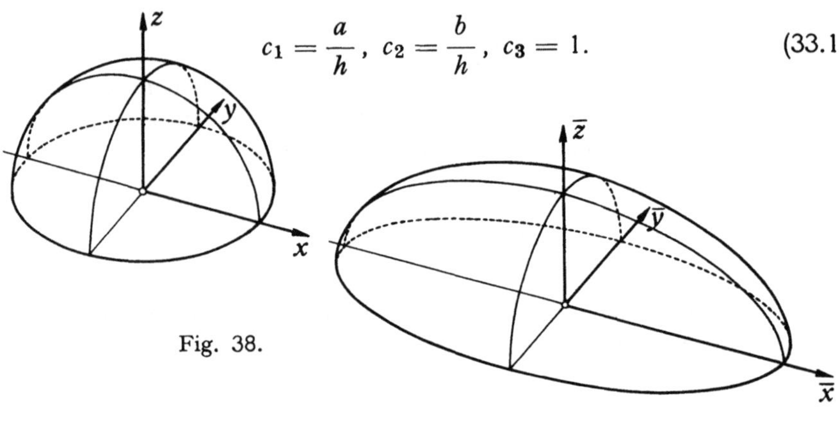

Fig. 38.

This result permits one to replace the calculation of the ellipsoidal cap by that of the hemispherical cap, loaded in a corresponding manner (Fig. 38).

Let the ellipsoidal cap be loaded by an external uniform normal pressure p and find the corresponding loading on the sphere.

The cosines of the angles between the normal to the ellipsoid and the axes x, y, z are given by

$$\cos(x, n) = \frac{c_2 \sin\theta \cos\varphi}{\sqrt{c_2^2 \sin^2\theta \cos^2\varphi + c_1^2 \sin^2\theta \sin^2\varphi + c_1^2 c_2^2 \cos^2\theta}},$$

$$\cos(y, n) = \frac{c_1 \sin\theta \sin\varphi}{\sqrt{c_2^2 \sin^2\theta \cos^2\varphi + c_1^2 \sin^2\theta \sin^2\varphi + c_1^2 c_2^2 \cos^2\theta}}, \tag{33.18}$$

$$\cos(z, n) = \frac{c_1 c_2 \cos\theta}{\sqrt{c_2^2 \sin^2\theta \cos^2\varphi + c_1^2 \sin^2\theta \sin^2\varphi + c_1^2 c_2^2 \cos^2\theta}}.$$

Multiplying the pressure by these direction cosines, one obtains $\bar{X}$, $\bar{Y}$, $\bar{Z}$, i.e., the components in the x, y, z directions of the external loading on the ellipsoidal cap. Now the corresponding components for the sphere may be calculated from (33.14), where the cosines n_1, n_2, n_3 may be determined from (33.18) by letting there $c_1 = c_2 = 1$. Thus one will have

$$X = -p\frac{c_2}{c_1} \sin\theta \cos\varphi, \quad Y = -p\frac{c_1}{c_2} \sin\theta \sin\varphi, \quad Z = -pc_1c_2 \cos\theta. \tag{33.19}$$

Taking into consideration that the cosines between the directions x, y, z and the unit vectors $\mathbf{e}_1$, $\mathbf{e}_2$, $\mathbf{e}_3$ (for the sphere) are given by (Fig. 39)

	x	y	z
$\mathbf{e}_1$	$\cos\theta \cos\varphi$	$\cos\theta \sin\varphi$	$-\sin\theta$
$\mathbf{e}_2$	$-\sin\varphi$	$\cos\varphi$	0
$\mathbf{e}_n$	$\sin\theta \cos\varphi$	$\sin\theta \sin\varphi$	$\cos\theta$

it is found that the uniform normal pressure on the ellipsoidal cap corresponds to a loading on the hemisphere with the components

$$q_1 = -\frac{p}{c_1c_2}(c_2^2 \cos^2\varphi + c_1^2 \sin^2\varphi - c_1^2c_2^2) \sin\theta \cos\theta,$$

$$q_2 = -\frac{p}{c_1c_2}(c_1^2 - c_2^2) \sin\theta \cos\varphi \sin\varphi, \tag{33.20}$$

$$q_n = -\frac{p}{c_1c_2}(c_2^2 \sin^2\theta \cos^2\varphi + c_1^2 \sin^2\theta \sin^2\varphi + c_1^2c_2^2 \cos^2\theta).$$

This loading may be written in the form

$$
\begin{aligned}
q_1 &= q_{1,0} + q_{1,2} \cos 2\varphi, \\
q_2 &= q_{2,0} + q_{2,2} \sin 2\varphi, \\
q_n &= q_{n,0} + q_{n,2} \cos 2\varphi,
\end{aligned}
\tag{33.21}
$$

where

$$
\begin{aligned}
q_{1,0} &= -\frac{p}{2c_1c_2}(c_1^2 + c_2^2 - 2c_1^2c_2^2)\cos\theta \sin\theta, \\
q_{2,0} &= 0, \\
q_{n,0} &= -\frac{p}{2c_1c_2}[(c_1^2 + c_2^2)\sin^2\theta + 2c_1^2c_2^2\cos^2\theta],
\end{aligned}
\tag{33.22}
$$

$$
\begin{aligned}
q_{1,2} &= \frac{p}{2c_1c_2}(c_1^2 - c_2^2)\sin\theta\cos\theta, \\
q_{2,2} &= -\frac{p}{2c_1c_2}(c_1^2 - c_2^2)\sin\theta, \\
q_{n,2} &= \frac{p}{2c_1c_2}(c_1^2 - c_2^2)\sin^2\theta.
\end{aligned}
\tag{33.23}
$$

Thus, the problem has been reduced to the analysis of a hemisphere subject to loading of the form (24.1) for the particular case when the series contains only those terms for which $k = 0$ and $k = 2$. These two cases will now be considered separately:

(a) Determine the stresses caused in a hemisphere by the symmetrical loading (33.22).

Using for this purpose directly the formulae (25.11), one will have

$$
\begin{aligned}
T_{1,0}^* &= -\frac{1}{h\sin^2\theta}\int_0^\theta ph^2c_1c_2\cos\theta\sin\theta\, d\theta = -\frac{phc_1c_2}{2}, \\
T_{2,0}^* &= q_{n,0}R_2 - T_{1,0}^*\frac{R_2}{R_1} = -\frac{ph}{2c_1c_2}[c_1^2c_2^2 + (c_1^2 + c_2^2 - 2c_1^2c_2^2)\sin^2\theta];
\end{aligned}
\tag{33.24}
$$

(b) Find the forces induced in a hemisphere by the loading (33.23).

In this case, the differential equation (31.1) may be written

$$
\frac{d^2U_2^*}{d\theta^2} + \operatorname{cotan}\theta\frac{dU_2^*}{d\theta} - \frac{4}{\sin^2\theta}U_2^* = -3ph^2\frac{c_1^2 - c_2^2}{c_1c_2}\sin^2\theta. \tag{33.25}
$$

Using Shpilrein's Tables of Spherical Functions (Part II, p. 91), it is

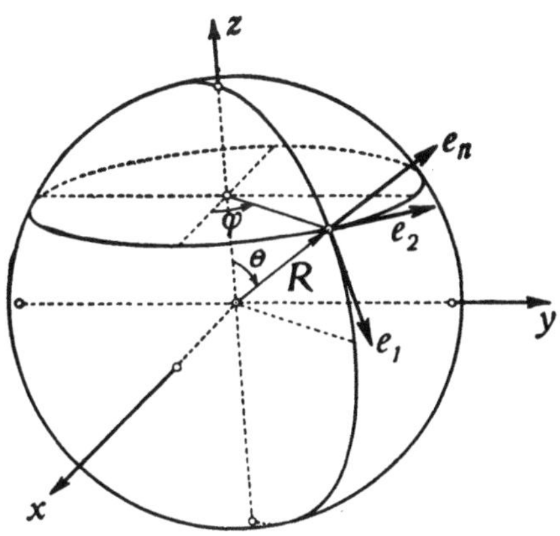

Fig. 39.

seen that,

$$P_{2,2}(\cos\theta) = 3\sin^2\theta, \quad (33.26)$$

where $P_{2,2}$ denotes the spherical harmonic of the first kind ($n = 2$, $k = 2$).

Hence, the general solution of the equation (33.25) may be written down immediately (§ 31)

$$U_2^* = B\tan^2\frac{\theta}{2} + \frac{ph^2}{2c_1c_2}(c_1^2 - c_2^2)\sin^2\theta. \quad (33.27)$$

It is now easy (by use of the formulae of § 23, specialized to the present particular case) to obtain expressions for the forces in a spherical shell, subject to the loading (33.23). The results are

$$\begin{aligned} T_1^* &= \left\{\frac{B}{h\sin^2\theta}\tan^2\frac{\theta}{2} + \frac{ph}{2c_1c_2}(c_1^2 - c_2^2)\right\}\cos 2\varphi, \\ T_2^* &= -\left\{\frac{B}{h\sin^2\theta}\tan^2\frac{\theta}{2} + \frac{ph}{2c_1c_2}(c_1^2 - c_2^2)\cos^2\theta\right\}\cos 2\varphi, \quad (33.28) \\ S^* &= \left\{\frac{-B}{h\sin^2\theta}\tan^2\frac{\theta}{2} + \frac{ph}{2c_1c_2}(c_1^2 - c_2^2)\cos\theta\right\}\sin 2\varphi. \end{aligned}$$

It will be assumed that the ellipsoidal end rests on a perfectly smooth support which cannot assimilate shear forces S^*. This being so, the same property must be imposed on the hemispherical end. In order that the formulae (33.28) satisfy this requirement, one has to put the constant B equal to zero.

Adding now the results obtained under a) and b) above, one finds that the forces in a hemispherical cap subject to the loading (33.21) are

$$\begin{aligned} T_1^* &= -\frac{phc_1c_2}{2}\left(1 - \frac{c_1^2 - c_2^2}{c_1^2c_2^2}\cos 2\varphi\right), \\ T_2^* &= -\frac{ph}{2c_1c_2}[c_1^2c_2^2 + (c_1^2 + c_2^2 - 2c_1^2c_2^2)\sin^2\theta + \\ &\qquad + (c_1^2 - c_2^2)\cos^2\theta\cos 2\varphi], \quad (33.29) \\ S^* &= \frac{ph}{2c_1c_2}(c_1^2 - c_2^2)\cos\theta\sin 2\varphi. \end{aligned}$$

The forces in the ellipsoidal cap may now be determined by means of the formulae (33.10) and (33.11) which, taking into consideration that in the present case

$$l_1 = \cos\theta\cos\varphi,\ m_1 = -\sin\varphi,$$
$$l_2 = \cos\theta\sin\varphi,\ m_2 = \cos\varphi, \tag{33.30}$$
$$l_3 = -\sin\theta, \qquad m_3 = 0,$$

give

$$\bar{T}_1{}^* = -\frac{phc_1c_2}{2}\left[1 - \frac{c_1{}^2 - c_2{}^2}{c_1{}^2c_2{}^2}\cos 2\varphi\right] \times$$
$$\times \frac{\sqrt{c_1{}^2\cos^2\theta\cos^2\varphi + c_2{}^2\cos^2\theta\sin^2\varphi + \sin^2\theta}}{\sqrt{c_1{}^2\sin^2\varphi + c_2{}^2\cos^2\varphi}},$$

$$\bar{T}_2{}^* = -\frac{ph}{2c_1c_2}[c_1{}^2c_2{}^2 + (c_1{}^2 + c_2{}^2 - 2c_1{}^2c_2{}^2)\sin^2\theta + (c_1{}^2 - c_2{}^2)\cos^2\theta\cos 2\varphi] \times$$
$$\times \frac{\sqrt{c_1{}^2\sin^2\varphi + c_2{}^2\cos^2\varphi}}{\sqrt{c_1{}^2\cos^2\theta\cos^2\varphi + c_2{}^2\cos^2\theta\sin^2\varphi + \sin^2\theta}}, \tag{33.31}$$

$$\bar{S}^* = \frac{ph}{2c_1c_2}(c_1{}^2 - c_2{}^2)\cos\theta\sin 2\varphi.$$

These last expressions completely determine the stresses in an ellipsoidal cap (on a smooth support) loaded by uniform external normal pressure p and they give the solution of this problem in closed form.

It is of the greatest interest to obtain the values of the forces at three points, namely, at the apex of the cap and at the points of intersection of the edge of the cap and the major and minor axes of the elliptic base. One finds

a) at the apex of the shell ($\theta = 0$, $\varphi = 0$):

$$\bar{T}_1{}^* = -\frac{phc_1{}^2}{2}\left(1 - \frac{c_1{}^2 - c_2{}^2}{c_1{}^2c_2{}^2}\right) = -\frac{pa^2}{2h}\left(1 - \frac{h^2}{b^2}\varepsilon^2\right),$$

$$\bar{T}_2{}^* = -\frac{ph}{2c_1{}^2}[c_1{}^2c_2{}^2 + (c_1{}^2 - c_2{}^2)] = -\frac{pb^2}{2h}\left(1 + \frac{h^2}{b^2}\varepsilon^2\right), \tag{33.32}$$

$$\bar{S}^* = 0,$$

where $\varepsilon^2 = 1 - \left(\frac{b}{a}\right)^2$ and $\bar{T}_1{}^*$ acts in the direction of the major, $\bar{T}_2{}^*$ in the direction of the minor axis of the base of the shell;

b) at the edge of the shell and at the end of the major axis:

$$\left(\theta = \frac{\pi}{2},\ \varphi = 0\right),$$

$$\bar{T}_1{}^* = -\frac{phc_1}{2}\left(1 - \frac{c_1{}^2 - c_2{}^2}{c_1{}^2 c_2{}^2}\right) = -\frac{pa}{2}\left(1 - \frac{h^2}{b^2}\varepsilon^2\right), \tag{33.33}$$

$$\bar{T}_2{}^* = \frac{phc_1 c_2{}^2}{2}\left(1 - \frac{c_1{}^2 + c_2{}^2}{c_1{}^2 c_2{}^2}\right) = \frac{pa}{2}\,\frac{b^2}{h^2}\left(1 - 2\frac{h^2}{a^2} - \frac{h^2}{b^2}\varepsilon^2\right),$$

$$\bar{S}^* = 0;$$

c) at the edge of the shell and at the end of the minor axis:

$$\left(\theta = \frac{\pi}{2},\ \varphi = \frac{\pi}{2}\right),$$

$$\bar{T}_1{}^* = -\frac{phc_2}{2}\left(1 + \frac{c_1{}^2 - c_2{}^2}{c_1{}^2 c_2{}^2}\right) = -\frac{pb}{2}\left(1 + \frac{h^2}{b^2}\varepsilon^2\right), \tag{33.34}$$

$$\bar{T}_2{}^* = \frac{phc_1{}^2 c_2}{2}\left[1 - \frac{c_1{}^2 + c_2{}^2}{c_1{}^2 c_2{}^2}\right] = \frac{pb}{2}\,\frac{a^2}{h^2}\left(1 - \frac{2h^2}{a^2} - \frac{h^2}{b^2}\varepsilon^2\right),$$

$$\bar{S}^* = 0.$$

It should be noted that the condition that the shear forces vanish at the edge of the shell can be replaced by any distribution of these forces.

In particular, by suitable choice of the constant B, one may arrive (for T_1 and S) at the case of a cap of the shape of half a triaxial ellipsoid attached to a tube of elliptic cross-section, the entire system being in a membrane state of stress.

§ 34. The equations of the membrane theory in Cartesian coordinates

A membrane analysis of roofs, the boundaries contours of which do not coincide with the lines of curvature, as a rule, is more conveniently performed in Cartesian coordinates. Let the middle surface of the shell be given by the equation

$$z = f(x, y). \tag{34.1}$$

The z-axis will be directed vertically upward, the plane xOy horizontally (cf. Fig. 11b). The direction of the axes Ox and Oy will be chosen for the convenience of the computations; for example, for a roof with rectangular plan form, they will be directed along the sides of the base of the proof.

The location of points on the middle surface of the shell can be fixed by the coordinates x and y of their projection on the plane xOy. The parameters x and y will then become the curvilinear coordinates of the middle surface. The lines x of this system of coordinates will be curves formed by the intersections of the middle surface with planes parallel to the plane xOz, the lines y, curves formed by the intersections of the middle surface with planes parallel to the plane yOz.

This system of curvilinear coordinates will not be orthogonal, except in the case when the middle surface of the roof is part of a cylinder with one of its axes, say Ox, parallel to the axis of the cylinder.

Consider the membrane conditions of equilibrium for an element, cut from the shell by two pairs of planes parallel to xOz, and yOz, respectively.

The projection of this element on the plane xOy will be a rectangle with sides dx and dy. Denote the force vector acting on the boundary AB of this element by $\mathbf{T}_x$ and the force vector acting on the boundary AD by $\mathbf{T}_y$. Then the resultant force acting on the side AB will be equal to

$$-\mathbf{T}_x ds_2 = -\mathbf{T}_x \sqrt{1+\left(\frac{\partial z}{\partial y}\right)^2}\, dy = -\mathbf{T}_{\check{x}}\, dy, \tag{34.2}$$

and the resultant force on the side AD

$$-\mathbf{T}_y ds_1 = -\mathbf{T}_y \sqrt{1+\left(\frac{\partial z}{\partial x}\right)^2}\, dx = -\mathbf{T}_{\check{y}}\, dx. \tag{34.3}$$

On the opposite sides CD and BC, one will have the resultant forces

$$\mathbf{T}_x ds_2 + \frac{\partial \mathbf{T}_x ds_2}{\partial x}\, dx = \left(\mathbf{T}_{\check{x}} + \frac{\partial \mathbf{T}_{\check{x}}}{dx}\, dx\right) dy,$$

$$\mathbf{T}_y ds_1 + \frac{\partial \mathbf{T}_y ds_1}{\partial y}\, dy = \left(\mathbf{T}_{\check{y}} + \frac{\partial \mathbf{T}_{\check{y}}}{dy}\, dy\right) dx.$$

Bearing in mind that the roof is to be stressed on the basis of its own weight, it will be assumed that the surface loading, acting on the shell, has only the vertical component q_z. Then the surface loading on the shell element $ABCD$ will be

$$q_z \sqrt{1+\left(\frac{\partial z}{\partial x}\right)^2+\left(\frac{\partial z}{\partial y}\right)^2}\, dx\, dy = q_{\check{z}}\, dx\, dy. \tag{34.4}$$

The condition that the resultant vector of all forces acting on this element is to vanish yields now

$$\frac{\partial \mathbf{T}_{\check{x}}}{\partial x} + \frac{\partial \mathbf{T}_{\check{y}}}{\partial y} + q_{\check{z}} \mathbf{i}_z = 0, \tag{34.5}$$

where $\mathbf{i}_z$ is the unit vector in the direction of the z-axis. Projecting on the coordinate axes, one obtains

$$\begin{aligned} &\frac{\partial T_{\overset{\vee}{xx}}}{\partial x} + \frac{\partial T_{\overset{\vee}{yx}}}{\partial y} = 0, \\ &\frac{\partial T_{\overset{\vee}{xy}}}{\partial x} + \frac{\partial T_{\overset{\vee}{yy}}}{\partial y} = 0, \\ &\frac{\partial T_{\overset{\vee}{xz}}}{\partial x} + \frac{\partial T_{\overset{\vee}{yz}}}{\partial y} = - q_{\overset{\vee}{z}}, \end{aligned} \tag{34.6}$$

where

$$\begin{gathered} T_{\overset{\vee}{xx}} = T_{xx}\sqrt{1+\left(\frac{\partial z}{\partial y}\right)^2}, \quad T_{\overset{\vee}{xy}} = T_{xy}\sqrt{1+\left(\frac{\partial z}{\partial y}\right)^2}, \\ T_{\overset{\vee}{xz}} = T_{xz}\sqrt{1+\left(\frac{\partial z}{\partial y}\right)^2}, \\ \mathbf{T}_{\overset{\vee}{yx}} = T_{yx}\sqrt{1+\left(\frac{\partial z}{\partial x}\right)^2}, \quad T_{\overset{\vee}{yy}} = T_{yy}\sqrt{1+\left(\frac{\partial z}{\partial x}\right)^2}, \\ T_{\overset{\vee}{yz}} = T_{yz}\sqrt{1+\left(\frac{\partial z}{\partial x}\right)^2}, \\ q_{\overset{\vee}{z}} = q_z\sqrt{1+\left(\frac{\partial z}{\partial x}\right)^2+\left(\frac{\partial z}{\partial y}\right)^2}. \end{gathered} \tag{34.7}$$

Next, consider the condition that the resultant moment of all forces acting on the element $ABCD$ of the middle surface is to vanish. Selecting as point of reference the centre of gravity of the element, constructing through this point three straight lines parallel to the coordinate axes and equating the sum of the moments about these straight lines of all the forces acting on the element to zero, one finds

$$\begin{gathered} T_{\overset{\vee}{xz}} - T_{\overset{\vee}{xx}}\frac{\partial z}{\partial x} - T_{\overset{\vee}{yx}}\frac{\partial z}{\partial y} = 0, \\ T_{\overset{\vee}{yz}} - T_{\overset{\vee}{yy}}\frac{\partial z}{\partial y} - T_{\overset{\vee}{xy}}\frac{\partial z}{\partial x} = 0, \\ T_{\overset{\vee}{xy}} = T_{\overset{\vee}{yx}}. \end{gathered} \tag{34.8}$$

The first two of these equalities express the condition that in a shell in a membrane state the forces must not have components normal to the middle surface. The third equation expresses the law of reciprocity of the shear forces in the adopted system of coordinates.

Substituting for $T_{\check{xz}}$, $T_{\check{yz}}$ from (34.8) into the third equation (34.6) and using the other two equations (34.6), one obtains

$$T_{\check{xx}} \frac{\partial^2 z}{\partial x^2} + 2T_{\check{xy}} \frac{\partial^2 z}{\partial x \partial y} + T_{\check{yy}} \frac{\partial^2 z}{\partial y^2} = -q_{\check{z}}. \tag{34.9}$$

Further, note that the first two equations (34.6) are identical in form to the equations of the plane problem of the theory of elasticity in terms of stresses. This makes it possible to introduce a stress function, related to the forces by the formulae

$$T_{\check{xx}} = \frac{\partial^2 \Phi}{\partial y^2}, \quad T_{\check{xy}} = T_{\check{yx}} = -\frac{\partial^2 \Phi}{\partial x \partial y}, \quad T_{\check{yy}} = \frac{\partial^2 \Phi}{\partial x^2}. \tag{34.10}$$

Substituting (34.10) into the first two equations (34.6), these equations are satisfied identically. Substitution from (34.10). into (34.9) yields the following equation for the stress function:

$$\frac{\partial^2 z}{\partial y^2} \frac{\partial^2 \Phi}{\partial x^2} - 2 \frac{\partial^2 z}{\partial x \partial y} \frac{\partial^2 \Phi}{\partial x \partial y} + \frac{\partial^2 z}{\partial x^2} \frac{\partial^2 \Phi}{\partial y^2} = -q_{\check{z}}, \tag{34.11}$$

which has been derived first by Pucher [56, 57].

Certain properties of the function Φ will now be studied. For this purpose, consider a linear element $d\sigma$ on the boundary of the middle surface of the shell, which may be an arbitrary, closed space curve. Through the ends of this element draw the coordinate lines x and y and consider the equilibrium of the triangular element ABC formed in this manner (Fig. 40).

The side BC will be acted upon by the tangential external force $\mathbf{f} d\sigma$, while the sides AB and AC will be subject to the internal forces

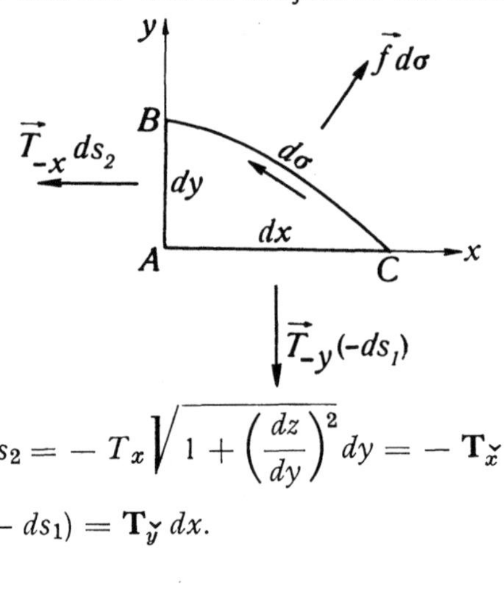

Fig. 40

$$\mathbf{T}_{-x} ds_2 = -T_x \sqrt{1 + \left(\frac{dz}{dy}\right)^2} dy = -\mathbf{T}_{\check{x}}\, dy,$$
$$\mathbf{T}_{-y}(-ds_1) = \mathbf{T}_{\check{y}}\, dx. \tag{34.12}$$

In the last formula ds_1 has a minus sign, since the positive increment along the contour of the shell (in the first quadrant of the plane xOy) corresponds to a negative increment of the coordinate x.

The requirement that the resultant vector of all forces, acting on the triangular shell element under consideration, should vanish yields

$$\mathbf{f}\,d\sigma = \mathbf{T}_{\overset{\vee}{x}}\,dy - \mathbf{T}_{\overset{\vee}{y}}\,dx - \tfrac{1}{2}q_{\overset{\vee}{z}}\,dx\,dy\,\mathbf{i}_z, \tag{34.13}$$

where the last term must be omitted as a small second order quantity. Hence (34.13) assumes the form

$$\mathbf{f}\,d\sigma = \mathbf{T}_{\overset{\vee}{x}}\,dy - \mathbf{T}_{\overset{\vee}{y}}\,dx. \tag{34.14}$$

Projecting this vector equation on the axes Ox and Oy, one obtains

$$\begin{aligned} f_x d\sigma &= T_{\overset{\vee}{x}x} dy - T_{\overset{\vee}{y}x} dx = \frac{\partial^2 \Phi}{\partial y^2} dy + \frac{\partial^2 \Phi}{\partial x \partial y} dx, \\ f_y d\sigma &= T_{\overset{\vee}{x}y} dy - T_{\overset{\vee}{y}y} dx = -\frac{\partial^2 \Phi}{\partial x \partial y} dy - \frac{\partial^2 \Phi}{\partial x^2} dx. \end{aligned} \tag{34.15}$$

These equations can be integrated along the boundary of the shell to render

$$\frac{\partial \Phi}{\partial x} = C_1 - F_y, \quad \frac{\partial \Phi}{\partial y} = C_2 + F_x \text{ on } L, \tag{34.16}$$

where

$$F_x = \int_0^\sigma f_x d\sigma, \quad F_y = \int_0^\sigma f_y d\sigma \tag{34.17}$$

and C_1, C_2 are arbitrary constants.

The symbol L denotes here the closed curve in the plane which is the projection of the edge of the middle surface on this plane. The origin of the arc σ for this edge can be chosen arbitrarily.

The integrals (34.17) are equal to the projections on the axes Ox and Oy of the resultant vector of all forces acting on the section of the boundary of the middle surface lying between the origin O and the point σ.

Multiplying the first equation (34.16) by dx and the second by dy, adding the two formulae and integrating along σ, one finds

$$\Phi = C + C_1 x + C_2 y + \mathfrak{M} \text{ on } L, \tag{34.18}$$

where

$$\mathfrak{M} = \int_0^\sigma (F_x dy - F_y dx) = \int_0^\sigma \{[y(\sigma) - y(t)]f_x(t) - [x(\sigma) - x(t)]f_y(t)\}dt \tag{34.19}$$

is the moment of all external forces, acting on the section of the edge of the shell, about a vertical axis through the point σ.

In the problem under consideration, as also in the plane problem of the theory of elasticity, one can add to the function Φ terms which are linear in x and y without influencing the stress field in the shell. Therefore, at least in the case when the region bounded by L is simply connected, one can set $C = C_1 = C_2 = 0$ in (34.17), (34.18), when these formulae reduce to the final form

$$\frac{\partial \Phi}{\partial x} = - F_y, \quad \frac{\partial \Phi}{\partial y} = F_x, \quad \Phi = \mathfrak{M} \text{ on } L. \tag{34.20}$$

The equalities (34.20) are analogous to the well known formulae of the plane theory of elasticity relating the stress function and its partial derivatives to the loading on the boundary.

The differential equation (34.11) is of the elliptic type, if the Gaussian curvature of the surface is positive, of the hyperbolic type, if it is negative, and of the parabolic type, if it is zero. The greatest interest attaches to the first case which will now be discussed in detail. The solution of (34.11) is then fully determined, if the function Φ or its normal derivative is known on the contour L.

The boundary condition

$$\Phi = 0 \text{ on } L \tag{34.21}$$

is of special interest. It corresponds to the case $\mathfrak{M} = 0$, i.e., the loading on the boundary of the shell is such that there is no moment about the vertical axis through an arbitrary point σ of the shell edge.

Hence it follows that, in order to practically realize the boundary condition (34.21), the basic contour of the shell must be made as flexible as possible in the horizontal plane. Simultaneously, it must be rigid in tension as well as in bending in a vertical plane, in order to make the given tangential loading on the shell possible and, in addition, to ensure that the shell will not be free to deform in pure bending.

It has been seen that this type of boundary condition is not only the simplest in a mathematical sense, but that it is also most advantageous from purely practical considerations. In fact, there is no difficulty in giving the edge of a roof rigidity in a vertical plane, i.e., in the plane of the wall of the building. Similarly, it is also easy to make the base contour of the roof rigid in tension.

In contrast, it is not very easy to achieve its rigidity in bending in the horizontal plane. For this purpose, the edge of the shell would have to be strengthened by powerful supports in the direction perpendicular to the wall of the building, which would create a number of constructional inconveniences, independently of the fact that these reinforcements would have to be placed inside or outside the roof.

In view of these facts, one usually seeks to realize in practice the type of boundary conditions for which the base contour does not bend in a horizontal plane, i.e., (34.21).

The problem under consideration can also be formulated as a variational principle. It turns out that among the infinite number of functions satisfying on the boundary curve L the condition (34.21), and, in addition, the equation (34.11), the correct function is that which gives a stationary value to the functional

$$V = \iint \Big[\frac{1}{2} \frac{\partial^2 z}{\partial y^2} \left(\frac{\partial \Phi}{\partial x} \right)^2 - \frac{\partial^2 z}{\partial x \partial y} \frac{\partial \Phi}{\partial x} \frac{\partial \Phi}{\partial y} + \\ + \frac{1}{2} \frac{\partial^2 z}{\partial x^2} \left(\frac{\partial \Phi}{\partial y} \right)^2 - q_{\check{z}} \Phi \Big] dx\, dy, \tag{34.22}$$

where the integrations extend over the area of the region bounded by L.

This result is readily verified by evaluating the variation of V and performing the usual manipulations of the calculus of variations. Since $\delta\Phi = 0$ on L, one arrives thus at

$$-\delta V = \iint \left[\frac{\partial^2 z}{\partial y^2} \frac{\partial^2 \Phi}{\partial x^2} - 2 \frac{\partial^2 z}{\partial x \partial y} \frac{\partial^2 \Phi}{\partial x \partial y} + \frac{\partial^2 z}{\partial x^2} \frac{\partial^2 \Phi}{\partial y^2} + q_{\check{z}} \right] \delta\Phi\, dx\, dy = 0, \tag{34.23}$$

whence follows, in view of the arbitrariness of $\delta\Phi$, the equation (36.11).

The formulae (34.22) and (34.23) reveal a possibility for solving (34.11) approximately for the boundary condition (34.21).

Let the contour L be determined by the equation

$$L(x, y) = 0; \tag{34.24}$$

then an approximation to the function Φ can be sought in the form

$$\Phi^* = L(x, y) \sum_{j=0}^{j=n} \sum_{i=0}^{i=j} A_{ij} x^i y^{j-i}, \tag{34.25}$$

where the A_{ij} are constants which can be determined by the method of Bubnov-Galerkin, for which (34.25) must be substituted into (34.23).

§ 35a. Analysis of a roof in the shape of an elliptical paraboloid

Let the middle surface of the roof be formed by the elliptical paraboloid

$$z = h - \frac{h}{2}\left(\frac{x^2}{a^2} + \frac{y^2}{b^2}\right). \tag{35.1a}$$

In this case, the equation (34.11) reduces to the simpler form

$$\frac{h}{b^2}\frac{\partial^2 \Phi}{\partial x^2} + \frac{h}{a^2}\frac{\partial^2 \Phi}{\partial y^2} = q_{\check{z}} = q_z\sqrt{1 + \left[\left(\frac{x}{a}\right)^2 + \left(\frac{y}{b}\right)^2\right]h^2}. \tag{35.2a}$$

If it is assumed that the loading is distributed uniformly over the area of the base of the roof and acts downwards, then

$$q_z\sqrt{1 + \left[\left(\frac{x}{a}\right)^2 + \left(\frac{y}{b}\right)^2\right]h^2} = -q_0 = \text{const.} \tag{35.3a}$$

Substituting (35.3a) into (35.2a) and introducing the new non-dimensional coordinates

$$\xi = \frac{x}{a}, \qquad \eta = \frac{y}{b} \tag{35.4a}$$

and the new auxiliary function

$$\Psi = \Phi : \frac{q_0 a^2 b^2}{2h}, \tag{35.5a}$$

one obtains the equation

$$\frac{\partial^2 \Psi}{\partial \xi^2} + \frac{\partial^2 \Psi}{\partial \eta^2} = -2, \tag{35.6a}$$

which, if it is assumed that the perimeter of the roof does not resist bending in a horizontal plane, must be solved for the boundary conditions

$$\Phi = \Psi = 0 \text{ on } L', \tag{35.7a}$$

where L' is the equation of the curve bounding the base of the roof structure in the ξ, η-plane.

Hence the analysis of a roof in the shape of an elliptical paraboloid under loading distributed uniformly over its base, mathematically speaking, is found to be identical to the problem of the torsion of a prismatic rod with a cross-section in the shape of L'.

This analogy was discovered by V. Z. Vlasov [3a]. It opens up the possibility of adopting certain results from the theory of torsion for use in the membrane theory of shells.

a) *Analysis of a dome with an elliptical plan form.*

Consider a shell in the form of an elliptical paraboloid (35.1a) with central rise h, bounded by the plane curve L

$$\frac{x^2}{a^2} + \frac{y^2}{b^2} = 1 \tag{35.8a}$$

in the ξ, η-plane; this curve can be transformed into the circle L'

$$\xi^2 + \eta^2 = 1. \tag{35.9a}$$

The torsion function Ψ for a rod with a circular cross-section is known to have the form

$$\Psi = -\tfrac{1}{2}(\xi^2 + \eta^2). \tag{35.10a}$$

Changing over from Ψ to Φ and from ξ, η to x, y, one obtains

$$\Phi = -\frac{q_0 a^2 b^2}{4h}\left(\frac{x^2}{a^2} + \frac{y^2}{b^2}\right), \tag{35.11a}$$

and hence the following expressions for the stresses in the roof:

$$T_{\check{x}\check{x}} = -\frac{q_0 a^2}{2h}, \quad T_{\check{y}\check{y}} = -\frac{q_0 b^2}{2h}, \quad T_{x\check{y}} = T_{y\check{x}} = 0. \tag{35.12a}$$

The roof is seen to be in compression in all cross-sections; its support ring is not under bending. However, one must subject the dome to a suitable thrust which ensures on its edge forces corresponding to (35.12a), when the support ring will be under tension. The condition of compatibility of strains in the ring and dome requires that at each point of the boundary the elongations of the ring and the dome must be equal. Thus, a contradiction arises which is analogous to that for the case of the dome in the form of a shell of revolution with parabolic meridian: the support ring is in tension, but the dome in compression.

b) *Analysis of a dome in the shape of a shell of revolution, having an equilateral, triangular plan form.*

Let the roof have the form of the paraboloid of revolution

$$z = h - \frac{h}{2R^2}(x^2 + y^2) \tag{35.13a}$$

and its base the form of an equilateral triangle of height $3a$ (Fig. 41). Then the edge of the shell is a space curve which has three angular points and is formed by three arched segments lying in vertical planes.

If it is assumed that the support ring is not subject to bending in a horizontal plane, this problem reduces to the problem of the torsion of a rod with an equilateral, triangular cross-section the side l of which is equal to

$$l = 2\sqrt{3}a = \sqrt{\tfrac{3}{2}}R.$$

As is known from the theory of elasticity, the torsion function Ψ in this case is given by

$$\Psi = -\left[\frac{1}{6a}(x^3 - 3xy^2) - \tfrac{2}{3}a^2 + \tfrac{1}{2}(x^2 + y^2)\right],$$

whence

$$\Phi = -\frac{1}{2}\frac{q_0R^2}{h}\left[\frac{1}{6a}(x^3 - 3xy^2) - \tfrac{2}{3}a^2 + \tfrac{1}{2}(x^2 + y^2)\right],$$

and therefore

$$T_{\check{x}x} = \frac{\partial^2\Phi}{\partial y^2} = -\frac{1}{2}\frac{q_0R^2}{h}\left(1 - \frac{x}{a}\right),$$

$$T_{\check{y}y} = \frac{\partial^2\Phi}{\partial x^2} = -\frac{1}{2}\frac{q_0R^2}{h}\left(1 + \frac{x}{a}\right),$$

$$T_{\check{x}y} = -\frac{\partial^2\Phi}{\partial x\partial y} = -\frac{1}{2}\frac{q_0R^2}{h}\frac{y}{a}.$$

Thus, the stress field in a dome with triangular plan form turns out to be sufficiently smooth. The maximum compressive stress acts at the corners and is three times as large as that at the vertex. It is worth noting that the stresses at the corners of the dome remain finite. Later on, when dealing with the analogous analysis for a dome with rectangular

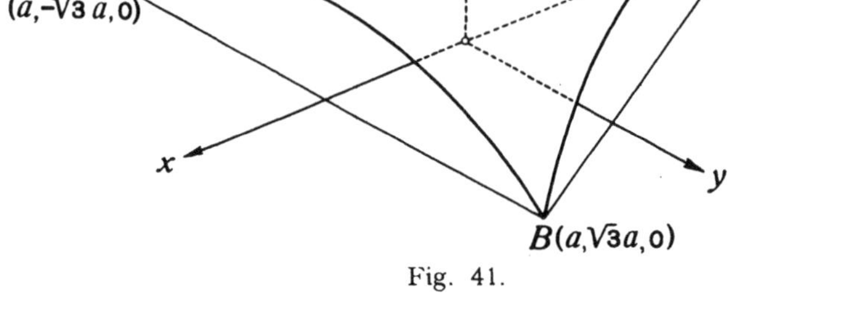

Fig. 41.

plan form, it will be shown that the stresses at the corners of the shell become infinite.

Finally, attention will be drawn to the fact that the loading on the roof under consideration is transmitted to the three supporting arcs through forces, tangential to the contour, whereas in the dome with circular plan form it is transmitted by means of normal forces.

The method of supporting a shell not through normal, but through tangential forces is often employed at the present time, since it has definite structural advantages in cases, where the plan form of the shell is not circular.

§ 35b. Analysis of an elliptical paraboloid with a rectangular base

If the surface of the roof is determined by (35.1a) and the plan form of the roof is a rectangle with sides 2α and 2β, then the edge of the shell is a space curve with four angular points consisting of four symmetric arched segments. This form of roof, though it may not necessarily be based on an elliptical paraboloid, is perhaps most typical for the present time; therefore it will be treated here in greater detail.

Assuming the roof loading to be uniformly distributed over the base, it follows from § 34 that the problem under consideration may be reduced to the problem of torsion of a rod with rectangular cross-section with sides

$$2\xi_0 = 2\frac{\alpha}{a}, \quad 2\eta_0 = 2\frac{\beta}{b}, \quad \frac{\xi_0}{\eta_0} \leqslant 1. \tag{35.1b}$$

However, the use of the corresponding solution in the form, in which it is given in treatises on the theory of elasticity, proves in the present case to be inconvenient, since, while in the problem of torsion interest centres around the function Ψ itself and its first partial derivatives, for the stress analysis of roofs the basic aim is the determination of the second partial derivatives of this function; in fact, near the edges of the shell, the usual series expansions for Ψ are too slowly convergent. Therefore, following a reasoning which applies in the torsion problem for a rod of rectangular cross-section, one can subject the solution to certain additional transformations for the purpose of separating from the series singularities which affect their convergence unfavourably.

Represent the right-hand side of (35.6a) in the form of the series

$$-2 = -\frac{8}{\pi}\sum_{n=0}^{\infty}\frac{(-1)^n}{2n+1}\cos\lambda_n\xi, \tag{35.2b}$$

where

$$\lambda_n = \frac{2n+1}{2}\frac{\pi}{\xi_0}, \tag{35.3b}$$

and seek Ψ in the form of the series

$$\Psi(\xi, \eta) = \sum_{n=0}^{\infty} \Psi_n(\eta) \cos \lambda_n \xi. \tag{35.4b}$$

Then one obtains for the coefficients $\Psi_n(\eta)$ the second order differential equation

$$\frac{d^2\Psi_n}{d\eta^2} - \lambda_n{}^2\Psi_n = -\frac{8}{\pi}\frac{(-1)^n}{2n+1}, \tag{35.5b}$$

whence

$$\Psi_n = A_n \cosh \lambda_n\eta + B_n \sinh \lambda_n\eta + \frac{8}{\pi}\frac{(-1)^n}{2n+1}\frac{1}{\lambda_n{}^2}. \tag{35.6b}$$

Subjecting this expression to the condition $\Psi_n = 0$ for $\eta = \pm\,\eta_0$, one obtains

$$\Psi_n = \frac{32\xi_0{}^2}{\pi^3}\frac{(-1)^n}{(2n+1)^3}\left(1 - \frac{\cosh \lambda_n\eta}{\cosh \lambda_n\eta_0}\right). \tag{35.7b}$$

and hence

$$\Psi = \frac{32\xi_0{}^2}{\pi^3}\sum_{n=0}^{\infty}\frac{(-1)^n}{(2n+1)^3}\left[1 - \frac{\cosh \lambda_n\eta}{\cosh \lambda_n\eta_0}\right]\cos \lambda_n\xi. \tag{35.8b}$$

It will be noted that

$$\frac{32\xi_0{}^2}{\pi^3}\sum_{n=0}^{\infty}\frac{(-1)^n}{(2n+1)^3}\cos \lambda_n\xi = \xi_0{}^2 - \xi^2, \tag{35.9b}$$

so that (35.9b) may be rewritten in the form

$$\Psi = \xi_0{}^2 - \xi^2 - \frac{32\xi_0{}^2}{\pi^3}\sum_{n=0}^{\infty}\frac{(-1)^n}{(2n+1)^3}\frac{\cosh \lambda_n\eta}{\cosh \lambda_n\eta_0}\cos \lambda_n\xi. \tag{35.10b}$$

In what follows, one will be interested in the second partial derivatives of this function

$$\begin{aligned}
\frac{\partial^2\Psi}{\partial\xi^2} &= -2 + \frac{8}{\pi}\sum_{n=0}^{\infty}\frac{(-1)^n}{2n+1}\frac{\cosh \lambda_n\eta}{\cosh \lambda_n\eta_0}\cos \lambda_n\xi,\\
\frac{\partial^2\Psi}{\partial\xi\partial\eta} &= \frac{8}{\pi}\sum_{n=0}^{\infty}\frac{(-1)^n}{2n+1}\frac{\sinh \lambda_n\eta}{\cosh \lambda_n\eta_0}\sin \lambda_n\xi,\\
\frac{\partial^2\Psi}{\partial\eta^2} &= -\frac{8}{\pi}\sum_{n=0}^{\infty}\frac{(-1)^n}{2n+1}\frac{\cosh \lambda_n\eta}{\cosh \lambda_n\eta_0}\cos \lambda_n\xi = -2 - \frac{\partial^2\Psi}{\partial\xi^2},
\end{aligned} \tag{35.11b}$$

which are related to the stresses in the shell by the formulae

$$T_{\check{x}\check{x}} = \frac{q_0 a^2}{2h} \frac{\partial^2 \Psi}{\partial \eta^2},$$

$$T_{\check{y}\check{y}} = \frac{q_0 b^2}{2h} \frac{\partial^2 \Psi}{\partial \xi^2}, \tag{35.12b}$$

$$T_{\check{x}\check{y}} = -\frac{q_0 ab}{2h} \frac{\partial^2 \Psi}{\partial \xi \partial \eta}.$$

Near the edge of the shell and in particular, near the corners, the series (35.11b) converge badly and one has to be concerned with improving their convergence. Following B. A. Palmov, consider for this purpose the series

$$W = \sum_{n=0}^{\infty} (-1)^n \frac{\cosh \lambda_n \eta}{\cosh \lambda_n \eta_0} \sin \lambda_n \xi \tag{35.13b}$$

and transform its terms, using the relations

$$\frac{\cosh \lambda_n \eta}{\cosh \lambda_n \eta_0} = \frac{\cosh \lambda_n \eta}{\cosh \lambda_n \eta_0} [(1 + e^{-2\lambda_n \eta_0}) - e^{-2\lambda_n \eta_0}] =$$

$$= e^{-\lambda_n(\eta_0 - \eta)} + e^{-\lambda_n(\eta_0 + \eta)} - e^{-2\lambda_n \eta_0} \frac{\cosh \lambda_n \eta}{\cosh \lambda_n \eta_0}. \tag{35.14b}$$

Substituting (35.14b) into (35.13b), one obtains

$$W = \sum_{n=0}^{\infty} (-1)^n e^{-\lambda_n(\eta_0 - \eta)} \sin \lambda_n \xi + \sum_{n=0}^{\infty} (-1)^n e^{-\lambda_n(\eta_0 + \eta)} \sin \lambda_n \xi -$$

$$- \sum_{n=0}^{\infty} (-1)^n e^{-2\lambda_n \eta_0} \frac{\cosh \lambda_n \eta}{\cosh \lambda_n \eta_0} \sin \lambda_n \xi. \tag{35.15b}$$

The series (35.15b) has been subdivided into three series the last of which converges very rapidly throughout the entire region under consideration, while the first two can be summed. In order to show this, express $\sin \lambda_n \xi$ by use of the formula

$$\sin \lambda_n \xi = \frac{e^{i\lambda_n \xi} - e^{-i\lambda_n \xi}}{2i}.$$

Then the first of the two series reduces to the form

$$I_1 = \sum_{n=0}^{\infty} (-1)^n e^{-\lambda_n(\eta_0 - \eta)} \sin \lambda_n \xi =$$

$$= \frac{1}{2i} \left\{ \sum_{n=0}^{\infty} (-1)^n e^{-\lambda_n[(\eta_0 - \eta) - i\xi]} - \sum_{n=0}^{\infty} (-1)^n e^{-\lambda_n[(\eta_0 - \eta) + i\xi]} \right\}. \tag{35.16b}$$

However, this is the sum of two infinite geometric progressions with the first terms

$$\frac{1}{2i}\,e^{-\lambda_0[(\eta_0-\eta)-i\xi]},\qquad \frac{1}{2i}\,e^{-\lambda_0[(\eta_0-\eta)+i\xi]}$$

and factors

$$-\,e^{-2\lambda_0[(\eta_0-\eta)-i\xi]},\qquad -\,e^{-2\lambda_0[(\eta_0-\eta)+i\xi]},$$

respectively. Hence one can write immediately

$$I_1=\sum_{n=0}^{\infty}(-1)^n e^{-\lambda_n(\eta_0-\eta)}\sin\lambda_n\xi=$$

$$=\frac{1}{2i}\left[\frac{e^{-\lambda_0(\eta_0-\eta-i\xi)}}{1+e^{-\lambda_0(\eta_0-\eta+i\xi)}}-\frac{e^{-\lambda_0(\eta_0-\eta-i\xi)}}{1+e^{-2\lambda_0(\eta_0-\eta+i\xi)}}\right].\tag{35.17b}$$

Similarly, one finds

$$I_2=\sum_{n=0}^{\infty}(-1)^n e^{-\lambda_n(\eta_0+\eta)}\sin\lambda_n\xi=$$

$$=\frac{1}{2i}\left[\frac{e^{-\lambda_0(\eta_0+\eta-i\xi)}}{1+e^{-2\lambda_0(\eta_0+\eta-i\xi)}}-\frac{e^{-\lambda_0(\eta_0+\eta+i\xi)}}{1+e^{-2\lambda_0(\eta_0+\eta+i\xi)}}\right].\tag{35.18b}$$

It follows from (35.11b), (35.13b) that

$$\frac{\partial^3\Psi}{\partial\xi\partial\eta^2}=\frac{4}{\xi_0}\,W.\tag{35.19b}$$

Substituting for W its expression (35.15b) and integrating with respect to ξ, one obtains

$$\begin{aligned}\frac{\partial^2\Psi}{\partial\eta^2}=f(\eta)+\frac{4}{\xi_0}\Bigg\{\int_{\xi_0}^{\xi}I_1d\xi+\int_{\xi_0}^{\xi}I_2d\xi+\\ +\frac{2\xi_0}{\pi}\sum_{n=1}^{\infty}\frac{(-1)^n}{2n+1}\,e^{-2\lambda_n\eta_0}\,\frac{\cosh\lambda_n\eta}{\cosh\lambda_n\eta_0}\cos\lambda_n\xi\Bigg\},\end{aligned}\tag{35.20b}$$

where

$$\begin{aligned}\int_{\xi_0}^{\xi}I_1d\xi&=\frac{1}{2\lambda_0}\{-\operatorname{artan}[e^{-\lambda_0(\eta_0-\eta-i\xi)}]-\operatorname{artan}[e^{-\lambda_0(\eta_0-\eta+i\xi)}]\}\\ &=-\frac{\xi_0}{\pi}\operatorname{artan}\left[\frac{\cos\lambda_0\xi}{\sinh\lambda_0(\eta_0-\eta)}\right],\end{aligned}\tag{35.21b}$$

$$\int_{\xi_0}^{\xi}I_2d\xi=-\frac{\xi_0}{\pi}\operatorname{artan}\left[\frac{\cos\lambda_0\xi}{\sinh\lambda_0(\eta_0+\eta)}\right].$$

The arbitrary function $f(\eta)$ entering into (35.20b) can be found from the condition that $\partial^2\Psi/\partial\eta^2 = 0$ for $\xi = \pm\,\xi_0$, as this follows from (35.11b). Letting for this reason $f(\eta) = 0$, one obtains

$$\frac{\partial^2\Psi}{\partial\eta^2} = -\frac{4}{\pi}\left\{\operatorname{artan}\left[\frac{\cos\lambda_0\xi}{\sinh\lambda_0(\eta_0-\eta)}\right] + \operatorname{artan}\left[\frac{\cos\lambda_0\xi}{\sinh\lambda_0(\eta_0+\eta)}\right]\right\} + \\ + \frac{8}{\pi}\sum_{n=0}^{\infty}\frac{(-1)^n}{2n+1}e^{-2\lambda_n\eta_0}\frac{\cosh\lambda_n\eta}{\cosh\lambda_n\eta_0}\cos\lambda_n\xi. \tag{35.22b}$$

This series converges extraordinarily quickly in view of the fact that

$$e^{-2\lambda_n\eta_0} = e^{-(2n+1)(\pi\eta_0/\xi_0)},$$

where, as has been pointed out previously, $\eta_0/\xi_0 \geqslant 1$ and the variation of the function

$$\frac{\cosh\lambda_n\eta}{\cosh\lambda_n\eta_0}\cos\lambda_n\xi$$

in the region under consideration does not exceed in absolute value unity. Therefore it will be sufficient to retain in this series only its first term, whence

$$\frac{\partial^2\Psi}{\partial\eta^2} \approx -\frac{4}{\pi}\left\{\operatorname{artan}\left[\frac{\cos\lambda_0\xi}{\sinh\lambda_0(\eta_0-\eta)}\right] + \operatorname{artan}\left[\frac{\cos\lambda_0\xi}{\sinh\lambda_0(\eta_0+\eta)}\right] - \\ - 2e^{-2\lambda_0\eta_0}\frac{\cosh\lambda_0\eta}{\cosh\lambda_0\eta_0}\cos\lambda_0\xi\right\}. \tag{35.23b}$$

This result differs from the exact value of $\partial^2\Psi/\partial\eta^2$ in the entire region of its definition by only a fraction of a percent. Using now the formula

$$\frac{\partial^2\Psi}{\partial\xi^2} = -2 - \frac{\partial^2\Psi}{\partial\eta^2}, \tag{35.24b}$$

one can also obtain an expression for $\partial^2\Psi/\partial\xi^2$.

In order to evaluate the mixed derivative $\partial^2\Psi/\partial\xi\partial\eta$, one must integrate (35.19b) with respect to η

$$\frac{\partial^2\Psi}{\partial\xi\partial\eta} = \frac{4}{\xi_0}\left\{f(\xi) + \int_0^\eta I_1 d\eta + \int_0^\eta I_2 d\eta - \\ - \frac{1}{\lambda_0}\sum_{n=0}^{\infty}\frac{(-1)^n}{2n+1}e^{-2\lambda_n\eta_0}\frac{\sinh\lambda_n\eta}{\cosh\lambda_n\eta_0}\sin\lambda_n\xi\right\}. \tag{35.25b}$$

The integrals in this formula are elementary; in fact, one has

$$\int_{\eta_0}^{\eta} I_1 d\eta = \frac{1}{2i}\left[\int_0^{\eta} \frac{e^{-\lambda_0(\eta_0-\eta-i\xi)}}{1+e^{-2\lambda_0(\eta_0-\eta-i\xi)}}\, d\eta - \right.$$

$$\left. - \int_0^{\eta} \frac{e^{-\lambda_0(\eta_0-\eta+i\xi)}}{1+e^{-2\lambda_0(\eta_0-\eta+i\xi)}}\, d\eta\right] = \frac{1}{2\lambda_0} \operatorname{artanh}\left[\frac{\sin \lambda_0\xi}{\cosh \lambda_0(\eta_0-\eta)}\right]\Bigg|_0^{\eta}, \quad (35.26b)$$

$$\int_0^{\eta} I_2 d\eta = -\frac{1}{2\lambda_0} \operatorname{artanh}\left[\frac{\sin \lambda_0\xi}{\cosh \lambda_0(\eta_0+\eta)}\right]\Bigg|_0^{\eta}.$$

The arbitrary function in (35.25b) must be determined from the condition that $\partial^2\Psi/\partial\xi\partial\eta = 0$ for $\eta = 0$, which follows from the second equation (35.11b). Hence one arrives at the final formula for $\partial^2\Psi/\partial\xi\partial\eta$

$$\frac{\partial^2\Psi}{\partial\xi\partial\eta} = \frac{4}{\pi}\left\{\operatorname{artanh}\left[\frac{\sin \lambda_0\xi}{\cosh \lambda_0(\eta_0-\eta)}\right] - \operatorname{artanh}\left[\frac{\sin \lambda_0\xi}{\cosh \lambda_0(\eta_0+\eta)}\right] - \right.$$

$$\left. - 2\sum_{n=0}^{\infty} \frac{(-1)^n}{2n+1} e^{-2\lambda_n\eta_0} \frac{\sinh \lambda_n\eta}{\cosh \lambda_n\eta_0} \sin \lambda_n\xi\right\}, \quad (35.27b)$$

where the series converge so quickly that there is no point in retaining more than one term.

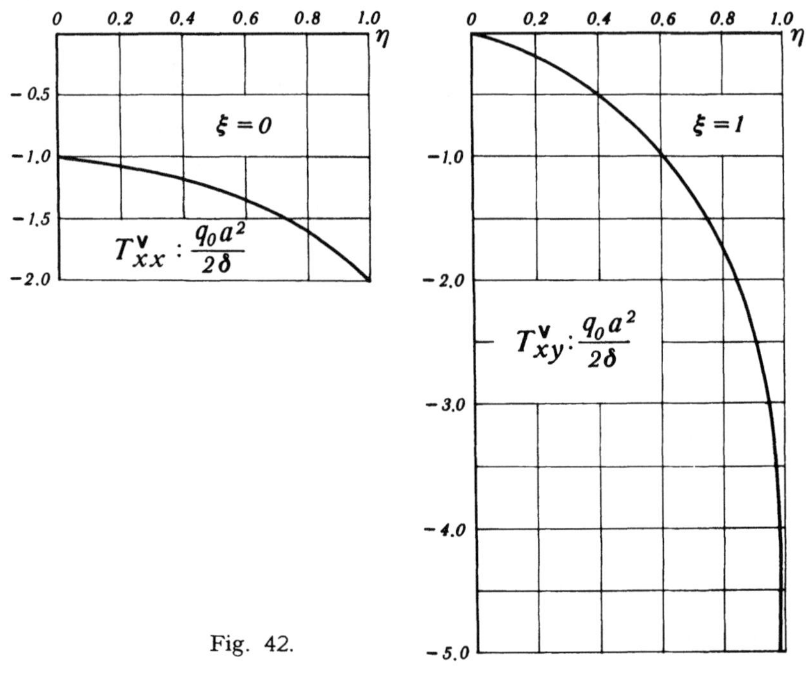

Fig. 42.

The three second partial derivatives of the function Ψ can now be substituted into (35.12b) to evaluate the stresses in the shell. Fig. 42 shows the values of $T_{\check{x}x}$ in the centre section of the shell $\xi = 0$ and $T_{\check{x}y}$ along one of the edges. The shear forces are seen to become infinite at the corners of the shell. However, this singularity of $\partial^2\Psi/\partial\xi\partial\eta$ is very weak, since it is of the order of artanh x for $x \to 1$ or $\pm \log r$ for $r \to 0$. The last result follows immediately, if one rewrites (35.24b) using

$$\operatorname{artanh} t = \tfrac{1}{2} \log \frac{1+t}{1-t}$$

in the form

$$\frac{\partial^2\Psi}{\partial\xi\partial\eta} = \frac{2}{\pi}\left[\log \frac{\cosh \lambda_0(\eta_0 - \eta) + \sin \lambda_0\xi}{\cosh \lambda_0(\eta_0 - \eta) - \sin \lambda_0\xi} - \log \frac{\cosh \lambda_0(\eta_0 + \eta) + \sin \lambda_0\xi}{\cosh \lambda_0(\eta_0 + \eta) - \sin \lambda_0\xi} - \right.$$
$$\left. - 4 \sum_{n=0}^{\infty} \frac{(-1)^n}{2n+1} e^{-2\lambda_n\eta_0} \frac{\sinh \lambda_n\eta}{\cosh \lambda_n\eta_0} \sin \lambda_n\xi \right]. \qquad (35.28b)$$

For an analysis of sections of the shell, far removed from the edge, it will be more convenient to use the more compact formula (35.27b), but for the corner regions (35.28b) will be more suitable. This is explained by the fact that the function $\tanh y = x$ to five decimals is equal to unity for $y \geqslant 7$. Therefore the inverse function $y = \operatorname{artanh} x$, in an approximate analysis, turns out to be undetermined for all values of x near to unity. However, use of (35.28b) allows to expose more clearly the character of the growth of the shear force in the proximity of the corners of the shell, and to locate in this sense the regions of the shell where the membrane theory is not applicable.

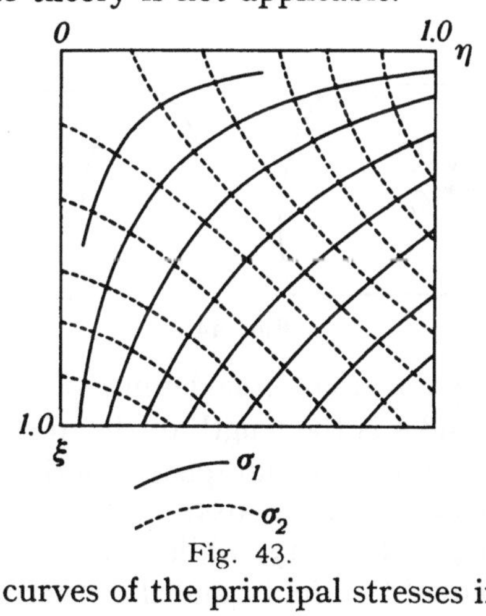

Fig. 43.

Fig. 43 shows the curves of the principal stresses in a roof with square plan form. The maximum principal stresses act at the corners of the shell, where they tend to infinity in accordance with the same law as the shear forces.

The fact that for the above boundary conditions the membrane theory yields such a rapidly changing force field suggests that it is, in essence, not possible to apply it in these regions. In order to obtain a realistic representation of the behaviour of a shell near its corners, one has to take into consideration in these areas the moment terms of the equations of shell theory. In a certain sense, this reduces the value of the results obtained above.

However, the significance of the formulae developed above lies in the fact that they describe everywhere sufficiently exactly the state of stress in the shell, except in four relatively small regions near the corners. Besides, they give quite reliable indications that there arise in the corners of a roof with rectangular plan form significant stress concentrations which are accompanied by bending moments and shear forces. As a rule, the collapse of a roof will commence in these areas, whence it will be necessary to take care with corresponding supports of the shell. It will be recalled that it has been shown in the last section that in a roof with triangular plan form the stresses in the corners remain finite.

§ 36. The Equations of the membrane theory for cylindrical shells of arbitrary shape

Consider a cylindrical shell of arbitrary shape. The position of a point on its middle surface can be given by the coordinates x and s, where x

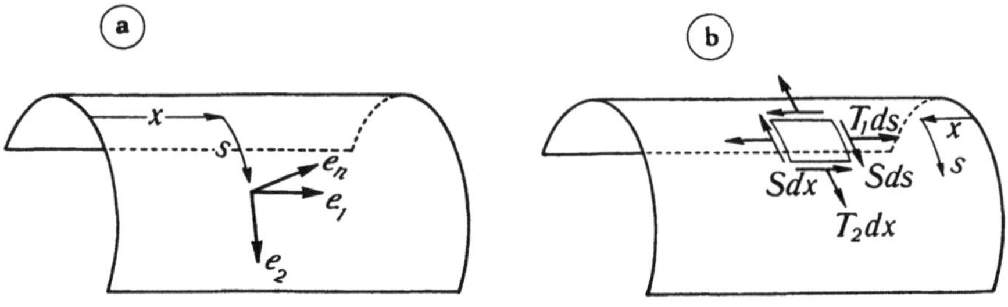

Fig. 44.

is the distance of the point from some definite plane perpendicular to the axis of the cylinder and s is the length of the arc in the circumferential direction measured from some definite generator (Fig. 44a). The coordinate lines of this system are the generators and the contours. Both Lamé parameters will be equal to unity, while one of the two principal radii of curvature, R_1, is equal to infinity and the other, $R_2 = r$, is a function of the arc length s only.

Using the general equations of the membrane theory, these results lead to the following differential equations for cylindrical shells of arbitrary shape

$$\frac{\partial T_1^*}{\partial x} + \frac{\partial S^*}{\partial s} + q_1 = 0,$$

$$\frac{\partial S^*}{\partial x} + \frac{\partial T_2^*}{\partial s} + q_2 = 0, \tag{36.1}$$

$$T_2^* = q_n r,$$

where q_1, q_2, q_n are the components of the surface loading in the directions x, s and the normal to the middle surface.

As can be seen from (36.1), the force T_2^* in this case can be expressed directly in terms of the surface loading. Substituting this expression into the second equation (36.1) and integrating with respect to x, one finds

$$S^* = S^0(s) - \int_{x_0}^{x} \left(\frac{\partial q_n r}{\partial s} + q_2 \right) dx, \tag{36.2}$$

where $S^0(s)$ is an arbitrary function of s. The lower limit of integration x_0 can likewise be selected arbitrarily.

Next, substituting (36.2) into the first equation (36.1) and integrating again with respect to x, one obtains

$$T_1^* = T_1^{(1)}(s) - (x - x_1) \frac{dS^0}{ds} + \int_{x_1}^{x} \left\{ \int_{x_0}^{x} \left(\frac{\partial^2 q_n r}{\partial s^2} + \frac{\partial q_2}{\partial s} \right) dx - q_1 \right\} dx, \tag{36.3}$$

where $T_1^{(1)}(s)$ is another arbitrary function of s, and x_1 is an arbitrarily chosen value of x. The third formula (36.1), (36.2) and (36.3) represent the general solution of the equations of the membrane theory of cylindrical shells. This solution involves two arbitrary functions which depend only on s. They have a simple significance which will be explained now by letting $x = x_0$ in (36.2) and $x = x_1$ in (36.3). Thus, one finds

$$S^*(x_0, s) = S^0(s), \quad T_1^*(x_1, s) = T_1^{(1)}(s), \tag{36.4}$$

whence it is clear that this arbitrariness in the general solutions of the equations of the membrane theory of cylindrical shells yields the possibility of prescribing laws of change for S^* and T_1^* on arbitrarily selected circles $x = x_0$ and $x = x_1$, respectively. In a particular case, x_0 and x_1 may coincide, i.e., S^* and T_1^* may be specified on the same circle.

Consequently, if a cylindrical shell is closed and bounded by two sections $x = x_0$, $x = x_1$, which are perpendicular to it, then the above solution makes it possible to prescribe either both shear forces on one of its two edges or the force T_1^* on one, S^* on the other edge. Side by side with S^* at the section x_0, one may give also $T_1^* = T_1^{(0)}(s)$, since (36.3) may be subjected to this condition for a corresponding choice of the arbitrary function S^0. It must be obtained from

$$\frac{dS^0}{dx} = \frac{1}{x_0 - x_1}\left\{T_1^{(1)}(s) - T_1^{(0)}(s) + \right.$$
$$\left. + \int_{x_1}^{x_0}\left[\int_{x_0}^{x}\left(\frac{\partial^2 q_n r}{\partial s^2} + \frac{\partial q_2}{\partial s}\right)dx - q_1\right]dx\right\}, \qquad (36.5)$$

determining it apart from an arbitrary constant which must be specified additionally, in order that the solution will be completely specified.

In the membrane analysis of tubes of arbitrary cross-section, the boundary conditions for the forces can be formulated in the same manner as in the problem of the analysis of shells of revolution of positive Gaussian curvature bounded by two parallel circles.

Next, consider an open cylindrical shell bounded by two generators $s = s_1$, $s = s_2$ and two circles $x = x_1$, $x = x_2$. It is readily seen that the membrane theory does not yield any freedom of choice for the tangential forces on the straight sections of the boundary $s = s_1$, $s = s_2$, since the general solution of its equations does not contain arbitrary functions in x. As regards the curvilinear parts of the boundary $x = x_1$, $x = x_2$, the boundary conditions appear to be the same as in the problem of the tube. Consequently, one finds that prescription of two tangential forces in the curvilinear sections of a curvilinear panel determines completely its membrane state of stress, including the values of both tangential forces on the straight sections of the boundary.

Eliminating from (36.1) consecutively the forces T_2^* and S^*, one arrives at the equation for T_1^*

$$\frac{\partial^2 T_1^*}{\partial x^2} = \frac{\partial^2 (q_n r)}{\partial s^2} + \frac{\partial q_2}{\partial s} - \frac{\partial q_1}{\partial x}. \qquad (36.6)$$

This is an equation of the parabolic type which determines the special features of the formulation of the boundary conditions for membrane problems of cylindrical shells.

§ 36a. The determination of membrane displacements

Applying the general formulae (4.23) to cylindrical shells, i.e., letting in these formulae $A_1 = A_2 = 1$, $\alpha_1 = x$, $\alpha_2 = s$, $R_1 = \infty$, $R_2 = r(s)$, one obtains

$$\varepsilon_1 = \frac{\partial u}{\partial x}, \quad \varepsilon_2 = \frac{\partial v}{\partial s} + \frac{w}{r}, \quad \omega = \frac{\partial u}{\partial s} + \frac{\partial v}{\partial x}, \tag{36.1a}$$

$$\kappa_1 = -\frac{\partial^2 w}{\partial x^2}, \quad \kappa_2 = -\frac{\partial}{\partial s}\left(\frac{\partial w}{\partial s} - \frac{v}{r}\right), \quad \tau = -\frac{\partial}{\partial x}\left(\frac{\partial w}{\partial s} - \frac{v}{r}\right). \tag{36.2a}$$

Next, using the formulae relating the deformations to the stresses, one can write

$$\begin{aligned} \frac{\partial u}{\partial x} &= \frac{1}{E\delta}(T_1^* - \mu T_2^*), \\ \frac{\partial v}{\partial s} + \frac{w}{r} &= \frac{1}{E\delta}(T_2^* - \mu T_1^*), \\ \frac{\partial u}{\partial s} + \frac{\partial v}{\partial x} &= \frac{2(1+\mu)}{E\delta} S^*. \end{aligned} \tag{36.3a}$$

Substituting on the right-hand sides of these formulae the general expressions for the membrane forces in a cylindrical shell, one finds

$$\begin{aligned} \frac{\partial u}{\partial x} &= t(s) - x\frac{d\eta}{ds} + F_1(x, s), \\ \frac{\partial v}{\partial s} + \frac{w}{r} &= -\mu\left[t(s) - x\frac{d\eta}{ds}\right] + F_2(x, s), \\ \frac{\partial u}{\partial s} + \frac{\partial v}{\partial x} &= 2(1+\mu)\eta(s) + F_3(x, s), \end{aligned} \tag{36.4a}$$

where $t(s)$, $\eta(s)$ are arbitrary functions related to $T_1^{(1)}(s)$ and $S^0(s)$ by the equalities

$$t(s) = \frac{1}{E\delta}\left[T_1^{(1)}(s) + x_1\frac{dS^0}{ds}\right], \quad \eta(s) = \frac{S^0(s)}{E\delta} \tag{36.5a}$$

and F_1, F_2, F_3 are known functions of the coordinates, expressible in terms of the components of the surface loading:

$$F_1(x, s) = \frac{1}{E\delta}\left\{-\mu q_n r + \int_{x_1}^{x}\left[\int_{x_0}^{x}\left(\frac{\partial^2 q_n r}{\partial s^2} + \frac{\partial q_2}{\partial s}\right)dx - q_1\right]dx\right\},$$

$$F_2(x, s) = \frac{1}{E\delta}\left\{q_n r - \mu \int_{x_1}^{x} \left[\int_{x_0}^{x} \left(\frac{\partial^2 q_n r}{\partial s^2} + \frac{\partial q_2}{\partial s}\right) dx - q_1\right] dx\right\}, \quad (36.6a)$$

$$F_3(x, s) = -\frac{2(1+\mu)}{E\delta} \int_{x_0}^{x} \left(\frac{\partial q_n r}{\partial s} + q_2\right) dx.$$

Integrating the first equation (36.4a) with respect to x, one obtains

$$u = u'(s) + (x - x')t(s) - \tfrac{1}{2}(x^2 - x'^2)\frac{d\eta}{ds} + \int_{x'}^{x} F_1(x, s)\, dx, \quad (36.7a)$$

where $u'(s)$ is an arbitrary function of s and x' is an arbitrarily assigned value of x.

Substituting (36.7a) into the third equation (36.4a), one arrives at the equation involving only the unknown function v

$$\frac{\partial v}{\partial x} = -\frac{du'}{ds} - (x - x')\frac{dt}{ds} + \tfrac{1}{2}(x^2 - x'^2)\frac{d^2\eta}{ds^2} - \int_{x'}^{x} \frac{\partial F_1}{\partial s} dx + $$
$$+ 2(1+\mu)\eta + F_3. \quad (36.8a)$$

Integrating (36.8a) with respect to x yields for v the formula

$$v = v''(s) - (x - x'')\frac{du'}{ds} - \tfrac{1}{2}(x - x'')(x + x'' - 2x')\frac{dt}{ds} +$$
$$+ \int_{x''}^{x} \left[\tfrac{1}{2}(x^2 - x'^2)\frac{d^2\eta}{ds^2} - \int_{x'}^{x} \frac{\partial F_1}{\partial s} dx + 2(1+\mu)\eta + F_3\right] dx. \quad (36.9a)$$

Besides $t(s)$ and $\eta(s)$, it contains two arbitrary functions of s, namely, u' and v''. In order to explain the significance of these functions, let in (36.7a) and (36.9a) $x = x'$ and $x = x''$, respectively. Then one finds

$$u(x', s) = u'(s), \quad v(x'', s) = v''(s). \quad (36.10a)$$

Thus, it is seen that $u'(s)$ expresses the law of change of the displacements u on the section $x = x'$ and $v''(s)$ is the law of change of the displacement v on the section $x = x''$.

Altogether the formulae for the displacements (36.7a) and (36.9a) contain four arbitrary functions of the arc length, namely, u', v'', t, η for the evaluation of which one can prescribe arbitrarily both tangential displacement components u and v on two edges of the shell which coincide with circles. However, if two boundary conditions on the edges are formulated in terms of stresses (cf. § 36), then the presence of the arbitrary functions $u'(s)$, $v''(s)$ makes it possible to impose on these edges two further conditions for the displacements. On a different type of boundary, the membrane theory of cylindrical shells does not admit prescription of boundary conditions either in terms of forces or in terms of displacements, since its corresponding general expressions for the forces and displacements do not contain arbitrary functions depending on the coordinate x.

§ 36b. Deformation of cylindrical shells without extension or shearing of the middle surface

For cylindrical shells, the general equations (20.5) reduce to the form

$$\frac{\partial u^{**}}{\partial x} = 0, \quad \frac{\partial v^{**}}{\partial s} + \frac{w^{**}}{r} = 0, \quad \frac{\partial u^{**}}{\partial s} + \frac{\partial v^{**}}{\partial x} = 0. \tag{36.1b}$$

This system can be obtained from (36.3a) for $T_1^* = T_2^* = S^* = 0$.

The first equation (36.1b) yields

$$u^{**} = u_0(s). \tag{36.2b'}$$

Substituting this result into the third equation (36.1b) and integrating with respect to x, one obtains

$$v^{**} = v_0(s) - x\frac{du_0}{ds}. \tag{36.2b''}$$

The second equation now gives

$$w^{**} = -r\left[\frac{dv_0}{ds} - x\frac{d^2u_0}{ds^2}\right]. \tag{36.2b'''}$$

The formulae (36.2b) contain two arbitrary functions $u_0(s)$, $v_0(s)$ of the arc length s. These formulae are the most general expressions for the displacements in cylindrical shells which are not resisted by stretching or shearing of the middle surface. This type of displacement comprises, firstly, displacements of the shell as a rigid body and, secondly, displacements of a shell in pure bending.

Rigid body displacements of a shell must satisfy the system (36.1b) as well as the equations $\kappa_1 = \kappa_2 = \tau = 0$, which, using the general formulae (4.23) adapted for cylindrical shells, assume in this particular case the form

$$\frac{\partial^2 w}{\partial x^2} = 0, \quad \frac{\partial}{\partial s}\left(\frac{\partial w}{\partial s} - \frac{v}{r}\right) = 0, \quad \frac{\partial}{\partial x}\left(\frac{\partial w}{\partial s} - \frac{v}{r}\right) = 0. \tag{36.3b}$$

The last two equations can be replaced by the single equation

$$\frac{\partial w}{\partial s} - \frac{v}{r} = \text{const.} = C. \tag{36.4b}$$

It is readily seen that the displacements (36.2b) satisfy the first equation (36.3b) identically.

The equation (36.4b), on substituting (36.2b), yields the differential equations for u_0, v_0

$$\frac{d}{ds}\left(r\frac{d^2u_0}{ds^2}\right) + \frac{1}{r}\frac{du_0}{ds} = 0, \qquad r\frac{d}{ds}\left(r\frac{dv_0}{ds}\right) + v_0 = -Cr \tag{36.5b}$$

with the general integrals

$$u_0 = A_1y + A_2z + A_3, \quad v_0 = C_1\cos\varphi + C_2\sin\varphi - C\frac{d\Omega}{ds}, \tag{36.6b}$$

where y, z are the coordinates of points of the cross-sections of the cylinder in the Cartesian system yOz, φ is the angle between the normal at an arbitrary point on the cross-section of the cylinder and the normal at the origin of the arc s and

$$\Omega = \int_0^s (y\,dz - z\,dy)$$

is twice the area of a sector. The origin of the coordinate system yOz can be chosen arbitrarily. The z-axis passes through the origin of the arc s (Fig. 44c).

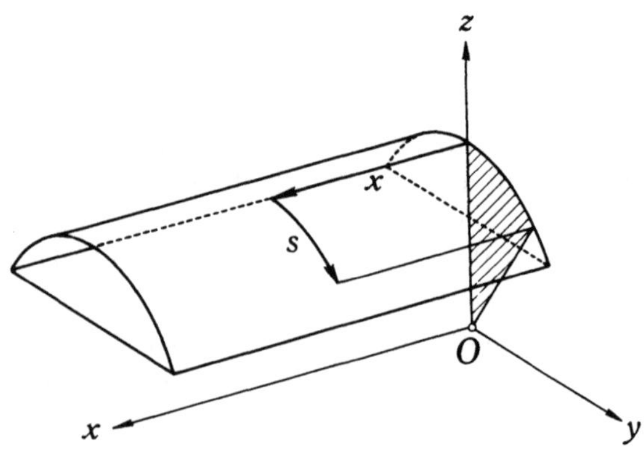

Fig. 44c

Substituting (36.6b) into (36.2b), one obtains the displacements of the middle surface of a cylindrical shell which correspond to rigid body motion in space. For any other choice of $u_0(s)$, $v_0(s)$, the formulae (36.2b) correspond to pure bending. It follows from these formulae that, in order to eliminate the possibility of pure bending, and also of rigid body displacements, it is sufficient to let

$$u = v = 0 \tag{36.7b}$$

for one of the cross-sections of the shell $x = x_0$.

In fact, in order to fulfill this condition, one has to set in (36.2b) $u_0(s) = v_0(s) = 0$, whence

$$u^{**} = v^{**} = w^{**} = 0. \tag{36.8b}$$

In order to eliminate pure bending, one may likewise clamp the shell edge on two circles $x = x_1$, $x = x_2$ against displacements v, since such a boundary condition leads to

$$v_0(s) - x_1 \frac{du_0}{ds} = 0, \qquad v_0(s) - x_2 \frac{du_0}{ds} = 0, \tag{36.9b}$$

whence

$$v_0 = 0, \quad u_0 = \text{const.} = C. \tag{36.10b}$$

Substituting these values for u_0 and v_0 into (36.2b), one arrives at the displacements

$$u^{**} = C, \quad v^{**} = w^{**} = 0. \tag{36.11b}$$

It is seen from (36.6b) that these displacements correspond to rigid body motion of the shell, in fact, to translation in the x-direction. Thus, these boundary conditions, while they ensure absence of pure bending of the shell, do not eliminate the possibility of rigid body motion.

However, it should be noted that for cylindrical shells of sufficiently large length the above boundary conditions may not prevent the occurrence of pure bending type of strains at large distances from the clamped edge (or edges) of the shell. This phenomenon will be studied in the next section.

§ 37. Determination of the stresses in tubes with arbitrary cross-section, loaded by uniform normal pressure

Using membrane theory, a complete cylindrical tube of arbitrary cross-section, loaded by uniform normal pressure p will now be studied. The ends of the shell will be two cross-sections perpendicular to the axis of the shell. The coordinate x (§ 36) will be measured from one of the end-sections. Then the other end-section of the tube will be determined by the coordinate L (the length of the tube). The radius of curvature of the cross-section of the shell will be assumed to be a continuous and at least four times differentiable function of the arc coordinate s.

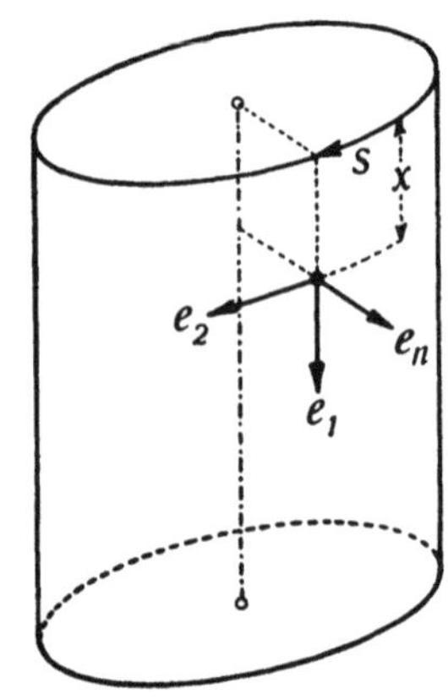

Fig. 45.

It will be required to solve the problem for two types of boundary conditions:

1) for $x = 0$, $x = L$: $T_1 = 0$, $v = 0$;

2) for $x = 0$, $x = L$: $u = 0$, $v = 0$.

The first type is statically determinate, while the second which corresponds to rigid fixing of the edges of the shell is statically indeterminate.

For the present problem the differential equations (36.1) take the form

$$\frac{\partial T_1^*}{\partial x} + \frac{\partial S^*}{\partial s} = 0, \quad \frac{\partial S^*}{\partial x} + \frac{\partial T_2^*}{\partial s} = 0, \quad T_2^* = pr. \tag{37.1}$$

The third of these equations directly determines the hoop stress in the tube

$$T_2^* = pr. \tag{37.2}$$

Substituting this value for T_2^* into the second equation (37.1) and integrating it with respect to x, one obtains

$$S^* = -\, px \frac{dr}{ds} + f_1(s), \tag{37.3}$$

where $f_1(s)$ is an arbitrary function of the arc length s. This function may be determined from the condition that at the centre section of the tube the shear stresses must be equal to zero, as a result of the symmetry of the above boundary conditions.

On this basis, one obtains the equation

$$S^*_{x=\frac{L}{2}} = 0 = -\, p \frac{L}{2} \frac{dr}{ds} + f_1(s), \tag{37.4}$$

whence

$$S^* = p\left(\frac{L}{2} - x\right)\frac{dr}{ds}. \tag{37.5}$$

Substituting this value for S^* into the first equation (37.1) and integrating again with respect to x, one will have

$$T_1{}^* = p\left[x\frac{x-L}{2}\frac{d^2r}{ds^2} + f_2(s)\right]. \tag{37.6}$$

From here on, the methods of solution of the above problems will differ, their course depending on the differences in the boundary conditions.

1. *The statically determinate type of boundary conditions.*

In this case, the force $T_1{}^*$ at $x = 0$ and $x = L$ will be equal to zero. Hence, it is found that in (37.6) $f_2(s) = 0$ and therefore

$$T_1{}^* = px\frac{x-L}{2}\frac{d^2r}{ds^2}. \tag{37.7}$$

The formulae (37.2), (37.5), (37.7) completely determine all the forces in the tube.

Now one may write down for the displacements u, v, w the system of differential equations

$$\begin{aligned}
\varepsilon_1 &= \frac{\partial u}{\partial x} = \frac{1}{E\delta}(T_1{}^* - \mu T_2{}^*) = \frac{p}{E\delta}\left[-\mu r + \frac{x(x-L)}{2}\frac{d^2r}{ds^2}\right],\\
\varepsilon_2 &= \frac{\partial v}{\partial s} + \frac{w}{r} = \frac{1}{E\delta}(T_2{}^* - \mu T_1{}^*) = \frac{p}{E\delta}\left[r - \mu\frac{x(x-L)}{2}\frac{d^2r}{ds^2}\right],\\
\omega &= \frac{\partial u}{\partial s} + \frac{\partial v}{\partial x} = \frac{2(1+\mu)}{E\delta}S^* = \frac{2(1+\mu)p}{E\delta}\left(\frac{1}{2}L - x\right)\frac{dr}{ds}.
\end{aligned} \tag{37.8}$$

Integrating the first of these equations with respect to x, one finds

$$u = -\frac{px}{E\delta}\left[\mu r - \frac{x}{2}\left(\frac{1}{3}x - \frac{1}{2}L\right)\frac{d^2r}{ds^2}\right] + f_3(s). \tag{37.9}$$

In order to evaluate the arbitrary function $f_3(s)$, use will be made of the fact that, by the symmetry of the boundary conditions, the displacement u must vanish at the section $x = \frac{L}{2}$. Hence

$$f_3(s) = -\frac{pL}{2E\delta}\left(\mu r + \frac{1}{12}L^2\frac{d^2r}{ds^2}\right) \tag{37.10}$$

and, consequently,

$$u = \mu \frac{pr}{E\delta}\left(\frac{1}{2} L - x\right) + \frac{1}{6} \frac{p}{E\delta}\left(x^3 - \frac{3}{2} Lx^2 + \frac{1}{4} L^3\right)\frac{d^2r}{ds^2}. \qquad (37.11)$$

Substituting (37.11) into the third equation (37.8) and integrating with respect to x, one will have

$$v = \frac{(2+\mu)}{2} \frac{px(L-x)}{E\delta} \frac{dr}{ds} - \frac{px}{24E\delta}(x^3-2Lx^2+L^3)\frac{d^3r}{ds^3} + f_4(s). \qquad (37.12)$$

In order to make this expression satisfy the boundary conditions $v = 0$ at $x = 0$ and $x = L$, it is sufficient to let $f_4(s) = 0$. Substituting the expression thus obtained for v into the second equation (37.8), one finds for the normal deflection (which is generally of greatest interest)

$$w = \frac{pr}{E\delta}\left\{r + x(x - L)\frac{d^2r}{ds^2} + \frac{x}{24}(x^3 - 2Lx^2 + L^3)\frac{d^4r}{ds^4}\right\}. \qquad (37.13)$$

2. *The statically indeterminate type of boundary conditions*

In this case, the arbitrary function $f_2(s)$ in (37.6) may be determined only after obtaining the displacement u.

The first equation (37.8) with $f_2(s) \neq 0$ takes the form

$$\frac{\partial u}{\partial x} = \frac{p}{E\delta}\left[-\mu r + \frac{x(x - L)}{2} \frac{d^2r}{ds^2} + f_2(s)\right]. \qquad (37.14)$$

Integrating this equation with respect to x, one finds

$$u = \frac{px}{E\delta}\left[-\mu r + \frac{x}{2}\left(\frac{x}{3} - \frac{L}{2}\right)\frac{d^2r}{ds^2} + f_2(s)\right] + f_3(s). \qquad (37.15)$$

The two arbitrary functions $f_2(s)$, $f_3(s)$ in this formula will be determined by the conditions that $u = 0$ for $x = 0$ and $x = L$, which lead to

$$f_3(s) = 0,$$
$$f_2(s) = \mu r + \frac{1}{12} L^2 \frac{d^2r}{ds^2}. \qquad (37.16)$$

Hence one obtains for u

$$u = \frac{px}{12E\delta}(2x^2 - 3Lx + L^2)\frac{d^2r}{ds^2}. \tag{37.17}$$

Now, $f_2(s)$ having been found, the force $T_1{}^*$ is completely determined by (37.6)

$$T_1{}^* = \frac{p}{2}\left(x^2 - Lx + \frac{1}{6}L^2\right)\frac{d^2r}{ds^2} + \mu pr. \tag{37.18}$$

Substituting (37.17) into the third equation (37.8) and integrating with respect to x, one finds

$$v = -\frac{px^2}{24E\delta}(x-L)^2\frac{d^3r}{ds^3} - \frac{(1+\mu)px}{E\delta}(x-L)\frac{dr}{ds} + f_4(s). \tag{37.19}$$

In this expression, the function $f_4(s)$, on the basis of the boundary conditions $v = 0$ for $x = 0$, $x = L$, must be set equal to zero.

Now one may obtain from the formula

$$w = \frac{r}{E\delta}(T_2{}^* - \mu T_1{}^*) - r\frac{\partial v}{\partial s} \tag{37.20}$$

for the normal displacement w

$$w = \frac{pr}{E\delta}\left\{r + \frac{2+\mu}{2}x(x-L)\frac{d^2r}{ds^2} + \frac{1}{24}x^2(x-L)^2\frac{d^4r}{ds^4}\right\}. \tag{37.21}$$

All the forces and displacements have now been determined for the second (statically indeterminate) type of problem under consideration.

The above results will now be applied to the particular case of a tube with an elliptical cross-section.

For an ellipse,

$$r = \frac{b^2}{a}\frac{1}{(1-\varepsilon^2\cos^2\varphi)^{3/2}}, \tag{37.22}$$

where $2a$ and $2b$ are the major and minor axes of the ellipse, ε — is the eccentricity $\left[\varepsilon^2 = 1 - \left(\frac{b}{a}\right)^2\right]$ and φ is the angle between the normal to the ellipse and its minor axis (Fig. 46).

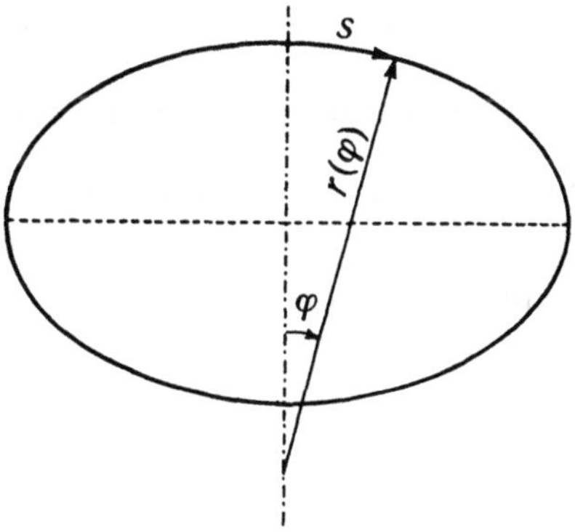

Fig. 46.

This angle is connected with the arc length s by the differential relation

$$ds = r\,d\varphi, \tag{37.23}$$

which permits one to replace differentiation with respect to the arc s by differentiation with respect to φ; this step is required because (37.22) gives r as a function of φ.

In applying the above formulae for the forces and deflections one has to know the first four derivatives of r with respect to s. On the basis of (37.22), (37.23), one finds

$$\frac{dr}{ds} = \frac{1}{r}\frac{dr}{d\varphi} = -\frac{3}{2}\varepsilon^2\frac{\sin 2\varphi}{1-\varepsilon^2\cos^2\varphi},$$

$$\frac{d^2r}{ds^2} = \frac{1}{r}\frac{d}{d\varphi}\left(\frac{dr}{ds}\right) = -\frac{3a}{b^2}\varepsilon^2\frac{\cos 2\varphi - \varepsilon^2\cos^2\varphi}{\sqrt{1-\varepsilon^2\cos^2\varphi}},$$

$$\frac{d^3r}{ds^3} = \frac{1}{r}\frac{d}{d\varphi}\left(\frac{d^2r}{ds^2}\right) = \frac{6a^2}{b^4}\varepsilon^2[1-\tfrac{3}{4}\varepsilon^2-\tfrac{1}{2}\varepsilon^2(1-\tfrac{1}{2}\varepsilon^2)\cos^2\varphi]\sin 2\varphi, \tag{37.24}$$

$$\frac{d^4r}{ds^4} = \frac{1}{r}\frac{d}{d\varphi}\left(\frac{d^3r}{ds^3}\right) = \frac{12a^3}{b^6}\varepsilon^2(1-\varepsilon^2\cos^2\varphi)^{3/2}[(1-\tfrac{3}{4}\varepsilon^2)\cos 2\varphi -$$
$$-\tfrac{1}{2}\varepsilon^2(1-\tfrac{1}{2}\varepsilon^2)(\cos^2\varphi - 3\sin^2\varphi)\cos^2\varphi].$$

With these expressions available, one may determine the maximum values of the stresses and deflections in elliptical tubes subject to uniform normal pressure. Consideration will here be limited to the statically determinate version of the problem.

By (37.2) the direct stress induced by the force $T_2{}^*$ reaches its maximum at the point where r has its maximum value, i.e., for the ellipse at the end of the minor axis. By (37.22), this maximum will be equal to

$$|\sigma_2|_{\max} = \frac{\max(T_2{}^*)}{\delta} = \frac{pa^2}{b\delta}. \tag{37.25}$$

By (37.5), the shearing stress will attain its maximum value at the ends of the shell for values of φ which correspond to the maxima of $\dfrac{dr}{ds}$. Equating $\dfrac{d^2r}{ds^2}$ to zero, one obtains for this angle the equation

$$\cos 2\varphi - \varepsilon^2\cos^2\varphi = 0, \tag{37.26}$$

whence

$$\cos\varphi = \frac{1}{\sqrt{2-\varepsilon^2}}, \quad \sin\varphi = \frac{\sqrt{1-\varepsilon^2}}{\sqrt{2-\varepsilon^2}}. \tag{37.27}$$

Introducing these values for $\sin\varphi$ and $\cos\varphi$ into the first formula (37.24) and substituting it thereafter into (37.5), one finds

$$\tau_{\max} = \frac{\max S^*}{\delta} = \frac{3}{4}\frac{pL}{\delta}\left(\frac{a}{b} - \frac{b}{a}\right). \tag{37.28}$$

By (37.7), the direct stress, due to the force T_1^*, will reach its maximum value at the centre section $\left(\text{for } x = \frac{L}{2}\right)$ at the end points of the major and minor axes of the ellipse. The values there will be

$$\begin{aligned} \sigma_1' &= \frac{T_1^*}{\delta} = \frac{3}{8}\frac{pL^2}{b\delta}\varepsilon^2 \quad \text{(at the end of the minor axis)}, \\ \sigma_1'' &= \frac{T_1^*}{\delta} = -\frac{3}{8}\frac{paL^2}{b^2\delta}\varepsilon^2 \quad \text{(at the end of the major axis)}. \end{aligned} \tag{37.29}$$

It should be noted that the stresses (37.29) must be augmented (or reduced) by the stresses due to axial tension (or compression) of the shell (if such should be present). As regards the displacements in the shell, only the values of the normal displacements at the centre section $\left(x = \frac{L}{2}\right)$ at the ends of the major and minor axes of the ellipse $\left(\varphi = 0,\ \varphi = \frac{\pi}{2}\right)$ will be given. In fact, at the end of the minor axis,

$$w_b = \frac{pa^4}{E\delta b^2}\left[1 + \frac{3}{4}\frac{L^2}{a^2}\varepsilon^2 + \frac{5}{32}\frac{L^4}{a^4}\varepsilon^2\left(1 - \frac{\varepsilon^2}{4}\right)\right], \tag{37.30}$$

while at the end of the major axis

$$w_a = -\frac{pa^4}{E\delta b^2}\left[-\frac{b^6}{a^6} + \frac{3}{4}\frac{L^2b^2}{a^4}\varepsilon^2 + \frac{5}{32}\frac{L^4}{a^2b^2}\varepsilon^2\left(1 - \frac{3}{4}\varepsilon^2\right)\right]. \tag{37.31}$$

In conclusion the reader's attention will be drawn to the fact that all the formulae of this section, giving the displacements and stresses in a tube subject to uniform normal pressure, will become infinite as the length of the tube tends to infinity (the only exception being the particular case when the tube has a circular plan form). Hence it is clear that the above formulae do not apply to tubes of sufficiently large length. For the analysis of long tubes, one has to take into consideration their bending stiffness in the transverse direction, which, in principle, determines also the strains.

The results, obtained above for a particular loading and for a particular shape of shell has general validity for the theory of cylindrical shells. In fact, the applicability of the membrane theory, as a rule, depends on the length of a shell. As soon as a cylindrical shell is sufficiently long, no formulation of the boundary conditions at its ends will ensure the proper boundedness of the pure bending displacements in the middle portion of the shell (unless the cross-section of the shell has been specially selected for a given type of loading to exclude the dependence of the transverse bending of the shell on its length. In particular, for uniform normal pressure, such a special cross-section is the circle).

§ 38. The membrane theory of cylindrical vaults

During the era of brick and stone constructions, the roofing of rectangular areas was often achieved by cylindrical vaults which, for the purpose of stress analysis, can be treated as arches. In order to ensure the proper functioning of the roof, one has to give it solid foundations along the straight edges, while the curvilinear edges need not be supported.

Such an approach practically indicates the need for supporting the straight edges of a vault on continuous walls which will divide the interior space of the building into a number of separate rooms (in the case that several vaults are involved). However, in many cases, such a subdivision of the interior of a building will be undesirable, and in some

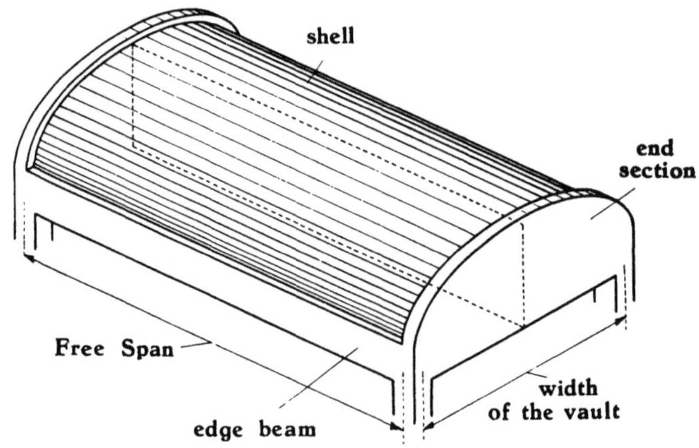

Fig. 47.

cases unacceptable. Thus, for example, the usefulness of factory premises, large stores or markets, hangars and garages, as a rule, increases with their size. In such cases, stone or brick vaults are still sometimes being used for roofs.

The progress in reinforced concrete technology has led to quite a different problem related to the cylindrical roof. The contemporary roof has small thickness and may be considered to be a shell, mainly supported along its curvilinear edges.

The advantage of transmitting the load on the roof along a curvilinear edge arises from the fact that the supports in this case have the form of arches and that they may be made sufficiently strong without their having to be solid walls. Further, it turns out that the distance between such supports may be made very large, several times the distance between straight supports of a vault. All this points towards the use of a number of adjoining cylindrical shells for the purpose of roofing large spaces with a minimum obstruction by supports. These supports then take the form of pillars with acceptably large spacing between them.

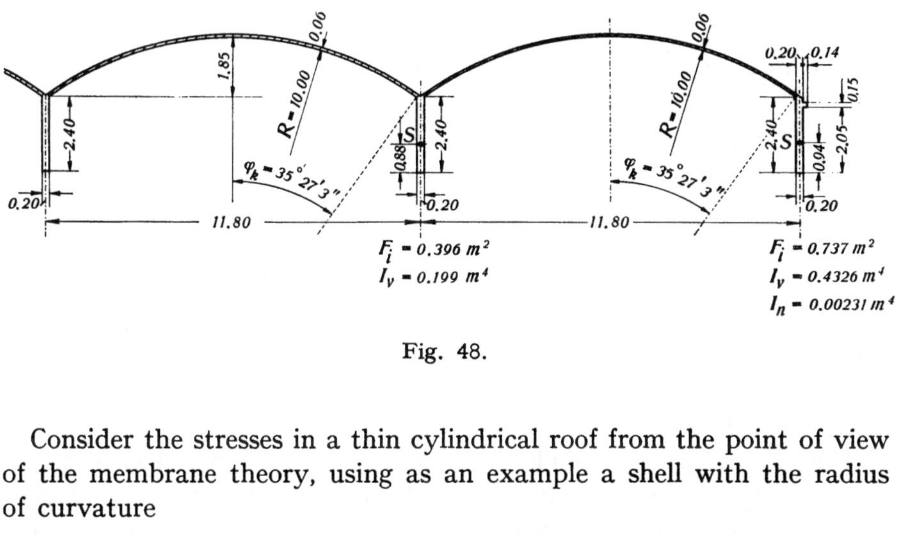

Fig. 48.

Consider the stresses in a thin cylindrical roof from the point of view of the membrane theory, using as an example a shell with the radius of curvature

$$r = \frac{r_0}{(1 + \gamma \sin^2 \varphi)^{3/2}}. \tag{38.1}$$

In this case, the contour of the cylinder will be a second order curve: for $\gamma = 0$, a circular, for $\gamma = -1$, a parabolic, for $\gamma > 1$, an elliptic, for $\gamma < -1$, a hyperbolic arc. Hence, using (38.1), one may study a family of roofs varying greatly in shape, and arrive at a number of results of a general nature.

Denoting the weight of the roof per unit area of its middle surface by q, the equations (36.1) applicable to the present problem may be written

$$\frac{\partial T_1^*}{\partial x} + \frac{1}{r}\,\frac{\partial S^*}{\partial \varphi} = 0,$$

$$\frac{\partial S^*}{\partial x} + \frac{1}{r}\,\frac{\partial T_2^*}{\partial \varphi} = -\,q \sin \varphi, \qquad (38.2)$$

$$T_2^* = -\,qr \cos \varphi,$$

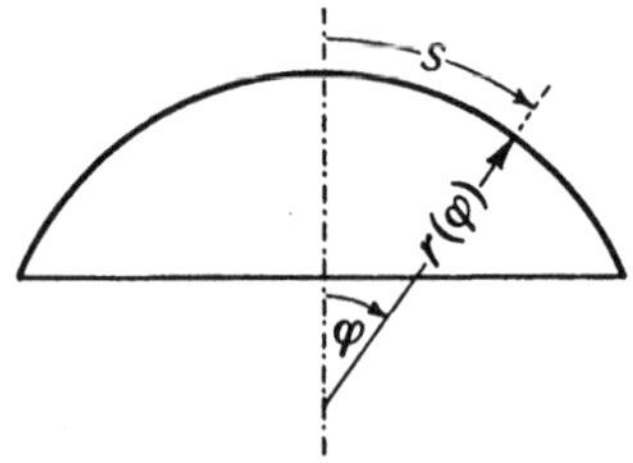

Fig. 49.

where φ is the angle between the normal to the roof and the vertical, related to the arc coordinate s by $r\,d\varphi = ds$ (Fig. 49) and r is given by (38.1).

Carrying out the integration of the system (38.2) with respect to x in the same manner as in the preceding section, one obtains

$$T_2^* = -\,qr_0 \frac{\cos \varphi}{(1 + \gamma \sin^2 \varphi)^{3/2}},$$

$$S^* = f_1(\varphi) - qx \frac{2 + \gamma(2 + \cos^2 \varphi)}{1 + \gamma \sin^2 \varphi} \sin \varphi,$$

$$\begin{aligned} T_1^* = f_2(\varphi) &- \frac{x}{r_0}(1 + \gamma \sin^2 \varphi)^{3/2} \frac{df_1}{d\varphi} - \\ &- \frac{qx^2}{2r_0}(1 + \gamma \sin^2 \varphi)^{3/2}\left[1 - 3(1+\gamma)\frac{1 - \gamma \sin^2 \varphi}{(1 + \gamma \sin^2 \varphi)^2}\right] \cos \varphi. \end{aligned} \qquad (38.3)$$

The two arbitrary functions $f_1(\varphi)$ and $f_2(\varphi)$ in these formulae will be determined by the condition that at the curvilinear boundaries, i.e., for $x = -\,l/2$, $x = +\,l/2$ (taking the origin for x the centre of the span), the force $T_1^* = 0$.

This condition corresponds to the case where the roof has no intermediate supports between its ends and the edge supports are analogous to those of a beam, freely supported at its ends. The more complicated, statically indeterminate case when, in addition to the terminal supports, there are intermediate supports and the roof is analogous to a beam with several supports will not be considered here.

Subjecting the last of the formulae (38.3) to the above boundary condition, one finds

$$\frac{df_1}{d\varphi} = 0,$$

$$f_2 = \frac{ql^2}{8r_0}(1 + \gamma \sin^2 \varphi)^{3/2}\left[1 - 3(1+\gamma)\frac{1 - \gamma \sin^2 \varphi}{(1 + \gamma \sin^2 \varphi)^2}\right] \cos \varphi. \qquad (38.4)$$

By the first of these equalities,

$$f_1 = C = \text{constant}.$$

Substituting this value of f_1 into the second of the expressions (38.3), it is seen that the constant C must be set equal to zero, since, in view of the symmetry of the loading about the plane $\varphi = 0$, the shear stress S^* must be an odd function of φ.

As a result, the formulae (38.3) assume the following final form:

$$T_2^* = -qr_0 \frac{\cos\varphi}{(1+\gamma\sin^2\varphi)^{3/2}},$$

$$S^* = -qx \frac{2+\gamma(2+\cos^2\varphi)}{1+\gamma\sin^2\varphi}\sin\varphi = \frac{2x}{l} S^*\left(\frac{l}{2}, \varphi\right), \tag{38.5}$$

$$T_1^* = -\frac{q(l^2-4x^2)}{8r_0}(1+\gamma\sin^2\varphi)^{3/2}\left[1-3(1+\gamma)\frac{1-\gamma\sin^2\varphi}{(1+\gamma\sin^2\varphi)^2}\right]\cos\varphi = \left(1-\frac{4x^2}{l^2}\right)T_1^*(0,\varphi),$$

where

$$S^*\left(\frac{l}{2}, \varphi\right) = -\frac{ql}{2}\frac{2+\gamma(2+\cos^2\varphi)}{1+\gamma\sin^2\varphi}\sin\varphi \tag{38.6}$$

is the value of S^* in the plane of the support $x = \frac{l}{2}$ and

$$T_1^*(0,\varphi) = \frac{ql^2}{8r_0}(1+\gamma\sin^2\varphi)^{3/2}\left[1-3(1+\gamma)\frac{1-\gamma\sin^2\varphi}{(1+\gamma\sin^2\varphi)^2}\right]\cos\varphi \tag{38.7}$$

is the value of the force T_1^* in the plane $x = 0$, i.e., at the centre section. Both these functions characterize the variation of the forces S^* and T_1^* across the width of the roof.

As regards the law of change of the forces along the roof, T_2^* remains constant, by (38.5), while S^* varies linearly and T_1^* parabolically. Thus, the changes of the forces S^* and T_1^* along the length of the roof turn out to be analogous to the laws of variation of the shear and bending stresses along the length of a freely supported and uniformly loaded beam on two supports.

Consider next the displacements of the roof, assuming that $\mu = 0$ (as is very often done for calculations referring to reinforced concrete structures). For this purpose, one may in the present case write the differential equations

$$\frac{\partial u}{\partial x} = \frac{T_1^*}{E\delta} = \frac{1}{E\delta}\left(1 - 4\frac{x^2}{l^2}\right) T_1^*(0, \varphi),$$
$$\frac{\partial v}{\partial \varphi} + w = \frac{r}{E\delta} T_2^*(\varphi), \tag{38.8}$$
$$\frac{1}{r}\frac{\partial u}{\partial \varphi} + \frac{\partial v}{\partial x} = \frac{4}{E\delta}\frac{x}{l} S^*\left(\frac{l}{2}, \varphi\right).$$

Integrating the first of these equations with respect to x, one finds

$$u = f_3(\varphi) + \frac{x}{E\delta}\left(1 - \frac{4}{3}\frac{x^2}{l^2}\right) T_1^*(0, \varphi). \tag{38.9}$$

But, in view of the symmetry of the supports of the roof at the curvilinear ends, the displacements u for $x = +\frac{l}{2}$ and $x = -\frac{l}{2}$ must be equal in magnitude and opposite in sign. Hence

$$u = \frac{x}{E\delta}\left(1 - \frac{4}{3}\frac{x^2}{l^2}\right) T_1^*(0, \varphi). \tag{38.10}$$

Now, the last of the equations (38.8), after integration with respect to x, gives

$$v = f_4(\varphi) + \frac{2x^2}{E\delta l} S^*\left(\frac{l}{2}, \varphi\right) - \frac{x^2}{2E\delta}\left(1 - \frac{2}{3}\frac{x^2}{l^2}\right)\frac{1}{r}\frac{dT_1^*(0, \varphi)}{d\varphi}. \tag{38.11}$$

Subjecting this displacement to the condition that it must vanish at the curvilinear edges, one has

$$f_4(\varphi) = \frac{5}{48}\frac{l^2}{E\delta}\frac{1}{r}\frac{dT_1^*(0, \varphi)}{d\varphi} - \frac{l}{2E\delta} S^*\left(\frac{l}{2}, \varphi\right), \tag{38.12}$$

and therefore

$$v = \frac{2}{E\delta l}\left(x^2 - \frac{l^2}{4}\right) S^*\left(\frac{l}{2}, \varphi\right) + $$
$$+ \frac{1}{3E\delta}\left(x^2 - \frac{l^2}{4}\right)\left(\frac{x^2}{l^2} - \frac{5}{4}\right)\frac{1}{r}\frac{dT_1^*(0, \varphi)}{d\varphi}. \tag{38.13}$$

Substituting this result into the second equation (38.8), one finds for the radial displacement

$$w = \frac{1}{E\delta}\left\{ rT_2^*(\varphi) - \left(x^2 - \frac{l^2}{4}\right)\left[\frac{2}{l}\frac{dS^*\left(\frac{l}{2}, \varphi\right)}{d\varphi} + \right.\right.$$
$$\left.\left. + \frac{1}{3}\left(\frac{x^2}{l^2} - \frac{5}{4}\right)\frac{1}{r}\frac{d^2T_1^*(0, \varphi)}{d\varphi^2}\right]\right\}. \tag{38.14}$$

The formulae (38.10), (38.13), (38.14) determine all the displacements of the cylindrical roof loaded by its own weight. The quantities $T_1^*(0, \varphi)$, $S^*\left(\frac{l}{2}, \varphi\right)$, $T_2^*(\varphi)$ in these expressions are determined by (38.6), (38.7) and (38.5). In view of the complexity of the above results, there is no purpose in giving explicit formulae for the displacements.

The obtained results will now be used to compare the four different roofs shown in Fig. 50 for which the height, length, width and weight per unit length are the same.

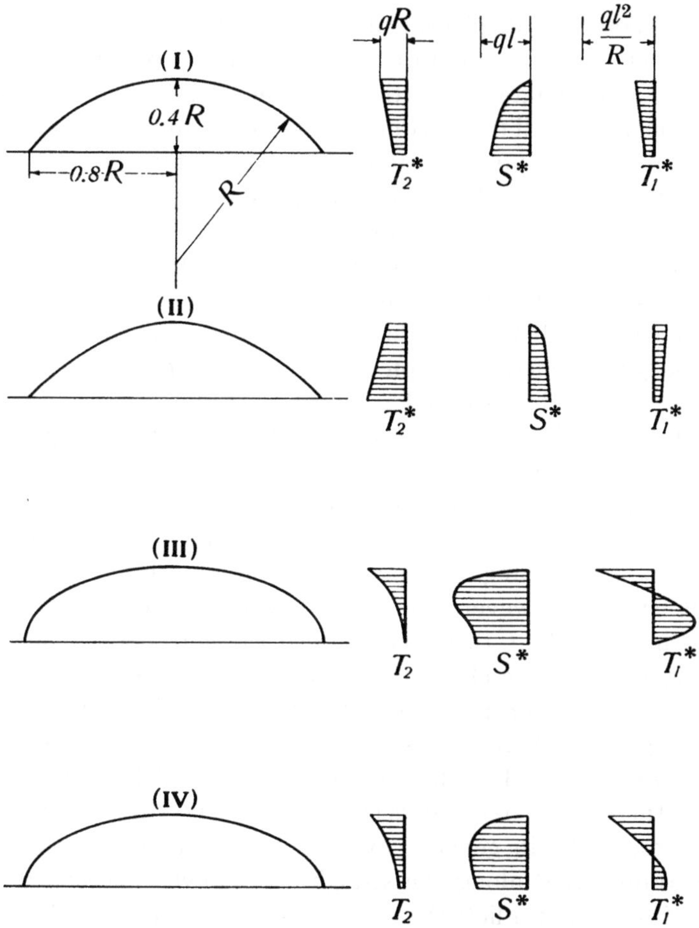

Fig. 50.

The cross-sections of these roofs will be the same curves which were investigated in § 26 as the meridian curves of domes. In fact, the section of roof (I) is an arc of a circle of radius R, roof (II) is an arc of a parabola, roof (III) is one half of the ellipse $\gamma = 3$, $r_0 = 1.60R$, roof IV is an arc of the ellipse $\gamma = 1.5$, $r_0 = 1.30R$. Fig. 50 shows to the same scale for all four shells the variations of the forces T_2^*, S^*, T_1^* with the height of each roof, calculated from the formulae obtained above. It is seen from these curves that normal and shear forces $T_2^*(x, \varphi_0)$, $S^*(x, \varphi_0)$ arise at the straight edges of roofs, subject to the conditions of the membrane state of stress. Both these forces must be absorbed by corresponding edge supports, so that along the straight boundaries there will be required edge girders which will be subject to bending and extension. For this purpose, the displacements of the edge beams under the loads acting on them must be equal to the displacements of the straight edges of the roof, as calculated from the formulae (38.10), (38.13), (38.14).

As it happens, not all these requirements can be fulfilled simultaneously and hence it will be impossible to design a roof which is in a membrane state of stress. In view of this situation, the analysis of thin skin roofs must take account of their bending stiffness.

The corresponding theory will be presented in Chapter III of this book. However, the membrane theory of roofs should not be considered useless, since (as will become clear in the further discussion), although it does not give a general solution of the problem, it leads in the simplest manner to one of the essential parts of the solution. In addition, the method permits the deduction of certain conclusions regarding the suitability of one or the other type of cross-section for a roof and regarding the appropriate design of the edge girders. For example, a study of the graphs in Fig. 50 shows that for the roofs (III) and (IV) the edge beams may have small bending stiffness, since the edge forces T_2^* for these roofs are small. However, the beams will require large extensional strength, in order that they may absorb the considerable shear forces at the edges, which are characteristic for this type of roof.

Hence the conclusion is reached that for the roofs (III) and (IV) the ideal edge girder will be a flexible, but very strong rope. In contrast, for the roofs (I) and (II), the edge forces T_2^* will be appreciable and therefore these roofs require deep edge girders. In this respect, the parabolic roof (II), in general, exhibits irrational features in that the pressure transmitted along the straight edges even exceeds the weight of the roof. This contradicts the basic principle accepted to-day by which the major part of the load should be transmitted to the curved boundaries.

CHAPTER III

THE ANALYSIS OF CYLINDRICAL SHELLS

§ 39. Historical introduction

In the theory of cylindrical shells, the fundamental problems are concerned with the analysis of complete cylindrical shells (tubes) and with the analysis of partial cylindrical shells, the boundaries of which are a pair of generators and a pair of cross-sections of the cylinder (cylindrical plates). The customary methods of solution of these two fundamental problems employ double and single trigonometric series respectively. Greater usefulness attaches to the method of single Fourier series which permits the solutions to be subjected to arbitrary boundary conditions along two edges of the shell. Utilization of one or the other method is made essentially more difficult by the complexity and the high order of the differential equations of the problems in view of which a great deal of attention must be given to the simplification of the initial formulae. In this connection, it has been found that the choice of one or the other system of simplifications depends on the ratios of the dimensions of the cylindrical shells.

Thus, it will be convenient to subdivide cylindrical shells into the following classes:

(a) Very long shells (thin walled rods) for which the length is many times greater than the average value of the radius of curvature (or the maximum dimension of the cross section);

(b) long shells for which the length is several times the average radius of curvature;

(c) medium length shells for which the length is comparable with the average radius of curvature;

(d) short shells for which the length is some fraction of the average radius of curvature.

For long and very long shells, V. Z. Vlasov [88] has proposed a special theory which may be called the "semi-membrane" theory. In this theory, the moments M_1, H and the normal shear force N_1 are neglected in the equilibrium equations for a element shell (which has led to the suggestion of the above name).

Using in addition to the above simplifications several additional assumptions (in particular, neglecting the strain ε_2 and the shear strain ω), Vlasov reduced the analysis of circular cylindrical shells to the solution of an equation of the form

$$\frac{\partial^4 \Phi_1}{\partial \eta^4} + \frac{\partial^2 \Phi_1}{\partial \eta^2} + i 2b^2 \frac{\partial^2 \Phi_1}{\partial \xi^2} = F_1(\xi, \eta), \tag{39.1}$$

where $2b^2$ is a constant and F_1 is a given function of ξ and η. Later on this equation was extended by Vlasov to apply to non-circular cylindrical shells. In this last case, it takes the form

$$\frac{\partial^2}{\partial \eta^2} \rho \frac{\partial^2 \Phi_1}{\partial \eta^2} + \frac{\partial}{\partial \eta}\left(\frac{1}{\rho} \frac{\partial \Phi_1}{\partial \eta}\right) + i 2b^2 \frac{\partial^2 \Phi_1}{\partial \xi^2} = F_2(\xi, \eta), \tag{39.2}$$

where $\rho(\eta)$ is the radius of curvature of the cross-section of the shell, divided by its mean value. In this way the foundations were laid for the general theory of thin-walled rods.

After adding the assumption that the cross-section of the shell remains undeformed, one may develop a semi-membrane theory which has proved very useful for the analysis of thin walled rods with open profiles [90]. Subsequently this approximate method was placed on a sound foundation in the works of G. I. Dzhanelidze [101] and A. L. Gol'denweizer [100]. In connection with the analysis of thin-walled rods, one should also mention the engineering theory of torsion of tubes of non-circular cross-section, developed by A. A. Umanski [112]. As regards long shells, it will be shown below that one may obtain even simpler equations than (39.1), (39.2), namely

$$\frac{\partial^4 \Phi_1}{\partial \eta^4} + i 2b^2 \frac{\partial^2 \Phi_1}{\partial \xi^2} = F_1(\xi, \eta), \tag{39.3}$$

$$\frac{\partial^2}{\partial \eta^2}\left(\rho \frac{\partial^2 \Phi_1}{\partial \eta^2}\right) + i 2b^2 \frac{\partial^2 \Phi_1}{\partial \xi^2} = F_2(\xi, \eta). \tag{39.4}$$

The first of these equations refers to circular, the second to non-circular shells.

It will be noted that for non-circular shells the transition from (39.2) to (39.4) simplifies the problem in form rather than in essence, since the omitted term is necessary for the transformation. However, when speaking of circular cylindrical shells, the replacement of the equation (39.1) by (39.3) gives an essential simplification. A theory arising from the equation (39.3) was proposed by S. Feinberg [113] and worked out in detail by the Author in the book [27].

For short and medium length shells, one has to employ a principally different scheme of simplification, based on the neglect of the tangential displacements in the formulae for the change of the curvature and twist, and likewise of the normal shear force in the second equilibrium equation (as has been mentioned in § 17, these two simplifications are interrelated). As a result of the above simplifying assumptions, the analysis of circular cylindrical shells under arbitrary loading may be reduced to the solution of the differential equation

$$\Delta\Delta\Phi + i2b^2 \frac{\partial^2\Phi}{\partial\xi^2} = F_2(\xi, \eta), \tag{39.5}$$

where $\Delta(\ldots) = \dfrac{\partial^2(\ldots)}{\partial\eta^2} + \dfrac{\partial^2(\ldots)}{\partial\xi^2}$.

This system of simplifications was first recommended by Kh. M. Mushtari (see [21] and earlier work by the same author). However, the final equation (39.5) seems to have appeared first in the work of Feinberg [113], devoted to a valuable criticism of the book by Vlasov [87]. The detailed method for the analysis of cylindrical plates, based on the equation (39.5), was developed by the Author of this book [27]. Later on, this equation was employed by A. I. Lur'e [18], Vlasov [91] and their pupils. Starting from equation (39.1), Lur'e succeeded in solving the problem of the stress concentration in a circular cylindrical shell, weakened by a small circular cut-out [102].

In connection with the appearance of the complex transformation of the differential equations of the theory of shells [32], it was found possible to obtain a sufficiently simple, general equation embracing all the types of cylindrical shells listed above [106]. Adapted to shells of arbitrary cross-section, this equation has the form

$$\Delta\rho\Delta\tilde{T} + \frac{\partial}{\partial\eta}\left(\frac{1}{\rho}\frac{\partial\tilde{T}}{\partial\eta}\right) + i2b^2\frac{\partial^2\tilde{T}}{\partial\xi^2} = F(\xi, \eta). \tag{39.6}$$

Neglecting in this equation the term

$$\frac{\partial}{\partial\eta}\left(\frac{1}{\rho}\frac{\partial\tilde{T}}{\partial\eta}\right)$$

and putting $\rho = 1$, one obtains an equation of the form (39.5).

Alternatively, neglecting in (39.6) the term $\dfrac{\partial^2\tilde{T}}{\partial\xi^2}$ in comparison with $\dfrac{\partial^2\tilde{T}}{\partial\eta^2}$ (with the exception of the last term on the left hand side of the equation (39.6), since it has the factor $i2b^2$ which has a very large modulus), one obtains an equation of the form (39.2). Finally, applying both the above simplifying assumptions, one finds equations of the form (39.3), (39.4). Thus, equation (39.6) actually comprises all the above fundamental variants of the approximate theory of cylindrical shells (circular as well as non-circular). Based on this equation, it has been possible to develop an effective method for the analysis of tubes with non-circular cross-sections and of arbitrary length [106].

In conclusion, it is necessary to dwell on the analysis of cylindrical shells reinforced by equidistant transverse ribs (a problem of considerable practical importance). When the ribs are sufficiently numerous, a significant simplification is achieved by treating the shell as structurally "anisotropic". This method is based on the replacement of the shell with ribs by an approximately equivalent shell without ribs, which has different bending stiffness in the two principal directions.

In such a case, the analysis of a circular cylindrical shell, reinforced by frequent ribs, reduces to the solution of an equation of the form

$$\frac{\partial^2}{\partial\eta^2}\Delta(\tilde{T}) + \frac{\partial^2\tilde{T}}{\partial\eta^2} + i2a^2\frac{\partial^2\tilde{T}}{\partial\xi^2} = F(\xi,\eta), \tag{39.7}$$

where $2a^2$ is a constant (see § 45).

§ 40. The equations of the theory of cylindrical shells in complex form

In this development of a general theory of cylindrical shells it is proposed to discuss the problem in terms of complex forces.

Determining the position of a point on the cylindrical surface by the coordinates x and s (i.e., the distance along the generator and the length of arc around a contour — see Fig. 51), it is convenient to introduce instead of these dimensional coordinates the two dimensionless quantities

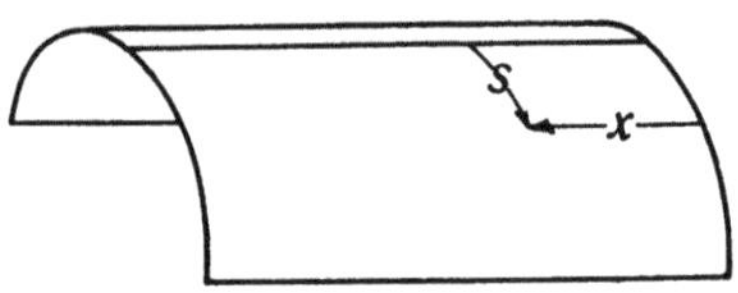

Fig. 51.

$$\xi = \frac{x}{r_0}, \quad \eta = \frac{s}{r_0}, \tag{40.1}$$

where r_0 is a constant quantity, characteristic for the shell and having the dimension of length. For example, one may take for r_0 the radius of the circle the circumference of which is equal to that of the cross-section of the shell, i.e.,

$$r_0 = \frac{2s_0}{\pi}$$

(s_0 — being a quarter of the length of arc of the transverse section of the cylinder). In the examples given below, the maximum value of the radius of curvature of the cross-section of the cylinder has been taken as r_0.

The Lamé parameters in the given case are

$$A_1 = A_2 = r_0, \tag{40.2}$$

whence the equations of shell theory in terms of complex forces [equations (16.10)], may for the present be written in the form

$$\begin{aligned}
&\frac{\partial \tilde{T}_1}{\partial \xi} + \frac{\partial \tilde{S}}{\partial \eta} = -q_1 r_0,\\
&\frac{\partial \tilde{S}}{\partial \xi} + \frac{\partial \tilde{T}_2}{\partial \eta} + \frac{i}{2b^2}\frac{1}{\rho}\frac{\partial \tilde{T}}{\partial \eta} = -q_2 r_0,\\
&\tilde{T}_2 - \frac{i}{2b^2}\rho \Delta(\tilde{T}) = \rho \tilde{q}_n r_0,
\end{aligned} \tag{40.3}$$

where

$$\begin{aligned}
\rho(\eta) &= \frac{r}{r_0},\\
4b^4 &= 12(1-\mu^2)\frac{r_0^2}{\delta^2},\\
\Delta(\ldots) &= \frac{\partial^2(\ldots)}{\partial \xi^2} + \frac{\partial^2(\ldots)}{\partial \eta^2},\\
\tilde{T} &= \tilde{T}_1 + \tilde{T}_2,\\
\tilde{q}_n &= q_n + i\frac{(1+\mu)}{2b^2}\left(\frac{\partial q_1}{\partial \xi} + \frac{\partial q_2}{\partial \eta}\right) \approx q_n.
\end{aligned} \tag{40.4}$$

The complex forces in these formulae are related to the forces and moments by means of the expressions (16.8), viz.,

$$\begin{aligned}
\tilde{T}_1 &= T_1 - i\frac{2b^2}{r_0}\frac{M_2 - \mu M_1}{1-\mu^2},\\
\tilde{T}_2 &= T_2 - i\frac{2b^2}{r_0}\frac{M_1 - \mu M_2}{1-\mu^2},\\
\tilde{S} &= S + i\frac{2b^2}{r_0}\frac{H}{1-\mu}.
\end{aligned} \tag{40.5}$$

Also q_1, q_2, q_n are the components of the surface loading in the directions x, s and along the normal to the middle surface.

Eliminating the unknown $\tilde{S}$ from the equations (40.3), one finds

$$\begin{gathered}\frac{\partial^2 \tilde{T}}{\partial \xi^2} - \Delta(\tilde{T}_2) - \frac{i}{2b^2}\frac{\partial}{\partial \eta}\left(\frac{1}{\rho}\frac{\partial \tilde{T}}{\partial \eta}\right) = r_0\left(\frac{\partial q_2}{\partial \eta} - \frac{\partial q_1}{\partial \xi}\right), \\ \tilde{T}_2 - \frac{i}{2b^2}\rho\Delta(\tilde{T}) = \tilde{q}_n r_0 \rho.\end{gathered} \tag{40.6}$$

Substituting now for $\tilde{T}_2$ from the second into the first equation (40.6), one arrives at the following differential equation of fourth order in a single unknown — the auxiliary function $\tilde{T}$:

$$\Delta[\rho\Delta(\tilde{T})] + \frac{\partial}{\partial \eta}\left(\frac{1}{\rho}\frac{\partial \tilde{T}}{\partial \eta}\right) + i2b^2\frac{\partial^2 \tilde{T}}{\partial \xi^2} = i2b^2 r_0\left[\Delta(\rho\tilde{q}_n) + \frac{\partial q_2}{\partial \eta} - \frac{\partial q_1}{\partial \xi}\right]. \tag{40.7}$$

The analysis of cylindrical shells of arbitrary form has thus been reduced to the solution of this equation, since, once $\tilde{T}$ is known, one may immediately find all the complex forces and then the complex displacements. It will now be shown how this may be accomplished.

The third of the equations (40.3) gives directly

$$\tilde{T}_2 = i\frac{\rho}{2b^2}\Delta(\tilde{T}) + \tilde{q}_n r_0 \rho, \tag{40.8}$$

and hence, since $\tilde{T}_1 = \tilde{T} - \tilde{T}_2$,

$$\tilde{T}_1 = \tilde{T} - i\frac{\rho}{2b^2}\Delta(\tilde{T}) - \tilde{q}_n \rho r_0. \tag{40.9}$$

Substituting this value for $\tilde{T}_2$ into the second of the equations (40.3), one has

$$\frac{\partial \tilde{S}}{\partial \xi} = -\frac{i}{2b^2}\left\{\frac{\partial}{\partial \eta}[\rho\Delta(\tilde{T})] + \frac{1}{\rho}\frac{\partial \tilde{T}}{\partial \eta}\right\} - r_0\left[\frac{\partial(\tilde{q}_n\rho)}{\partial \eta} + q_2\right]. \tag{40.10}$$

Further, on the basis of the first of the equations (40.3) and the formula (40.9), one may write

$$\frac{\partial \tilde{S}}{\partial \eta} = -\frac{\partial \tilde{T}}{\partial \xi} + i\frac{\rho}{2b^2}\frac{\partial}{\partial \xi}[\Delta(\tilde{T})] + r_0\left(\rho\frac{\partial \tilde{q}_n}{\partial \xi} - q_1\right). \tag{40.11}$$

The formulae (40.10) and (40.11) together determine the complex force $\tilde{S}$ in terms of $\tilde{T}$. Thus all the complex forces have been expressed in terms of this auxiliary function $\tilde{T}$.

There remains to express the complex displacements in terms of $\tilde{T}$. Using the formulae (16.11), (16.12) and (4.23) and applying them to cylindrical shells (taking into consideration that for this case $A_1 = A_2 = r_0$, $R_1 = \infty$, $R_2 = \rho r_0$), one obtains for the determination of the complex displacements the relations

$$
\begin{aligned}
r_0\tilde{\varepsilon}_1 &= \frac{\partial \tilde{u}}{\partial \xi} = \frac{r_0}{E\delta}(\tilde{T}_1 - \mu \tilde{T}_2),\\
r_0\tilde{\varepsilon}_2 &= \frac{\partial \tilde{v}}{\partial \eta} + \frac{\tilde{w}}{\rho} = \frac{r_0}{E\delta}(\tilde{T}_2 - \mu \tilde{T}_1),\\
r_0\tilde{\omega} &= \frac{\partial \tilde{u}}{\partial \eta} + \frac{\partial \tilde{v}}{\partial \xi} = \frac{2(1+\mu)r_0}{E\delta}\tilde{S}.
\end{aligned} \tag{40.12}
$$

$$
\begin{aligned}
r_0^2\tilde{\kappa}_1 &= -\frac{\partial^2 \tilde{w}}{\partial \xi^2} = i\frac{2b^2 r_0}{E\delta}(\tilde{T}_2 - T_2^*),\\
r_0^2\tilde{\kappa}_2 = r_0\frac{\partial \tilde{\psi}}{\partial \eta} &= -\frac{\partial}{\partial \eta}\left(\frac{\partial \tilde{w}}{\partial \eta} - \frac{\tilde{v}}{\rho}\right) = i\frac{2b^2 r_0}{E\delta}(\tilde{T}_1 - T_1^*),\\
r_0^2\tilde{\tau} = r_0\frac{\partial \tilde{\psi}}{\partial \xi} &= -\frac{\partial^2 \tilde{w}}{\partial \xi \partial \eta} + \frac{1}{\rho}\frac{\partial \tilde{v}}{\partial \xi} = -i\frac{2b^2 r_0}{E\delta}(\tilde{S} - S^*).
\end{aligned} \tag{40.13}
$$

in which T_1^*, T_2^*, S^* are any solutions of the membrane equations, i.e., in the given case solutions of the equations

$$
\begin{aligned}
\frac{\partial T_1^*}{\partial \xi} + \frac{\partial S^*}{\partial \eta} &= -q_1 r_0,\\
\frac{\partial S^*}{\partial \xi} + \frac{\partial T_2^*}{\partial \eta} &= -q_2 r_0,\\
T_2^* &= q_n \rho r_0.
\end{aligned} \tag{40.14}
$$

But, as has been established in Chapter II, the system of equations of the membrane theory applicable to cylindrical shells can always be integrated. For example, in the particular case when q_1, q_2 and q_n do not depend on ξ, one has for the solutions of the equations (40.14) the expressions

$$
\begin{aligned}
T_2^* &= q_n r_0 \rho,\\
S^* &= -\xi r_0\left(\frac{dq_n\rho}{d\eta} + q_2\right),\\
T_1^* &= \frac{\xi^2}{2} r_0\left[\frac{d^2 q_n\rho}{d\eta^2} + \frac{dq_2}{d\eta}\right] - q_1 r_0 \xi,
\end{aligned} \tag{40.15}
$$

which must be introduced in the above case on the right hand sides of the formulae (40.13).

The two systems (40.12) and (40.13) together determine the complex displacements in terms of the complex forces or, since the complex forces can be expressed in terms of the auxiliary function $\tilde{T}$, it may also be said that they determine the complex displacements in terms of $\tilde{T}$.

It must be emphasized that the systems (40.12) and (40.13) will usually be only approximate, — exact to terms of order $\dfrac{\delta}{r_0}$ compared with unity, since terms of that order were repeatedly neglected in the derivation of the equations (16.11), (16.12). Therefore, in the process of transformation and solution of the equations (40.12) and (40.13), there will be discovered inconsistencies in small terms which must be removed by corresponding arguments.

Having found the general solution of the systems (40.12) and (40.13), i.e., having determined $\tilde{u}$, $\tilde{v}$, $\tilde{w}$, one may immediately find also the displacements u, v, w, since they are equal to the real parts of the former. On the basis of the system (40.12) and taking into consideration that the complex forces can be expressed in terms of the auxiliary function $\tilde{T}$ which is a solution of the equation (40.7), one may obtain by (40.8), (40.9) and (40.10) the following expressions for the derivatives of the complex displacements:

$$\frac{\partial \tilde{u}}{\partial \xi} = \frac{r_0}{E\delta}\left[\tilde{T} - i\,\frac{1+\mu}{2b^2}\,\rho\Delta(\tilde{T})\right] - (1+\mu)\,\frac{r_0^2}{E\delta}\,\rho\tilde{\sigma}$$

$$\frac{\partial^2 \tilde{v}}{\partial \xi^2} = -\,\frac{r_0}{E\delta}\left\{\left(1 + i\,\frac{1+\mu}{\rho b^2}\right)\frac{\partial \tilde{T}}{\partial \eta} + i\,\frac{1+\mu}{2b^2}\,\frac{\partial}{\partial \eta}\,[\rho\Delta(\tilde{T})]\right\} \cdot$$

$$-\,\frac{(1+\mu)r_0^2}{E\delta}\,\frac{\partial(\rho\tilde{q}_n)}{\partial \eta} - \frac{2(1+\mu)}{E\delta}\,r_0^2 q_2, \qquad (40.16)$$

$$\frac{\partial^2 \tilde{w}}{\partial \xi^2} = \frac{r_0}{E\delta}\,\rho\left[\Delta(\tilde{T}) + i\,\frac{1+\mu}{2b^2}\,\frac{\partial}{\partial \eta}\left(\frac{1}{\rho}\,\frac{\partial \tilde{T}}{\partial \eta}\right)\right] + \frac{(1+\mu)r_0^2\rho}{E\delta}\left(\frac{\partial q_1}{\partial \xi} + \frac{\partial q_2}{\partial \eta}\right).$$

These expressions, directly linking the derivatives of the complex displacements with respect to ξ with the auxiliary function $\tilde{T}$, will repeatedly be used in what follows.

§ 41. The theory of cylindrical shells, reinforced by transverse ribs

In order to ensure the strength and stability of cylindrical shells, it is often convenient to reinforce them by closely spaced transverse stiffeners (Fig. 52). The analysis of such shells turns out to be a very difficult problem, if it is desired to satisfy exactly the conditions which prevail at the junction between the rings and the actual skin. However, at the present time, no use will be made of this complicated approach. Instead, an approximate method will be employed which consists of distributing the bending stiffness and tensile strength of each rib uniformly over the width of the bay, supported by the rib, and of adding it to the corresponding stiffness of the shell in the circumferential direction. Under these conditions, the shell with ribs is replaced in the analysis by an approximately equivalent shell without ribs, where this last shell exhibits different elastic properties in the longitudinal and circumferential directions, i.e., it is anisotropic (or better "orthotropic"). If the ribs are sufficiently closely spaced and their number is sufficiently large, then the described method involves relatively small errors. At the same time, the method greatly simplifies the problem. These two properties are the reason for the wide application which the method has found recently. Anisotropic behaviour of the above type may be called constructional, since it does not result from a property of the material, but from the manner in which it is distributed.

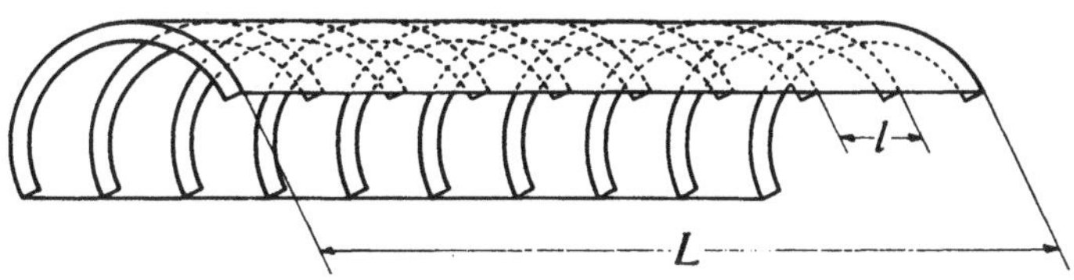

Fig. 52.

The relations between the forces, moments and strains of the middle surface for cylindrical shells, reinforced by transverse equidistant ribs and conceived as constructionally anisotropic will now be established. For this purpose, it will be assumed that the ribs possess bending and extensional stiffness in their own plane only and that the resistance to displacements perpendicular to their plane and to torsion may be neglected.

Then it is obvious that the ribs will not influence the shear displacement or the torsion of the shell, as a result of which one will have for S and H the usual formulae

$$S = \frac{E\delta}{2(1+\mu)}\,\omega, \quad H = \frac{E\delta^3}{12(1+\mu)}\,\tau \tag{41.1}$$

The remaining forces and moments require a special study. The formulae for their determination can be deduced, if one takes into account that for transverse bending the axis of rotation of the cross-section will pass through the centre of gravity of the profile, formed by the rib and the associated shell section. Let the distance of this point from the middle surface be denoted by z_0 (Fig. 53). Then the strains

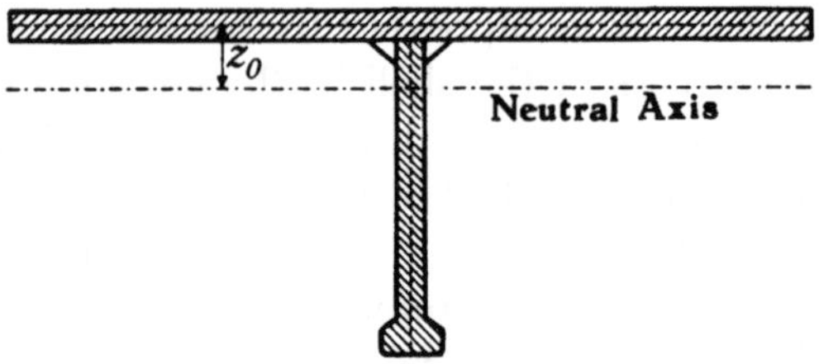

Fig. 53.

of a fibre of the shell parallel to the middle surface may be represented by

$$\begin{aligned} e_{xx} &= \varepsilon_1 + z\kappa_1, \\ e_{\eta\eta} &= \frac{1}{1 + \dfrac{z'}{r - z_0}} (\varepsilon_2' + z'\kappa_2') \approx \varepsilon_2' + z'\kappa_2', \end{aligned} \tag{41.2}$$

where z is the distance measured from the middle surface (in the direction of the convex side) and $z' = z + z_0$ is the distance from the centre of gravity of the profile measured in the same direction. Also, ε_2' and $\kappa_2' \approx \kappa_2$ are the direct strain and parameter for the change of curvature of the neutral axis of the profile. Hence the stresses in the shell can be written in the form

$$\begin{aligned} \sigma_{xx} &= \frac{E}{1-\mu^2}[\varepsilon_1 + z\kappa_1 + \mu(\varepsilon_2' + z'\kappa_2')], && -\frac{\delta}{2} \leqslant z \leqslant \frac{\delta}{2}, \\ \sigma_{\eta\eta} &= \frac{E}{1-\mu^2}[\varepsilon_2' + z'\kappa_2' + \mu(\varepsilon_1 + z\kappa_1)], && -\frac{\delta}{2} \leqslant z \leqslant \frac{\delta}{2}, \\ \sigma_{\eta\eta} &= E(\varepsilon_2' + z\kappa_2') && z_0 - \frac{\delta}{2} \geqslant z \geqslant -\left(h - z_0 + \frac{\delta}{2}\right). \end{aligned} \tag{41.3}$$

In the brackets on the right the limits are indicated between which these formulae hold true, where h is the height of the profile.

The first of the formulae (41.3) will now be integrated with respect to z over the thickness of the shell. Then one obtains

$$T_1 = \int_{-\frac{\delta}{2}}^{+\frac{\delta}{2}} \sigma_{xx} dz = \frac{E\delta}{1-\mu^2} [\varepsilon_1 + \mu(\varepsilon_2{}' + z_0\kappa_2{}')]. \tag{41.4}$$

But $\varepsilon_2{}' + z_0\kappa_2{}' = \varepsilon_2$, where ε_2 is the extension of the middle surface. Hence one obtains for T_1 the usual expression

$$T_1 = \frac{E\delta}{1-\mu^2} (\varepsilon_1 + \mu\varepsilon_2). \tag{41.5}$$

Similarly, multiplying the first of the formulae (41.3) by z and integrating over the thickness, one obtains the usual formula for M_1

$$M_1 = \frac{E\delta^3}{12(1-\mu^2)} (\kappa_1 + \mu\kappa_2). \tag{41.6}$$

Next, the stress $\sigma_{\eta\eta}$ will be integrated over the entire area of the profile (which comprises the area of the section of the rib Ω and the area of the associated shell section $l\delta$) taking into consideration for this purpose both the second and the third of the formulae (41.3). This gives

$$T_2 l = E\left[\left(\Omega + \frac{l\delta}{1-\mu^2}\right)\varepsilon_2 + \mu\frac{l\delta}{1-\mu^2}\varepsilon_1\right], \tag{41.7}$$

or

$$T_2 = \frac{E\delta}{1-\mu^2}\left[\left(1 + \frac{(1-\mu^2)\Omega}{l\delta}\right)\varepsilon_2 + \mu\varepsilon_1\right]. \tag{41.8}$$

Finally, multiplying $\sigma_{\eta\eta}$ by z' and integrating again over the entire area of the profile, one finds

$$M_2 l = E\left[J\kappa_2 + \mu\frac{\delta^3 l}{12(1-\mu^2)}\kappa_1\right], \tag{41.9}$$

where J is the moment of inertia of the profile (including the associated shell portion).

It will be noted that the bending stiffness of the rib per unit length of bay between ribs, i.e., the quantity $\frac{EJ}{l}$, is obviously considerably larger than the stiffness of the cylindrical panel $\frac{E\delta^3}{12(1-\mu^2)}$. The area of the cross-section of the rib per unit length of bay between ribs, i.e., the quantity $\frac{\Omega}{l}$, obviously turns out to be only a fraction of the thickness δ of the shell.

The above observations will now be used to simplify the formulae (41.8) and (41.9) to the following forms:

$$T_2 \approx \frac{E\delta}{1-\mu^2}(\varepsilon_2 + \mu\varepsilon_1),$$
$$M_2 \approx \frac{EJ}{l}\kappa_2. \tag{41.10}$$

Formulae (41.1), (41.5), (41.6) and (41.10) express the dependence between the forces, moments and strains of the middle surface for cylindrical shells, reinforced by transverse ribs.

Now the equations of equilibrium for the element of a constructionally anisotropic cylindrical shell will be established. Applying the general equations of the first chapter to shells of the present shape and taking into consideration that the bending stiffness of the panel is to be neglected in comparison with the bending stiffness of the rib, one has to neglect in the equilibrium equations the moments M_1 and H in comparison with M_2 and will thus obtain

$$\frac{\partial T_1}{\partial \xi} + \frac{\partial S}{\partial \eta} = -q_1 r_0,$$
$$\frac{\partial S}{\partial \xi} + \frac{\partial T_2}{\partial \eta} + \frac{N_2}{\rho} = -q_2 r_0,$$
$$\frac{\partial N_2}{\partial \eta} - \frac{T_2}{\rho} + q_n r_0 = 0, \tag{41.11}$$
$$r_0 N_2 = \frac{\partial M_2}{\partial \eta}.$$

Here, the fifth equilibrium equation, as a result of the omission of M_1 and H, has not to be retained and the sixth equation reduces to

$$T_{12} = T_{21} = S.$$

The system (41.11) consists of four equations for five unknowns. The missing equation may be obtained from the compatibility relations for the middle surface (5.1). These last, adapted to cylindrical shells, have the form

$$\frac{\partial \kappa_2}{\partial \xi} - \frac{\partial \tau}{\partial \eta} = 0,$$

$$\frac{\partial \kappa_1}{\partial \eta} - \frac{\partial \tau}{\partial \xi} + \frac{1}{r_0 \rho}\left(\frac{\partial \omega}{\partial \xi} - \frac{\partial \varepsilon_1}{\partial \eta}\right) = 0, \tag{41.12}$$

$$\frac{\kappa_1}{\rho} + \frac{1}{r_0}\left(\frac{\partial^2 \varepsilon_1}{\partial \eta^2} + \frac{\partial^2 \varepsilon_2}{\partial \xi^2} - \frac{\partial^2 \omega}{\partial \xi \partial \eta}\right) = 0.$$

[Cf. the formulae (5.1) and (40.1) and (40.2)] Eliminating between these equations κ_1 and τ, one finds

$$r_0 \frac{\partial^2 \kappa_2}{\partial \xi^2} + \frac{\partial^2}{\partial \eta^2}\left[\rho\left(\frac{\partial^2 \varepsilon_1}{\partial \eta^2} + \frac{\partial^2 \varepsilon_2}{\partial \xi^2} - \frac{\partial^2 \omega}{\partial \xi \partial \eta}\right)\right] + \frac{\partial}{\partial \eta}\left[\frac{1}{\rho}\left(\frac{\partial \varepsilon_1}{\partial \eta} - \frac{\partial \omega}{\partial \xi}\right)\right] = 0. \tag{41.13}$$

There still remains to introduce in these formulae for the strains their expressions in terms of the forces and moments (41.1), (41.5), (41.10). In this way one arrives at the relation

$$\frac{r_0^2 l \delta}{J} \frac{\partial^2 M_2}{\partial \xi^2} + r_0 \left\{\frac{\partial^2}{\partial \eta^2} \rho \left[\frac{\partial^2 (T_1 - \mu T_2)}{\partial \eta^2} + \frac{\partial^2 (T_2 - \mu T_1)}{\partial \xi^2} - 2(1+\mu)\frac{\partial^2 S}{\partial \xi \partial \eta}\right] + \right.$$

$$\left. + \frac{\partial}{\partial \eta} \frac{1}{\rho}\left[\frac{\partial (T_1 - \mu T_2)}{\partial \eta} - 2(1+\mu)\frac{\partial S}{\partial \xi}\right]\right\} = 0. \tag{41.14}$$

This last formula may be transformed to a more convenient form, if one takes into consideration that by (41.11)

$$2\frac{\partial^2 S}{\partial \xi \partial \eta} = -\frac{\partial^2 T_1}{\partial \xi^2} - \frac{\partial^2 T_2}{\partial \eta^2} - \frac{1}{r_0}\frac{\partial}{\partial \eta}\left(\frac{1}{\rho}\frac{\partial M_2}{\partial \eta}\right) - r_0\left(\frac{\partial q_1}{\partial \xi} + \frac{\partial q_2}{\partial \eta}\right),$$

$$T_2 = \frac{\rho}{r_0}\frac{\partial^2 M_2}{\partial \eta^2} + q_n \rho r_0, \quad \frac{\partial S}{\partial \xi} = -\frac{\partial T_2}{\partial \eta} - \frac{1}{r_0 \rho}\frac{\partial M_2}{\partial \eta} - q_2 r_0. \tag{41.15}$$

Substituting from (41.15) into (41.14), onc has

$$\frac{r_0 l \delta}{J}\frac{\partial^2 M_2}{\partial \xi^2} + \frac{\partial^2}{\partial \eta^2}[\rho \Delta (T_1 + T_2)] + \frac{\partial}{\partial \eta}\left[\frac{1}{\rho}\frac{\partial (T_1 + T_2)}{\partial \eta}\right] +$$

$$+ \frac{(1+\mu)}{r_0}\left\{\frac{\partial}{\partial \eta}\frac{1}{\rho}\frac{\partial}{\partial \eta}\rho\frac{\partial^2 M_2}{\partial \eta^2} + \frac{\partial^2}{\partial \eta^2}\rho\frac{\partial}{\partial \eta}\frac{1}{\rho}\frac{\partial M_2}{\partial \eta} + 2\frac{\partial}{\partial \eta}\left(\frac{1}{\rho^2}\frac{\partial M_2}{\partial \eta}\right)\right\} +$$

$$+ (1+\mu) r_0 \left[\frac{\partial^2}{\partial \eta^2}\rho\left(\frac{\partial q_1}{\partial \xi} + \frac{\partial q_2}{\partial \eta}\right) + \frac{\partial}{\partial \eta}\frac{1}{\rho}\left(\frac{\partial q_n \rho}{\partial \eta} + q_2\right)\right] = 0. \tag{41.16}$$

The relation so obtained is complicated enough. However, fortunately the most cumbersome of its terms (in fact, the term involving M_2 and contained between braces) is unimportant and may be neglected, as will be shown somewhat later when applying the above theory to circular cylinders (§ 45).

Another equation involving the same two unknowns (M_2 and $T = T_1 + T_2$) may be obtained from (41.11). Eliminating from these equations all unknowns except T_1 and M_2, one finds

$$r_0 \frac{\partial^2 T_1}{\partial \xi^2} - \frac{\partial^2}{\partial \eta^2}\left[\rho \frac{\partial^2 M_2}{\partial \eta^2}\right] - \frac{\partial}{\partial \eta}\left(\frac{1}{\rho}\frac{\partial M_2}{\partial \eta}\right) = r_0^2\left[\frac{\partial^2 (q_n \rho)}{\partial \eta_2} + \frac{\partial q_2}{\partial \eta} - \frac{\partial q_1}{\partial \xi}\right]. \tag{41.17}$$

Adding this equation to the equation

$$r_0 \frac{\partial^2 T_2}{\partial \xi^2} - \rho \frac{\partial^4 M_2}{\partial \xi^2 \partial \eta^2} = \rho r_0^2 \frac{\partial^2 q_n}{\partial \xi^2}, \tag{41.18}$$

obtained from the last two equations of (41.11), one has

$$r_0 \frac{\partial^2 (T_1 + T_2)}{\partial \xi^2} - \Delta\left(\rho \frac{\partial^2 M_2}{\partial \eta^2}\right) - \frac{\partial}{\partial \eta}\left(\frac{1}{\rho}\frac{\partial M_2}{\partial \eta}\right) = r_0^2\left[\Delta(\rho q_n) + \frac{\partial q_2}{\partial \eta} - \frac{\partial q_1}{\partial \xi}\right]. \tag{41.19}$$

The equations (41.16) and (41.19) will now be rewritten and at the same time the first of these equations will be simplified in the above stated manner:

$$\begin{aligned} &\frac{\partial^2 (T_1 + T_2)}{\partial \xi^2} - \frac{1}{r_0}\left\{\Delta\left[\rho \frac{\partial^2 M_2}{\partial \eta^2}\right] + \frac{\partial}{\partial \eta}\left(\frac{1}{\rho}\frac{\partial M_2}{\partial \eta}\right)\right\} = r_0 F_1(\xi, \eta), \\ &\frac{4a^4}{r_0}\frac{\partial^2 M_2}{\partial \xi^2} + \frac{\partial^2}{\partial \eta^2}[\rho \Delta (T_1 + T_2)] + \frac{\partial}{\partial \eta}\left[\frac{1}{\rho}\frac{\partial (T_1 + T_2)}{\partial \eta}\right] = r_0 F_2(\xi, \eta), \end{aligned} \tag{41.20}$$

where

$$4a^4 = \frac{r_0^2 l \delta}{J},$$

$$F_1(\xi, \eta) = \Delta(\rho q_n) + \frac{\partial q_2}{\partial \eta} - \frac{\partial q_1}{\partial \xi}, \tag{41.21}$$

$$F_2(\xi, \eta) = -(1 + \mu)\left\{\frac{\partial^2}{\partial \eta^2}\rho\left(\frac{\partial q_1}{\partial \xi} + \frac{\partial q_2}{\partial \eta}\right) + \frac{\partial}{\partial \eta}\frac{1}{\rho}\left(\frac{\partial q_n \rho}{\partial \eta} + q_2\right)\right\}.$$

The analysis of cylindrical shells of arbitrary shape, reinforced by equidistant transverse ribs of equal strength also leads to the solution of the system (41.20).

Unfortunately, a complex transformation of the equations (41.20) is not possible or, more correctly, is only possible in the case when $\rho = 1$, i.e., when the cylinder is circular. In this last case, one may introduce the complex auxiliary function

$$\tilde{V} = T_1 + T_2 - i\frac{2a^2}{r_0}M_2, \tag{41.22}$$

which satisfies the differential equation

$$\frac{\partial^2}{\partial\eta^2}\Delta(\tilde{V}) + \frac{\partial^2\tilde{V}}{\partial\eta^2} + i2a^2\frac{\partial^2\tilde{V}}{\partial\xi^2} = i2a^2F_1. \tag{41.23}$$

This particular case is considered in detail in § 45.

§ 42. The analysis of cylindrical plates

The results of § 40 will now be applied to the analysis of curved plates, removed from circular cylindrical shells by two cuts along generators and two cuts perpendicular to the axis of the shell.

For circular cylindrical shells

$$\rho = \frac{r}{r_0} = 1,$$

and the fundamental differential equation (40.7) takes the form

$$\Delta\Delta(\tilde{T}) + \frac{\partial^2\tilde{T}}{\partial\varphi^2} + i2b^2\frac{\partial\tilde{T}}{\partial\xi^2} = i2b^2r_0\left[\Delta(q_n) + \frac{\partial q_2}{\partial\varphi} - \frac{\partial q_1}{\partial\xi}\right], \tag{42.1}$$

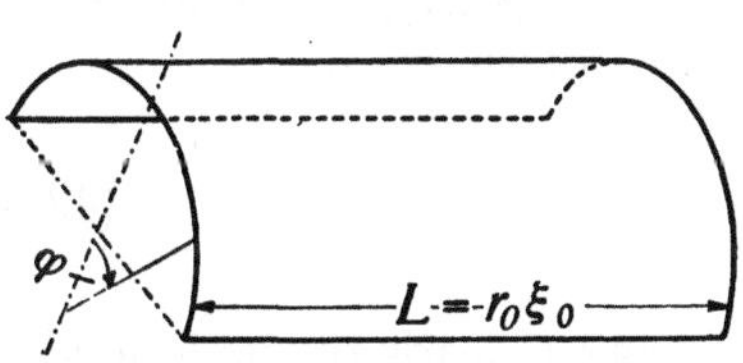

Fig. 54.

where by (40.4)

$$\Delta(\ldots) = \frac{\partial^2(\ldots)}{\partial\xi^2} + \frac{\partial^2(\ldots)}{\partial\varphi^2},$$

$$4b^4 = \frac{12(1-\mu^2)r_0^2}{\delta^2}.$$

In the present particular case, it is more customary to take instead of η the symbol φ, where φ has the simple geometrical meaning of the angle corresponding to the arc length s. This angle will always be measured from the plane of symmetry of the cylindrical plate (see Fig. 54).

In practical problems, it is ordinarily possible to assume that the curvilinear edges of cylindrical plates rest freely on supports which are absolutely rigid in their planes. The boundary conditions at the curvilinear edges thus reduce to the equalities

$$v = w = T_1 = M_1 = 0, \tag{42.2}$$

which hold true for $\xi = 0$, $\xi_0 = \dfrac{L}{r_0}$ (where L is the length of the plate and the origin for ξ has been taken at one of the curvilinear edges). As regards the boundary conditions at the straight edges ($\varphi = \varphi_0$ and $\varphi = -\varphi_0$), they may vary greatly, so that in solving the equations (42.1) one has, if possible, to preserve the possibility of leaving them arbitrary.

A solution, satisfying all the above mentioned requirements, can be obtained by expanding the components of the surface loading in ordinary trigonometric series in terms of the coordinate ξ. Let these expansions have the form

$$\begin{aligned} q_1 &= \sum_{m=1}^{\infty} q_1^{(m)} \cos \lambda_m \xi, \\ q_2 &= \sum_{m=1}^{\infty} q_2^{(m)} \sin \lambda_m \xi, \\ q_n &= \sum_{m=1}^{\infty} q_n^{(m)} \sin \lambda_m \xi, \end{aligned} \tag{42.3}$$

where

$$\lambda_m = m \frac{\pi r_0}{L}, \tag{42.4}$$

and $q_1^{(m)}$, $q_2^{(m)}$, $q_n^{(m)}$ are functions of φ only which may be determined by use of the general formulae for calculating Fourier coefficients.

Substituting (42.3) on the right hand side of (42.1), the solution of this equation will itself be sought in the form of the series

$$\tilde{T} = \sum_{m=1}^{\infty} \tilde{T}^{(m)} \sin \lambda_m \xi. \tag{42.5}$$

In this way, the partial differential equation (42.1) leads to the following set of independent ordinary differential equations

$$\Delta_m[\Delta_m(\tilde{T}^{(m)})] + \frac{d^2 \tilde{T}^{(m)}}{d\varphi^2} - i2b^2 \lambda_m^2 \tilde{T}^{(m)} = \\ = i2b^2 r_0 \left[\Delta_m(q_n^{(m)}) + \frac{dq_2^{(m)}}{d\varphi} + \lambda_m q_1^{(m)} \right], \tag{42.6}$$

where

$$\Delta_m(\ldots) = \frac{d^2(\ldots)}{d\varphi^2} - \lambda_m{}^2(\ldots) \tag{42.7}$$

and there will be as many equations as there were terms retained in the series (42.3).

Since the system of equations (42.6) is linear, its general solution may be obtained by adding any of its particular integrals to the general solution of the homogeneous equation

$$\Delta_m[\Delta_m(\tilde{T}^{(m)})] + \frac{d^2\tilde{T}^{(m)}}{d\varphi^2} - i2b^2\lambda_m{}^2\tilde{T}^{(m)} = 0. \tag{42.8}$$

If the surface loading is a sufficiently smooth function of the angle φ, one may take for this purpose as a close approximation to a particular integral the expression

$$\tilde{T}_{part}{}^{(m)} = -\frac{r_0}{\lambda_m{}^2}\left\{\Delta_m[q_n{}^{(m)}] + \frac{dq_2{}^{(m)}}{d\varphi} + \lambda_m q_1{}^{(m)}\right\}, \tag{42.9}$$

which will be obtained, if one omits on the left hand side of (42.6) all but the last term.

Employment of this formula is justified on the basis that the magnitude of the components of the surface loading have been assumed to be non-increasing for differentiation with respect to φ, so that on substitution of (42.9) on the left hand side of (42.6) one may neglect all terms except those with the very large factor $2b^2\lambda_m{}^2$.

By the way, the formula (42.9) coincides with the solution which would be obtained for the problem on the basis of the membrane theory. In fact, the equations of the membrane theory in the given case can be written

$$\begin{gathered}\frac{\partial T_1{}^*}{\partial \xi} + \frac{\partial S^*}{\partial \varphi} = -r_0 \sum_{m=1}^{\infty} q_1{}^{(m)} \cos \lambda_m \xi,\\ \frac{\partial S^*}{\partial \xi} + \frac{\partial T_2{}^*}{\partial \varphi} = -r_0 \sum_{m=1}^{\infty} q_2{}^{(m)} \sin \lambda_m \xi,\\ T_2{}^* = r_0 \sum q_n{}^{(m)} \sin \lambda_m \xi.\end{gathered} \tag{42.10}$$

Hence, if $T_1{}^*$, $T_2{}^*$, S^* are sought in the form

$$\begin{gathered}T_1{}^* = \sum T_{1,m}{}^* \sin \lambda_m \xi, \; T_2{}^* = \sum T_{2,m}{}^* \sin \lambda_m \xi,\\ S^* = \sum S_m{}^* \cos \lambda_m \xi,\end{gathered} \tag{42.11}$$

one will have for the coefficients in these series the system

$$\lambda_m T_{1,m}^* + \frac{dS_m^*}{d\varphi} = -r_0 q_1^{(m)},$$
$$-\lambda_m S_m^* + \frac{dT_{2,m}^*}{d\varphi} = -r_0 q_2^{(m)}, \tag{42.12}$$
$$T_{2,m}^* = r_0 q_n^{(m)},$$

from which follows

$$T_{2,m}^* = r_0 q_n^{(m)},$$
$$S_m^* = \frac{r_0}{\lambda_m}\left[\frac{dq_n^{(m)}}{d\varphi} + q_2^{(m)}\right], \tag{42.13}$$
$$T_{1,m}^* = -\frac{r_0}{\lambda_m^2}\left[\frac{d^2 q_n^{(m)}}{d\varphi^2} + \frac{dq_2^{(m)}}{d\varphi}\right] - \frac{r_0}{\lambda_m} q_1^{(m)},$$

and

$$T_m^* = T_{1,m}^* + T_{2,m}^* = -\frac{r_0}{\lambda_m^2}\left\{\Delta_m[q_n^{(m)}] + \frac{dq_2^{(m)}}{d\varphi} + \lambda_m q_1^{(m)}\right\}, \tag{42.14}$$

which coincides with (42.9), if one takes into consideration that

$$\text{for } M_1 = M_2 = 0,\ T_m^* = \tilde{T}^{(m)}.$$

One thus arrives at the important conclusion that (at least in the case under consideration) one may take as the particular integral of the differential equations of the theory of elastic shells the solution of the membrane theory.

In general, the following will be a completely obvious rule:

if one can state for a shell under consideration in the case of prescribed surface loading boundary conditions fulfillment of which will ensure a membrane state of stress, or one very close to it, the membrane solution can be selected in the capacity of a particular solution corresponding to the moment problem.

Following this rule, and likewise the conditions formulated in Chapter II ensuring a membrane state of stress in a shell, one can always find out whether the above method of approximate determination of a particular solution is admissible or not. Thus, for shells with continuous or, may be, rapidly changing radii of curvature or thickness and for shells loaded by surface forces which vary sufficiently rapidly, it is impossible to adopt a particular solution from membrane theory. An estimate of the error involved in the replacement of a particular solution of the moment equations by a membrane solution can always be obtained by substitution of this solution in the moment equations, as has been shown above in the example of cylindrical plates.

Later on, it will be repeatedly verified that this method turns out to be admissible in many cases, so that the determination of particular integrals of the equations of the theory of shells does not as a rule cause any difficulties and reduces to a range of problems which have been dealt with in the preceding chapter.

These results make it possible to consider the present problem of the stress analysis of cylindrical plates as a homogeneous problem, i.e., to put in all formulae

$$q_1^{(m)} = q_2^{(m)} = q_n^{(m)} = 0.$$

Having obtained the homogeneous solution, it is always possible to generalize it to the non-homogeneous problem. For this purpose, it is sufficient to add to the expressions found for the forces, moments and displacements the corresponding solution of the membrane theory.

Thus, attention will be concentrated on the integration of the homogeneous equation (42.8) and in what follows $\tilde{T}_1$, $\tilde{T}_2$, $\tilde{S}$, $\tilde{u}$, $\tilde{v}$, $\tilde{w}$ will not denote the general expressions for these complex forces and complex displacements, but merely the expressions corresponding to the homogeneous problem.

The solution of (42.8) must be sought in the form

$$\tilde{T}^{(m)} = Ce^{\gamma\varphi}, \tag{42.15}$$

where C and γ are constants.

Substituting (42.15) into (42.8) and dividing the final expression by $Ce^{\gamma\varphi}$, one arrives at the following bi-quadratic equation for γ:

$$(\gamma^2 - \lambda_m^2)^2 + \gamma^2 - i2b^2\lambda_m^2 = 0, \tag{42.16}$$

whence

$$\gamma^2 = \lambda_m^2 - \tfrac{1}{2} \pm \sqrt{i2b^2\lambda_m^2 + (\lambda_m^2 - \tfrac{1}{2})^2}. \tag{42.17}$$

In this last expression, one may neglect under the root sign the second term on comparison with the first. This step is justified, since in the problem under consideration the parameter $2b^2 \gg \lambda_m^2$ and λ_m^2 is usually a quantity of the same order as unity.

After this simplification, one has

$$\gamma^2 = \lambda_m^2 - \tfrac{1}{2} \pm (1 + i)b\lambda_m = b\lambda_m[f_m \pm (1 + i)], \tag{42.18}$$

where

$$f_m = \frac{\lambda_m^2 - \frac{1}{2}}{b\lambda_m}. \tag{42.19}$$

Taking the square root of the expression (42.18), one finds for γ the four values

$$\gamma_{1,2} = \pm (c_1 + id_1),$$
$$\gamma_{3,4} = \pm (c_2 - id_2), \tag{42.20}$$

where

$$c_1 = \sqrt{\frac{b\lambda_m}{2}} \sqrt{\sqrt{(1+f_m)^2+1}+1+f_m},$$
$$d_1 = \sqrt{\frac{b\lambda_m}{2}} \sqrt{\sqrt{(1+f_m)^2+1}-1-f_m},$$
$$c_2 = \sqrt{\frac{b\lambda_m}{2}} \sqrt{\sqrt{(1-f_m)^2+1}-1+f_m}, \tag{42.21}$$
$$d_2 = \sqrt{\frac{b\lambda_m}{2}} \sqrt{\sqrt{(1-f_m)^2+1}+1-f_m}.$$

Hence one may write for the auxiliary function $\tilde{T}^{(m)}$ the expression

$$\tilde{T}^{(m)} = \tilde{C}_1 e^{\gamma_1 \varphi} + \tilde{C}_2 e^{\gamma_2 \varphi} + \tilde{C}_3 e^{\gamma_3 \varphi} + \tilde{C}_4 e^{\gamma_4 \varphi}, \tag{42.22}$$

where $\tilde{C}_1$, $\tilde{C}_2$, $\tilde{C}_3$, $\tilde{C}_4$ are constants of integration which have been specially identified to indicate that they are complex quantities. Strictly speaking, the constants $\tilde{C}_j$ as well as the exponents γ_j should carry an index m; however, in order to avoid formulae with double indices, this has not been done and instead the reader will be asked to keep this circumstance in mind.

Now, having constructed the solution of the homogeneous problem in the form of the series (42.5), one may obtain, in an analogous form, the expressions for all the complex forces. Substituting (42.5) into (40.8), (40.9), (40.10), taking (42.22) into consideration and putting there $p = 1$, $q_1 = q_2 = q_n = 0$, one has

$$\tilde{T}_1 = \sum \tilde{T}_1^{(m)} \sin \lambda_m \xi, \quad \tilde{T}_2 = \sum \tilde{T}_2^{(m)} \sin \lambda_m \xi, \quad \tilde{S} = \sum \tilde{S}^{(m)} \cos \lambda_m \xi, \tag{42.23}$$

where

$$\tilde{T}_1^{(m)} = \sum_{j=1}^{j=4} \tilde{C}_j \left[1 - \frac{i}{2b^2}(\gamma_j^2 - \lambda_m^2)\right] e^{\gamma_j \varphi} \approx \sum_{j=1}^{j=4} \tilde{C}_j e^{\gamma_j \varphi}. \tag{42.24}$$

$$\tilde{T}_2^{(m)} = \frac{i}{2b^2} \sum_{j=1}^{j=4} (\gamma_j^2 - \lambda_m^2) \tilde{C}_j e^{\gamma_j \varphi},$$
$$\tilde{S}^{(m)} = -\lambda_m \sum_{j=1}^{j=4} \frac{1}{\gamma_j} \tilde{C}_j \left[1 - \frac{i}{2b^2}(\gamma_j^2 - \lambda_m^2)\right] e^{\gamma_j \varphi} \approx -\lambda_m \sum_{j=1}^{j=4} \frac{1}{\gamma_j} \tilde{C}_j e^{\gamma_j \varphi}. \tag{42.25}$$

Introducing next these expressions in (40.12) and (40.13), one obtains equations for the determination of the complex displacements. For this purpose, the last will be sought in the form of the series

$$\tilde{u} = \sum_m \tilde{u}_m \cos \lambda_m \xi, \quad \tilde{v} = \sum \tilde{v}_m \sin \lambda_m \xi, \quad \tilde{w} = \sum \tilde{w}_m \sin \lambda_m \xi, \tag{42.26}$$

the coefficients of which, by virtue of (40.12), (40.13) and (42.23), will be subject to the relations

$$\begin{aligned}
-\lambda_m \tilde{u}_m &= \frac{r_0}{E\delta}(\tilde{T}_1^{(m)} - \mu \tilde{T}_2^{(m)}), \\
\frac{d\tilde{v}_m}{d\varphi} + \tilde{w}_m &= \frac{r_0}{E\delta}(\tilde{T}_2^{(m)} - \mu \tilde{T}_1^{(m)}), \\
\frac{d\tilde{u}_m}{d\varphi} + \lambda_m \tilde{v}_m &= \frac{2(1+\mu) r_0}{E\delta} \tilde{S}^{(m)}, \\
\lambda_m^2 \tilde{w}_m &= i 2 b^2 \frac{r_0}{E\delta} \tilde{T}_2^{(m)}, \\
-\frac{d^2 \tilde{w}_m}{d\varphi^2} + \frac{d\tilde{v}_m}{d\varphi} &= i 2 b^2 \frac{r_0}{E\delta} \tilde{T}_1^{(m)}, \\
\lambda_m \left(\frac{d\tilde{w}_m}{d\varphi} - \tilde{v}_m \right) &= i 2 b^2 \frac{r_0}{E\delta} \tilde{S}^{(m)},
\end{aligned} \tag{42.27}$$

the first and fourth of which immediately give

$$\begin{aligned}
\tilde{w}_m &= i \frac{2b^2}{\lambda_m^2} \frac{r_0}{E\delta} \tilde{T}_2^{(m)}, \\
\tilde{u}_m &= -\frac{1}{\lambda_m} \frac{r_0}{E\delta} (\tilde{T}_1^{(m)} - \mu \tilde{T}_2^{(m)}) \approx -\frac{1}{\lambda_m} \frac{r_0}{E\delta} \tilde{T}_1^{(m)}.
\end{aligned} \tag{42.28}$$

After this, one obtains from the third equation

$$\tilde{v}_m = \frac{1}{\lambda_m} \frac{2(1+\mu) r_0}{E\delta} \tilde{S}^{(m)} + \frac{1}{\lambda_m^2} \frac{r_0}{E\delta} \frac{d}{d\varphi} [\tilde{T}_1^{(m)} - \mu \tilde{T}_2^{(m)}]. \tag{42.29}$$

Substituting in (42.28) and (42.29) for $\tilde{T}_1^{(m)}$, $\tilde{T}_2^{(m)}$, $\tilde{S}^{(m)}$ their expressions given by (42.25), one has

$$\begin{aligned}
\tilde{w}_m &= -\frac{r_0}{E\delta} \frac{1}{\lambda_m^2} \sum_{j=1}^{j=4} (\gamma_j^2 - \lambda_m^2) \tilde{C}_j e^{\gamma_j \varphi}, \\
\tilde{u}_m &\approx -\frac{r_0}{E\delta} \frac{1}{\lambda_m} \sum_{j=1}^{j=4} \tilde{C}_j e^{\gamma_j \varphi}, \\
\tilde{v}_m &\approx \frac{r_0}{E\delta} \frac{1}{\lambda_m^2} \sum_{j=1}^{j=4} \frac{\gamma_j^2 - (2+\mu)\lambda_m^2}{\gamma_j} \tilde{C}_j e^{\gamma_j \varphi}.
\end{aligned} \tag{42.30}$$

One is now faced with the problem of deducing from the complex forces and displacements the real forces, moments and displacements respectively. This can be done by separating real and imaginary parts in the above expressions, where the constants $\tilde{C}_1$, $\tilde{C}_2$, $\tilde{C}_3$, $\tilde{C}_4$ must be assumed to be complex.

No space will be devoted here to the purely formal manipulations which by now should be sufficiently clear. However, the final results will be written down. They give for the displacements, the angle of rotation ψ, the stress and moment resultants the series

$$\begin{aligned} &u = \sum u_m \cos \lambda_m \xi, \qquad v = \sum v_m \sin \lambda_m \xi, \qquad w = \sum w_m \sin \lambda_m \xi, \\ &\psi = \sum \psi_m \cos \lambda_m \xi, \\ &T_1 = \sum T_1{}^{(m)} \sin \lambda_m \xi,\ T_2 = \sum T_2{}^{(m)} \sin \lambda_m \xi,\ S = \sum S^{(m)} \cos \lambda_m \xi, \\ &N_1 = \sum N_1{}^{(m)} \cos \lambda_m \xi,\ N_2 = \sum N_2{}^{(m)} \sin \lambda_m \xi, \\ &M_1 = \sum M_1{}^{(m)} \sin \lambda_m \xi,\ M_2 = \sum M_2{}^{(m)} \sin \lambda_m \xi,\ H = \sum H^{(m)} \cos \lambda_m \zeta, \end{aligned} \tag{42.31}$$

where the coefficients will expressed by formulae of one of the following two types:

$$\begin{aligned} \text{(I)} = {} & (\alpha C_1+\beta C_2)\Phi_1+(\alpha C_2-\beta C_1)\Phi_2+(\alpha C_3-\beta C_4)\Phi_3+(\alpha C_4+\beta C_3)\Phi_4+ \\ & +(\alpha' C_1'+\beta' C_2')\Psi_1+(\alpha' C_2'-\beta' C_1')\Psi_2+ \\ & +(\alpha' C_3'-\beta' C_4')\Psi_3+(\alpha' C_4'+\beta' C_3')\Psi_4, \\ \text{(II)} = {} & (\alpha C_4+\beta C_3)\Phi_1+(\alpha C_3-\beta C_4)\Phi_2+(\alpha C_2-\beta C_1)\Phi_3+(\alpha C_1+\beta C_2)\Phi_4+ \\ & +(\alpha' C_4'+\beta' C_3')\Psi_1+(\alpha' C_3'-\beta' C_4')\Psi_2+ \\ & +(\alpha' C_2'-\beta' C_1')\Psi_3+(\alpha' C_1'+\beta' C_2')\Psi_4. \end{aligned}$$

In fact, the coefficients of u_m, w_m, $T_1{}^{(m)}$, $T_2{}^{(m)}$, $N_1{}^{(m)}$, $M_1{}^{(m)}$, $M_2{}^{(m)}$ are determined by formulae of the type (I), the coefficients of v_m, $S^{(m)}$, $H^{(m)}$, $N_2{}^{(m)}$, ψ_m by those of the type (II). In those formulae:

C_j, C_j' are eight, real, arbitrary constants by means of which one can satisfy on each of the two straight edges of the cylindrical plate four edge conditions;

α, β, α', β' are parameters which take for each of the coefficients a special value;

Φ_j, Ψ_j are the following combinations of trigonometric and hyperbolic functions:

$$\begin{aligned} \Phi_1 &= \cosh c_1\varphi \cos d_1\varphi, & \Psi_1 &= \cosh c_2\varphi \cos d_2\varphi, \\ \Phi_2 &= \sinh c_1\varphi \sin d_1\varphi, & \Psi_2 &= \sinh c_2\varphi \sin d_2\varphi, \\ \Phi_3 &= \cosh c_1\varphi \sin d_1\varphi, & \Psi_3 &= \cosh c_2\varphi \sin d_2\varphi, \\ \Phi_4 &= \sinh c_1\varphi \cos d_1\varphi, & \Psi_4 &= \sinh c_2\varphi \cos d_2\varphi. \end{aligned} \tag{42.32}$$

Table 3 gives the formulae for the calculation of the parameters α, β, α', β' for all the coefficients of the series (42.31).

In the derivation of these formulae, it has been assumed that for the displacement w the parameters $\alpha = \alpha' = 1$, $\beta = \beta' = 0$.

The above information makes it possible to construct the series (42.31) for given shell dimensions. The number of terms in these series will correspond to the number of terms in the series for the surface loading, i.e., in the series (42.3). For this, the terms corresponding to each value of m must be subjected individually to the boundary conditions at the straight edges of the cylindrical plate (in a similar manner this must be done in the method of Maurice Lévy for ordinary plates). Hence it is clear that the complexity of the calculations increases proportionally to the number of terms retained in the series (42.3).

Mostly, cylindrical plates are analyzed for loadings which do not vary along their length (for example, roofs under their own weight or shells subject to uniform normal pressure).

The expansion of the constant quantity B in a trigonometric series for the interval $0 \leqslant \xi \leqslant \xi_0$ is known to have the form

$$B = \frac{4}{\pi} B \sum_{m=1}^{\infty} \frac{1}{m} \sin \lambda_m \xi, \tag{42.33}$$

where the summation extends only over the odd integers m and the λ_m are determined by (42.4). This series does not converge fast (especially not near the ends of the interval). Nevertheless, in order to simplify the calculations, it is customary to approximate the loading by a single term of the given series, i.e., to replace the rectangular load distribution along the length of the shell by a sinusoidal distribution of area $0.81pl$.

Although the total loading on the shell has been somewhat altered, this unfavorable situation (from the point of view of the accuracy of the analysis) is compensated by the fact that the loading is redistributed from the support to the middle of the span. This method is justified by the fact that the series for the displacements, forces and moments, at least at the sections where they attain maxima, converge better than the series for the loading, so that the error inherent in the analysis turns out to be smaller than the error incurred by retaining in (42.33) only the first term.

Thus, in what follows, it will be assumed that the series (42.31) contain only first terms corresponding to $m = 1$. The coefficients of these terms will be distinguished by primes (for example, u', T_1', ..., etc.). In addition, the notation

$$\lambda_1 = \frac{\pi r_0}{L} = \lambda, \tag{42.34}$$

will be adopted, i.e., the subscript of λ_1 will be omitted.

Table 3

Formulae for determining α, β, α', β'

	α	β	α'	β'	Type of formula
u_m	$\frac{1}{2b}$	$-\frac{1+(1+\mu)f_m}{2b}$	$-\frac{1}{2b}$	$-\frac{1-(1+\mu)f_m}{2b}$	I
v_m	$-\frac{d_1}{b\lambda_m}\left[1-\frac{\mu}{2}f_m+\frac{1}{2}\sqrt{1+(1+f_m)^2}\right]$	$\frac{c_1}{b\lambda_m}\left[1-\frac{\mu}{2}f_m-\frac{1}{2}\sqrt{1+(1+f_m)^2}\right]$	$-\frac{d_2}{b\lambda_m}\left[1+\frac{\mu}{2}f_m-\frac{1}{2}\sqrt{1+(1-f_m)^2}\right]$	$\frac{c_2}{b\lambda_m}\left[1+\frac{\mu}{2}f_m+\frac{1}{2}\sqrt{1+(1-f_m)^2}\right]$	II
w_m	1, 0	0, 0	1, 0	0, 0	I
$r_0\psi_m$	$-c_1+\alpha_{v_m}$	$-d_1+\beta_{v_m}$	$-c_2+\alpha'_{v_m}$	$-d_2+\beta'_{v_m}$	II
$\frac{r_0}{E\delta}T_1^{(m)}$	$-\frac{\lambda_m}{2b}$	$\frac{\lambda_m}{2b}(1+f_m)$	$\frac{\lambda_m}{2b}$	$\frac{\lambda_m}{2b}(1-f_m)$	I
$\frac{r_0}{E\delta}T_2^{(m)}$	0, 0	$-\frac{\lambda_m^2}{2b^2}$	0, 0	$\frac{\lambda_m^2}{2b^2}$	I
$\frac{r_0}{E\delta}S^{(m)}$	$\frac{\lambda_m}{2b^2}d_1$	$-\frac{\lambda_m}{2b^3}c_1$	$-\frac{\lambda_m}{2b^2}d_2$	$\frac{\lambda_m}{2b^2}c_2$	II
$\frac{r_0}{E\delta}N_1^{(m)}$	$\frac{\lambda_m}{4b^4}(\lambda_m^2-c_1^2+d_1^2)$	$-\frac{c_1d_1\lambda_m}{2b^4}$	$\frac{\lambda_m}{4b^4}(\lambda_m^2-c_2^2+d_2^2)$	$-\frac{c_2d_2\lambda_m}{2b^4}$	I
$\frac{r_0}{E\delta}N_2^{(m)}$	$\frac{c_1}{4b^4}(\lambda_m^2-c_1^2+3d_1^2)$	$\frac{d_1}{4b^4}(\lambda_m^2-3c_1^2+d_1^2)$	$\frac{c_2}{4b^4}(\lambda_m^2-c_2^2+3d_2^2)$	$\frac{d_2}{4b^4}(\lambda_m^2-3c_2^2+d_2^2)$	II
$\frac{1}{E\delta}M_1^{(m)}$	$-\frac{\lambda_m}{4b^3}[\mu-(1-\mu)f_m]$	$-\mu\frac{\lambda_m}{4b^3}$	$\frac{\lambda_m}{4b^3}[\mu+(1-\mu)f_m]$	$-\mu\frac{\lambda_m}{4b^3}$	I
$\frac{1}{E\delta}M_2^{(m)}$	$-\frac{\lambda_m}{4b^3}[1+(1-\mu)f_m]$	$-\frac{\lambda_m}{4b^3}$	$\frac{\lambda_m}{4b^3}[1-(1-\mu)f_m]$	$-\frac{\lambda_m}{4b^3}$	I
$\frac{1}{E\delta}H^{(m)}$	$-\frac{(1-\mu)c_1\lambda_m}{4b^4}$	$-\frac{(1-\mu)d_1\lambda_m}{4b^4}$	$-\frac{(1-\mu)c_2\lambda_m}{4b^4}$	$-\frac{(1-\mu)d_2\lambda_m}{4b^4}$	II

Further, one has to design a convenient scheme for the computations which in spite of the above stated simplifications will still be very involved. Such a practical scheme will now be given [98], having in mind the case which, while being simplest, occurs most often in practice, namely when the boundary conditions at the straight edges are the same and the loading is symmetrical with respect to the plane $\varphi = 0$.

Under these circumstances, the coefficients u_m, w_m, $T_1^{(m)}$, $T_2^{(m)}$, $N_1^{(m)}$, $M_1^{(m)}$, $M_2^{(m)}$ will be even and the coefficients v_m, ψ_m, $H^{(m)}$, $S^{(m)}$, $N_2^{(m)}$ odd functions of φ, as a consequence of which one must put

$$C_3 = C_4 = C_3' = C_4' = 0.$$

With the help of the remaining four integration constants the solution may be subjected to boundary conditions at one of the rectilinear edges (for example, at the edge $\varphi = \varphi_0$), while, by virtue of the symmetry of the deformation, they will be automatically satisfied at the other edge.

It is convenient to consider beforehand the following four types of boundary conditions:

$$\left.\begin{aligned} u' &= 1, & v' = w' = \psi' &= 0 \\ v' &= 1, & u' = w' = \psi' &= 0 \\ w' &= 1, & u' = v' = \psi' &= 0 \\ r_0\psi' &= 1, & u' = v' = w' &= 0 \end{aligned}\right\} \text{ for } \varphi = \varphi_0 \tag{42.35}$$

by determining for each of them the values of the constants C_1, C_2, C_1', C_2'.

Substituting these four constants into the formulae for the forces S, T_2, N_2 and the moment M_2, the boundary values of these quantities, corresponding to the above types of boundary conditions at the straight edges, will now be obtained. The results may be presented in the form of the following table:

	$u' = 1$	$v' = 1$	$w' = 1$	$r_0\psi' = 1$	(*)
$\frac{r_0}{E\delta} S'$	a_{11}	a_{12}	a_{13}	a_{14}	
$\frac{r_0}{E\delta} T_2'$	a_{21}	a_{22}	a_{23}	a_{24}	
$\frac{r_0}{E\delta} N_2'$	a_{31}	a_{32}	a_{33}	a_{34}	
$\frac{1}{E\delta} M_2'$	a_{41}	a_{42}	a_{43}	a_{44}	

The coefficients a_{jk} in this table completely characterize the elastic properties of the straight edges. Once they are known, the forces and moments at the edges can be presented as linear functions of the edge displacements and the angle of rotation (strictly speaking this may be done with regard to the coefficients u', v', ..., T_2', S_1', ...).

The corresponding expressions will have the form

$$\begin{aligned}
\frac{r_0}{E\delta} S' &= a_{11}u' + a_{12}v' + a_{13}w' + a_{14}(r_0\psi') + \frac{r_0}{E\delta} s, \\
\frac{r_0}{E\delta} T_2' &= a_{21}u' + a_{22}v' + a_{23}w' + a_{24}(r_0\psi') + \frac{r_0}{E\delta} t, \\
\frac{r_0}{E\delta} N_2' &= a_{31}u' + a_{32}v' + a_{33}w' + a_{34}(r_0\psi') + \frac{r_0}{E\delta} n, \\
\frac{1}{E\delta} M_2' &= a_{41}u' + a_{42}v' + a_{43}w' + a_{44}(r_0\psi') + \frac{1}{E\delta} m,
\end{aligned} \tag{42.36}$$

where the quantities s, t, n and m must be equal to the boundary values of S', T_2', N_2' and M_2' for the boundary conditions

$$u = v = w = \psi = 0,$$

i.e., for rigidly clamped straight edges. They will only differ from zero when the surface loading on the shell is distinct from zero, whence it is clear in the formulae (42.36) that they represent the particular integrals of the problem which were to be taken from the membrane theory.

It will now be assumed that the membrane problem for the given surface loading has been solved and that the corresponding values of the displacements and of the angle of rotation u^*, v^*, w^*, ψ^* are known. In order to arrive at the general solution, one must add these results to the solution of the homogeneous problem which consequently will have the form

$$u = u_0 + u^*, \quad v = v_0 + v^*, \quad w = w_0 + w^*, \quad \psi = \psi_0 + \psi^*, \tag{42.37}$$

where u_0, v_0, w_0, ψ_0 are the expressions for the displacements (involving the constants of integration) the determination of which has been explained above. Thus the conditions for the rigid clamping of the straight edges reduce to

$$u_0' = - u^{*\prime}, \quad v_0' = - v^{*\prime}, \quad w_0' = - w^{*\prime}, \quad \psi_0' = - \psi^{*\prime} \tag{42.38}$$

which are to hold at $\varphi = \varphi_0$. [As before, the primes indicate the coefficients of the first terms in the expansions (42.31)].

Hence it follows that the quantities s, t, n and m are given by the formulae

$$\begin{aligned}
\frac{r_0}{E\delta}s &= -a_{11}u^{*\prime} - a_{12}v^{*\prime} - a_{13}w^{*\prime} - a_{14}(r\psi^{*\prime}) + \frac{r_0}{E\delta}S^{*\prime},\\
\frac{r_0}{E\delta}t &= -a_{21}u^{*\prime} - a_{22}v^{*\prime} - a_{23}w^{*\prime} - a_{24}(r\psi^{*\prime}) + \frac{r_0}{E\delta}T_2^{*\prime},\\
\frac{r_0}{E\delta}n &= -a_{31}u^{*\prime} - a_{32}v^{*\prime} - a_{33}w^{*\prime} - a_{34}(r\psi^{*\prime}),\\
\frac{r_0}{E\delta}m &= -a_{41}u^{*\prime} - a_{42}v^{*\prime} - a_{43}w^{*\prime} - a_{44}(r\psi^{*\prime}),
\end{aligned} \tag{42.39}$$

where $u^{*\prime}$, $v^{*\prime}$, $w^{*\prime}$, $S^{*\prime}$, $T_2^{*\prime}$ are understood to be the boundary values of the particular integrals taken from the membrane theory.

By means of these formulae, one may fill in the fifth column (*) of the table on page 236 which was intended for the free terms in the expressions (42.36).

The availability of the expressions (42.36) makes it much simpler to reveal the statically indeterminate nature of composite shells with supports at their straight edges. This fact justifies the effort of setting up the stiffness table. Further, the execution of the stress analysis of cylindrical plates by use of the intermediate stiffness table has one more important advantage. In fact, by the theorem of virtual work, the coefficients a_{jk} must be symmetrical so that

$$a_{jk} = a_{kj}. \tag{42.40}$$

The accuracy of the computations can be assessed by the degree to which these conditions are fulfilled.

Thus, by constructing tables of the above type, one is in a position to check the computations which is very desirable in view of their great complexity. However, it should be noted that, since the solution of this section has been obtained by neglecting a number of small terms, the conditions (42.40) may be fulfilled exactly and nevertheless this will not ensure the same degree of accuracy of the computations. Therefore a 10—15% violation of (42.40) must be considered legitimate and only disagreements exceeding this limit must be interpreted as indications of arithmetical errors in the computations.

§ 43. The evaluation of the stresses in oval shells composed of four cylindrical plates. I.

A thin-walled cylinder with its cross-section formed by four smoothly joined arcs of circles two of which have the radius r_1 and two the radius r_2 (Fig. 55) will now be considered and the strains and stresses due to uniform normal pressure evaluated.

The radius of curvature of the shell whose section is shown in Fig. 55 varies discontinuously and therefore the formulae deduced in § 37 are not applicable. Its stress analysis may be approached by dividing the shell into four interconnected cylindrical plates. In this way, the problem is reduced to the analysis of these cylindrical plates subject to uniform normal pressure with special consideration of the conditions at their common boundaries.

All four plates will be deformed symmetrically with respect to their planes $\varphi = 0$, and the deformations of opposite plates will obviously be equal. Hence, only two cylindrical plates have to be considered (for example, the plates I and II of Fig. 56) with statically indeterminate conditions along their junctions. Below, this analysis will be performed for a specific shell and the calculations will be carried through in the same order which has been indicated in the preceding section.

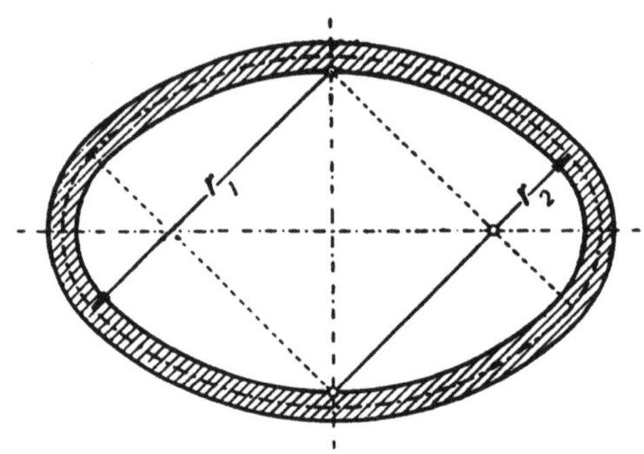

Fig. 55.

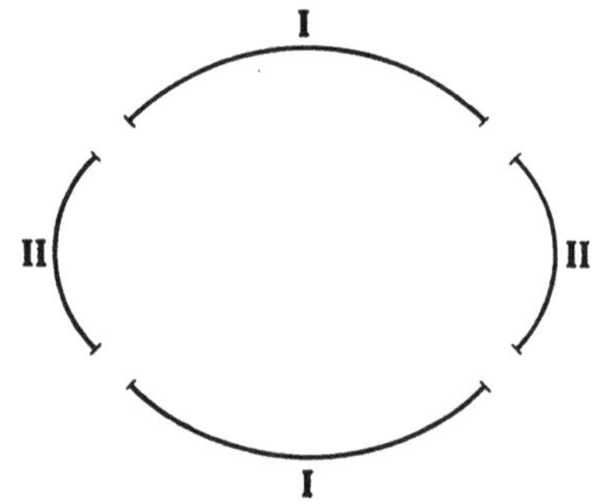

Fig. 56.

As a first step, the membrane solution for a cylindrical shell loaded by uniform normal pressure $q_n = p =$ constant will be determined. Representing this load distribution in the form of the series (42.33) and retaining in the series only the first term, one has

$$q_n = \frac{4}{\pi} p \sin \lambda\xi = q_n' \sin \lambda\xi, \quad q_1 = q_2 = 0. \tag{43.1}$$

Then the equations (42.10) give

$$T_2^* = \frac{4}{\pi} p r_0 \sin \lambda\xi, \quad T_1^* = S^* = 0. \tag{43.2}$$

Forming next the equations for the displacements

$$\frac{\partial u}{\partial \xi} = \frac{r_0}{E\delta}(T_1^* - \mu T_2^*) = -\mu \frac{4}{\pi} \frac{p r_0^2}{E\delta} \sin \lambda\xi,$$

$$\frac{\partial v}{\partial \varphi} + w = \frac{r_0}{E\delta}(T_2^* - \mu T_1^*) = \frac{4}{\pi} \frac{p r_0^2}{E\delta} \sin \lambda\xi, \tag{43.3}$$

$$\frac{\partial u}{\partial \varphi} + \frac{\partial v}{\partial \xi} = 0,$$

one finds

$$u^* = u_1^* \cos \lambda\xi = \frac{4}{\pi} \frac{\mu}{\lambda} \frac{p r_0^2}{E\delta} \cos \lambda\xi,$$

$$v^* = 0, \tag{43.4}$$

$$w^* = \frac{4}{\pi} \frac{p r_0^2}{E\delta} \sin \lambda\xi.$$

The expressions (43.2) and (43.4) will be used later on as particular integrals of this problem, where one will have for the shell I the radius $r_0 = r_1$ and for the shell II the radius $r_0 = r_2$.

I. Calculation of the stiffness table for the edges of the first cylindrical plate

Given data:

$$\frac{r_1}{\delta} = 150, \quad \frac{r_1}{L} = 0.625, \quad \varphi_0 = 45°, \quad \lambda = \frac{\pi r_1}{L} = 1.964.$$

$$4b^4 = 12(1 - \mu^2)\frac{r_1^2}{\delta^2} = 2.45 . 10^5, \quad b = 15.72.$$

Calculation of the roots of the characteristic Equation (42.16)

$$f = \frac{\lambda^2 - \frac{1}{2}}{b\lambda} = 0.108,$$

$$c_1 = 6.37, \quad d_1 = 2.42, \quad c_2 = 2.65, \quad d_2 = 5.84,$$

$$c_1\varphi_0 = 5.01, \quad d_1\varphi_0 = 1.90, \quad c_2\varphi_0 = 2.08, \quad d_2\varphi_0 = 4.59.$$

The boundary values of the functions Φ and Ψ:

$$\Phi_1(\varphi_0) = -24.2, \quad \Phi_2(\varphi_0) = 70.9, \quad \Phi_3(\varphi_0) = 7.09, \quad \Phi_4(\varphi_0) = -24.2,$$

$$\Psi_1(\varphi_0) = -0.496, \quad \Psi_2 = -3.91, \quad \Psi_3 = -4.04, \quad \Psi_4(\varphi_0) = -0.481.$$

On the basis of the following table and the boundary values of the functions Φ and Ψ, the boundary values of the displacements and of the angle of rotation will be expressed in terms of the constants of integration by the following formulae:

$$\begin{aligned} u' &= 1.86C_1 + 3.16C_2 - 0.088C_1' + 0.138C_2', \\ v' &= -0.05C_1 - 10.75C_2 + 0.612C_1' + 0.201C_2', \\ w' &= -24.2C_1 + 70.9C_2 - 0.496C_1' - 3.91C_2', \\ r_1\psi' &= -326C_1 + 414C_2 + 21.7C_1' - 13.74C_2'. \end{aligned} \tag{43.5}$$

Table 4

Values of the coefficients α, β, α', β'

	α	β	α'	β'
u'	$3.19\cdot10^{-2}$	$-3.71\cdot10^{-2}$	$-3.19\cdot10^{-2}$	$-2.67\cdot10^{-2}$
v'	$-1.37\cdot10^{-1}$	$4.70\cdot10^{-2}$	$-6.70\cdot10^{-2}$	$1.44\cdot10^{-2}$
w'	1.00	0.00	1.00	0.00
$r_1\Psi'$	-6.51	-2.37	-2.72	-5.70
$\frac{r_1}{E\delta}S'$	$9.63\cdot10^{-3}$	$-25.4\cdot10^{-3}$	$-23.3\cdot10^{-3}$	$10.55\cdot10^{-3}$
$\frac{r_1}{E\delta}T_2'$	0.00	$-7.82\cdot10^{-3}$	0.00	$7.82\cdot10^{-3}$
$\frac{r_1}{E\delta}N_2'$	$-4.97\cdot10^{-4}$	$-1.105\cdot10^{-3}$	$1.072\cdot10^{-3}$	$4.03\cdot10^{-4}$
$\frac{1}{E\delta}M_2'$	$-1.37\cdot10^{-4}$	$-1.26\cdot10^{-4}$	$1.15\cdot10^{-4}$	$-1.26\cdot10^{-4}$
$\frac{r_1}{E\delta}T_1'$	$-6.25\cdot10^{-2}$	$7.03\cdot10^{-2}$	$6.25\cdot10^{-2}$	$5.46\cdot10^{-2}$

Employing the given expressions, one may set up the four systems of equations corresponding to the case for $\varphi = \varphi_0$:

$$\begin{aligned} u' &= 1, & v' &= w' = \psi' = 0, \\ v' &= 1, & u' &= w' = \psi' = 0, \\ w' &= 1, & u' &= v' = \psi' = 0, \\ r_1\psi' &= 1, & u' &= v' = w' = 0. \end{aligned} \tag{43.6}$$

Thus, the following table of values of the constants of integration is obtained; the fifth column (*) gives the constants, calculated on the basis of the assumption that the edges of the shell are rigidly clamped, so that there

$$u' = v' = w' = \psi' = 0. \tag{43.7}$$

This column may be easily calculated from the preceding ones. In fact, the conditions (43.7) reduce to the system of equations, obtained from (42.36), (42.37),

$$u_0' = -u^{*\prime},\ v_0' = -v^{*\prime},\ w_0' = -w^{*\prime},\ \psi_0' = -\psi^{*\prime}, \tag{43.8}$$

Table 5

	$u' = 1$	$v' = 1$	$w' = 1$	$r_1\psi' = 1$	(*)
C_1	0.381	0.216	0.0380	0.00374	$-0.0961\,q'$
C_2	0.162	0.00318	0.00354	$-6.63\cdot 10^{-4}$	$-0.0283\,q'$
C_1'	2.82	2.22	0.214	$2.3\cdot 10^{-4}$	$-0.644\,q'$
C_2'	0.206	-1.56	-0.455	-0.0351	$0.0037\,q'$

where u_0 .... refer to the solutions of the homogeneous problem and u^* ...., to the particular integrals. For the loading case under consideration, the last are given by (43.4)

$$\begin{aligned} v^* &= \psi^* = 0, \\ u^{*\prime} &= \frac{\mu}{\lambda}\frac{4}{\pi}\frac{pr_1^2}{E\delta} = \frac{\mu}{\lambda}q', \\ w^{*\prime} &= \frac{4}{\pi}\frac{pr_1^2}{E\delta} = q'. \end{aligned} \tag{43.9}$$

Hence, it is seen that the fifth column (*) may be obtained by multiplying the first column by $-\frac{\mu}{\lambda}q'$ $\left(\text{where } q' = \frac{4}{\pi}\frac{pr_1^2}{E\delta}\right)$, the third by $-q'$ and by adding these results.

On the basis of Table 5 and of the boundary values of the auxiliary functions Φ and Ψ, one may write the following expressions for the boundary forces and moments in terms of the integration constants

$$\frac{r_1}{E\delta}S' = 1.562C_1 + 1.295C_2 + 0.0537C_1' + 0.0889C_2',$$

$$\begin{aligned} \frac{r_1}{E\delta} T_2' &= 0.555C_1 + 0.1895C_2 + 0.0306C_1' - 0.00388C_2', \\ \frac{r_1}{E\delta} N_2' &= [90.3C_1 - 8.5C_2 + 1.06C_1' - 4.30C_2']10^{-3}, \\ \frac{1}{E\delta} M_2' &= [12.24C_1 - 6.66C_2 - 0.549C_1' - 0.388C_2']10^{-3}. \end{aligned} \tag{43.10}$$

Substituting here the values of C_1, C_2, C_1', C_2' from Table 5, one obtains the stiffness table for the edge of the shell:

Table 6

	$u' = 1$	$v' = 1$	$w' = 1$	$r_1\psi = 1$	(*)
$\frac{r_1}{E\delta} S'$	$9.76 \cdot 10^{-1}$	$3.20 \cdot 10^{-1}$	$3.46 \cdot 10^{-2}$	$1.88 \cdot 10^{-3}$	$-1.83 \cdot 10^{-1} q'$
$\frac{r_1}{E\delta} T_2'$	$3.27 \cdot 10^{-1}$	$1.94 \cdot 10^{-1}$	$3.02 \cdot 10^{-2}$	$2.09 \cdot 10^{-3}$	$9.20 \cdot 10^{-1} q'$
$\frac{r_1}{E\delta} N_2'$	$3.51 \cdot 10^{-2}$	$2.85 \cdot 10^{-2}$	$5.6 \cdot 10^{-3}$	$4.87 \cdot 10^{-4}$	$-1.1 \cdot 10^{-2} q'$
$\frac{1}{E\delta} M_2'$	$1.93 \cdot 10^{-3}$	$2.01 \cdot 10^{-3}$	$5.02 \cdot 10^{-4}$	$6.37 \cdot 10^{-5}$	$-8.0 \cdot 10^{-4} q'$

The fifth column (*) of this table contains the values of the boundary forces and moments for the boundary conditions (43.7). It has been computed in the manner stated in the preceding section, where, as before,

$$q' = \frac{4}{\pi} p \frac{r_1^2}{E\delta}.$$

A study of the first four columns of this table shows that the terms in it possess the necessary symmetry about the principal diagonal, thus confirming the accuracy of the computations.

It is true that this symmetry is not complete, but (considering that the formulae which have been employed entail certain errors and also that all calculations have been carried out on a slide rule) the deviations are within the limits admissible in any technical calculations.

II. Calculation of the stiffness table for the edges of the second cylindrical plate

In an analogous manner, one can obtain the influence coefficients for the second cylindrical plate.

The results will be given here without explanation:

$$\frac{r_2}{\delta} = 40, \quad \frac{r_2}{L} = \frac{1}{6}, \quad \lambda = \frac{\pi}{6} = 0.524, \quad \varphi_0 = 45°,$$

$$4b^4 = 12(1-\mu^2)\frac{r_2{}^2}{\delta^2} = 1.745.10^4, \quad b = 8.12,$$

$$f = -0.053,$$

$$\begin{array}{llll} c_1 = 2.31, & d_1 = 0.915, & c_2 = 0.959, & d_2 = 2.21, \\ c_1\varphi_0 = 1.820, & d_1\varphi_0 = 0.719, & c_2\varphi_0 = 0.753, & d_2\varphi_0 = 1.735, \\ \Phi_1 = 2.38, & \Phi_2 = 1.98, & \Phi_3 = 2.09, & \Phi_4 = 2.26, \\ \Psi_1 = -0.211, & \Psi_2 = 0.815, & \Psi_3 = 1.278, & \Psi_4 = -0.1345. \end{array}$$

Table 7

	α	β	α'	β'
u'	$6.15\cdot10^{-2}$	$-6.66\cdot10^{-2}$	$-6.15\cdot10^{-2}$	$-5.63\cdot10^{-2}$
v'	$3.68\cdot10^{-1}$	$1.38\cdot10^{-1}$	$-1.74\cdot10^{-1}$	$3.84\cdot10^{-1}$
w'	1.00	0.00	1.00	0.00
$r_2\psi'$	-2.683	-0.777	-1.132	-1.826
$\frac{r_2}{E\delta}S'$	$3.63\cdot10^{-3}$	$-9.19\cdot10^{-3}$	$-8.76\cdot10^{-3}$	$3.81\cdot10^{-3}$
$\frac{r_2}{E\delta}T_2'$	0.00	$-2.08\cdot10^{-3}$	0.00	$2.08\cdot10^{-3}$
$\frac{r_2}{E\delta}N_2'$	$-3.16\cdot10^{-4}$	$-7.74\cdot10^{-4}$	$7.8\cdot10^{-4}$	$3.26\cdot10^{-4}$
$\frac{1}{E\delta}M_2'$	$-2.54\cdot10^{-4}$	$-2.44\cdot10^{-4}$	$2.33\cdot10^{-4}$	$-2.44\cdot10^{-4}$
$\frac{r_2}{E\delta}T_1'$	$-3.22\cdot10^{-2}$	$3.43\cdot10^{-2}$	$3.22\cdot10^{-2}$	$3.02\cdot10^{-2}$

$$\begin{aligned} u' &= 0.279C_1 - 0.037C_2 + 0.0596C_1' - 0.0378C_2', \\ v' &= -1.12C_1 - 0.456C_2 - 0.476C_1' - 0.273C_2', \\ w' &= 2.38C_1 + 1.98C_2 - 0.211C_1' + 0.815C_2', \\ -r_2\psi' &= 4.45C_1 + 7.36C_2 - 2.49C_1' + 1.20C_2'. \end{aligned} \tag{43.11}$$

Table 8

	$u' = 1$	$v' = 1$	$w' = 1$	$r_2\psi' = 1$	(*)
C_1	3.17	0.561	0.430	0.065	$-2.24\, q^{(2)}$
C_2	-1.97	-1.475	-1.070	-0.317	$2.20\, q^{(2)}$
C_1'	-2.64	-2.765	-1.278	-0.160	$2.79\, q^{(2)}$
C_2'	-5.19	1.225	$+2.18$	$+0.537$	$0.79\, q^{(2)}$

$$q^{(2)} = \frac{4}{\pi} \frac{pr_2^2}{E\delta}. \tag{43.12}$$

$$\begin{aligned} \frac{r_2}{E\delta} S' &= [2.79C_1 - 1.318C_2 - 0.349C_1' - 1.17C_2']10^{-2}, \\ \frac{r_2}{E\delta} T_2' &= [0.412C_1 - 0.496C_2 - 0.170C'_1 - 0.043C_2']10^{-2}, \\ \frac{r_2}{E\delta} N_2' &= [9.0C_1 - 24.1C_2 - 5.21C_1' + 9.56C_2']10^{-4}, \\ \frac{1}{E\delta} M_2' &= [-1.23C_1 - 10.85C_2 + 1.5C_1' + 2.42C_2']10^{-4}. \end{aligned} \tag{43.13}$$

Table 9

	$u' = 1$	$v' = 1$	$w' = 1$	$r_2\psi' = 1$	(*)
$\frac{r_2}{E\delta} S_1$	$1.83 \cdot 10^{-1}$	$3.07 \cdot 10^{-2}$	$5.1 \cdot 10^{-3}$	$2.6 \cdot 10^{-4}$	$-1.1 \cdot 10^{-1} q^{(2)}$
$\frac{r_2}{E\delta} T_2'$	$2.27 \cdot 10^{-2}$	$1.38 \cdot 10^{-2}$	$8.3 \cdot 10^{-3}$	$1.87 \cdot 10^{-3}$	$9.75 \cdot 10^{-1} q^{(2)}$
$\frac{r_2}{E\delta} N_2'$	$4.0 \cdot 10^{-3}$	$6.7 \cdot 10^{-3}$	$5.72 \cdot 10^{-3}$	$1.42 \cdot 10^{-3}$	$-8 \cdot 10^{-3} q^{(2)}$
$\frac{1}{E\delta} M_2'$	$3.5 \cdot 10^{-4}$	$1.42 \cdot 10^{-3}$	$1.42 \cdot 10^{-3}$	$4.38 \cdot 10^{-4}$	$-1.66 \cdot 10^{-3} q^{(2)}$

III. The joints between the cylindrical plates

Along a common straight edge of the plates I and II the following equalities must hold (Fig. 57):

$$\begin{aligned} &u_I' = u_{II}', \quad v_I' = -v_{II}', \quad w_I' = w_{II}', \quad \psi_I' = -\psi_{II}', \\ &S_I' = -S_{II}', \quad \underset{(I)}{T_2'} = \underset{(II)}{T_2'}, \quad \underset{(I)}{N_2'} = -\underset{(II)}{N_2'}, \quad \underset{(I)}{M_2'} = \underset{(II)}{M_2'}. \end{aligned} \tag{43.14}$$

(Since the positive directions of u, w, S and N_2 for both plates are the same and the positive directions for v, ψ, T_2, M_2 are opposite).

With the stiffness tables for both plates being available, one may write down on the basis of the eight relations above the following four equations with the four unknowns u_1, v_1, w_1, ψ_1 (where, for the sake of

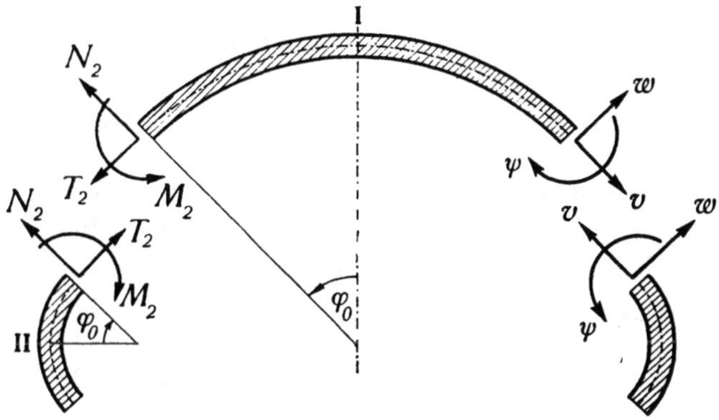

Fig. 57.

brevity, the subscript I will in future be omitted):

$$(1)\quad 0.976u' + 0.320v' + 0.0346w' + 0.00188(r_1\psi') - 0.183q' =$$
$$= -0.686u' + 0.115v' - 0.0191w' + 0.00026(r_1\psi') + 0.029q',$$

$$(2)\quad 0.327u' + 0.194v' + 0.302w' + 0.0021(r_1\psi') + 0.920q' =$$
$$= 0.111u' - 0.0518v' + 0.0312w' - 0.00187(r_2\psi') + 0.260q',$$

$$(3)\quad 0.035u' + 0.0285v' + 0.0056w' + 0.00049(r_2\psi') - 0.011\,q' =$$
$$= -0.0150u' + 0.0252v' - 0.021w' + 0.00142(r_2\psi') + 0.0021q',$$

$$(4)\quad 0.00193u' + 0.00201v' + 0.000502w' + 0.0000637(r_1\psi') - \qquad (43.15)$$
$$- 0.0008q' = 0.00035u' - 0.00142v' + 0.00142w' -$$
$$- 0.000117(r_1\psi') - 0.00019q'.$$

Transferring all the terms to the left hand side of these equations, one has

$$1.662u' + 0.205v' + 0.0537w' + 0.00162(r_1\psi') - 0.212q' = 0,$$
$$0.216u' + 0.246v' - 0.0010w' + 0.00397(r_1\psi') + 0.660q' = 0, \quad (43.16)$$
$$0.050u' + 0.0038v' + 0.0267w' - 0.00093(r_1\psi') - 0.132q' = 0,$$
$$0.00158u' + 0.00343v' - 0.00092w' + 0.000181(r_1\psi') - 0.00068q' = 0.$$

Hence one obtains for the displacements and the rotation at the edges of the first plate

$$u' = 0.475q', \; v' = -4.93q', \; w' = 4.24q', \; (r_1\psi') = 115q'. \qquad (43.17)$$

Similarly, the boundary values of the displacements and rotation for the second plate are

$$u_{II}' = u_I' = 0.475q', \; v_{II} = -v_I' = 4.93q', \; w_{II}' = w_I' = 4.24q',$$
$$(r_2\psi')_{II} = -(r_2\psi')_I = -\frac{r_2}{r_1} 115q' = -30.7q'. \qquad (43.18)$$

From the displacements at the edges of both plates one may determine the constants C_1, C_2, C_1', C_2' for each plate (taking advantage here of the tables which have already been prepared for these constants):

a) first plate

$$\begin{aligned} C_1 &= (0.181 - 1.061 + 0.161 + 0.430 - 0.096)q' = -0.385q', \\ C_2 &= (0.077 - 0.057 + 0.015 - 0.076 - 0.028)q' = -0.028q', \\ C_1' &= (1.34 - 10.95 + 0.906 + 0.026 - 0.64)q' = -9.32q', \\ C_2' &= (0.098 + 7.69 - 1.93 - 4.03 + 0.004)q' = 1.83q'; \end{aligned} \qquad (43.19)$$

b) second plate:

$$\begin{aligned} C_1 &= (1.51 + 2.77 + 1.82 - 1.99 - 0.159)q' = 3.85q', \\ C_2 &= (-0.935 - 7.27 - 4.54 + 9.73 - 0.156)q' = -2.86q', \\ C_1' &= (-1.255 - 13.62 - 5.41 + 4.91 + 0.198)q' = -15.17q', \\ C_2' &= (-2.47 + 6.05 + 9.24 - 16.45 + 0.056)q' = -3.57q'. \end{aligned} \qquad (43.20)$$

Having obtained all the constants, one can now write down the expressions for any of the displacements, forces and moments in an oval shell. Thus, for example, for the first plate:

$$w' = (-0.385\Phi_1 - 0.028\Phi_2 - 9.31\Psi_1 + 1.83\Psi_2 + 1)q',$$

$$\frac{r_1}{E\delta} S' = (-0.01\Phi_3 - 0.00301\Phi_4 + 0.056\Psi_3 + 0.236\Psi_4)q',$$

$$\frac{r_1}{E\delta} T' = (0.0221\Phi_1 + 0.0288\Phi_2 - 0.481\Psi_1 + 0.623\Psi_2)q', \qquad (43.21)$$

$$\sigma'_{M_2} = \frac{6M_2'}{\delta^2} = (0.0505\Phi_1 - 0.0402\Phi_2 - 1.17\Psi_1 + 0.623\Psi_2)\frac{4}{\pi}\frac{pr_1}{\delta};$$

for the second plate:

$$
\begin{aligned}
w' &= (3.85\Phi_1 - 2.86\Phi_2 - 15.17\Psi_1 - 3.57\Psi_2 + 0.071)q', \\
\sigma_{T_1}{}' &= \frac{T_1{}'}{\delta} = (-0.833\Phi_1 - 0.210\Phi_2 - 2.24\Psi_1 + 1.286\Psi_2)\frac{4}{\pi}\frac{pr_1}{\delta}, \\
\sigma_S{}' &= \frac{S'}{\delta} = (0.0937\Phi_3 + 0.151\Phi_4 + 0.333\Psi_3 + 0.446\Psi_4)\frac{4}{\pi}\frac{pr_1}{\delta}, \\
\sigma'_{M_2} &= \frac{6M_2{}'}{\delta^2} = (0.252\Phi_1 + 1.50\Phi_2 - 2.39\Psi_1 - 4.08\Psi_2)\frac{4}{\pi}\frac{pr_1}{\delta},
\end{aligned}
\qquad (43.22)
$$

where Φ_j and Ψ_j are the functions of the angle φ, defined by (42.32), the angle being measured from the plane of symmetry of each plate.

The formulae (43.21) and (43.22) give the variation of w', S', $T_1{}'$, $M_2{}'$ over the transverse section of the oval shell or, more accurately, the coefficients of the terms in the series (42.31), corresponding to $w = 1$. On this basis, the following table may be evaluated:

Table 10

φ	0	15°	25°	35°	45°	60°	75°	90°
$w' : \frac{pr_1^2}{E\delta}$	−11.1	+0.87	+17.7	+25.4	$\frac{+6.1}{+5.1}$	−4.3	−12.3	−14.4
$v' : \frac{pr_1^2}{E\delta}$	0.00	2.19	0.68	−3.32	$\frac{-6.9}{-6.85}$	−6.3	−4.0	0.00
$\sigma_{T_1}{}' : \frac{pr_1}{\delta}$	−0.585	+0.663	+1.86	+1.63	$\frac{-0.87}{-1.11}$	−2.41	−3.52	−3.91
$\sigma_S{}' : \frac{pr_1}{\delta}$	0.00	+0.05	−0.36	−1.06	$\frac{-1.25}{-1.23}$	−0.92	−0.57	0.00
$\sigma_{T_2}{}' : \frac{pr_1}{\delta}$	1.29	1.34	1.30	1.05	$\frac{0.625}{0.665}$	0.470	0.370	0.330
$\sigma_{M_2}{}' : \frac{pr_1}{\delta}$	−1.42	−0.88	+0.60	+2.32	$\frac{-0.14}{-0.32}$	−1.94	−2.71	−2.72

In using this table, one must take into consideration that here the same origin has been taken for the angle φ in both plates, namely, the angle $\varphi = 0$ corresponds to the plane of symmetry of the first plate and the angle $\varphi = 90°$ to the apex of the second plate.

The results in Table 10 are shown graphically in Figs. 58, 59.

First consider Fig. 58, showing on an exaggerated scale the deformation of the oval cross-section of the cylindrical shell under discussion when loaded by an internal pressure. This curve, first calculated in 1938, was for a long time considered incorrect. It shows naturally that an internal pressure must induce at the ends of the minor and major axes of the oval deformations of different signs, so that the eccentricity of the cross-section of the shell will tend to decrease. This is the same as would be the case if the radius of curvature of the oval were a continuous function of the arc length (see § 35). However, for shells with discontinuously changing radii of curvature, the deformation of the cross-section, as shown by Fig. 58, is more complicated. Experimental results (shown dotted in

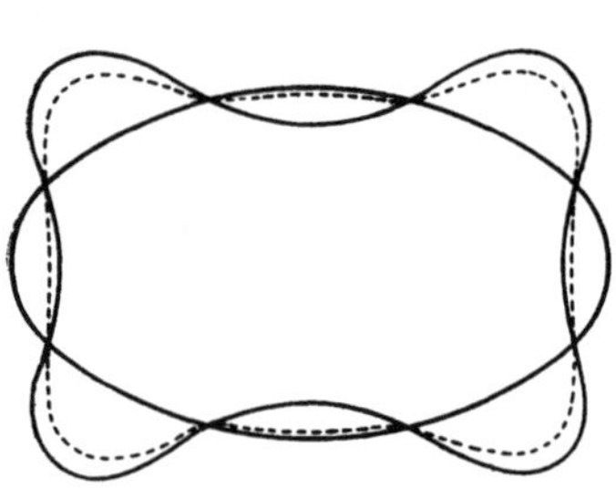

Fig. 58.

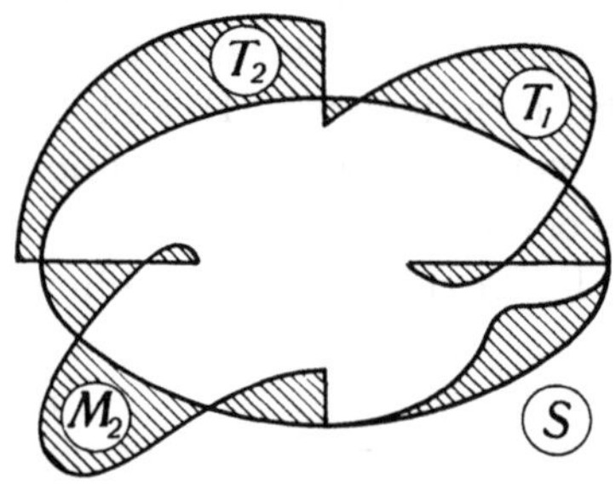

Fig. 59.

Fig. 58) fully confirm the predictions of the theory, and not only qualitatively, but also quantitatively. It is essential to further note that, as can be seen from Table 10, the bending stresses in the problem under consideration are not at all less significant than the membrane stresses (i.e., the stresses, uniformly distributed through the thickness of the shell.)

Hence it is clear that the membrane theory does not apply to this problem.

§ 43a. The evaluation of the stresses in oval shells, composed of four cylindrical plates. II.

By the example of the problem solved in § 43, it has been shown that one has to compute stiffness tables of the edges of the cylindrical plates and to use them for satisfying the boundary conditions. Such an approach is general and applicable whether the conditions are specified on rectilinear edges of a cylindrical plate or not. For example, it is convenient in the analysis of cylindrical roofs supported along rectilinear edges by longitudinal beams (cf. § 44). However, the problem of § 43 can also be solved in a different manner, since it belongs to a class of problems for which the boundary conditions can be formulated in complex form (cf. § 15a). The corresponding analysis reduces to the solution of a system of four, linear, algebraic equations the coefficients of which are complex constants. This approach has its advantages and disadvantages. The computations are reduced, the formulae are more lucid, but one has to work with complex quantities. It will be instructive for the reader to become acquainted with this type of analysis.

Consider a cylindrical plate, freely supported along curvilinear edges and loaded normally to its middle surface in accordance with (43.1). The membrane theory yields for this problem the solution (43.2) which, as has been established in the last section, can be considered to be a particular solution of the equations of the moment theory of shells. Hence one can take as particular solutions of the complex equations of shell theory for this problem the expressions

$$\begin{gathered} \tilde{t}_1 = T_1^* = \frac{4}{\pi} pr \sin \lambda\xi, \quad \tilde{t}_2 = T_2^* = 0, \\ \tilde{S} = S^* = 0, \quad \tilde{t} = T_1^* + T_2^* = \frac{4}{\pi} pr \sin \lambda\xi. \end{gathered} \tag{43.1a}$$

For the loading (43.1) in the series (42.5) one must retain only the first term. In correspondence with this, the complex auxiliary function for the problem under consideration becomes

$$\begin{aligned} \tilde{T} &= \tilde{T}_{(1)} \sin \lambda\xi = \\ &= (\tilde{C}_1 e^{\gamma_1\varphi} + \tilde{C}_2 e^{-\gamma_2\varphi} + \tilde{C}_3 e^{\gamma_2\varphi} + \tilde{C}_4 e^{-\gamma_2\varphi} + (4/\pi)pr) \sin \lambda\xi, \end{aligned} \tag{43.2a}$$

where γ_1, γ_2 are to be determined from (42.20), (42.21).

In what follows, as in the last section, the coordinate φ will be measured from the plane of symmetry of the cylindrical plate and it will be assumed that the deformation is symmetric with respect to this plane. Then $\tilde{T}$ must be even in φ, and hence one must take in (43.2a)

$$\tilde{C}_1 = \tilde{C}_2, \quad \tilde{C}_3 = \tilde{C}_4,$$

so that

$$\tilde{T} = \left(\tilde{A}_1 \cosh \gamma_1\varphi + \tilde{A}_2 \cosh \gamma_2\varphi + \frac{4}{\pi} pr\right) \sin \lambda\xi. \tag{43.3a}$$

Using the formulae relating the complex auxiliary function $\tilde{T}$ to the complex forces, one finds

$$\tilde{T}_1 = \tilde{T}_{1,1}(\varphi) \sin \lambda\xi, \quad \tilde{T}_2 = \tilde{T}_{2,1}(\varphi) \sin \lambda\xi, \quad \tilde{S} = \tilde{S}_1(\varphi) \cos \lambda\xi, \tag{43.4a}$$

where

$$\tilde{T}_{1,1} = \left\{\left[1 - \frac{i}{2b^2}(\gamma_1^2 - \lambda^2)\right]\tilde{A}_1 \cosh \gamma_1\varphi + \right.$$

$$\left. + \left[1 - \frac{i}{2b^2}(\gamma_2^2 - \lambda^2)\right] \times \tilde{A}_2 \cosh \gamma_2\varphi \approx \tilde{A}_1 \cosh \gamma_1\varphi + \tilde{A}_2 \cosh \gamma_2\varphi,\right.$$

$$\tilde{T}_{2,1} = \frac{i}{2b^2}\{(\gamma_1^2 - \lambda^2)\tilde{A}_1 \cosh \gamma_1\varphi + (\gamma_2^2 - \lambda^2)\tilde{A}_2 \cosh \gamma_2\varphi\} + \frac{4}{\pi} pr,$$

$$\tilde{S}_1 = -\lambda\left\{\frac{1}{\gamma_1}\left[1 - \frac{i}{2b^2}(\gamma_1^2 - \lambda^2)\right]\tilde{A}_1 \sinh \gamma_1\varphi + \right. \tag{43.5a}$$

$$\left. + \frac{1}{\gamma_2}\left[1 - \frac{i}{2b^2}(\gamma_2^2 - \lambda^2)\right]\tilde{A}_2 \sinh \gamma_2\varphi\right\} \approx$$

$$\approx -\lambda\left[\frac{1}{\gamma_1}\tilde{A}_1 \sinh \gamma_1\varphi + \frac{1}{\gamma_2}\tilde{A}_2 \sinh \gamma_2\varphi\right].$$

These formulae will now be applied to the analysis of a cylindrical tube of oval cross-section for the same formulation of the problem and the same data as in § 43. The general solution (43.4a) for the plate I, which will contain two complex arbitrary constants A_1, A_2 will be written down as well as the analogous solution for the plate II which involves two other arbitrary constants B_1, B_2. Next, construct the four equations expressing the conditions of compatibility of stresses and deformations of the plates I and II along their common rectilinear boundary. These conditions have been shown in § 15a to reduce to the requirement of continuity for transition through the common boundary of the two shells of four complex functions which adapted to the present problem, taking into consideration that the two contiguous cylindrical plates have the same thickness, are expressed by the four complex forces

$$\tilde{T}_2, \quad \tilde{S}, \quad \tilde{T}_1, \quad \frac{1}{r_0}\left(\frac{\partial \tilde{T}_1}{\partial \varphi} - 2\frac{\partial \tilde{S}}{\partial \xi}\right).$$

The last leads to the equalities

$$\tilde{T}_{21}' = \tilde{T}_{21}'', \quad \tilde{S}_1' = \tilde{S}_1'',$$

$$\tilde{T}_{1,1}' = \tilde{T}_{1,1}'', \quad \frac{1}{r_1}\frac{d\tilde{T}_{1,1}'}{d\varphi} = \frac{1}{r_2}\frac{d\tilde{T}_{1,1}''}{d\varphi}, \tag{43.6a}$$

where one and two strokes indicate quantities, relating to the plates I and II, respectively. For this purpose, one has to take for both plates

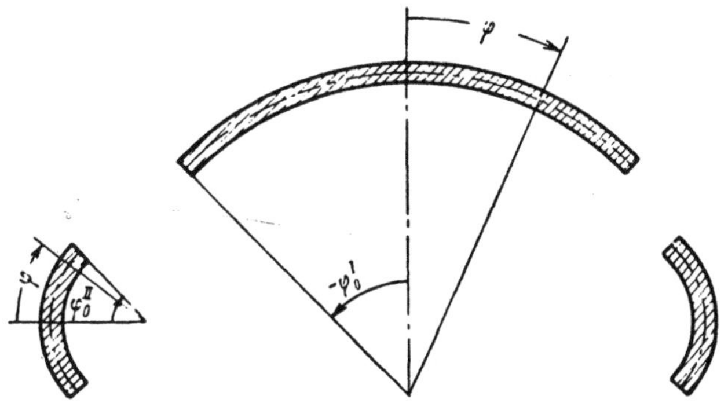

Fig. 59a.

the same direction of increase of angle [Fig. 59a], in correspondence with which the value of φ on their common edge will be $\varphi_0^{\mathrm{I}} = -\pi/4$ for the plate I and $\varphi_0^{\mathrm{II}} = \pi/4$ for the plate II.

The data for the analysis have been given in § 43. Hence one finds

a) for the plate I:

$$\lambda = 1.964, \quad 2b^2 = 496, \quad b = 15.7,$$
$$\gamma_1 = 6.37 + 2.42i, \quad \gamma_1^2 - \lambda^2 = 30.9 + 30.9i,$$
$$\gamma_2 = 2.65 + 5.84i, \quad \gamma_2^2 - \lambda^2 = -30.95 + 30.95i.$$

Substituting these quantities into (43.4a), one finds for $\varphi_0 = -\pi/4$

$$\begin{aligned} \tilde{T}_{2,1} &= \tfrac{1}{2}(-5.79 - 11.9i)\tilde{A}_1 + \tfrac{1}{2}(0.425 + 0.553i)\tilde{A}_2 + 1.273pr_1, \\ \tilde{T}_{1,1} &= \tfrac{1}{2}(-48.9 + 142i)\tilde{A}_1 + \tfrac{1}{2}(-0.97 + 7.86i)\tilde{A}_2, \\ \tilde{S}_1 &= \tfrac{1}{2}(1.34 + 43.2i)\tilde{A}_1 + \tfrac{1}{2}(-2.38 + 0.763i)\tilde{A}_2, \end{aligned} \tag{43.7a}$$

$$\frac{1}{r_1}\frac{d\tilde{T}_{1,1}}{d\varphi} = \frac{1}{2r_1}\{(654 - 784i)\tilde{A}_1 + (-44.8 - 27.0i)\tilde{A}_2\}.$$

b) for the plate II:

$$\lambda = 0.524, \quad 2b^2 = 132, \quad b = 8.12,$$
$$\gamma_1 = 2.31 + 0.915i, \quad \gamma_1^2 - \lambda^2 = 4.22 + 4.23i,$$
$$\gamma_2 = 0.959 - 2.21i, \quad \gamma_2^2 - \lambda^2 = -4.24 - 4.24i.$$

Substituting these results into (43.5a), one finds for $\varphi_0 = \pi/4$

$$\begin{aligned} \tilde{T}_{2,1} &= \tfrac{1}{2}(-0.279 + 0.025i)\tilde{B}_1 + \tfrac{1}{2}(-0.0663 - 0.00386i)\tilde{B}_2 + 1.273pr_2, \\ \tilde{T}_{1,1} &= \tfrac{1}{2}(4.75 + 3.95i)\tilde{B}_1 + \tfrac{1}{2}(-0.432 - 1.63i)\tilde{B}_2, \\ \tilde{S}_1 &= \tfrac{1}{2}(-1.21 - 0.467i)\tilde{B}_1 + \tfrac{1}{2}(-0.487 + 0.277i)\tilde{B}_2, \end{aligned} \tag{43.8a}$$

$$\frac{1}{r_2}\frac{d\tilde{T}_{1,1}}{d\varphi} = \frac{1}{2r_2}\{(6.59 + 13.74i)\tilde{B}_1 + (-5.92 - 1.85i)\tilde{B}_2\}.$$

Introducing (43.7a) and (43.8a) into (43.6a) and taking into consideration that $r_2/r_1 = 0.267$, one arrives at the linear algebraic system of four equations in four unknowns

$$\begin{gathered}(5.79 + 11.9i)\tilde{A}_1 + (-0.425 - 0.553i)\tilde{A}_2 + (-0.279 + 0.025i)\tilde{B}_1 \\ + (-0.0663 + 0.0386i)\tilde{B}_2 = 1.868pr_1, \\ (48.9 - 142i)\tilde{A}_1 + (0.97 - 7.86i)\tilde{A}_2 + (4.75 + 3.95i)\tilde{B}_1 + \\ + (-0.432 - 1.63i)\tilde{B}_2 = 0, \\ (+1.34 - 43.2i)\tilde{A}_1 + (2.38 - 0.763i)\tilde{A}_2 + (-1.21 - 0.467i)\tilde{B}_1 + \\ + (-0.487 + 0.277i)\tilde{B}_2 = 0, \\ (-654 + 784i)\tilde{A}_1 + (44.8 + 27.0i)\tilde{A}_2 + (24.7 + 51.5i)\tilde{B}_1 + \\ + (-22.2 - 6.92i)\tilde{B}_2 = 0.\end{gathered} \tag{43.9a}$$

Its solutions yield for the unknown constants the values

$$\begin{aligned}\tilde{A}_1 &= 2(0.0140 - 0.0186i)pr_1, & \tilde{B}_1 &= 2(-0.554 + 0.093i)pr_1, \\ \tilde{A}_2 &= 2(-0.286 + 0.420i)pr_1, & \tilde{B}_2 &= 2(-1.425 + 0.803i)pr_1.\end{aligned} \tag{43.10a}$$

Substituting these values into (43.4a), (43.5a) and separating real and imaginary parts, one finds the distribution of forces and moments in the shell and the corresponding stresses. Table 10a gives the values of σ_{T_1}, σ_{T_2}, σ_{M_2} in the centre section of the shell $x = l/2$ and the values of the tangential stress σ_S in the plane of the supporting section ($x = 0$).

Table 10a

φ_{I}	0°	−15°	−25°	−35°	−45°			
φ^{II}					45°	30°	15°	0°
$w : \frac{pr_1}{E\delta}$	−10.6	0.5	16.8	27.5	6.16			
					4.97	−4.8	−10.6	−13.0
$\sigma_{T_1} : \frac{pr_1}{\delta}$	−0.545	0.72	1.87	1.50	−1.07			
					−1.07	−2.45	−3.48	−3.96
$\sigma_S : \frac{pr_1}{\delta}$	0	0.05	−0.42	−1.07	−1.18			
					−1.18	−0.94	−0.52	0
$\sigma_{T_1} : \frac{pr_1}{\delta}$	1.29	1.34	1.29	1.05	0.618			
					0.618	0.46	0.36	0.33
$\sigma_{\mu_2} : \frac{pr_1}{\delta}$	−1.46	−0.79	1.29	3.95	−0.416			
					−0.416	−2.93	−3.32	−3.30

Comparison of Tables 10 and 10a verifies that the methods of §§ 43 and 43a yield similar results, as was, of course, to be expected, since both are based on the same theory. Certain differences between corresponding values of these tables must be due mainly to the approximate nature of the computations. The complexity of the computations can lead to certain accumulations of error.

The above method is undoubtedly clearer and briefer with regard to manipulations than that of § 43. However, in its application, one has to operate on complex numbers which requires some practice before it becomes convenient.

§ 44. The simplified theory of long cylindrical plates

The above formulae for the analysis of cylindrical plates may be used to investigate short as well as relatively long cylindrical plates.

If one has only the study of long cylindrical plates in mind, the theory can be essentially simplified; since this case is of considerable practical interest in connection with the stress analysis of thin-walled reinforced roofs, it should be treated here in detail. For this purpose, as before, only the first terms corresponding to $m = 1$ in the expansions of the load distribution (42.3) will be retained.

If the length L of the cylindrical plate is several times its radius r_0, the parameter

$$f = \frac{\lambda^2 - \frac{1}{2}}{b\lambda} \tag{44.1}$$

will be small and may be neglected compared with unity. Introducing this simplification into the formulae (42.21), one has

$$\begin{aligned} c_1 = d_2 &= \sqrt{\frac{b\lambda}{2}}\sqrt{\sqrt{2}+1} = \nu\cos\frac{\pi}{8}, \\ c_2 = d_1 &= \sqrt{\frac{b\lambda}{2}}\sqrt{\sqrt{2}-1} = \nu\sin\frac{\pi}{8}, \end{aligned} \tag{44.2}$$

where

$$\nu = \sqrt[4]{2}\sqrt{b\lambda}. \tag{44.3}$$

The adoption of these formulae corresponds to writing equation (42.1) in the form

$$\frac{\partial^4 \tilde{T}}{\partial \varphi^4} + i2b^2 \frac{\partial^2 \tilde{T}}{\partial \xi^2} = 0, \tag{44.4}$$

i.e., it corresponds to neglecting $\frac{\partial^2 \tilde{T}}{\partial \xi^2}$ and $\tilde{T}$ compared with $\frac{\partial^2 \tilde{T}}{\partial \varphi^2}$. (In equation (44.4) as well as in later formulae, terms depending on the surface loading have been omitted, since the homogeneous problem is of prime interest). Introducing analogous simplifications into the formulae (40.8), (40.9), (40.11) and (40.16) expressing the complex stresses and displacements in terms of the auxiliary function $\tilde{T}$, one finds

$$\tilde{T}_1 \approx \tilde{T},\ \tilde{T}_2 \approx \frac{i}{2b^2}\frac{\partial^2 \tilde{T}}{\partial \varphi^2},\quad \frac{\partial \tilde{S}}{\partial \xi} \approx -\frac{i}{2b^2}\frac{\partial^3 \tilde{T}}{\partial \varphi^3}, \tag{44.5}$$

$$\frac{\partial \tilde{u}}{\partial \xi} \approx \frac{r_0}{E\delta}\tilde{T},\ \frac{\partial^2 \tilde{v}}{\partial \xi^2} \approx -\frac{r_0}{E\delta}\frac{\partial \tilde{T}}{\partial \varphi},\ \frac{\partial^2 \tilde{w}}{\partial \xi^2} \approx \frac{r_0}{E\delta}\frac{\partial^2 \tilde{T}}{\partial \varphi^2}. \tag{44.6}$$

The solution corresponding to the first term of the series (42.31) will now be sought in the form

$$\begin{gathered}\tilde{T} = \tilde{T}' \sin \lambda\xi,\ \tilde{T}_1 = \tilde{T}_1' \sin \lambda\xi,\ \tilde{T}_2 = \tilde{T}_2' \sin \lambda\xi,\ \tilde{S} = \tilde{S}' \cos \lambda\xi,\\ \tilde{u} = \tilde{u}' \cos \lambda\xi,\ \tilde{v} = \tilde{v}' \sin \lambda\xi,\ \tilde{w} = \tilde{w}' \sin \lambda\xi.\end{gathered} \tag{44.7}$$

Substitution of these expressions in (44.5) and (44.6) gives

$$\begin{gathered}\tilde{T}_1' \approx \tilde{T}',\ \tilde{T}_2' \approx \frac{i}{2b^2}\frac{d^2\tilde{T}'}{d\varphi^2},\ \tilde{S}' \approx \frac{i}{2b^2}\frac{1}{\lambda}\frac{d^3\tilde{T}'}{d\varphi^3},\\ \tilde{u}' \approx -\frac{r_0}{E\delta}\frac{1}{\lambda}\tilde{T}',\ \tilde{v}' \approx \frac{r_0}{E\delta}\frac{1}{\lambda^2}\frac{d\tilde{T}'}{d\varphi},\ \tilde{w}' \approx -\frac{r_0}{E\delta}\frac{1}{\lambda^2}\frac{d^2\tilde{T}'}{d\varphi^2}.\end{gathered} \tag{44.8}$$

Consideration will now be limited to deformations of cylindrical plates which are symmetrical with respect to the plane $\varphi = 0$. Hence one needs only retain in the general expression for $\tilde{T}'$ even functions of φ, i.e., one may write

$$\tilde{T}' = \tilde{C}_1 \cosh (c_1 + id_1)\varphi + \tilde{C}_2 \cosh (c_2 - id_2)\varphi. \tag{44.9}$$

This formula follows immediately from the formulae (42.20) and (42.22). Separating it into real and imaginary parts and letting

$$\tilde{C}_1 = C_1 - iC_2,\ \tilde{C}_2 = C_3 + iC_4, \tag{44.10}$$

where C_1, C_2, C_3, C_4 are real arbitrary constants, one obtains by use of known formulae of the theory of hyperbolic functions

$$\tilde{T}' = R_e + iI_m, \tag{44.11}$$

where

$$\begin{aligned}R_e &= C_1\Phi_1 + C_2\Phi_2 + C_3\Psi_1 + C_4\Psi_2,\\ I_m &= C_1\Phi_2 - C_2\Phi_1 - C_3\Psi_2 + C_4\Psi_1,\end{aligned} \tag{44.12}$$

and the functions Φ and Ψ are defined by (42.32).

The present principal aim is the deduction of formulae for the direct evaluation of the coefficients a_{jk} (i.e., of the stiffness coefficients for the edges of cylindrical plates). Therefore, in what follows, the functions Φ and Ψ will be considered mainly at the straight edges, i.e., for $\varphi = \varphi_0$. But then one may write as an approximation

$$\begin{aligned} \cosh c_1\varphi &\approx \sinh c_1\varphi \approx \tfrac{1}{2}e^{c_1\varphi}, \\ \cosh c_2\varphi &\approx \sinh c_2\varphi \approx \tfrac{1}{2}e^{c_2\varphi}, \end{aligned} \tag{44.13}$$

admitting an error (near the straight boundary) of order $e^{-2c_2\varphi_0}$ in comparison with unity. As a rule, this error is relatively small (approx. 10—15%). Applying this simplification to the formulae (44.12), it is found that

$$\begin{aligned} R_e &\approx e^{c_1\varphi}(C_1 \cos d_1\varphi + C_2 \sin d_1\varphi) + e^{c_2\varphi}(C_3 \cos d_2\varphi + C_4 \sin d_2\varphi), \\ I_m &\approx e^{c_1\varphi}(C_1 \sin d_1\varphi - C_2 \cos d_1\varphi) - e^{c_2\varphi}(C_3 \sin d_2\varphi - C_4 \cos d_2\varphi), \end{aligned} \tag{44.14}$$

where the common factors $\frac{1}{2}$ have been included in the arbitrary constants.

The formulae (44.14) may likewise be represented in the form

$$\begin{aligned} R_e &\approx B_1'e^{c_1\varphi} \cos(d_1\varphi + \beta_1') + B_2'e^{c_2\varphi} \cos(d_2\varphi + \beta_2'), \\ I_m &\approx B_1'e^{c_1\varphi} \sin(d_1\varphi + \beta_1') - B_2'e^{c_2\varphi} \sin(d_2\varphi + \beta_2'), \end{aligned} \tag{44.15}$$

where B_1', B_2', β_1', β_2' are new arbitrary constants.

Differentiating R_e with respect to φ and taking for this purpose (44.2) into consideration, one finds

$$\frac{dR_e}{d\varphi} = \nu \left[B_1'e^{c_1\varphi} \cos\left(d_1\varphi + \beta_1' + \frac{\pi}{8}\right) + \right. \\ \left. + B_2'e^{c_2\varphi} \cos\left(d_2\varphi + \beta_2' + \frac{3}{8}\pi\right)\right]. \tag{44.16}$$

Hence one arrives at the following simple rule for the differentiation of R_e: for each differentiation, one has to multiply the entire expression by ν, increase the argument of the first cosine by $\frac{\pi}{8}$ and of the second by $\frac{3\pi}{8}$.

Next, differentiating I_m, one finds

$$\frac{dI_m}{d\varphi} = \nu \left[B_1'e^{c_1\varphi} \sin\left(d_1\varphi + \beta_1' + \frac{\pi}{8}\right) + \right. \\ \left. - B_2'e^{c_2\varphi} \sin\left(d_2\varphi + \beta_2' + \frac{3}{8}\pi\right)\right], \tag{44.17}$$

whence one obtains for the differentiation of I_m a rule which is quite analogous to that above for differentiating R_e. Since the displacements are equal to the real parts of the corresponding complex displacements, it follows from (44.8), taking (44.11) into consideration, that

$$u' = -\frac{r_0}{E\delta}\frac{1}{\lambda}R_e, \quad v' = \frac{r_0}{E\delta}\frac{1}{\lambda^2}\frac{dR_e}{d\varphi}, \qquad w' = -\frac{r_0}{E\delta}\frac{1}{\lambda^2}\frac{d^2R_e}{d\varphi^2}. \tag{44.18}$$

Using (44.15) and (44.16), one arrives at the following expressions for the coefficients u', v', w':

$$u' = -\frac{1}{\lambda}\frac{r_0}{E\delta}\left[B_1'e^{c_1\varphi}\cos(d_1\varphi+\beta_1') + B_2'e^{c_2\varphi}\cos(d_2\varphi+\beta_2')\right],$$

$$v' = \frac{\nu}{\lambda^2}\frac{r_0}{E\delta}\left[B_1'e^{c_1\varphi}\cos\left(d_1\varphi+\beta_1'+\frac{\pi}{8}\right) + B_2'e^{c_2\varphi}\cos\left(d_2\varphi+\beta_2'+\frac{3}{8}\pi\right)\right], \tag{44.19}$$

$$w' = -\frac{\nu^2}{\lambda^2}\frac{r_0}{E\delta}\left[B_1'e^{c_1\varphi}\cos\left(d_1\varphi+\beta_1'+\frac{\pi}{4}\right) - B_2'e^{c_2\varphi}\sin\left(d_2\varphi+\beta_2'+\frac{\pi}{4}\right)\right].$$

To these may be added the expressions for the coefficient of the rotation ψ:

$$r_0\psi' = -\frac{dw'}{d\varphi}+v' \approx -\frac{dw'}{d\varphi} = \frac{\nu^3}{\lambda^2}\frac{r_0}{E\delta}\left[B_1'e^{c_1\varphi}\cos\left(d_1\varphi+\beta_1'+\frac{3}{8}\pi\right) - B_2'e^{c_2\varphi}\cos\left(d_2\varphi+\beta_2'+\frac{\pi}{8}\right)\right]. \tag{44.20}$$

From (44.8) and likewise from (40.5) follows that

$$T_1' \approx R_e, \qquad -\frac{2b^2}{r_0}\frac{M_2-\mu M_1}{1-\mu^2} \approx I_m,$$

$$T_2' \approx -\frac{1}{2b^2}\frac{d^2I_m}{d\varphi^2}, \qquad -\frac{2b^2}{r_0}\frac{M_1-\mu M_2}{1-\mu^2} \approx \frac{1}{2b^2}R_e \approx 0, \tag{44.21}$$

$$S' = -\frac{1}{2b^2}\frac{1}{\lambda}\frac{d^3I_m}{d\varphi^3}, \qquad \frac{2b^2}{r_0}\frac{H}{1-\mu} \approx \frac{1}{2b^2}\frac{1}{\lambda}\frac{d^3R_e}{d\varphi^3}.$$

Substituting here the expressions for R_e and I_m and using their rules of differentiation, one finds

$$T_1' \approx B_1'e^{c_1\varphi}\cos(d_1\varphi+\beta_1') + B_2'e^{c_2\varphi}\cos(d_2\varphi+\beta_2'),$$

$$T_2' \approx -\frac{\nu^2}{2b^2}\left[B_1'e^{c_1\varphi}\sin\left(d_1\varphi+\beta_1'+\frac{\pi}{4}\right) - B_2'e^{c_2\varphi}\cos\left(d_2\varphi+\beta_2'+\frac{1}{4}\pi\right)\right],$$

$$S' \approx -\frac{1}{2b^2}\frac{\nu^3}{\lambda}\left[B_1'e^{c_1\varphi}\sin\left(d_1\varphi+\beta_1'+\frac{3}{8}\pi\right) + B_2'e^{c_2\varphi}\sin\left(d_2\varphi+\beta_2'+\frac{\pi}{8}\right)\right],$$

$$M_1' \approx \mu M_2', \tag{44.22}$$
$$M_2' \approx -\frac{r_0}{2b^2}\left[B_1'e^{c_1\varphi}\sin(d_1\varphi+\beta_1') - B_2'e^{c_2\varphi}\sin(d_2\varphi+\beta_2')\right],$$
$$H' \approx \frac{(1-\mu)r_0}{4b^4}\frac{\nu^3}{\lambda}\left[B_1'e^{c_1\varphi}\cos\left(d_1\varphi+\beta_1'+\frac{3}{8}\pi\right) - B_2'e^{c_2\varphi}\cos\left(d_2\varphi+\beta_2'+\frac{\pi}{8}\right]\right).$$

It is now possible to determine also the normal shear resultants, using the formulae (16.6), viz.,

$$(1+\mu)r_0N_1 \approx \frac{\partial(M_1+M_2)}{\partial\xi}, \quad (1+\mu)r_0N_2 \approx \frac{\partial(M_1+M_2)}{\partial\varphi}. \tag{44.23}$$

Substituting here

$$N_1 = N_1'\cos\lambda\xi, \quad M_1 = M_1'\sin\lambda\xi,$$
$$N_2 = N_2'\sin\lambda\xi, \quad M_2 = M_2'\sin\lambda\xi, \tag{44.24}$$

one obtains

$$(1+\mu)r_0N_1' = \lambda(M_1'+M_2'),$$
$$(1+\mu)r_0N_2' = \frac{d(M_2'+M_1')}{d\varphi}. \tag{44.25}$$

Hence, by (44.22),

$$N_1' \approx -\frac{\lambda}{2b^2}\left[B_1'e^{c_1\varphi}\sin(d_1\varphi+\beta_1') - B_2'e^{c_2\varphi}\sin(d_2\varphi+\beta_2')\right],$$
$$N_2' \approx -\frac{\nu}{2b^2}\left[B_1'e^{c_1\varphi}\sin\left(d_1\varphi+\beta_1'+\frac{\pi}{8}\right) - \right. \tag{44.26}$$
$$\left. - B_2'e^{c_2\varphi}\sin\left(d_2\varphi+\beta_2'+\frac{3\pi}{8}\right)\right].$$

The set of formulae (44.19), (44.20), (44.22), and (44.26) make it possible to write down general expressions for the displacements, the angle of rotation, the forces and moments for the case when only the first terms have been retained in the series (42.31).

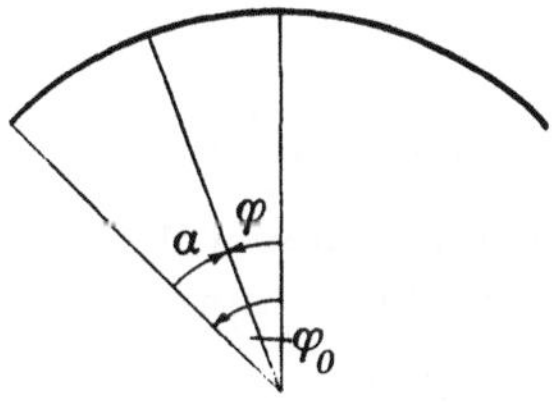

Fig. 60.

In what follows, it is more convenient to measure the angle φ not from the apex of the shell (as this has been done so far), but from one of its edges (see Fig. 60). For this purpose, the angle φ in the formulae (44.19), (44.22) and (44.26) will be replaced by

$$\alpha = \varphi_0 - \varphi. \tag{44.27}$$

One thus finds

$$u' = \frac{\lambda}{\nu^2}\left[B_1e^{-c_1\alpha}\cos\left(d_1\alpha+\beta_1+\frac{\pi}{4}\right) - B_2e^{-c_2\alpha}\cos\left(d_2\alpha+\beta_2-\frac{\pi}{4}\right)\right],$$

$$v' = -\frac{1}{\nu}\left[B_1 e^{-c_1\alpha}\cos\left(d_1\alpha+\beta_1+\frac{\pi}{8}\right)-B_2 e^{-c_2\alpha}\sin\left(d_2\alpha+\beta_2-\frac{\pi}{8}\right)\right],$$

$$w' = B_1 e^{-c_1\alpha}\cos(d_1\alpha+\beta_1) + B_2 e^{-c_2\alpha}\cos(d_2\alpha+\beta_2), \tag{44.28}$$

$$r_0\psi' = -\nu\left[B_1 e^{-c_1\alpha}\cos\left(d_1\alpha+\beta_1-\frac{\pi}{8}\right)+B_2 e^{-c_2\alpha}\sin\left(d_2\alpha+\beta_2+\frac{\pi}{8}\right)\right],$$

$$\frac{r_0}{E\delta}T_1' = -\lambda u',$$

$$\frac{r_0}{E\delta}T_2' = -\frac{\lambda^4}{\nu^4}\left[B_1 e^{-c_1\alpha}\sin(d_1\alpha+\beta_1) - B_2 e^{-c_2\alpha}\sin(d_2\alpha+\beta_2)\right],$$

$$\frac{r_0}{E\delta}S' = -\frac{\lambda^3}{\nu^3}\left[B_1 e^{-c_1\alpha}\sin\left(d_1\alpha+\beta_1-\frac{\pi}{8}\right)+\right.$$
$$\left.+B_2 e^{-c_2\alpha}\cos\left(d_2\alpha+\beta_2+\frac{\pi}{8}\right)\right],$$

$$\frac{r_0}{E\delta}N_2' = -\frac{\nu^3}{4b^4}\left[B_1 e^{-c_1\alpha}\sin\left(d_1\alpha+\beta_1+\frac{\pi}{8}\right)-\right.$$
$$\left.-B_2 e^{-c_2\alpha}\cos\left(d_2\alpha+\beta_2-\frac{\pi}{8}\right)\right], \tag{44.29}$$

$$\frac{1}{E\delta}M_2' = -\frac{\nu^2}{4b^4}\left[B_1 e^{-c_1\alpha}\cos\left(d_1\alpha+\beta_1-\frac{\pi}{4}\right)-\right.$$
$$M_1' \approx \mu M_2', \qquad \left.-B_2 e^{-c_2\alpha}\cos\left(d_2\alpha+\beta_2+\frac{\pi}{4}\right)\right],$$

$$\frac{1}{E\delta}H' = -\frac{(1-\mu)\lambda\nu}{4b^4}\left[B_1 e^{-c_1\alpha}\cos\left(d_1\alpha+\beta_1-\frac{\pi}{8}\right)+\right.$$
$$\left.+B_2 e^{-c_2\alpha}\sin\left(d_2\alpha+\beta_2+\frac{\pi}{8}\right)\right],$$

where B_1, B_2, β_1, β_2 are new constants of integration. These very convenient expressions represent the simplest formulation of the approximate solution of the homogeneous problem for cylindrical plates.

On the basis of the conditions underlying their deduction, this solution may only be used near the edges of a cylindrical plate. However, it will be seen from the example below that it gives satisfactory agreement even near the apex of the shell. Thus, it may be used in the interval $0 \leqslant \alpha \leqslant \varphi_0$, i.e., over half the cross-section of the shell. This is also sufficient, since it has been assumed that the deformation of the cylindrical plate is symmetrical with respect to the plane $\varphi = 0$.

Using the formulae (44.28) and (44.29), the influence coefficients for the edges of cylindrical plates will now be obtained. For this purpose, first of all the constants of integration corresponding to the four sets of boundary conditions (42.35) will be evaluated. These constants are given in the following table (the intermediate calculations having been omitted in view of their elementary nature):

Table 11

	$u' = 1$	$v' = 1$	$w' = 1$	$r_0\psi' = 1$	(*)
$B_1 \cos \beta_1$	0.00	-0.924ν	-0.707	$-\dfrac{0.924}{\nu}$	—
$B_1 \sin \beta_1$	$-2.41\dfrac{\nu^2}{\lambda}$	-4.08ν	-4.12	$-\dfrac{2.32}{\nu}$	—
$B_2 \cos \beta_2$	0.00	0.924ν	1.707	$\dfrac{0.924}{\nu}$	—
$B_2 \sin \beta_2$	$\dfrac{\nu^2}{\lambda}$	2.23ν	1.707	$\dfrac{0.383}{\nu}$	—

Substituting these values of the constants into the formulae for the forces and moments (44.29), the following table of stiffness coefficients for the edges of the cylindrical plates is found:

Table 12

	$u' = 1$	$v' = 1$	$w' = 1$	$r_0'\psi = 1$
$\dfrac{r_0}{E\delta} S'$	$a_1 \dfrac{\lambda^2}{\nu}$	$a_2 \dfrac{\lambda^3}{\nu^2}$	$a_1 \dfrac{\lambda^3}{\nu^3}$	$\dfrac{\lambda^3}{\nu^4}$
$\dfrac{r_0}{E\delta} T_2'$	$a_2 \dfrac{\lambda^3}{\nu^2}$	$a_3 \dfrac{\lambda^4}{\nu^3}$	$a_4 \dfrac{\lambda^4}{\nu^4}$	$a_1 \dfrac{\lambda^4}{\nu^5}$
$\dfrac{r_0}{E\delta} N_2'$	$a_1 \dfrac{\lambda^3}{\nu^3}$	$a_4 \dfrac{\lambda^4}{\nu^4}$	$a_3 \dfrac{\lambda^4}{\nu^5}$	$a_2 \dfrac{\lambda^4}{\nu^6}$
$\dfrac{1}{E\delta} M_2'$	$\dfrac{\lambda^3}{\nu^4}$	$a_1 \dfrac{\lambda^4}{\nu^5}$	$a_2 \dfrac{\lambda^4}{\nu^6}$	$a_1 \dfrac{\lambda^4}{\nu^7}$

In this table, the quantities a_j are

$$a_1 = \frac{1}{\sin \dfrac{\pi}{8}} \approx 2.61,$$

$$a_2 = \frac{\sqrt{2}}{\sqrt{2} - 1} \approx 3.41, \tag{44.30}$$

$$a_3 = \frac{1}{(\sqrt{2}-1)\sin\frac{\pi}{8}} \approx 6.31,$$
$$a_4 = \frac{\sqrt{2}+1}{\sqrt{2}-1} \approx 5.84. \tag{44.30}$$

Thus, the expressions (44.28) and (44.29) offer the means of deducing simple formulae for the direct evaluation of stiffness tables the computation of which by the method of the preceding section would have represented a tedious problem.

Using the above simplified theory, a cylindrical roof the edges of which are reinforced by vertical beams will now be studied. Let the length of the roof be

$$L = \tfrac{4}{3}\pi r_0$$

and let it be loaded by its own weight

$$q = 144\ [\text{kg/m}^2].$$

These data have been taken from A. L. Gol'denweizer's paper [98] in which the stress analysis of this roof has been executed using more accurate equations and formulae. Thanks to this circumstance, the simplified theory of this section may be compared with a theory which corresponds approximately to that of § 42. The auxiliary parameters of the shell are $\lambda = 0.75$, $4b^4 = 1.2 \times 10^5$, $\nu = 3.73$.

Substituting these values into Table 12, one finds the following values for the stiffness coefficients of the cylindrical plate under consideration:

Table 13

	$u' = 1$	$v' = 1$	$w' = 1$	$r_0\psi' = 1$	(*)
$\frac{r_0}{E\delta} S'$	0.3485 0.393 12.6%	0.09051 0.103 13%	0.01912 0.0212 11%	0.001899 0.00217 14%	$43\frac{qr_0}{E}$
$\frac{r_0}{E\delta} T_2'$	0.09051 0.103 13%	0.03391 0.0382 13%	0.008448 0.00948 12%	0.0009377 0.00114 21%	$-142\frac{qr_0}{E}$
$\frac{r_0}{E\delta} N_2'$	0.01912 0.0212 11%	0.008448 0.00948 12%	0.002469 0.00274 11%	0.0003456 0.000399 15%	$-18\frac{qr_0}{E}$
$\frac{1}{E\delta} M_2'$	0.001899 0.00217 14%	0.0009377 0.00114 21%	0.0003456 0.000399 15%	0.00007325 0.0000815 11%	$-2.0\frac{qr_0}{E}$

The upper numbers in this table were obtained by Gol'denweizer, the lower ones by use of the formulae in Table 12. The percentage errors are given on the right hand side of each column.

It is thus seen that in the given particular case the approximate formulae determine the influence coefficients for the edge of the shell with an average error of about 13% and, further, that this error is everywhere positive. In other words, the approximate formulae somewhat overestimate the stiffness of the edge of the shell. The fifth column (*) of the table contains the values of the forces and moments for the case when the edge of the shell is clamped. These values were calculated from (42.39), where the particular solutions (determined by the formulae of the membrane theory), on the basis of the given data, are

$$\begin{aligned} u^{*\prime} &= 603\,\frac{qr_0}{E}\cos\varphi, & S^{*\prime} &= 340q\delta\sin\varphi, \\ v^{*\prime} &= 1710\,\frac{qr_0}{E}\sin\varphi, & T_2^{*\prime} &= -\,127q\delta\cos\varphi. \\ w^{*\prime} &= 1840\,\frac{qr_0}{E}\cos\varphi, & N_2^{*} &= M_2^{*} = 0. \\ r_0\psi^{*\prime} &= -\,130\,\frac{qr_0}{E}\sin\varphi, \end{aligned} \tag{44.31}$$

Hence one finds the boundary values of these quantities [setting in (44.31) $\varphi = 60°$]

$$\begin{aligned} &u_0^{*\prime} = 301\,\frac{qr_0}{E}, \quad v_0^{*\prime} = 1480\,\frac{qr_0}{E}, \quad w_0^{*\prime} = -\,920\,\frac{qr_0}{E}, \\ &r_0\psi_0^{*\prime} = -\,112\,\frac{qr_0}{E}, \quad S_0^{*\prime} = 294q\delta, \quad T_{2,0}^{*\prime} = -\,63.5q\delta. \end{aligned} \tag{44.32}$$

It should be noted that these values (taken from the membrane theory) differ only very slightly from the results of Gol'denweizer, obtained by solution of the differential equation which take account of bending

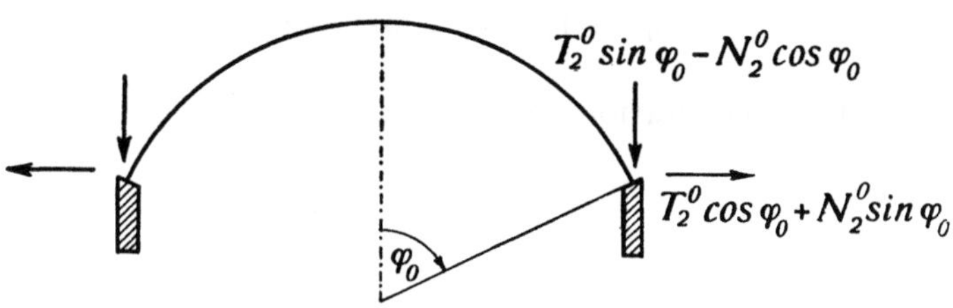

Fig. 61.

moments [98]. Substituting (44.32) into (42.39), one finds the numbers in the fifth column (*) of Table 13.

The next stage in the calculations is the determination of the influence coefficients for the edge beams (referring to the same displacements and forces which have been evaluated in Table 13). For this calculation, the resistance of the beams to torsion and to bending in the plane of minimum stiffness will be neglected. As regards bending in the plane of maximum stiffness of the beam, it will be assumed that the law of plane sections applies.

Hence one may write down the following formulae, relating the displacements of the upper edges of the beams to the forces acting on them (which are to vary along the beam like $\sin\lambda\xi$ or $\cos\lambda\xi$ — see Figs. 61 and 62),

$$u_0 = \frac{r_0^2}{\lambda^2}\left(1 + \frac{h^2\Omega}{4J}\right)\frac{S^0}{E\Omega} - \frac{r_0^3 h}{2\lambda^3 EJ}(N_2^0\cos\varphi_0 - T_2^0\sin\varphi_0),$$

$$w_0\cos\varphi_0 - v_0\sin\varphi_0 = \frac{r_0^4}{\lambda^4 EJ}\left(N_2^0\cos\varphi_0 - T_2^0\sin\varphi_0 - \frac{h}{2r_0}S^0\right). \qquad (44.33)$$

In these formulae h, Ω, J are the height, the area of the cross-section and the moment of inertia about the neutral axis of the section of the beam (which will be assumed to be rectangular).

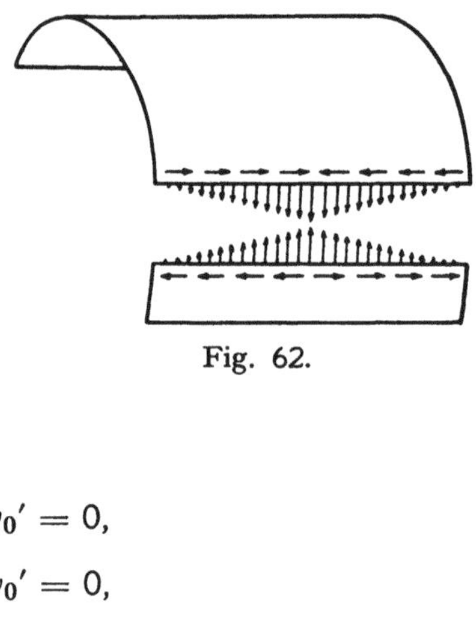

Fig. 62.

The following equalities will be added to these equations:

$$N_2^0\sin\varphi_0 + T_2^0\cos\varphi_0 = 0, \quad M_2^0 = 0.$$

Setting now

$$u_0' = 1, \quad v_0' = w_0' = 0,$$

$$v_0' = 1, \quad u_0' = w_0' = 0,$$

$$w_0' = 1, \quad v_0' = u_0' = 0,$$

and taking into consideration that

$$\frac{J}{h^2\Omega} = \frac{1}{12}, \qquad (44.34)$$

one obtains the following table, characterizing the stiffness of the beam:

Table 14

	$u' = 1$	$v' = 1$	$w' = 1$	$r_0\psi' = 1$	(*)
$\frac{r_0}{E\delta} S'$	$\frac{\Omega}{r_0\delta}\lambda^2$	$-\frac{\lambda^3 h\Omega}{2r_0^2\delta}\sin\varphi_0$	$\frac{\lambda^3 h\Omega}{2r_0^2\delta}\cos\varphi_0$	0	0
$\frac{r_0}{E\delta} T_2'$	$-\frac{\lambda^3 h\Omega}{2r_0^2\delta}\sin\varphi_0$	$\frac{\lambda^4 h^2\Omega}{3r_0^3\delta}\sin^2\varphi_0$	$-\frac{1}{6}\frac{\lambda^4 h^2\Omega}{r_0^3\delta}\sin 2\varphi_0$	0	$-\frac{4}{\pi}\frac{qr_0}{E\delta^2}\Omega\sin\varphi_0$
$\frac{r_0}{E\delta} N_2'$	$\frac{\lambda^3 h\Omega}{2r_0^2\delta}\cos\varphi_0$	$-\frac{1}{6}\frac{\lambda^4 h^2\Omega}{r_0^3\delta}\sin 2\varphi_0$	$\frac{1}{3}\frac{\lambda^4 h^2\Omega}{r_0^3\delta}\cos^2\varphi_0$	0	$\frac{4}{\pi}\frac{qr_0}{E\delta^2}\Omega\cos\varphi_0$
$\frac{1}{E\delta} M_2'$	0	0	0	0	0

The fifth column (*) of this table contains the values of the forces and moments, corresponding to clamping of the upper edge of the beam.

Substituting into this table the dimensions of the beam and the shell, one finds

Table 15

	$u_0' = 1$	$v_0' = 1$	$w_0' = 1$	$r_0\psi_0' = 1$	(*)
$\frac{r_0}{E\delta} S'$	0.244	− 0.017	0.0094	0	0
$\frac{r_0}{E\delta} T_2'$	− 0.017	0.0016	− 0.00093	0	$-47.8\,\frac{qr_0}{E}$
$\frac{r_0}{E\delta} N_2'$	0.0099	− 0.00093	0.00054	0	$27.6\,\frac{qr_0}{E}$
$\frac{1}{E\delta} M_2'$	0	0	0	0	0

The numbers in this table agree with those which were given by Gol'denweizer, since the calculations for the beams have not been modified.

Having obtained Tables 13 and 15 with the characteristic stiffnesses of the cylindrical plate and its reinforcing beams, one may now proceed to the determination of the edge displacements. For this purpose, the equations expressing the compatibility of the displacements of the beams and the edges of the shell may be written:

$$\begin{gathered} u_s = u_b, \quad v_s = v_b, \quad w_s = w_b, \quad \psi_s = \psi_b, \\ S_s = -S_b, \; T_{2s} = -T_{2b}, \; N_{2s} = -N_{2b}, \; M_{2s} = -M_{2b}, \end{gathered} \tag{44.35}$$

where the subscripts s, b refer to the shell and the beams respectively. Using the above mentioned tables, these conditions lead to the following system of simultaneous equations:

$$
\begin{aligned}
&0.637u' + 0.086v' + 0.0311w' + 0.00217(r\psi') + 43\frac{qr_0}{E} 0,\\
&0.086u' + 0.0398v' + 0.0086w' + 0.00114(r\psi') - 190\frac{qr_0}{E} = 0,\\
&0.0311u' + 0.0086v' + 0.00328w' + 0.000400(r\psi') + 10\frac{qr_0}{E} = 0,\\
&0.0217u' + 0.00114v' + 0.000400w' + 0.0000815(r\psi') - 2\frac{qr_0}{E} = 0.
\end{aligned}
\tag{44.36}
$$

Solving this algebraic system, one finds

$$
u_0' = 380\frac{qr_0}{E}, \quad v_0' = 11\,900\frac{qr_0}{E}, \quad w_0' = -48\,000\frac{qr_0}{E}, \quad r\psi_0' = 84\,200\frac{qr_0}{E},
\tag{44.37}
$$

while Gol'denweizer obtained the values

$$
u_0' = 480.9\frac{qr_0}{E}, \quad v_0' = 12\,200\frac{qr_0}{E}, \quad w_0' = -48\,580\frac{qr_0}{E}, \quad r\psi_0' = 83\,000\frac{qr_0}{E}.
\tag{44.38}
$$

These two sets of results agree very closely except for u' which exhibits an appreciable difference. However, as will be clear from the further discussion, the displacement u tends to zero not far from the edge of the shell, so that the encountered appreciable error in the value of u' at the edge is quite natural. It will be verified below that in determining the maximum values of the displacement u' (or of the force T_2' proportional to it) the error of the approximate solution does not exceed the error incurred during the determination of the remaining displacements and forces.

Having obtained the values of the edge displacements, one can calculate all the constants of integration by use of Table 11. Substituting into this table the parameters of the shell and evaluating the fifth column (*) by the rules established in § 43, one obtains the following values of the constants:

Table 16

	$u_0' = 1$	$v_0' = 1$	$w_0' = 1$	$r\psi_0' = 1$	(*)
$B_1 \cos \beta_1$	0.00	− 3.44	− 0.707	− 0.248	$4\,410 \frac{qr_0}{E}$
$B_1 \sin \beta_1$	− 44.8	− 15.2	− 4.12	− 0.597	$32\,100 \frac{qr_0}{E}$
$B_2 \cos \beta_2$	0.00	3.44	1.707	0.248	$-\,3\,490 \frac{qr_0}{E}$
$B_2 \sin \beta_2$	18.6	8.30	1.707	0.103	$-\,16\,300 \frac{qr_0}{E}$

It follows from this table that

$$B_1 \cos \beta_1 = (0-40800+33900-20800+4410)\frac{qr_0}{E} = -23300\frac{qr_0}{E},$$

$$B_1 \sin \beta_1 = (-17000-181000+198000-50200+32100)\frac{qr_0}{E} =$$

$$= -\,18100\frac{qr_0}{E}, \tag{44.39}$$

$$B_2 \cos \beta_2 = (0+40800-82000+20800-3490)\frac{qr_0}{E} = -23900\frac{qr_0}{E},$$

$$B_2 \sin \beta_2 = (7050+98600-82000+8670-16000)\frac{qr_0}{E} = 16000\frac{qr_0}{E}.$$

Introducing these constants into the formulae (44.29) and taking into consideration that in this example $q\delta = 0.0865$ kg/cm, one deduces the following expressions for the forces and moments:

$$\begin{aligned}
T_2' &— e^{-c_1\alpha}[2.56 \cos d_1\alpha + 3.27 \sin d_1\alpha] + \\
&\quad + e^{-c_2\alpha}[2.26 \cos d_2\alpha - 3.38 \sin d_2\alpha] - 11 \cos \varphi, \\
M_2' &= e^{-c_1\alpha}[175 \cos d_1\alpha + 22 \sin d_1\alpha] - \\
&\quad - e^{-c_2\alpha}[175 \cos d_2\alpha - 33.5 \sin d_2\alpha], \\
T_1' &= e^{-c_1\alpha}[12.8 \cos d_1\alpha - 100 \sin d_1\alpha] - \\
&\quad - e^{-c_2\alpha}[19.5 \cos d_2\alpha + 100 \sin d_2\alpha] - 39.2 \cos \varphi, \\
S' &= e^{-c_1\alpha}[5.48 \cos d_1\alpha + 20 \sin d_1\alpha] + \\
&\quad + e^{-c_2\alpha}[19.7 \cos d_2\alpha + 4.01 \sin d_2\alpha] + 29.4 \sin \varphi.
\end{aligned} \tag{44.40}$$

These formulae contain the auxiliary functions

$$\begin{aligned} f_1(z) &= e^{-0,924z}\cos(0.383z), & f_3(z) &= e^{-0,383z}\cos(0.924z), \\ f_2(z) &= e^{-0,924z}\sin(0.383z), & f_4(z) &= e^{-0,383z}\cos(0.924z) \end{aligned} \tag{44.41}$$

with $z = \nu\alpha = \nu(\varphi_0 - \varphi)$ which are shown in Fig. 63.

Table 17 contains T_1', T_2', S', M_2', as calculated from the formulae (44.40) (lower numbers) for $\varphi = 0$, 15°, 30°, 45°, 60°. The same table also gives the results calculated by Gol'denweizer (upper numbers).

Table 17

	0°	15°	30°	40°	60°
T_1' kg/cm	− 18.24 − 28	− 37.49 − 51.2	− 79.52 − 87.4	− 99.23 − 99.7	− 31.16 − 26.3
S' kg/cm	0 − 0.7	− 4.86 − 3.6	− 16.28 − 17.3	− 36.69 − 36.4	− 49.51 − 49.6
T_2' kg/cm	− 10.84 − 11.1	− 10.79 − 11.5	− 9.94 − 10.6	− 6.78 − 7.2	− 0.74 − 0.68
M_1' kg	60.61 33	58.80 62.0	42.21 43.4	14.10 12.4	0 0

Figures 64 to 67 exhibit (on the basis of this table) the distributions of the forces and moments; the results of Gol'denweizer correspond to the full lines, the approximate results to the dotted lines.

Comparison of the full and dotted lines shows that the approximate theory gives quite a correct representation of the magnitude of the forces and moments as well as of the character of their distributions. The greatest disagreements occur for $\varphi = 0$, i.e., at the apex of the shell, as had to be expected, since the formulae (44.28), (44.29) were deduced by neglecting terms causing errors which increase with the distance of the section under consideration from the edge of the shell. In view of this, the achieved accuracy can be accepted as completely satisfactory. Thus, this analysis indicates the suitability of the formulae (44.28) and (44.29) and of the corresponding Tables 11 and 12 for the evaluation of shells of the type used for thin-walled cylindrical roofs. This circumstance should be brought to the attention of engineers engaged on the design of roof structures who at the present time use more complicated methods which, however, do not lead to essentially greater accuracy than the formulae (44.28) and (44.29).

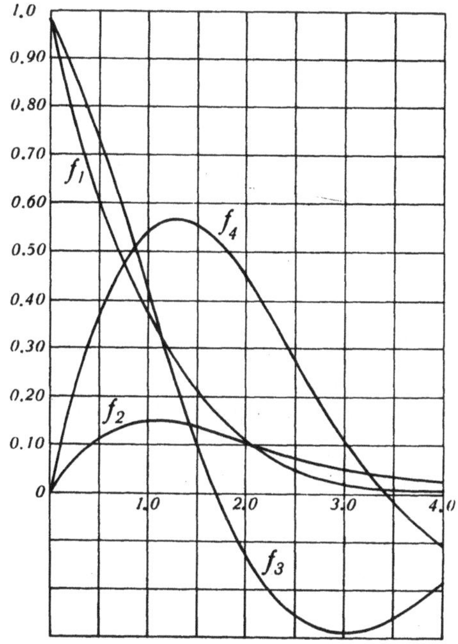

Fig. 63.

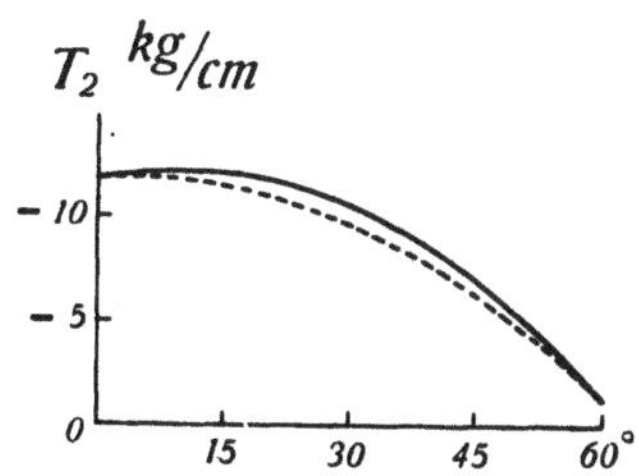

Fig. 64.

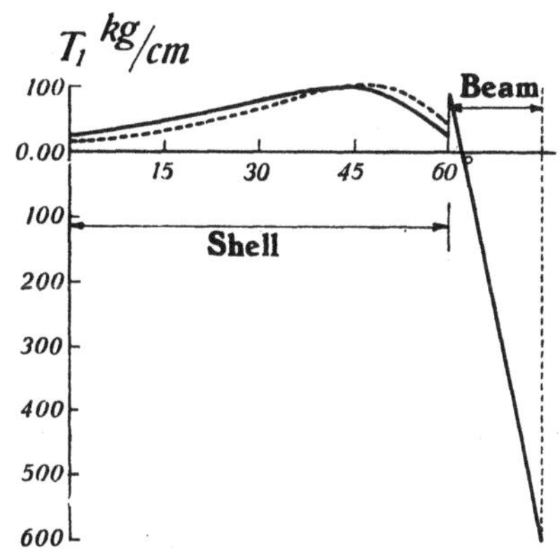

Fig. 65.

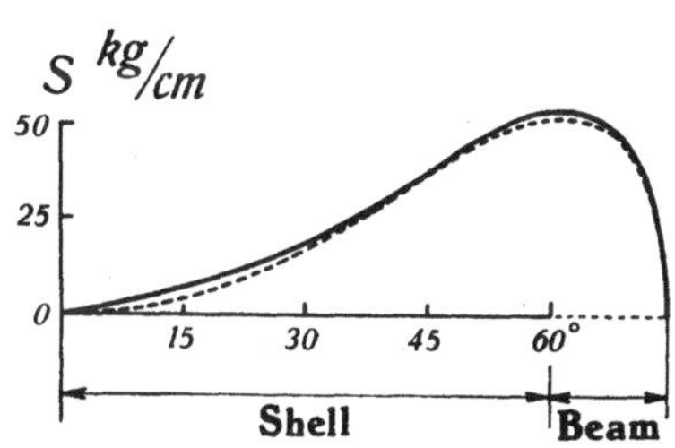

Fig. 66.

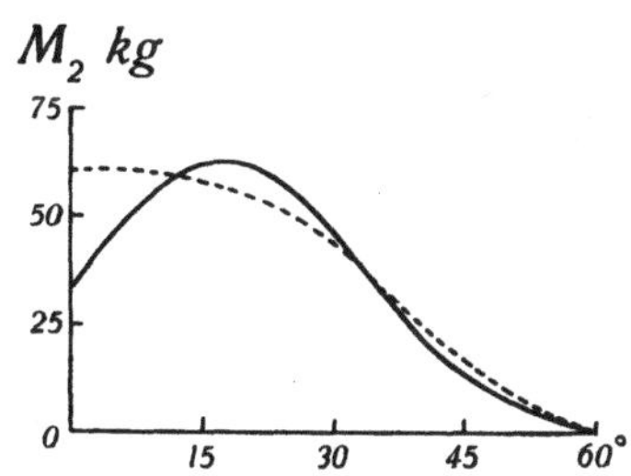

Fig. 67.

§ 45. The analysis of cylindrical plates, reinforced by transverse elastic ribs

The theory developed in § 41 will now be applied to the analysis of cylindrical plates reinforced by equidistant, transverse, elastic ribs. For this purpose, consider first of all the second equation of the system (41.20) with all its terms, i.e., in the form (41.16); in this way, the errors will be exhibited which arise from those simplifications employed to reduce this equation to the form it has in the system (41.20).

One thus has the system

$$\frac{\partial^2(T_1+T_2)}{\partial\xi^2}-\frac{1}{r_0}\left[\frac{\partial^2}{\partial\varphi^2}\Delta(M_2)+\frac{\partial^2 M_2}{\partial\varphi^2}\right]=r_0F_1(\xi,\eta),$$

$$\frac{4a^4}{r_0}\frac{\partial^2 M_2}{\partial\xi^2}+\frac{2(1+\mu)}{r_0}\left[\frac{\partial^4 M_2}{\partial\varphi^4}+\frac{\partial^2 M_2}{\partial\varphi^2}\right]+\frac{\partial^2}{\partial\varphi^2}\Delta(T_1+T_2)+ \tag{45.1}$$

$$+\frac{\partial^2(T_1+T_2)}{\partial\varphi^2}=r_0F_2(\xi,\varphi),$$

where F_1 and F_2 are the functions (41.21) of the external loading. Having in mind application of this system to the analysis of cylindrical plates, the functions F_1 and F_2 will be expressed in the form of the series

$$F_1=\sum_{m=1}^{\infty}F_{1,m}(\varphi)\sin\lambda_m\xi,\quad F_2=\sum_{m=1}^{\infty}F_{2,m}(\varphi)\sin\lambda_m\xi \tag{45.2}$$

$\left(\text{where, as before, } \lambda_m=m\dfrac{\pi r_0}{L}\right)$. Expressions for T_1, T_2 and M_2 will be sought in the form of similar series

$$T_1=\sum_{m=1}^{\infty}T_{1,m}(\varphi)\sin\lambda_m\xi,\quad T_2=\sum_{m=1}^{\infty}T_{2,m}(\varphi)\sin\lambda_m\xi, \tag{45.3}$$

$$M_2=\sum_{m=1}^{\infty}M_{2,m}\sin\lambda_m\xi.$$

After substituting (45.2) and (45.3) in (45.1), one obtains for the coefficients of (45.3) the following system of two ordinary differential equations:

$$\lambda_m{}^2T_m+\frac{1}{r_0}\left[\frac{d^2}{d\varphi^2}\Delta_m(M_{2,m})+\frac{d^2M_{2,m}}{d\varphi^2}\right]=-r_0F_{1,m}(\varphi),$$

$$\frac{1}{r_0}\left\{4a^4\lambda_m{}^2M_{2,m}-2(1+\mu)\left[\frac{d^4M_{2,m}}{d\varphi^4}+\frac{d^2M_{2,m}}{d\varphi^2}\right]\right\}- \tag{45.4}$$

$$-\frac{d^2}{d\varphi^2}\Delta_m(T_m)-\frac{d^2T_m}{d\varphi^2}=-r_0F_{2,m}(\varphi),$$

where

$$T_m=T_{1,m}+T_{2,m},\quad \Delta_m(\ldots)=\frac{d^2(\ldots)}{d\varphi^2}-\lambda_m{}^2(\ldots).$$

For reasons which have already been explained attention will be concentrated on the solution of the homogeneous problem, i.e., it will be assumed that $F_{1,m} = F_{2,m} = 0$. The solution of the corresponding system of homogeneous differential equations must be sought in the form

$$T_m = A e^{\gamma_m \varphi}, \quad M_{2,m} = r_0 B e^{\gamma_m \varphi}, \tag{45.5}$$

where A, B and γ_m are constants.

Substitution of (45.5) into (45.4) leads to the linear homogeneous system in the unknowns A and B:

$$\begin{gathered} \lambda_m{}^2 A + \gamma_m{}^2(\gamma_m{}^2 + 1 - \lambda_m{}^2)B = 0, \\ -\gamma_m{}^2(\gamma_m{}^2 + 1 - \lambda_m{}^2)A + [4a^4\lambda_m{}^2 - 2(1+\mu)\gamma_m{}^2(\gamma_m{}^2+1)]B = 0. \end{gathered} \tag{45.6}$$

Hence, one obtains for the parameter γ_m the determinantal equation

$$\begin{vmatrix} \lambda_m{}^2, & \gamma_m{}^2(\gamma_m{}^2 + 1 - \lambda_m{}^2) \\ -\gamma_m{}^2(\gamma_m{}^2 + 1 - \lambda_m{}^2), & 4a^4\lambda_m{}^2 - 2(1+\mu)\gamma_m{}^2(\gamma_m{}^2+1) \end{vmatrix} = 0, \tag{45.7}$$

which in expanded form leads to the following fourth degree polynomial in $\gamma_m{}^2$:

$$\gamma_m{}^4(\gamma_m{}^2 + 1 - \lambda_m{}^2)^2 - \underline{2(1+\mu)\gamma_m{}^2(\gamma_m{}^2+1)\lambda_m{}^2} + 4a^4\lambda_m{}^4 = 0. \tag{45.8}$$

As will be seen from what follows, this equation has eight complex roots, the moduli of which are of order $\sqrt{a\lambda_m}$. This may likewise be verified on the basis of the results of the preceding section.

But if this is so, then the underlined centre term of the equation (45.8) will be small in comparison with the two outer terms (which are quantities of the same order). Indeed, the modulus of the ratio of this term to the last term of the equation will be a quantity of the order:

$$\left| \frac{2(1+\mu)\gamma_m{}^2(\gamma_m{}^2+1)\lambda_m{}^2}{4a^4\lambda_m{}^4} \right| \approx \frac{2(1+\mu)a^2\lambda_m{}^4}{4a^4\lambda_m{}^4} \approx \frac{1+\mu}{2a^2},$$

i.e., it will be essentially less than unity (since $2a^2 \gg 1$).

Thus the underlined term in (45.8) can be omitted without significant error.

Tracing the deduction of (45.8) back to (45.1), it is easy to show that the neglect of the above term is equivalent to omitting the underlined term in the system (45.1). Reverting still further to the equations (41.16) and (41.20), it is seen that the simplification of the present section is equivalent to that which leads from equation (41.16) to the second equation (41.20). In this way, the present simplification has been based in a certain manner on that in § 41 by showing that at least for the particular case of the circular cylinder it is completely admissible. After rejecting in equation (45.8) the middle term, one obtains the approximate equation

$$\gamma_m^4(\gamma_m^2 + 1 - \lambda_m^2)^2 + 4a^4\lambda_m^4 = 0, \tag{45.9}$$

which can be resolved into the two bi-quadratic equations

$$\gamma_m^4 + (1 - \lambda_m^2)\gamma_m^2 \pm i2a^2\lambda_m^2 = 0, \tag{45.10}$$

the eight roots of which may be expressed by the two formulae

$$\begin{aligned} (\gamma_m)_{1,2,3,4} &= \pm c_1 \pm id_1, \\ (\gamma_m)_{5,6,7,8} &= \pm c_2 \pm id_2, \end{aligned} \tag{45.11}$$

where in the present case

$$\begin{aligned} c_1 &= \sqrt{\frac{a\lambda_m}{2}}\sqrt{\sqrt{(h+g_1)^2+g_2^2}+h+g_1}, \\ d_1 &= \sqrt{\frac{a\lambda_m}{2}}\sqrt{\sqrt{(h+g_1)^2+g_2^2}-h-g_1}, \\ c_2 &= \sqrt{\frac{a\lambda_m}{2}}\sqrt{\sqrt{(h-g_1)^2+g_2^2}+h-g_1}, \\ d_2 &= \sqrt{\frac{a\lambda_m}{2}}\sqrt{\sqrt{(h-g_1)^2+g_2^2}-h+g_1}, \end{aligned} \tag{45.12}$$

and

$$h = \frac{\lambda_m^2 - 1}{2a\lambda_m}, \tag{45.13}$$

$$g_1 = \frac{1}{\sqrt{2}}\sqrt{\sqrt{4+h^4}+h^2}, \quad g_2 = \frac{1}{\sqrt{2}}\sqrt{\sqrt{4+h^4}-h^2}.$$

After evaluation of the roots of the characteristic equation, one can write the following general expression for T_m:

$$T_m = A_1\Phi_1 + A_2\Phi_2 + A_3\Phi_3 + A_4\Phi_4 + A_1'\Psi_1 + A_2'\Psi_2 + A_3'\Psi_3 + A_4'\Psi_4 \tag{45.14}$$

[where the functions Φ_j and Ψ_j are defined by the formulae (42.32)]. The coefficients in the series (42.31), may be determined in terms of T_m, using the formulae of § 41 with $\rho = 1$. As a result, the coefficients u_m, v_m, w_m, ψ_m, $T_{1,m}$, $T_{2,m}$, S_m, $N_{2,m}$, $M_{2,m}$ can be represented by

Table 18

	α	β	α'	β'
$\frac{E\delta}{r_0} u_m$	$-\frac{1}{\lambda_m}\left[1 + (1+\mu)\frac{c_1 d_1}{a^2}\right]$	$(1+\mu)\frac{c_1^2 - d_1^2}{2a^2\lambda_m}$	$-\frac{1}{\lambda_m}\left[1 - (1+\mu)\frac{c_2 d_2}{a^2}\right]$	$-(1+\mu)\frac{c_2^2 - d_2^2}{2a^2\lambda_m}$
$\frac{E\delta}{r_0} v_m$	$\frac{c_1}{\lambda_m^2}\left[1 - \frac{(1+\mu)\lambda_m^2}{c_1^2 + d_1^2}\right]$	$\frac{d_1}{\lambda_m^2}\left[1 + \frac{(1+\mu)\lambda_m^2}{c_1^2 + d_1^2}\right]$	$\frac{c_2}{\lambda_m^2}\left[1 - \frac{(1+\mu)\lambda_m^2}{c_2^2 + d_2^2}\right]$	$\frac{d_2}{\lambda_m^2}\left[1 + \frac{(1+\mu)\lambda_m^2}{c_2^2 + d_2^2}\right]$
$\frac{E\delta}{r_0} w_m$	$-\frac{1}{\lambda_m^2}(c_1^2 - d_1^2 - \lambda_m^2)$	$-\frac{2c_1 d_1}{\lambda_m^2}$	$-\frac{1}{\lambda_m^2}(c_2^2 - d_2^2 - \lambda_m^2)$	$-\frac{2c_2 d_2}{\lambda_m^2}$
$E\delta\psi_m$	$\frac{d_1}{c_1^2 + d_1^2} 2a^2$	$\frac{c_1}{c_1^2 + d_1^2} 2a^2$	$-\frac{d_2}{c_2^2 + d_2^2} 2a^2$	$-\frac{c_2}{c_2^2 + d_2^2} 2a^2$
$T_{1,m}$	$1 + \frac{c_1 d_1}{a^2}$	$-\frac{c_1^2 - d_1^2}{2a^2}$	$1 - \frac{c_2 d_2}{a^2}$	$\frac{c_2^2 - d_2^2}{2a^2}$
$T_{2,m}$	$-\frac{c_1 d_1}{a^2}$	$\frac{c_1^2 - d_1^2}{2a^2}$	$\frac{c_2 d_2}{a^2}$	$-\frac{c_2^2 - d_2^2}{2a^2}$
S_m	$-\frac{d_1}{2a^2\lambda_m}(3c_1^2 - d_1^2 + 1)$	$-\frac{c_1}{2a^2\lambda_m}(3d_1^2 - c_1^2 - 1)$	$\frac{d_2}{2a^2\lambda_m}(3c_2^2 - d_2^2 + 1)$	$\frac{c_2}{2a^2\lambda_m}(3d_2^2 - c_2^2 - 1)$
$N_{2,m}$	$-\frac{d_1}{2a^2}$	$\frac{c_1}{2a^2}$	$\frac{d_2}{2a^2}$	$-\frac{c_2}{2a^2}$
$\frac{M_{2,m}}{r_0}$	0	$\frac{1}{2a^2}$	0	$-\frac{1}{2a^2}$

formulae of one of the following two types:

$$=(\alpha A_1+\beta A_2)\Phi_1+(\alpha A_2-\beta A_1)\Phi_2+(\alpha A_3-\beta A_4)\Phi_3+(\alpha A_4+\beta A_3)\Phi_4+$$
$$+(\alpha' A_1'+\beta' A_2')\Psi_1+(\alpha' A_2'-\beta' A_1')\Psi_2+(\alpha' A_3'-\beta' A_4')\Psi_3+(\alpha' A_4'+\beta' A_3')\Psi_4,$$
$$=(\alpha A_4+\beta A_3)\Phi_1+(\alpha A_3-\beta A_4)\Phi_2+(\alpha A_2-\beta A_1)\Phi_3+(\alpha A_1+\beta A_2)\Phi_4+$$
$$+(\alpha' A_4'+\beta' A_3')\Psi_1+(\alpha' A_3'-\beta' A_4')\Psi_2+(\alpha' A_2'-\beta' A_1')\Psi_3+(\alpha' A_1'+\beta' A_2')\Psi_4,$$
$$(45.15)$$

where u_m, $T_{1,m}$, $T_{2,m}$, $M_{2,m}$ will be expressed by formulae of the first type and v_m, $N_{2,m}$, S_m, ψ_m by formulae of the second type.

Table 18 gives the values of the coefficients α, β, α', β', corresponding to the different coefficients of the series (42.31).

No further explanation is required, since the sequence of the computations should again correspond with the recommendations of § 43. The example treated there represents a sufficiently detailed introduction to the computations involved.

§ 46. The analysis of tubes of arbitrary cross-section with freely supported ends

The preceding sections have only been concerned with cylindrical plates, i.e., with shells having four edges which are portions of circular cylinders. Proceeding now to the analysis of complete cylindrical shells (i.e., tubes), this problem will be solved in relation to tubes of arbitrary shape, assuming, however, that the cross-section has the form of a regular oval (i.e., an oval with two axes of symmetry, Fig. 68). Further, it should be emphasized that this limitation is not compulsory for the following reasoning (from the point of view of possible application of the method below) and that it has only been adopted since shells of that form are of the greatest practical interest. The origin of the coordinate x (and therefore also the origin of the non-dimensional coordinate ξ) will be taken at one of the ends of the tube. Hence the beginning and the end of the tube will be determined by $\xi = 0$, $\xi = \xi_0 = \dfrac{L}{r_0}$ (where L is the length of the tube). The origin of the arc s (and therefore also

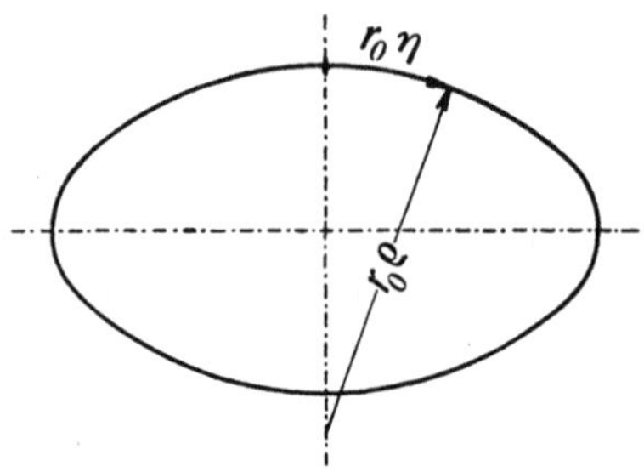

Fig. 68.

the origin of $\eta = \dfrac{s}{r_0}\Big)$ will be placed at one of the ends of the minor axis of the oval.

In view of the hypothesis referring to the shape of the cross-section of the tube, the ratio $\rho = \dfrac{r}{r_0}$ will be an even periodic function of η with period $\eta_0 = \dfrac{2s_0}{r_0}$ (where s_0 is a quarter of the length of the circumference of the tube).

Let the external loading be due to normal pressure q_n with the other two components being identically zero. Then the fundamental differential equation of the theory of cylindrical shells (40.7) may be written

$$\Delta[\rho\Delta(\tilde{T})] + \frac{\partial}{\partial\eta}\left(\frac{1}{\rho}\frac{\partial\tilde{T}}{\partial\eta}\right) + i2b^2\frac{\partial^2\tilde{T}}{\partial\xi^2} = i2b^2r_0\Delta(\rho q_n), \tag{46.1}$$

where

$$\Delta(\ldots) = \frac{\partial^2(\ldots)}{\partial\xi^2} + \frac{\partial^2(\ldots)}{\partial\eta^2},$$
$$4b^4 = 12(1-\mu^2)\frac{r_0^2}{\delta^2}. \tag{46.2}$$

For the problem to be fully determined, certain boundary conditions must be given at the edges of the tube. In this context, there is an essential difference between a tube and a cylindrical plate, in that the tube will always have two edges and that it will be closed with respect to the coordinate η. Hence, the solution of the equations (46.1) must in this case be a periodic function of η. In this section, consideration will be restricted to one type of boundary condition, namely, it will be assumed that the edges of the tube rest on supports which are absolutely rigid in their own planes but offer no resistance to edge moments and edge forces acting in the direction of the axis of the shell.

Mathematically, these boundary conditions take the form

$$v = w = T_1 = M_1 = 0, \tag{46.3}$$

for $\xi = 0$ and $\xi = \xi_0$.

In order to obtain the solution of the equation (46.1), satisfying the boundary conditions (46.3), one has to represent the loading in the form of the series

$$q_n = \sum_{m=1}^{\infty} q_n^{(m)} \sin \lambda_m\xi, \tag{46.4}$$

where, as before, $\lambda_m = m \dfrac{\pi r_0}{L}$ and the $q_n^{(m)}$ are functions of η only which may always be determined with the help of known formulae of the theory of Fourier series.

In the particular case of uniform pressure $q_n = p =$ constant, one will have the expansion

$$q_n = p = \frac{4}{\pi} p \sum_{m=1}^{\infty} \frac{1}{m} \sin \lambda_m \xi, \tag{46.5}$$

in which the summation extends only over odd m. The series (46.5) will now be substituted into the equation (46.1) and the solution will be sought in the form

$$\tilde{T} = \sum_{m=1}^{\infty} \tilde{T}^{(m)} \sin \lambda_m \xi, \tag{46.6}$$

where the coefficients $\tilde{T}^{(m)}$ are functions of η only.

Then the partial differential equation (46.1) reduces to a set of ordinary differential equations of the general type

$$\Delta_m[\rho \Delta_m(\tilde{T}^{(m)})] + \frac{d}{d\eta}\left(\frac{1}{\rho}\,\frac{d\tilde{T}^{(m)}}{d\eta}\right) - i2b^2\lambda_m{}^2\tilde{T}^{(m)} = i2b^2 q_m \Delta_m(\rho), \tag{46.7}$$

where

$$\left.\begin{aligned} q_m &= \frac{4}{\pi}\,\frac{p r_0}{m}, \\ \Delta_m(\ldots) &= \frac{d^2(\ldots)}{d\eta^2} - \lambda_m{}^2(\ldots). \end{aligned}\right\} \tag{46.8}$$

On the right hand side of (46.7) one has an even periodic function of η (since ρ is such a function). Hence $\tilde{T}^{(m)}$ must also be an even periodic function which may thus be sought in the form

$$\tilde{T}^{(m)} = \sum_{j=0}^{n} \tilde{T}^{(m,j)} \cos (j\nu\eta), \tag{46.9}$$

where $\nu = \dfrac{\pi r_0}{s_0}$ and the coefficients $\tilde{T}^{(m,j)}$ are as yet unknown constants which will be determined by use of Galerkin's method. Applying this method to the equation (46.7), one may write

$$\int_0^{\eta_0}\Big\{\Delta_m[\rho\Delta_m(\tilde{T}^{(m)})] + \frac{d}{d\eta}\Big(\frac{1}{\rho}\,\frac{d\tilde{T}^{(m)}}{d\eta}\Big) - i2b^2\lambda_m{}^2\tilde{T}^{(m)} - $$

$$- 2ib^2q_m\Delta_m(\rho)\Big\}\,\delta(\tilde{T}^{(m)}) = 0, \qquad (46.10)$$

where $\delta(\ldots)$ is the symbol for a variation compatible with the boundary conditions (in this case with the periodicity condition of the solution). Taking $\tilde{T}^{(m)}$ in the form (46.9) and varying (independently of each other) the coefficients $\tilde{T}^{(m,k)}$, one obtains the following system of $(n+1)$ algebraic equations with the unknowns $\tilde{T}^{(m,k)}$:

$$\int_0^{\eta_0}\Big\{\Delta_m[\rho\Delta_m(\tilde{T}^{(m)})] + \frac{d}{d\eta}\Big(\frac{1}{\rho}\,\frac{d\tilde{T}^{(m)}}{d\eta}\Big) - i2b^2\lambda_m{}^2\tilde{T}^{(m)} - $$

$$- i2b^2q_m\Delta_m(\rho)\Big\}\cos(k\nu\eta)\,d\eta = C, \qquad (46.11)$$

where $k = 0, 1, 2 \ldots n$ and by $\tilde{T}^{(m)}$ is understood the expression (46.9). Before substituting it in (46.11), it is expedient to reduce in this formula the order of the derivatives by an integration by parts which gives

$$-\int_0^{\eta_0}\Big\{\rho(k^2\nu^2 + \lambda_m{}^2)\Big[\frac{d^2\tilde{T}^{(m)}}{d\eta^2} - \lambda_m{}^2\tilde{T}^{(m)} - i2b^2q_m\Big] + $$

$$+ i2b^2\lambda_m{}^2\tilde{T}^{(m)}\Big\}\cos(k\nu\eta)\,d\eta + k\nu\int_0^{\eta_0}\frac{1}{\rho}\,\frac{d\tilde{T}^{(m)}}{d\eta}\sin(k\nu\eta)\,d\eta = 0. \qquad (46.12)$$

Introducing now instead of $\tilde{T}^{(m)}$ the expression (46.9), one has

$$\sum_{j=0}^{n}\{[(k^2\nu^2 + \lambda_m{}^2)(j^2\nu^2 + \lambda_m{}^2)I_1 - jk\nu^2I_2 - i2b^2\lambda_m{}^2I_3]\tilde{T}^{(m,j)} + $$

$$+ i2b^2q_m(k^2\nu^2 + \lambda_m{}^2)I_4\} = 0. \qquad (46.13)$$

where

$$I_1 = \int_0^{\eta_0}\rho\cos(k\nu\eta)\cos(j\nu\eta)\,d\eta, \quad I_2 = \int_0^{\eta_0}\frac{1}{\rho}\sin(k\nu\eta)\sin(j\nu\eta)\,d\eta,$$

$$I_3 = \int_0^{\eta_0}\cos(k\nu\eta)\cos(j\nu\eta)\,d\eta, \quad I_4 = \int_0^{\eta_0}\rho\cos(k\nu\eta)\,d\eta. \qquad (46.14)$$

Replacing in these integrals η by the new variable $z = \nu\eta$ and noting that $\eta_0 = \frac{\pi}{\nu}$, one has

$$I_1 = \frac{1}{\nu}\int_0^\pi \rho \cos kz \cos jz\, dz = \frac{1}{2\nu}\int_0^\pi \rho \cos (k+j)z\, dz + \frac{1}{2\nu}\int_0^\pi \rho \cos (k-j)z\, dz,$$

$$I_2 = \frac{1}{\nu}\int_0^\pi \frac{1}{\rho} \sin kz \sin jz\, dz = \frac{1}{2\nu}\int_0^\pi \frac{1}{\rho} \cos (k-j)z\, dz - \frac{1}{2\nu}\int_0^\pi \frac{1}{\rho} \cos (k+j)z\, dz,$$

$$I_3 = \frac{1}{\nu}\int_0^\pi \cos kz \cos jz\, dz, \tag{46.15}$$

$$I_4 = \frac{1}{\nu}\int_0^\pi \rho \cos kz\, dz.$$

Hence it follows, first of all, that

$$\begin{aligned} &\text{for } k \neq j : \quad I_3 = 0, \\ &\text{for } k = j > 0 : I_3 = \frac{\pi}{2\nu}, \\ &\text{for } k = j = 0 : I_3 = \frac{\pi}{\nu}. \end{aligned} \tag{46.16}$$

Further, since ρ is an even periodic function of η with period $2\eta_0$, it will also be an even periodic function of z (with period 2π). Hence the integral I_4 will be equal to

$$I_4 = \frac{\pi}{2\nu}\, \rho_k, \tag{46.17}$$

where ρ_k is a coefficient of the expansion

$$\rho = \tfrac{1}{2}\rho_0 + \rho_1 \cos \nu\eta + \rho_2 \cos 2\nu\eta + \ldots + \rho_k \cos k\nu\eta + \ldots, \tag{46.18}$$

i.e., a coefficient of the expansion of the function $\rho(\eta)$ in a Fourier series in terms of the arc length of the circumferential section.

A similar series applies also for the inverse function, namely

$$\frac{1}{\rho} = \frac{1}{2}\left(\frac{1}{\rho}\right)_0 + \left(\frac{1}{\rho}\right)_1 \cos \nu\eta + \left(\frac{1}{\rho}\right)_2 \cos 2\nu\eta + \ldots + \left(\frac{1}{\rho}\right)_k \cos k\nu\eta + \ldots, \tag{46.19}$$

where

$$\left(\frac{1}{\rho}\right)_k = \frac{2}{\pi}\int_0^\pi \frac{1}{\rho}\cos kz\,dz. \tag{46.20}$$

On the basis of the above discussion, the integrals I_1 and I_2 can be expressed in terms of the coefficients of the series (46.18) and (46.19) by the formulae

$$\begin{aligned} I_1 &= \frac{\pi}{4\nu}(\rho_{k+j} + \rho_{k-j}), \\ I_2 &= \frac{\pi}{4\nu}\left[\left(\frac{1}{\rho}\right)_{k-j} - \left(\frac{1}{\rho}\right)_{k+j}\right]. \end{aligned} \tag{46.21}$$

Now all the coefficients of the system (46.13) have been determined and it can be written in the form

$$\begin{aligned} &\beta_{0,0}^{(m)}T^{(m,0)}+\beta_{0,1}^{(m)}T^{(m,1)}+\beta_{0,2}^{(m)}T^{(m,2)}+\ldots+\beta_{0,n}^{(m)}T^{(m,n)}=f_0^{(m)}, \\ &\beta_{1,0}^{(m)}T^{(m,0)}+\beta_{1,1}^{(m)}T^{(m,1)}+\beta_{1,2}^{(m)}T^{(m,2)}+\ldots+\beta_{1,n}^{(m)}T^{(m,n)}=f_1^{(m)}, \\ &\cdots\cdots\cdots\cdots\cdots\cdots\cdots\cdots \\ &\cdots\cdots\cdots\cdots\cdots\cdots\cdots\cdots \\ &\cdots\cdots\cdots\cdots\cdots\cdots\cdots\cdots \\ &\beta_{n,0}^{(m)}T^{(m,0)}+\beta_{n,1}^{(m)}T^{(m,1)}+\beta_{n,2}^{(m)}T^{(m,2)}+\ldots+\beta_{n,n}^{(m)}T^{(m,n)}=f_n^{(m)}, \end{aligned} \tag{46.22}$$

where for $\beta_{jk}^{(m)}$ and $f_k^{(m)}$ the following formulae have been obtained:

$$\begin{aligned} \beta_{kk} &= \frac{\rho_0+\rho_{2k}}{2}(k^2\nu^2+\lambda_m^2)^2 + \frac{k^2\nu^2}{2}\left[\left(\frac{1}{\rho}\right)_{2k} - \left(\frac{1}{\rho}\right)_0\right] - i2b^2\lambda_m^2, \\ \beta_{jk} &= \tfrac{1}{2}(\rho_{k+j}+\rho_{k-j})\,(k^2\nu^2+\lambda_m^2)\,(j^2\nu^2+\lambda_m^2)+ \\ &\qquad + \frac{kj\nu^2}{2}\left[\left(\frac{1}{\rho}\right)_{k+j} - \left(\frac{1}{\rho}\right)_{k-j}\right], \\ f_k^{(m)} &= -\,i2b^2(k^2\nu^2+\lambda_m^2)\rho_k q_m, \qquad \beta_{jk} = \beta_{kj}. \end{aligned} \tag{46.23}$$

These general expressions do not apply to $\beta_{0,0}^{(m)}$ which is given by

$$\beta_{0,0}^{(m)} = (\lambda_m^2\rho_0 - i4b^2)\lambda_m^2. \tag{46.24}$$

Thus, when the Fourier series for the functions ρ and $\frac{1}{\rho}$ are known as functions of the arc length of the circumferential section of the tube, the construction of the system (46.22) will present no difficulty. It should be noted that the same system could be derived without the use of

Galerkin's method by substituting the series (46.9), (46.18), (46.19) in the equations (46.7) and by comparing the coefficients of the corresponding trigonometric functions.

The obtained solution must therefore be considered a Fourier type solution of the equation (46.7). Since the above direct method involves multiplication of trigonometric series and subsequent replacement of the indices, it has been considered preferable to follow Galerkin's method which yields the same results but simplifies the computations. It is then not an accident that the method of Galerkin in this case leads to a solution which completely coincides with the Fourier type solution. The fact is that for the solution of linear problems the above mentioned method satisfies the differential equations with least square accuracy (for given series of approximating functions). But it is known that this same property attaches to solutions of equations in the form of trigonometric series. Hence, if in applying the method of Galerkin the solution is sought in the form of a finite trigonometric series (as this has been done above), then the coefficients of this series (if the problem is linear) will always coincide with the coefficients of the Fourier series.

There remains to find out how many terms must be retained in the series (46.6) and (46.9) respectively. This question is very important, because inclusion of any extra term complicates significantly the computations, arising in the analysis of tubes.

In the analysis of cylindrical plates, only the first terms in the series (46.6) corresponding to $m = 1$ have been retained, i.e., the solution was taken in the form

$$\tilde{T} = \tilde{T}' \sin \lambda\xi, \tag{46.25}$$

corresponding to replacement of the uniform normal pressure loading by

$$q_n = \frac{4}{\pi} p \sin \lambda\xi. \tag{46.26}$$

It has been shown that in spite of this sweeping simplification the maximum values of the deflections and displacement components involve errors ordinarily not exceeding 10—15% and directed towards the safe side. Since the above problem has been analyzed by use of a method analogous to that applied to cylindrical plates and since it is closely related to it, there is no reason to assume that more significant errors will arise, if one makes use of the degree of approximation expressed by the formulae (46.25) and (46.26).

Thus, in the series (46.6), one may only retain the first term. As regards the series (46.9) the number of terms in it must be selected depending on the shape of the cross-section of the tube.

Namely, if the series (46.18) and (46.19) for the form of the shell converge rapidly, one may retain in the series (46.9) only two or three terms. Conversely, if the series for ρ and $\dfrac{1}{\rho}$ converge slowly, the series (46.9) may require four or five terms, but never a larger number of terms. In this context, the algebraic system (46.22), in as much as the solution is taken in the form (46.25), will only contain as many equations as the number of terms retained in the series (46.9).

In the following section, an unfavourable example for the application of this method will be presented, namely the case of a tube whose cross-section is formed by four circular arcs. Then the functions ρ and $\dfrac{1}{\rho}$ will be discontinuous, in that they exhibit step changes, and therefore the series (46.18) and (46.19) will converge slowly.

Further, since it is impossible to differentiate these series, the mathematical legality of applying the above method to this problem may, in general, be questioned. Nevertheless, the results presented below for the case when in the series (46.9) six terms have been retained are in good agreement with those for the same shell in § 43 by use of another method (by subdividing it into cylindrical plates) as well as with given experiments.

§ 47. The analysis of a cylindrical shell with abruptly changing radius of curvature

As an example of the application of the theory of the previous section, the analysis of an oval shaped cylindrical shell will be considered the shape of which is close to that studied in § 43 (by a different method). In fact, a thin-walled cylinder will be investigated which is subject to uniform internal pressure and the cross-section of which is formed by two pairs of circular arcs. Fig. 69 shows this cross-section together with an ellipse of the same eccentricity.

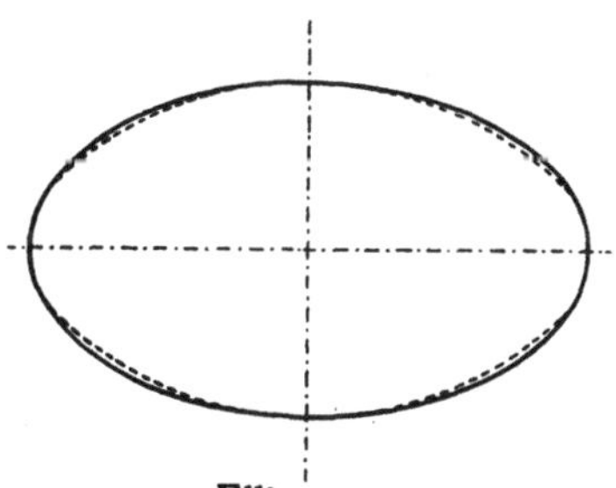

Fig. 69.

The analysis will be based on the following data:

$$\frac{r_2}{r_1} = 0.318, \qquad \frac{r_1}{r_2} = 3.14, \qquad \varphi_0 = 45°,$$

$$L = 1.47r_1, \qquad \lambda = \frac{\pi r_0}{L} = \frac{\pi r_1}{L} = 2.14, \qquad \lambda^2 = 4.58,$$

$$\frac{r_0}{\delta} = \frac{r_1}{\delta} = 175, \qquad 2b^2 = \sqrt{12(1-\mu^2)}\,\frac{r_1}{\delta} = 578, \tag{47.1}$$

$$s_0 = \frac{\pi}{4}(r_1 + r_2) = 1.035r_1, \qquad \eta_0 = 1.035, \qquad \nu = \frac{\pi}{\eta_0} = 3.04,$$

$$\nu^2 = 9.25, \qquad s_1 = \frac{\pi}{4} r_1, \quad \frac{s_1}{s_0} = \frac{r_1}{r_1 + r_2} = 0.758 = \alpha.$$

First of all, the coefficients in the expansions of the functions ρ and $\frac{1}{\rho}$ in Fourier series in terms of the arc length of the oval will be calculated. In this case, ρ will be a step function of the arc length, namely, for

$$\begin{aligned} 0 \leqslant s \leqslant s_1, \quad & \rho = 1, \qquad \frac{1}{\rho} = 1, \\ s_1 \leqslant s \leqslant s_0, \quad & \rho = 0.318, \qquad \frac{1}{\rho} = 3.14. \end{aligned} \tag{47.2}$$

Hence

$$\rho_k = \frac{2}{\pi}\int_0^\pi \rho \cos kz\,dz = \frac{2}{\pi}\left[\int_0^{\alpha\pi} \cos kz\,dz + 0.318\int_{\alpha\pi}^{\pi} \cos kz\,dz\right] =$$

$$= \frac{1}{k}\,0.434 \sin(0.758k\pi), \tag{47.3}$$

$$\rho_0 = \frac{2}{\pi}\left[\int_0^{\alpha\pi} dz + 0.318\int_{\alpha\pi}^{\pi} dz\right] = 2[\alpha + 0.318(1-\alpha)] = 1.67.$$

$$\left(\frac{1}{\rho}\right)_k = \frac{2}{\pi}\int_0^\pi \frac{1}{\rho}\cos kz\,dz = \frac{2}{\pi}\left[\int_0^{\alpha\pi} \cos kz\,dz + 3.14\int_{\alpha\pi}^{\pi} \cos kz\,dz\right] =$$

$$= -\frac{1.362}{k}\sin \alpha k\pi = -\frac{1.362}{k}\sin(0.758k\pi), \tag{47.4}$$

$$\left(\frac{1}{\rho}\right)_0 = 2[\alpha + 3.14(1-\alpha)] = 3.04.$$

On the basis of these formulae, the following table of values of ρ_k

and $\left(\frac{1}{\rho}\right)_k$ (up to $k = 8$) may be evaluated:

Table 19

k	0	1	2	3	4	5	6	7	8
ρ_k	1.67	0.299	−0.216	0.110	−0.011	−0.053	+0.071	−0.050	0.010
$\left(\frac{1}{\rho}\right)_k$	3.04	−0.937	0.679	−0.343	0.034	0.167	−0.224	0.158	−0.032

Substituting these values in the formulae (46.23) and (46.24), one finds the coefficients β_{jk} (corresponding to $m = 1$) and also the free terms f_k of the algebraic system (46.22).

Table 20

j, k	0.0	0.1	0.2	0.3	0.4	1.1	1.2	1.3
$\beta_{j,k}$	$35-i5300$	18.9	−41.1	44.2	−7.52	$128-i2650$	123	−146

j, k	1.4	2.2	2.3	2.4	3.3	3.4	4.4
$\beta_{j,k}$	69	$1380-i2650$	480	−493	$6560-i2650$	1720	$19200-i2650$

Table 21

k	0	1	2	3	4
f_k	$-5620ipr_1$	$-3040ipr_1$	$6620ipr_1$	$-7120ipr_1$	$1210ipr_1$

Hence, in this case, the algebraic system (46.22) may be written in the following manner (retaining in it five equations):

$$
\begin{aligned}
&(35-5300i)\tilde{T}^{(0)}+18.9\tilde{T}^{(1)}-41.1\tilde{T}^{(2)}+44.2\tilde{T}^{(3)}-7.52\tilde{T}^{(4)}=-i5620pr_1,\\
&18.9\tilde{T}^{(0)}+(128-2650i)\tilde{T}^{(1)}+123\tilde{T}^{(2)}-146\tilde{T}^{(3)}+69\tilde{T}^{(4)}=-i3040pr_1,\\
&-41.1\tilde{T}^{(0)}+123\tilde{T}^{(1)}+(1380-2650i)\tilde{T}^{(2)}+480\tilde{T}^{(3)}-493\tilde{T}^{(4)}=\ i6620pr_1,\\
&44.2\tilde{T}^{(0)}-146\tilde{T}^{(1)}+480\tilde{T}^{(2)}+(6560-2650i)\tilde{T}^{(3)}+1720\tilde{T}^{(4)}=-i7120pr_1,\\
&-7.52\tilde{T}^{(0)}+69\tilde{T}^{(1)}-493\tilde{T}^{(2)}+1720\tilde{T}^{(3)}+(19200-2650i)\tilde{T}^{(4)}=\ i1210pr_1.
\end{aligned}
\tag{47.5}
$$

The solutions of this problem are

$$\begin{aligned} \tilde{T}^{(0)} &= (1.042 - i0.034)pr_1, \\ \tilde{T}^{(1)} &= (1.23 + i0.078)pr_1, \\ \tilde{T}^{(2)} &= (-2.28 + i0.798)pr_1, \\ \tilde{T}^{(3)} &= (0.600 - i0.930)pr_1, \\ \tilde{T}^{(4)} &= (-0.138 + i0.148)pr_1. \end{aligned} \tag{47.6}$$

Thus, retaining in the series (46.6) only one term, one has

$$\begin{aligned} \tilde{T} = \tilde{T}' \sin \lambda\xi = pr_1[&1.042 - 0.034i + (1.23 + 0.078i) \cos \nu\eta + \\ &+ (-2.28 + 0.798i) \cos 2\nu\eta + (0.600 - 0.930i) \cos 3\nu\eta + \\ &+ (-0.138 + 0.148i) \cos 4\nu\eta] \sin \lambda\xi \end{aligned} \tag{47.7}$$

and

$$\begin{aligned} \Delta(\tilde{T}) = -pr_1[&4.77 + 0.157i + (17.1 + 1.079i) \cos \nu\eta + \\ &+ (-94.8 + 33.2i) \cos 2\nu\eta + (52.7 - 81.6i) \cos 3\nu\eta + \\ &+ (-21.0 + 22.6i) \cos 4\nu\eta] \sin \lambda\xi. \end{aligned}$$

The formulae of § 40 now permit the determination of all the complex forces and complex displacements. In this context, in correspondence with the accepted degree of accuracy of the calculations, the quantities ρ and $\frac{1}{\rho}$ in these formulae refer to their expressions in the form of the series (46.18) and (46.19), the first nine coefficients of which are given in Table 19.

Further, the quantity q_n in the formulae must be interpreted as

$$q_n = \frac{4}{\pi} p \sin \lambda\xi,$$

with the remaining components of the loading having been set equal to zero.

Reasoning in this manner, one obtains

$$\begin{aligned} \tilde{T}_1' = \tilde{T}' - \frac{i}{2b^2} \rho \left(\frac{d^2\tilde{T}'}{d\eta^2} - \lambda^2 \tilde{T}' \right) - \frac{4}{\pi} pr_1\rho = \\ = pr_1\{[&-0.018 - 0.034i + (0.855 + 0.078i) \cos \nu\eta + \\ &+ (-2.00 + 0.798i) \cos 2\nu\eta + (0.464 - 0.93i) \cos 3\nu\eta - \\ &- (0.124 - 0.148i) \cos 4\nu\eta] + \rho[-0.0003 + 0.0083i - \\ &- (0.0019 - 0.029i) \cos \nu\eta - (0.057 + 0.164i) \cos 2\nu\eta + \\ &+ (0.141 + 0.091i) \cos 3\nu\eta - (0.039 + 0.036i) \cos 4\nu\eta]\}; \end{aligned}$$

$$\tilde{T}_2' = \rho\left[\frac{i}{2b^2}\Delta_m(\tilde{T}') + \frac{4}{\pi}pr_1\right] = [1.27 - 0.008i +$$

$$+ (0.0019 - 0.029i)\cos\nu\eta + (0.057 + 0.164i)\cos 2\nu\eta -$$

$$- (0.141 + 0.092i)\cos 3\nu\eta + (0.039 + 0.036i)\cos 4\nu\eta]pr_1\rho,$$

$$\tilde{w}' = \frac{pr_1^2}{E\delta}\rho[1.042 - 0.034i + (3.72 + 0.24i)\cos\nu\eta +$$

$$+ (-20.7 + 7.25i)\cos 2\nu\eta + (11.1 - 17.8i)\cos 3\nu\eta + \tag{47.8}$$

$$+ (-4.6 + 4.9i)\cos 4\nu\eta],$$

where $\tilde{T}_1' \ldots, w'$ are the coefficients in the formulae

$$\tilde{T}_1 = \tilde{T}_1'\sin\lambda\xi,\ \tilde{T}_2 = \tilde{T}_2'\sin\lambda\xi,\ \tilde{S} = \tilde{S}'\cos\lambda\xi,\ \tilde{w} = \tilde{w}'\sin\lambda\xi.$$

Separating in these expressions real and imaginary parts, one obtains the following formulae for the forces, moments and the normal displacement w:

$$w' = |\tilde{w}'|_{\text{Real}} = \frac{pr_1^2}{E\delta}\rho[1.042 + 3.72\cos\nu\eta - 20.7\cos 2\nu\eta +$$

$$+ 11.5\cos 3\nu\eta - 4.6\cos 4\nu\eta],$$

$$T_1' = |\tilde{T}_1'|_{\text{Real}} = pr_1[-(0.018 + 0.0003\rho) + (0.855 - 0.002\rho)\cos\nu\eta -$$

$$- (2.00 + 0.06\rho)\cos 2\nu\eta + (0.46 + 0.141\rho)\cos 3\nu\eta -$$

$$- (0.124 + 0.039\rho)\cos 4\nu\eta], \tag{47.9}$$

$$T_2' = |\tilde{T}_2'|_{\text{Real}} = pr_1[1.27 + 0.002\cos\nu\eta + 0.057\cos 2\nu\eta -$$

$$- 0.141\cos 3\nu\eta + 0.039\cos 4\nu\eta]\rho,$$

$$\frac{M_2' - \mu M_1'}{1-\mu^2} = \frac{pr_1^2}{2b^2}[(0.034 - 0.0083\rho) - (0.078 + 0.029\rho)\cos\nu\eta -$$

$$- (0.798 - 0.164\rho)\cos 2\nu\eta + (0.93 - 0.09\rho)\cos 3\nu\eta -$$

$$- (0.148 - 0.036\rho)\cos 4\nu\eta],$$

$$\frac{M_1' - \mu M_2'}{1-\mu^2} = \frac{pr_1^2}{2b^2}\rho[0.008 + 0.029\cos\nu\eta - 0.164\cos 2\nu\eta +$$

$$+ 0.092\cos 3\nu\eta - 0.036\cos 4\nu\eta].$$

Multiplying the last of the formulae (47.9) by Poisson's ratio μ and adding it to the second last expression, one obtains the most interesting of the internal moments in the shell, namely the moment M_2, on which depends the deformation of the cross-section of the cylinder.

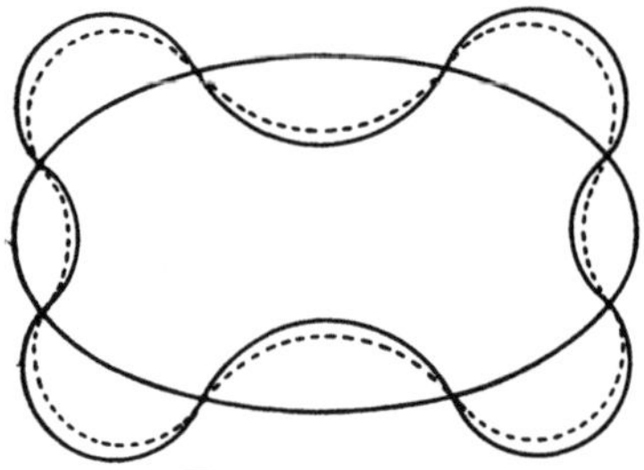

Fig. 70.

If one takes $\mu = 0.3$, one finds in this way

$$M_2' = \frac{pr_1^2}{2b^2}[0.034 - 0.0107\rho - (0.078 + 0.0207\rho)\cos\nu\eta - (0.798 - 0.115\rho)\cos 2\nu\eta + (0.936 - 0.064\rho)\cos 3\nu\eta - (0.148 - 0.026\rho)\cos 4\nu\eta].$$

Now one may evaluate the displacements and stresses at any point of the middle surface. Fig. 70 shows the graph of the deformation of the cross-section of the tube, as given by the first of the formulae (47.9).

Table 22 below presents a comparison of the deduced results with those obtained by experiment (the theoretical and experimental values of the radial displacement w at the three most characteristic points of the shell, lying in the plane $x = L/2$).

Table 22

	$\nu\eta = 0$	$\nu\eta = \frac{\pi}{2}$	$\nu\eta = \pi$
Theoretical	$-9.0\frac{pr_1^2}{E\delta}$	$17.2\frac{pr_1^2}{E\delta}$	$-12.6\frac{pr_1^2}{E\delta}$
Experimental	$-6.5\frac{pr_1^2}{E\delta}$	$15.9\frac{pr_1^2}{E\delta}$	$-11.9\frac{pr_1^2}{E\delta}$

A study of this table will lead to the conclusion that the method developed above is completely acceptable . The study of the capabilities of this method has purposely been based on an unfavourable example for which the trigonometric series for ρ and $\frac{1}{\rho}$ converge slowly. Nevertheless, by retaining in the expressions (46.9) only five terms, results were obtained which exhibit satisfactory accuracy. Hence there is no doubt that for tubes with smoother cross-sections it will be sufficient to retain in the series (46.9) three to four terms with a corresponding simplification of the calculations.

It is important to note in conclusion that the method of § 46 in comparison with the method of § 43 has the advantage that it is general and may be applied to tubes of arbitrary cross-section, while the method of § 43 is specialized and may only be used to analyze tubes whose cross-sections may be assumed to be formed by a number of cylindrical plates.

§ 48. The analysis of tubes for arbitrary end conditions

The method of § 46 for the analysis of tubes is based on the representation of the solution in the form of double Fourier series (in terms of the coordinates ξ and η), thus permitting only the study of one type of boundary condition in the problem, namely, the case when the edges of the tube are freely supported and when the supports do not resist displacements along the tube.

In order to obtain a solution of the same problem for arbitrary boundary conditions, one has to seek it in the form of an ordinary Fourier series (using the circumstance that for closed cylindrical shells the solution will always be a periodic function of the coordinate η). As in § 46, it will then be assumed that the cross-section of the tube has two axes of symmetry and that the loading on the tube is uniform, normal, internal pressure. Then the differential equation (46.1) takes the form

$$\Delta[\rho\Delta(\tilde{T})] + \frac{\partial}{\partial\eta}\left(\frac{1}{\rho}\,\frac{\partial\tilde{T}}{\partial\eta}\right) + i2b^2\frac{\partial^2\tilde{T}}{\partial\xi^2} = i2b^2pr_0\frac{d^2\rho}{d\eta^2}, \tag{48.1}$$

and its general solution for the problem under consideration can be written in the form of the series

$$\tilde{T} = \sum_{k=0}^{\infty} \tilde{T}^{(k)} \cos k\nu\eta, \tag{48.2}$$

where the coefficients $\tilde{T}^{(k)}$ are functions of ξ only and $\nu = \dfrac{\pi r_0}{s_0}$ (as in § 46, s_0 is a quarter of the length of the circumference of the shell).

The series (48.2) must be substituted into (48.1) to obtain [after expanding the right hand side of (48.1) in a series analogous to (48.2)] the system of ordinary differential equations for the determination of the $\tilde{T}^{(k)}(\xi)$. It is readily seen that the calculations necessary for the deduction of this system of equations will be quite analogous to those in § 46, the only difference being that the role played in § 46 by the constant multiplier λ^2 will now be taken by the differential operator $-\dfrac{d^2(\ldots)}{d\xi^2}$.

Hence one can immediately write down the following infinite system of linear differential equations with constant coefficients for the $\bar{T}^{(k)}$:

$$\begin{aligned}
&D_{0,0}[\bar{T}^{0}] + D_{0,1}[\bar{T}^{(1)}] + D_{0,2}[\bar{T}^{(2)}] + \ldots + D_{0,n}[\bar{T}^{(n)}] + \ldots = f_0,\\
&D_{1,0}[\bar{T}^{(0)}] + D_{1,1}[\bar{T}^{(1)}] + D_{1,2}[\bar{T}^{(2)}] + \ldots + D_{1,n}[\bar{T}^{(n)}] + \ldots = f_1,\\
&\cdots\cdots\cdots\cdots\cdots\cdots\cdots\cdots\cdots\cdots\\
&\cdots\cdots\cdots\cdots\cdots\cdots\cdots\cdots\cdots\cdots\\
&D_{n,0}[\bar{T}^{(0)}] + D_{n,1}[\bar{T}^{(1)}] + D_{n,2}[\bar{T}^{(2)}] + \ldots + D_{n,n}[\bar{T}^{(n)}] + \ldots = f_n,
\end{aligned} \tag{48.3}$$

in which $D_{kk}(\ldots)$ and $D_{kj}(\ldots)$ are the differential operators

$$\begin{aligned}
D_{k,k}[\ldots] &= \tfrac{1}{2}(\rho_0 + \rho_{2k})\left[\frac{d^4(\ldots)}{d\xi^4} - 2k^2\nu^2\frac{d^2(\ldots)}{d\xi^2} + k^4\nu^4(\ldots)\right] - \\
&\quad - \frac{1}{2}\left[\left(\frac{1}{\rho}\right)_0 - \left(\frac{1}{\rho}\right)_{2k}\right]k^2\nu^2(\ldots) + i2b^2\frac{d^2(\ldots)}{d\xi^2},\\
D_{k,j}[\ldots] &= D_{j,k}[\ldots] = -\frac{1}{2}\left[\left(\frac{1}{\rho}\right)_{k-j} - \left(\frac{1}{\rho}\right)_{k+j}\right]kj\nu^2(\ldots) + \\
&\quad + \tfrac{1}{2}(\rho_{k+j} + \rho_{k-j})\left[\frac{d^4(\ldots)}{d\xi^4} - \nu^2(k^2 + j^2)\frac{d^2(\ldots)}{d\xi^2} + k^2j^2\nu^4(\ldots)\right],\\
D_{0,0}[\ldots] &= \rho_0\frac{d^4(\ldots)}{d\xi^4} + i4b^2\frac{d^2(\ldots)}{d\xi^2},
\end{aligned} \tag{48.4}$$

and the free terms are the constants

$$f_k = -\, i2b^2 p r_0 k^2\nu^2\rho_k. \tag{48.5}$$

For practical utilization of the proposed general method one must retain in the series (48.2) several of its early terms (up to $k = n$), as a result of which (48.3) becomes a system of $(n + 1)$ differential equations of order $4(n + 1)$. The particular integrals of this system may be taken in the form

$$\bar{T}^{(0)} = c_0, \quad \bar{T}^{(1)} = c_1, \; \ldots, \; \bar{T}^{(n)} = c_n, \tag{48.6}$$

where $c_0, c_1 \ldots c_n$ are the constants solving the system of linear algebraic equations obtained after substitution of (48.6) into (48.3).

In this manner, the system (48.3) may be made homogeneous and it may then be integrated by the well-known methods of the theory of linear ordinary differential systems with constant coefficients.

Having found the auxiliary function $\bar{T}$ and subsequently employing the formulae of §40, one may write down expressions for all the displacements, forces and moments in the form of finite series, similar to (48.2). The coefficients of all these series will involve $8(n + 1)$ real arbitrary constants [since, as a result of the integration of the system (48.3), there occur $4(n + 1)$ complex arbitrary constants]. For the determination of these constants, one must give at each of the two ends of the tube up to four boundary conditions (in terms of displacements and rotations or forces and moments) and, in agreement with the general type of solution under consideration, all the quantities given at the edges of the shell must be expanded in Fourier series of the form

$$F(\eta) = \sum_{k=0}^{\infty} a_k \cos k\nu\eta, \tag{48.7}$$

where the a_k are constant coefficients.

Further, in order to obtain the degree of accuracy accepted above, i.e., retaining in the series (48.7) only the first $(n + 1)$ terms and comparing the coefficients in these series with the coefficients in the corresponding series of the deduced solution (for $\xi = 0$ and $\xi = \xi_0$, that is, for the values of ξ corresponding to the ends of the shell), a system of $8(n + 1)$ linear algebraic equations is obtained which permits the determination of all the constants of integration in the solution. Hence the problem under consideration will have been solved completely.

It will be seen from the above discussion that, although the method of solution of the problem encounters no fundamental difficulties, nevertheless, for $n = 3$ to 5, it will lead to very involved computations.

§ 49. On the semi-membrane theory of cylindrical shells

From 1932 onwards, V. Z. Vlasov [87, 4] has proposed a number of versions of a theory of cylindrical shells which differed more or less from each other. The most valuable of these, for a reason which will become clear from what follows, will be called the semi-membrane theory of cylindrical shells. The necessity for such a theory follows from the following considerations:

As has been shown previously (Chap. II), the accuracy of the solution given by the momentless theory, generally speaking, depends on the smoothness of the shape of the shell, on the smoothness of its loading and on the formulation of the boundary conditions. However, for cylindrical shells, these three factors must be supplemented by an additional one, namely the length of the shell (§ 37); as the length increases, the accuracy of the membrane theory diminishes without dependence on the degree to which the other conditions for the existence of the membrane state of stress are satisfied. In physical terms, this indicates that for a sufficiently great length the type of support of the ends of the tube has little influence on the stresses and strains in its centre section and it is not possible to avoid the appearance of significant bending stresses (if they are only due to the external loading).

It is essential to note in this context that the bending under discussion here will be caused mainly by the deformation of the cross-section of the tube, i.e., by the moment M_2. The bending due to deformation of the generators (i.e., by the moment M_1) and due to torsion oft the middle surface (i.e., by the moment H), will be insignificant provided the supports of the ends of the shell obey the requirements of the membrane theory.

Since for long (and very long) cylindrical shells the membrane theory is inadequate, there arises the necessity of constructing a theory of these shells which might occupy the central position between the membrane and the general theory, starting from the equations (40.7). As is clear from the above discussion, a first step towards the development of such an intermediate theory must be to neglect the moments M_1 and H (and therefore also the force N_1) in the equilibrium equations for a shell element.

As a result of this simplification, one will have

$$\begin{aligned}
&\frac{\partial T_1}{\partial \xi} + \frac{\partial S}{\partial \eta} + q_1 r_0 = 0,\\
&\frac{\partial S}{\partial \xi} + \frac{\partial T_2}{\partial \eta} + \frac{N_2}{\rho} + q_2 r_0 = 0,\\
&\frac{\partial N_2}{\partial \eta} - \frac{T_2}{\rho} + q_n r_0 = 0,\\
&N_2 = \frac{1}{r_0}\frac{\partial M_2}{\partial \eta}.
\end{aligned} \tag{49.1}$$

This system coincides with the system (41.11), which is natural, since in § 41 exactly the same assumptions have been made for shells reinforced by ribs.

The relationships between the forces and strains will be expressed by the usual formulae

$$\begin{gathered}
T_1 = \frac{E\delta}{1-\mu^2}(\varepsilon_1 + \mu\varepsilon_2), \quad T_2 = \frac{E\delta}{1-\mu^2}(\varepsilon_2 + \mu\varepsilon_1), \quad S = \frac{E\delta}{2(1+\mu)}\,\omega,\\
M_2 = \frac{E\delta^3}{12(1-\mu^2)}(\kappa_2 + \mu\kappa_1),
\end{gathered} \tag{49.2}$$

where

$$\varepsilon_1 = \frac{1}{r_0}\frac{\partial u}{\partial \xi}, \quad \varepsilon_2 = \frac{1}{r_0}\left(\frac{\partial v}{\partial \eta} + \frac{w}{\rho}\right), \quad \omega = \frac{1}{r_0}\left(\frac{\partial u}{\partial \eta} + \frac{dv}{\partial \xi}\right),$$
$$\kappa_2 = -\frac{1}{r_0^2}\frac{\partial}{\partial \eta}\left(\frac{\partial w}{\partial \eta} - \frac{v}{\rho}\right). \tag{49.3}$$

It also follows as a result of the neglecting of the moment

$$M_1 = \frac{E\delta^3}{12(1-\mu^2)}(\kappa_1 + \mu\kappa_2)$$

that

$$\kappa_1 \approx -\mu\kappa_2, \tag{49.4}$$

which approximately reduces the last of the formulae (49.2) to the form

$$M_2 = \frac{E\delta^3}{12}\kappa_2. \tag{49.5}$$

Eliminating from the system (49.1) all the unknowns except T_1 and M_2, one obtains the equation

$$r_0\frac{\partial^2 T_1}{\partial \xi^2} - \frac{\partial^2}{\partial \eta^2}\left(\rho\frac{\partial^2 M_2}{\partial \eta^2}\right) - \frac{\partial}{\partial \eta}\left(\frac{1}{\rho}\frac{\partial M_2}{\partial \eta}\right) =$$
$$= r_0^2\left[\frac{\partial^2(q_n\rho)}{\partial \eta^2} + \frac{\partial q_2}{\partial \eta} - \frac{\partial q_1}{\partial \xi}\right], \tag{49.6}$$

which is identical to (41.17).

Vlasov obtained a second equation involving the same unknowns from the condition of compatibility of strains (41.13)

$$r_0\frac{\partial^2 \kappa_2}{\partial \xi^2} + \frac{\partial^2}{\partial \eta^2}\left[\rho\left(\frac{\partial^2 \varepsilon_1}{\partial \eta^2} + \frac{\partial^2 \varepsilon_2}{\partial \xi^2} - \frac{\partial^2 \omega}{\partial \xi \partial \eta}\right)\right] + \frac{\partial}{\partial \eta}\left[\frac{1}{\rho}\left(\frac{\partial \varepsilon_1}{\partial \eta} - \frac{\partial \omega}{\partial \xi}\right)\right] = 0, \tag{49.7}$$

which he simplified beforehand by letting (cf. [4], p. 482)

$$\omega = \varepsilon_2 = 0. \tag{49.8}$$

After this simplification, equation (49.7) may be rewritten in the form

$$r_0\frac{\partial^2 \kappa_2}{\partial \xi^2} + \frac{\partial^2}{\partial \eta^2}\left(\rho\frac{\partial^2 \varepsilon_1}{\partial \eta^2}\right) + \frac{\partial}{\partial \eta}\left(\frac{1}{\rho}\frac{\partial \varepsilon_1}{\partial \eta}\right) = 0. \tag{49.9}$$

Replacing here the strains by their expressions in terms of the forces and moments, one finds

$$\frac{12r_0}{\delta^2}\frac{\partial^2 M_2}{\partial \xi^2} + W(T_1) - \mu W(T_2) = 0, \tag{49.10}$$

where $W(\ldots)$ is Vlasov's operator

$$W(\ldots) = \frac{\partial^2}{\partial \eta^2}\left[\rho \frac{\partial^2(\ldots)}{\partial \eta^2}\right] + \frac{\partial}{\partial \eta}\left(\frac{1}{\rho}\frac{\partial(\ldots)}{\partial \eta}\right). \tag{49.11}$$

Further, neglecting in equation (49.10) the term multiplied by Poisson's ratio μ, one obtains finally from (49.6) and (49.10) the differential equations

$$\begin{gathered} \frac{12r_0}{\delta^2}\frac{\partial^2 M_2}{\partial \xi^2} + W(T_1) = 0, \\ r_0\frac{\partial^2 T_1}{\partial \xi^2} - W(M_2) = r_0^2\left[\frac{\partial^2 \rho q_n}{\partial \eta^2} + \frac{\partial q_2}{\partial \eta} - \frac{\partial q_1}{\partial \xi}\right]. \end{gathered} \tag{49.12}$$

A general defect of the above discussion is the large number of arbitrary assumptions leading to the five conditions

$$M_1 = H = \varepsilon_2 = \omega = \mu = 0.$$

Several of these conditions are superfluous from the point of view of the fundamental idea of the semi-membrane theory (as formulated at the beginning of this section).

Thus, for example, the condition $\omega = 0$ is unacceptable for the solution of torsion problems of tubes and therefore this important problem of the theory of long (and very long) thin walled rods will fall outside the range of the developed theory.

Now it is not difficult to show that the defects indicated above are not at all natural to the equations (49.12), but rather to the above method of their derivation. In fact, it is not necessary to invoke the conditions (49.8) for the transition from the formula (49.7) to the formula (49.9) and it is sufficient for this purpose to assume that the derivatives $\frac{\partial^2 \omega}{\partial \xi \partial \eta}$, $\frac{\partial^2 \varepsilon_2}{\partial \xi^2}$ and $\frac{\partial \omega}{\partial \xi}$ may be neglected in comparison with $\frac{\partial^2 \varepsilon_1}{\partial \eta^2}$ and $\frac{\partial \varepsilon_1}{\partial \eta}$ (respectively) which are, of course, not at all equivalent to the conditions (49.8) and which represent considerably weaker restrictions. Since the derivatives of the strains are connected with each other by the equations (49.1), (49.2), it is clear (in this context) that the deduction of the system (49.12) need not be based on a number of unrelated assumptions, referring to different forces, moments and strains, but can be deduced from a certain deeper and more general principle. It may be shown that this single principle, sufficient for establishing equations of the form (49.12), involves the neglect of the term $\frac{\partial^2 f}{\partial \xi^2}$ in comparison with $\frac{\partial^2 f}{\partial \eta^2}$ (where f, any function characteristic for the problem under consideration, may be a displacement, a force or a moment).

Indeed, reverting to the general equations (40.7) and performing there the indicated simplification (with the exception of the term with the very large factor: $i2b^2$), one will have

$$W(\tilde{T}) + i2b^2 \frac{\partial^2 \tilde{T}}{\partial \xi^2} = i2b^2 r_0 \left[\frac{\partial^2 p q_n}{\partial \eta^2} + \frac{\partial q_2}{\partial \eta} - \frac{\partial q_1}{\partial \xi} \right]. \tag{49.13}$$

Separating real and imaginary parts, a system of two equations is obtained, the structure of which is identical to that of the system (49.12). The unknowns in this system are $T_1 + T_2$, instead of T_1 in the system (49.12), and M_2, as in the system (49.12). This last result follows from the fact that, in agreement with the accepted principle of simplification, one has

$$M_1 \approx \mu M_2. \tag{49.14}$$

Having determined the auxiliary function $\tilde{T}$ from the equation (49.13), all the complex forces may be found using the formulae (40.8) to (40.11). For this purpose, following the basic idea of the simplification, one has to put in these formulae

$$\Delta(\ldots) = \frac{\partial^2(\ldots)}{\partial \eta^2} + \frac{\partial^2(\ldots)}{\partial \xi^2} \approx \frac{\partial^2(\ldots)}{\partial \eta^2}.$$

Further, the equations (40.12) and (40.13) may be employed for the determination of the complex displacements.

Thus it is seen that the principal difference between the usual treatment of the semi-membrane theory and its treatment above is due to the fact that in the usual discussion certain strains (ε_2, ω) are arbitrarily put equal to zero, whereas in the above discussion all the strains, forces and moments enter, generally speaking, as non-zero quantities the determination of which, however, is subsequently achieved by use of a single simplification. Such an approach retains all the advantages of the semi-membrane theory (since the same fundamental equations are obtained) and it removes a number of its serious defects.

It should not be concluded from this that the proposed semi-membrane theory is any less 'physical' than the usual theory. The neglect of $\frac{\partial^2(\ldots)}{\partial \xi^2}$

in comparison with $\frac{\partial^2(\ldots)}{\partial\eta^2}$, translated into the language of physical ideas, indicates that the variations of all displacements and stresses in the direction along the cylindrical shell are assumed to be smoother than those in the circumferential direction of the shell. If the length of the shell under consideration is sufficiently large, the state of stress in it will usually satisfy this requirement (since the local state of stress of the edge effect type will either not be taken into account or it may be assumed to have been eliminated by the formulation of membrane boundary conditions at the ends of the tube).

In conclusion, it may be noted that from the above discussion follows a simple criterion for an error estimate of the solution of the semi-membrane theory which involves a comparison of the values of $\frac{\partial^2 f}{\partial\xi^2}$ and $\frac{\partial^2 f}{\partial\eta^2}$ (where f may be any displacement or force). This remark refers, of course, only to the semi-membrane theory generalized in the above stated manner. The error of a theory based on the condition $\omega = 0$ may in individual problems considerably exceed the stated estimate.

CHAPTER IV

SHELLS OF REVOLUTION

§ 50. Historical remarks.

Shells of revolution, including the circular cylindrical shells considered in the preceding chapter, have many technical applications. Hence many investigations have been devoted to this class of shells and the greatest advances have been made in the analysis of circular cylindrical and spherical shells. However, circular cylindrical shells have been discussed in sufficient detail in the last chapter and therefore they will receive no further attention here. The first significant success in the analysis of spherical shells is to be found in the work of H. Reissner [159] who reduced the differential equations for symmetrical deformations of such shells to a convenient form and then applied the asymptotic method for their integration [145]. He also discovered the possibility of reducing the order of the differential equations of this problem by means of their complex transformation, so that the analysis of spherical shells for symmetrical loading was found to reduce to the integration of a single second order differential equation.

A short time after the appearance of the article [159], E. Meissner [155, 154] was able to generalize Reissner's result to symmetrical deformations of shells of revolution of arbitrary shape (and even of variable thickness). In this way, the difficulties arising in the analysis of shells of revolution under symmetrical loading were largely overcome, particularly since the asymptotic method was shown to be a simple and sufficiently accurate process for the integration of the relevant differential equations. However, a long time after the appearance of the above results, efforts were still directed not so much towards approximate but towards mathematically exact solutions of these equations [146, 148, 162 et al.]. These results of the Zürich school (a name which in the history of shell theory refers to the group of investigators led by Meissner and Reissner) are to a great extent useless, since they are inconvenient for practical applications. In particular, in the analysis of spherical shells by use of hypergeometric series [155, 146], one has to retain more than twenty terms. The lack of purpose in seeking mathematically accurate solutions of the equations of shell theory becomes quite clear if one takes into consideration that these equations (as a result of their deduction) involve errors of order δ/R in comparison with unity. Hence there is little sense in retaining in their solutions terms of smaller order than δ/R. However, it may be shown that in the integration of the equations for the symmetrical deformation of shells of revolution this degree of accuracy is already achieved by the asymptotic method which must therefore be employed irrespective of the difference between the solutions obtained by this method and the so-called "exact solutions". This idea was apparently stated explicitly for the first time in the Author's doctoral dissertation (1945). The same conclusion was reached independently by A. I. Lur'e who employed it subsequently in his monograph [18].

As regards the development of simplified methods for the analysis of shells of revolution under symmetrical loadings, special mention should be made of I. Ia. Shtaerman [140] and P. L. Pasternak [157]. Shtaerman in his work gave, in addition, a very clear deduction of the equations of the problem by establishing the analogy between the problems of the symmetrical deformation of shells of revolution and of the bending of arches on elastic foundations [142, 143].

The simplest and most frequently applied approximate method of integration of the equations for symmetrical deformations of shells of revolution, based on the omission of terms of order up to $\sqrt{\delta/R}$ (compared with unity) inclusively, is referred to, in practice, as the "method of Geckeler". However, it is impossible to acknowledge this title as justified, because, in fact, it is the result of the efforts of a number of the above-mentioned authors. Thus, at the present time, the analysis of shells of revolution under symmetrical loading has been studied in sufficient detail and the complex transformation of the equations halving the order of the differential equations of the problem has in this respect played an important role.

The analysis of shells of revolution under unsymmetrical loading is more complicated. The most important loading of this type is anti-symmetrical and is sometimes referred to as 'wind' loading. For spherical shells, the corresponding problem was solved in the dissertation of E. Schwerin [160] who (evidently to please his teacher and opponent H. Reissner) transformed the differential equations in the manner typical of the Zürich school. He thus obtained the solution in the form of a badly converging hyper-geometric series, a result which was also reached by the Author of this book. In this connection, two quadratures and also the possibility of a complex transformation were discovered so that the analysis of spherical shells under 'wind' type loading was found to reduce to the integration of a single second order equation. This last result has recently been generalized in the work [134] to shells of revolution of arbitrary shape.

As regards spherical shells with arbitrary loading, this general problem was solved first by A. Havers [151] (or, more correctly, by A. van der Neut [156], since Havers merely applied to the problem of stress analysis the transformation which van der Neut employed for the stability analysis of spherical shells). This problem was solved by use of other methods by V. V. Sokolovskii [136], Iu. V. Repman [135], A. L. Gol'denweizer [129] and the Author [133].

In this context, it should be noted that both Havers as well as Gol'denweizer gave the transformation for the homogeneous problem only, whence their results may not be applied directly to the stability and vibration analysis of spherical shells. If this is taken into consideration, the transformation proposed in the paper [133] may be given preference, since it may be found to be more convenient for the solution of problems of the above two types. However, if one restricts discussion to the homogeneous problem, the solution by Gol'denweizer [129] is of greater interest. Based on the works of the preceding writer, I. N. Vekua [125, 126] formulated the homogeneous problem of the theory of spherical shells, expressing its solutions in terms of four analytic functions of one complex variable.

Thus, the analysis of spherical shells under arbitrary loading, in its final form developed in detail by the Russian school, has reached at the present time a stage of completeness and, if one is orientated towards the asymptotic integration of the differential equations, the problem of computations is less cumbersome than might be expected.

As regards results for other types of shells of revolution, the solution for conical shells due to A. E. H. Love [49] and applicable for arbitrary loading should be mentioned. This same author indicated that for arbitrary shells of revolution, bounded by parallel circles, a first approximation to a general solution may be obtained in the form of the sum of the solutions of the membrane equations and of the equations for the so-called 'edge effect'. Integrals of the latter equations are expressed in terms of rapidly changing functions of the distance from the edge of the shell. This idea was subsequently used by Love in his book [49], however, without justification and likewise without taking notice of possible simplifications of the equations of the edge effect.

Both these questions have been studied in great detail in a number of papers by A. L. Gol'denweizer [8, 9] and also in the work of Yu. N. Rabotnov [34] and Kh. M. Mushtari [25]. These papers by the Russian scientists essentially defined the conditions for which rapidly decaying solutions of the equations of the theory of shells may be obtained and studied a number of possible special cases.

It is more convenient to consider the problem of the analysis of shells of revolution under arbitrary load in complex form. It then turns out that the corresponding differential equations may be transformed in much the same way as this was possible in the membrane theory. As a result, the analysis of shells of revolution reduces to the solution of a system of two differential equations of fourth order. Firstly, one may obtain from this system the known results for symmetrical deformations of shells of revolution, i.e., the solution of this problem may be reduced to the integration of a single equation of second order. Further, analogous results can be obtained from it also for the so called 'wind type' loading which, as has already been mentioned, was hitherto only known to be applicable to spherical shells.

For the family of shells the radii of curvature of which are given by

$$\left(\frac{1}{R_1} - \frac{1}{R_2}\right) \cdot \frac{1}{\sin^2 \theta} = \text{const.},$$

the basic system of differential equations for shells of revolution disintegrates by means of a repeated application of a complex transformation into two independent second order equations one of which is close to the equation of the membrane theory, while the other is close to the equation for symmetrical deformations, and this result is true for any loading. These results may be extended approximately also to shells of revolution of arbitrary shape.

Thus, the complex transformation of the differential equations of shell theory, as applied to shells of revolution, is found to be useful and it leads to an essential extension of the knowledge referring to such shells.

In this survey the valuable work of B. G. Galerkin [128], A. I. Lur'e [131] and G. S. Shapiro [139] has not been mentioned, since their contributions do not refer to the theory of thin, but to the theory of thick shells.

§ 51. The differential equations for shells of revolution in "complex forces" and their transformation

The position of points on the middle surface of the shell of revolution of arbitrary shape will be determined by the angles θ and φ, shown in Fig. 71, in the same way as this was done in Chapter II. Further, let R_1

be the radius of curvature of the meridian and R_2 the radius of curvature of the normal section, tangential to the parallel circle. This second radius is equal to the segment of the perpendicular to the middle surface between this surface and the axis of the shell. Obviously both these radii will be functions of the angle θ only.

The Lamé parameters in this case are determined by the expressions

$$A_1 = R_1, \; A_2 = R_2 \sin\theta, \tag{51.1}$$

and the Codazzi-Gauss conditions reduce to the relation

$$\frac{dR_2 \sin\theta}{d\theta} = R_1 \cos\theta. \tag{51.2}$$

In correspondence with the above, the differential equations of the theory of shells in terms of complex forces [equations (16.10)] for shells of revolution take the form

$$\begin{gathered}
\frac{1}{R_1}\frac{\partial \tilde{T}_1}{\partial\theta} + \frac{\cotan\theta}{R_2}(\tilde{T}_1 - \tilde{T}_2) + \frac{1}{R_2\sin\theta}\frac{\partial\tilde{S}}{\partial\varphi} + i\frac{c}{R_1{}^2}\frac{\partial\tilde{T}}{\partial\theta} = -q_1, \\
\frac{1}{R_1}\frac{\partial\tilde{S}}{\partial\theta} + 2\frac{\cotan\theta}{R_2}\tilde{S} + \frac{1}{R_2\sin\theta}\frac{\partial\tilde{T}_2}{\partial\varphi} + i\frac{c}{R_2{}^2\sin\theta}\frac{\partial\tilde{T}}{\partial\varphi} = -q_2, \\
\frac{\tilde{T}_1}{R_1} + \frac{\tilde{T}_2}{R_2} - ic\Delta(\tilde{T}) = q_n,
\end{gathered} \tag{51.3}$$

where q_1, q_2, q_n are the components of the surface loading in the directions θ, φ and along the normal to the middle surface,

$$\Delta(\;) = \frac{1}{R_1{}^2}\frac{\partial^2(\ldots)}{\partial\theta^2} + \left(\frac{\cotan\theta}{R_1R_2} - \frac{1}{R_1{}^3}\frac{dR_1}{d\theta}\right)\frac{\partial(\ldots)}{\partial\theta} + \frac{1}{R_2{}^2\sin^2\theta}\frac{\partial^2(\ldots)}{\partial\varphi^2}, \tag{51.4}$$

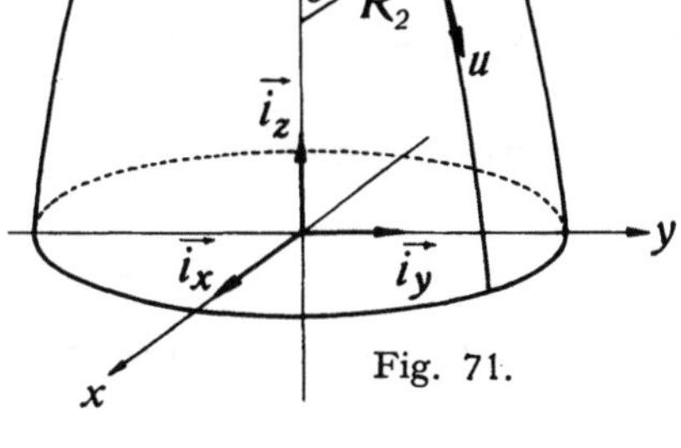

Fig. 71.

$$\begin{aligned}
c &= \frac{\delta}{\sqrt{12(1-\mu^2)}}, \\
\tilde{T}_1 &= T_1 - \frac{i}{c}\,\frac{M_2 - \mu M_1}{1-\mu^2}, \\
\tilde{T}_2 &= T_2 - \frac{i}{c}\,\frac{M_1 - \mu M_2}{1-\mu^2}, \\
\tilde{S} &= S + \frac{i}{c}\,\frac{H}{1-\mu}, \\
\tilde{T} &= \tilde{T}_1 + \tilde{T}_2.
\end{aligned} \tag{51.5}$$

Letting $c = 0$ in the system (51.3), the equations reduce to the system

(23.4), i.e., to the differential equations of the membrane theory of shells of revolution. Thus these two systems have quite a lot in common. This suggests that a transformation of the system (51.3) might be accomplished in an analogous manner as this has been done in § 23 during transformation of the membrane equations.

In fact, introduce two new auxiliary functions

$$\begin{aligned} \tilde{U} &= \tilde{T}_1 R_2 \sin^2\theta - ic\frac{R_2}{R_1}\sin\theta\cos\theta\frac{\partial\tilde{T}}{\partial\theta}, \\ \tilde{V} &= \tilde{S}R_2{}^2\sin^2\theta, \end{aligned} \tag{51.6}$$

which for $c = 0$ will be identical with the functions U and V of the membrane theory of shells of revolution [formulae (23.6)]. Replacing $\tilde{T}_2$ in the second equation (51.3) by the expression

$$\tilde{T}_2 = \tilde{T} - \tilde{T}_1 \tag{51.7}$$

and eliminating from this equation $\tilde{T}_1$ and $\tilde{S}$ with the help of (51.6), one has, on the basis of the condition (51.2),

$$\begin{aligned} \frac{\partial\tilde{V}}{\partial\theta} - \frac{R_1}{\sin\theta}\frac{\partial\tilde{U}}{\partial\varphi} - icR_2\cos\theta\frac{\partial^2\tilde{T}}{\partial\theta\,\partial\varphi} + R_1(R_2 + ic)\sin\theta\frac{\partial\tilde{T}}{\partial\varphi} = \\ = -q_2R_1R_2{}^2\sin^2\theta. \end{aligned} \tag{51.8}$$

Further, introduce in the first and third equations of the system (51.3) for $\tilde{T}_1$ the function $\tilde{U}$. After certain manipulations involving the formulae (51.2) and (51.4), this gives

$$\begin{aligned} &\frac{1}{R_1R_2\sin^2\theta}\frac{\partial\tilde{U}}{\partial\theta} - \frac{1}{R_1R_2}\frac{\cos\theta}{\sin^3\theta}\tilde{U} - \frac{\cotan\theta}{R_2}\tilde{T}_2 + \frac{1}{R_2\sin\theta}\frac{\partial\tilde{S}}{\partial\varphi} + \\ &\quad + ic\cotan\theta\Delta(\tilde{T}) - ic\frac{\cotan\theta}{R_2{}^2\sin^2\theta}\frac{\partial^2\tilde{T}}{\partial\varphi^2} - ic\frac{\cotan^2\theta}{R_1{}^2}\frac{\partial\tilde{T}}{\partial\theta} = -q_1 \end{aligned} \tag{51.9}$$

and

$$\frac{\tilde{U}}{R_1R_2\sin^2\theta} + ic\frac{\cotan\theta}{R_1{}^2}\frac{\partial\tilde{T}}{\partial\theta} + \frac{\tilde{T}_2}{R_2} - ic\Delta(\tilde{T}) = q_n. \tag{51.10}$$

Multiplying (51.10) by cotan θ and adding it to (51.9), one finds

$$\frac{\partial\tilde{U}}{\partial\theta} + \frac{R_1}{R_2{}^2\sin\theta}\frac{\partial\tilde{V}}{\partial\varphi} - ic\frac{R_1}{R_2}\cotan\theta\frac{\partial^2\tilde{T}}{\partial\varphi^2} = (q_n\cos\theta - q_1\sin\theta)R_1R_2\sin\theta. \tag{51.11}$$

Next, the function $\tilde{V}$ will be eliminated from the equations (51.8) and

(51.11) by multiplying (51.11) by $\dfrac{R_2^2 \sin\theta}{R_1}$ and by subsequently differentiating (51.8) with respect to φ and (51.11) with respect to θ. As a result one arrives at the following differential equation in two unknowns

$$G(\tilde{U}) - \left[1 - ic\left(\frac{1}{R_1} - \frac{1}{R_2}\right)\frac{1}{\sin^2\theta}\right]\frac{\partial^2 \tilde{T}}{\partial\varphi^2} = f(\theta, \varphi), \tag{51.12}$$

in which

$$f(\theta, \varphi) = \frac{1}{R_1 R_2 \sin\theta}\left\{\frac{\partial}{\partial\theta}\left[(q_n \cos\theta - q_1 \sin\theta) R_2^3 \sin^2\theta\right] + \right.$$

$$\left. + \frac{\partial q_2}{\partial\varphi} R_1 R_2^2 \sin^2\theta\right\}, \tag{51.13}$$

$$G(\ldots) = \frac{1}{R_1 R_2 \sin\theta}\frac{\partial}{\partial\theta}\left[\frac{R_2^2 \sin\theta}{R_1}\frac{\partial(\ldots)}{\partial\theta}\right] + \frac{1}{R_2 \sin^2\theta}\frac{\partial^2(\ldots)}{\partial\varphi^2}. \tag{51.14}$$

It will be noted that the operators $\Delta(\ldots)$ and $G(\ldots)$ are connected by the relation

$$G(\ldots) = R_2\Delta(\ldots) + \left(\frac{1}{R_1} - \frac{R_2}{R_1^2}\right)\cot\theta\,\frac{\partial(\ldots)}{\partial\theta}, \tag{51.15}$$

as a result of which the equation (51.10) can be written in the form

$$-icG(\tilde{T}) + \tilde{T} + \left(\frac{1}{R_1} - \frac{1}{R_2}\right)\frac{1}{\sin^2\theta}\tilde{U} = q_n R_2. \tag{51.16}$$

The expressions (51.12) and (51.16) represent a system of two linear partial differential equations in two unknowns — the auxiliary functions $\tilde{T}$ and $\tilde{U}$.

Thus, as the result of the above discussion, the analysis of shells of revolution of arbitrary shape subject to any type of loading has been reduced to the solution of the system

$$\begin{aligned} G(\tilde{U}) - \left[1 - ic\left(\frac{1}{R_1} - \frac{1}{R_2}\right)\frac{1}{\sin^2\theta}\right]\frac{\partial^2 \tilde{T}}{\partial\varphi^2} &= f(\theta, \varphi), \\ -icG(\tilde{T}) + \tilde{T} + \left(\frac{1}{R_1} - \frac{1}{R_2}\right)\frac{1}{\sin^2\theta}\tilde{U} &= q_n R_2, \end{aligned} \tag{51.17}$$

which for $c = 0$ takes the form

$$\begin{aligned} G(\tilde{U}) - \frac{\partial^2 \tilde{T}}{\partial\varphi^2} &= f(\theta, \varphi), \\ \tilde{T} + \left(\frac{1}{R_1} - \frac{1}{R_2}\right)\frac{1}{\sin^2\theta}\tilde{U} &= q_n R_2. \end{aligned} \tag{51.18}$$

Eliminating $\tilde{T}$, one arrives at the equation

$$G(\tilde{U}) + \left(\frac{1}{R_1} - \frac{1}{R_2}\right)\frac{1}{\sin^2\theta}\,\frac{\partial^2 \tilde{U}}{\partial \varphi^2} = f(\theta, \varphi) + R_2\frac{\partial^2 q_n}{\partial \varphi^2}, \qquad (51.19)$$

which completely coincides with the equation (23.8). This was to be expected, since for $c = 0$ the original equations (51.3) reduce to the system of equations of the membrane theory. Thus, the system (51.17) includes as a particular case the fundamental differential equations of the membrane theory of shells of revolution. Hence it may be considered a generalization of these equations to shells of revolution which are not only subject to membrane stresses, but also to bending stresses.

§ 52. General method for solution of the system (51.17)

If a shell of revolution is bounded by one or two parallel circles, as is frequently the case in practical problems, it will be complete with respect to the coordinate φ and hence the components of the external surface and edge loadings will be periodic functions of this coordinate with period 2π. As a consequence, the components of the surface loading may be expanded in Fourier series in terms of φ and the functions $\tilde{U}$, $\tilde{V}$ and $\tilde{T}$ may be sought in the form of such series. Then the system (51.17) breaks up into a series of systems of ordinary differential equations. This method is of general importance and applies to any type of loading. As has already been noted in § 22, a study of loadings with the components

$$q_1 = q_{1,k} \cos k\varphi, \quad q_2 = q_{2,k} \sin k\varphi, \quad q_n = q_{n,k} \cos k\varphi, \qquad (52.1)$$

where $q_{1,k}, q_{2,k}, q_{n,k}$ are functions of θ, will thus lead to the solution of the problem in its general form, since any loading can be expressed as a sum (finite or infinite) of loadings of the form (52.1). Hence, in what follows consideration will be restricted to loadings of this type.

After substituting (52.1) into the system (51.17), its solution will be sought in the form

$$\tilde{U} = \tilde{U}^{(k)} \cos k\varphi, \qquad \tilde{T} = \tilde{T}^{(k)} \cos k\varphi, \qquad (52.2)$$

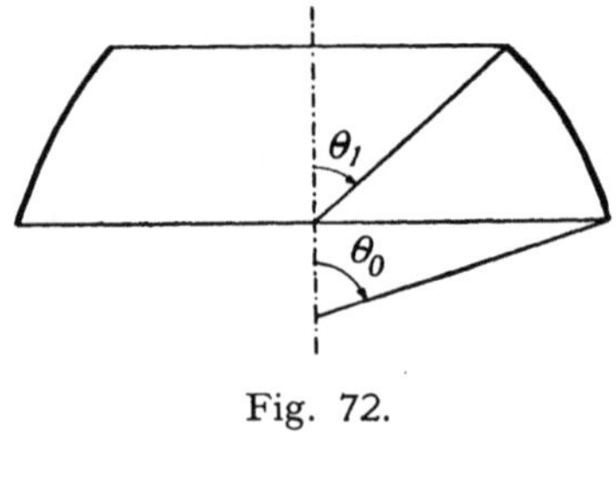

Fig. 72.

where $\tilde{U}^{(k)}$, $\tilde{T}^{(k)}$ are functions of θ only.

In this way one obtains, after elimination of common trigonometric terms, the following system of two ordinary differential equations for $\tilde{U}^{(k)}$ and $\tilde{T}^{(k)}$:

$$G_k[\tilde{U}^{(k)}] + k^2\left[1 - ic\left(\frac{1}{R_1} - \frac{1}{R_2}\right)\frac{1}{\sin^2\theta}\right]\tilde{T}^{(k)} = f_k(\theta),$$
$$- icG_k[\tilde{T}^{(k)}] + \tilde{T}^{(k)} + \left(\frac{1}{R_1} - \frac{1}{R_2}\right)\frac{1}{\sin^2\theta}\tilde{U}^{(k)} = R_2 q_{n,k}, \tag{52.3}$$

where

$$f_k(\theta) = \frac{1}{R_1R_2\sin\theta}\left\{\frac{d}{d\theta}[(q_{n,k}\cos\theta - q_{1,k}\sin\theta)R_2{}^3\sin^2\theta] + \right.$$
$$\left. + kq_{2,k}R_1R_2{}^2\sin^2\theta\right\}. \tag{52.4}$$

$$G_k(\ldots) = \frac{1}{R_1R_2\sin\theta}\frac{d}{d\theta}\left[\frac{R_2{}^2\sin\theta}{R_1}\frac{d(\ldots)}{d\theta}\right] - \frac{k^2}{R_2\sin^2\theta}(\ldots). \tag{52.5}$$

Integration of the system (52.3) leads to the determination of the auxiliary functions $\tilde{U}$ and $\tilde{T}$ for the loading (52.1). Since, on the one hand, any loading can be obtained in terms of loadings of the type (52.1) and, on the other hand, having found $\tilde{U}$ and $\tilde{T}$ one can immediately evaluate all the complex forces by the help of the formulae (51.6), (51.11) and (51.7), it may be stated that the problem under consideration has been reduced to the integration of the system (52.3).

§ 53. Determination of the displacements

The relations between the complex forces and complex displacements (§ 34) for shells of revolution are expressed by the formulae

$$R_1\tilde{\varepsilon}_1 = \frac{\partial\tilde{u}}{\partial\theta} + \tilde{w} = \frac{R_1}{E\delta}(\tilde{T}_1 - \mu\tilde{T}_2),$$
$$R_2\tilde{\varepsilon}_2 = \frac{1}{\sin\theta}\frac{\partial\tilde{v}}{\partial\varphi} + \tilde{u}\,\mathrm{cotan}\,\theta + \tilde{w} = \frac{R_2}{E\delta}(\tilde{T}_2 - \mu\tilde{T}_1), \tag{53.1}$$
$$R_2\tilde{\omega} = \frac{R_2}{R_1}\frac{\partial\tilde{v}}{\partial\theta} - \tilde{v}\,\mathrm{cotan}\,\theta + \frac{1}{\sin\theta}\frac{\partial\tilde{u}}{\partial\varphi} = \frac{2(1+\mu)R_2}{E\delta}\tilde{S}.$$

$$\tilde{\kappa}_1 = -\frac{1}{R_1}\frac{\partial}{\partial\theta}\left[\frac{1}{R_1}\left(\frac{\partial\tilde{w}}{\partial\theta} - \tilde{u}\right)\right] = \frac{i}{cE\delta}(\tilde{T}_2 - T_2{}^*),$$
$$\tilde{\kappa}_2 = -\frac{1}{R_2{}^2\sin\theta}\frac{\partial}{\partial\varphi}\left(\frac{1}{\sin\theta}\frac{\partial\tilde{w}}{\partial\varphi} - \tilde{v}\right) - \frac{\mathrm{cotan}\,\theta}{R_1R_2}\left(\frac{\partial\tilde{w}}{\partial\theta} - \tilde{u}\right) = \frac{i}{cE\delta}(\tilde{T}_1 - T_1{}^*), \tag{53.2}$$

$$\tilde{\tau} = -\frac{1}{R_1}\frac{\partial}{\partial\theta}\left[\frac{1}{R_2}\left(\frac{1}{\sin\theta}\frac{\partial\tilde{w}}{\partial\varphi} - \tilde{v}\right)\right] + \\ + \frac{1}{R_1 R_2 \sin\theta}\left(\frac{\partial\tilde{u}}{\partial\varphi} - \tilde{v}\cos\theta\right) = -\frac{i}{cE\delta}(\tilde{S} - S^*), \tag{53.2}$$

where T_1^*, T_2^*, S^* are the general or arbitrary particular integrals of the equations of the membrane theory of shells of revolution, i.e., of the equations (23.4).

If the surface loading is given by the components (52.1), the complex forces and solutions of the membrane theory are obtained in the form

$$\begin{aligned} \tilde{T}_1 &= \tilde{T}_{1,k}\cos k\varphi, & \tilde{T}_2 &= \tilde{T}_{2,k}\cos k\varphi, & \tilde{S} &= \tilde{S}_k \sin k\varphi, \\ T_1^* &= T^*_{1,k}\cos k\varphi, & T_2^* &= T^*_{2,k}\cos k\varphi, & S^* &= S_k^* \sin k\varphi, \end{aligned} \tag{53.3}$$

where $\tilde{T}_{1,k} \ldots \tilde{T}_{1,k}$ are functions of θ only.

It follows then that also the complex displacements must be taken in a similar form, viz.,

$$\tilde{u} = \tilde{u}_k \cos k\varphi, \quad \tilde{v} = \tilde{v}_k \sin k\varphi, \quad \tilde{w} = \tilde{w}_k \cos k\varphi. \tag{53.4}$$

Substituting (53.3) and (53.4) into the formulae (53.1) and (53.2), one finds

$$\begin{aligned} &\frac{d\tilde{u}_k}{d\theta} + \tilde{w}_k = \frac{R_1}{E\delta}(\tilde{T}_{1,k} - \mu\tilde{T}_{2,k}), \\ &\frac{k}{\sin\theta}\tilde{v}_k + \tilde{u}_k \operatorname{cotan}\theta + \tilde{w}_k = \frac{R_2}{E\delta}(\tilde{T}_{2,k} - \mu\tilde{T}_{1,k}), \\ &\frac{R_2}{R_1}\frac{d\tilde{v}_k}{d\theta} - \tilde{v}_k \operatorname{cotan}\theta - \frac{k}{\sin\theta}\tilde{u}_k = \frac{2(1+\mu)R_2}{E\delta}\tilde{S}_k. \end{aligned} \tag{53.5}$$

$$\begin{aligned} &-\frac{1}{R_1}\frac{d}{d\theta}\left[\frac{1}{R_1}\left(\frac{d\tilde{w}_k}{d\theta} - \tilde{u}_k\right)\right] = \frac{i}{cE\delta}(\tilde{T}_{2,k} - T^*_{2,k}), \\ &\frac{k}{R_2^2\sin\theta}\left(\frac{k}{\sin\theta}\tilde{w}_k + \tilde{v}_k\right) - \frac{\operatorname{cotan}\theta}{R_1R_2}\left(\frac{d\tilde{w}_k}{d\theta} - \tilde{u}_k\right) = \frac{i}{cE\delta}(\tilde{T}_{1,k} - T^*_{1,k}), \\ &\frac{1}{R_1}\frac{d}{d\theta}\left[\frac{1}{R_2}\left(\frac{k}{\sin\theta}\tilde{w}_k + \tilde{v}_k\right)\right] - \frac{1}{R_1R_2\sin\theta}(k\tilde{u}_k + \tilde{v}_k\cos\theta) = \\ &\qquad = -\frac{i}{cE\delta}(\tilde{S}_k - S_k^*). \end{aligned} \tag{53.6}$$

Determination of the coefficients in the formulae (53.4) reduces to the simultaneous solution of the systems (53.5) and (53.6).

In this context it should be noted that the formulae (53.1) and (53.2) are only approximately compatible with each other with an error of order δ/R compared with unity. In view of this fact, a similar reasoning holds true with regard to the integration of the equations (53.5), (53.6).

§ 54. Symmetrical deformations of shells of revolution

A particularly important practical problem is the special case when $k = 0$ in the formulae (52.1). Then

$$q_1 = q_{1,0}(\theta), \quad q_2 = 0, \quad q_n = q_{n,0}(\theta), \tag{54.1}$$

i.e., the surface loading is symmetrical with respect to the axis of the shell and the component $q_2(\theta)$ which would twist the shell about its axis is absent.

It will be noted however, that this circumstance does not ensure that the deformations of the shell will be functions of θ only. For example, if a dome loaded by its own weight which is of the form (54.1) is not resting on a continuous foundation, but on a series of columns, then the strains in the dome will be functions of θ as well as of φ. Therefore, in order that the deformations of shells of revolution are axially symmetric, not only the surface loadings but also the edge loadings must exhibit the given symmetry.

It will be assumed that this condition is satisfied and that, in addition, there are no edge forces twisting the shell. Then the loading (54.1) produces in the shell only the forces and moments T_1, T_2, M_1, M_2, while the shear force S and the twisting moment H will be equal to zero. Hence the state of stress will be completely characterized by the two complex forces $\tilde{T}_1$ and $\tilde{T}_2$ (as functions of the angle θ only).

In this case, the equation (51.8) is identically satisfied and the equation (51.11) takes the form

$$\frac{d\tilde{U}}{d\theta} = (q_n \cos\theta - q_1 \sin\theta) R_2 R_1 \sin\theta. \tag{54.2}$$

(Here and in what follows the index ($_0$) will be omitted in order to avoid confusion in the formulae).

Integrating (54.2) with respect to θ, one finds

$$\tilde{U} = \tilde{C}_1 + \int_{\theta''}^{\theta} (q_n \cos\theta - q_1 \sin\theta)\, R_1 R_2 \sin\theta\, d\theta, \tag{54.3}$$

where the lower limit of integration θ'' can be chosen arbitrarily (for example, at one of the edges of the shell).

Later on it will be shown that the constant $\tilde{C}_1$ in (54.3) is real, so that in the problem under consideration the auxiliary function $\tilde{U}$ may be identified with the auxiliary function U of the membrane theory; this follows from a comparison of the formula (54.3) with (25.3).

Substituting the expression obtained for $\tilde{U}$ into the second of the equations (51.17), one obtains

$$-icG(\tilde{T}) + \tilde{T} = q_n R_2 - \\ -\left(\frac{1}{R_1} - \frac{1}{R_2}\right)\frac{1}{\sin^2\theta}[\tilde{C}_1 + \int_{\theta''}^{\theta}(q_n \cos\theta - q_1 \sin\theta)R_1 R_2 \sin\theta\, d\theta], \quad (54.4)$$

where in this case

$$G(\ldots) = \frac{1}{R_1 R_2 \sin\theta}\frac{d}{d\theta}\left[\frac{R_2^2 \sin\theta}{R_1}\frac{d(\ldots)}{d\theta}\right], \quad (54.5)$$

since $\tilde{T}$ does not depend on φ.

Expanding equation (54.4), it may be written in the form

$$\frac{d^2\tilde{T}}{d\theta^2} + \left[\left(2\frac{R_1}{R_2} - 1\right)\operatorname{cotan}\theta - \frac{1}{R_1}\frac{dR_1}{d\theta}\right]\frac{d\tilde{T}}{d\theta} + i\frac{R_1^2}{R_2 c}\tilde{T} = i\frac{R_1^2}{R_2 c}F(\theta), \quad (54.6)$$

where

$$F(\theta) = q_n R_2 - \left(\frac{1}{R_1} - \frac{1}{R_2}\right)\frac{1}{\sin^2\theta}[\tilde{C}_1 + \int_{\theta''}^{\theta}(q_n \cos\theta - q_1 \sin\theta)R_1 R_2 \sin\theta\, d\theta]. \quad (54.7)$$

The equation (54.6) is linear and as a result its general solution may be expressed by the formula

$$\tilde{T} = \mathfrak{T} + \tilde{t}, \quad (54.8)$$

where $\mathfrak{T}$ is the general solution of the homogeneous equation

$$\frac{d^2\mathfrak{T}}{d\theta^2} + \left[\left(2\frac{R_1}{R_2} - 1\right)\operatorname{cotan}\theta - \frac{1}{R_1}\frac{dR_1}{d\theta}\right]\frac{d\mathfrak{T}}{d\theta} + i\frac{R_1^2}{R_2 c}\mathfrak{T} = 0, \quad (54.9)$$

and $\tilde{t}$ is any particular integral of the non-homogeneous equation (54.6).

As is known from any textbook on differential equations, the determination of a particular integral does not encounter any principal difficulties, once the solution of the homogeneous equation has been found.

In the present problem, in many cases, an approximation to a particular integral of the non-homogeneous equation (54.6) may be obtained by the following reasoning. It will be noted that the coefficient $i\frac{R_1^2}{R_2 c}$ of the last term on the left hand side of the equation (54.6) has a large absolute value of the order $\frac{R}{\delta}$, whereas the coefficients of the remaining terms are quantities of order unity. Thus a particular integral of the equation (54.6) may be obtained by neglecting all but the last term on the left hand side. In this way one will have

$$\tilde{t} \approx F(\theta). \tag{54.10}$$

The error in this formula may be estimated by substituting (54.10) into (54.6) and by comparing the sole remaining term on the left hand side with the omitted terms, i.e., by comparing

$$\frac{cR_2}{R_1^2}\frac{1}{F}\left\{\frac{d^2F}{d\theta^2}+\left[\left(2\frac{R_1}{R_2}-1\right)\cot\theta-\frac{1}{R_1}\frac{dR_1}{d\theta}\right]\frac{dF}{d\theta}\right\} \tag{54.11}$$

with unity. In the majority of practical problems the error incurred is not great. In those cases where it turns out to be substantial, a particular integral of the equation (54.6) can be obtained by the ordinary method of variation of parameters.

Once the general solution of the equation (54.6) has been found, the complex forces and displacements can be readily determined. It thus follows from (51.6) and (51.7) that

$$\begin{aligned}\tilde{T}_1 &= i\frac{c}{R_1}\cot\theta\frac{d\tilde{T}}{d\theta}+\frac{1}{R_2\sin^2\theta}\tilde{U},\\ \tilde{T}_2 &= \tilde{T}-\tilde{T}_1=\tilde{T}-i\frac{c}{R_1}\cot\theta\frac{d\tilde{T}}{d\theta}-\frac{1}{R_2\sin^2\theta}\tilde{U}.\end{aligned} \tag{54.12}$$

Substituting these expressions into the equations (53.1), (53.2) and taking into consideration the assumed independence of the solution from φ, one obtains the following four equations for the complex displacements

$$\begin{aligned}R_1\tilde{\varepsilon}_1 &= \frac{d\tilde{u}}{d\theta}+\tilde{w}=\frac{R_1}{E\delta}\left\{(1+\mu)i\frac{c}{R_1}\cot\theta\frac{d\tilde{T}}{d\theta}-\mu\tilde{T}+\frac{1+\mu}{R_2\sin^2\theta}\tilde{U}\right\},\\ \Delta_x &= \tilde{u}\cos\theta+\tilde{w}\sin\theta=\frac{R_2\sin\theta}{E\delta}\left\{\tilde{T}-(1+\mu)i\frac{c}{R_1}\cot\theta\frac{d\tilde{T}}{d\theta}-\frac{1+\mu}{R_2\sin^2\theta}\tilde{U}\right\},\end{aligned} \tag{54.13}$$

$$\frac{1}{R_1}\frac{d\tilde{\vartheta}}{d\theta} = -\frac{1}{R_1}\frac{d}{d\theta}\frac{1}{R_1}\left(\frac{d\tilde{w}}{d\theta} - \tilde{u}\right) = \tag{54.13}$$

$$= \frac{1}{cE\delta}\left\{i\tilde{T} + \frac{c}{R_1}\operatorname{cotan}\theta\frac{d\tilde{T}}{d\theta} - i\left[\frac{1}{R_2\sin^2\theta}\tilde{U} + T_2{}^*\right]\right\},$$

$$\tilde{\vartheta} = -\frac{1}{R_1}\left(\frac{d\tilde{w}}{d\theta} - \tilde{u}\right) = \frac{1}{E\delta}\left\{-\frac{R_2}{R_1}\frac{d\tilde{T}}{d\theta} + i\frac{R_2\tan\theta}{c}\left(\frac{1}{R_2\sin^2\theta}\tilde{U} - T_1{}^*\right)\right\}.$$

In order that the last two equations of this system will be compatible, the operator $\dfrac{1}{R_1}\dfrac{d(\ldots)}{d\theta}$ on the right hand side of the fourth equation must be equal to the right hand side of the third equation. Hence it is seen that equation (54.6) will be obtained only if

$$T_1{}^* = \frac{1}{R_2\sin^2\theta}\tilde{U}. \tag{54.14}$$

Thus this last equality is the condition of compatibility of these two equations. Comparing (54.3) and (54.14) with (25.3), it is seen that

$$\tilde{C}_1 = C_1. \tag{54.15}$$

In other words, as has already been mentioned, the constant $\tilde{C}_1$ in (54.3) is real. It now follows from (54.7), (54.14), (54.15) and the formulae of § 25 that the function $F(\theta)$ on the right hand side of the equation (54.6) may be written in the form

$$F(\theta) = T_1{}^* + T_2{}^*, \tag{54.16}$$

whence it is quite clear that in determining the particular integral of the equation (54.6) by the approximate formula (54.10) one is, in effect, invoking the membrane theory.

On the basis of (54.14) and the last of the formulae (54.13) one has for the complex function $\tilde{\vartheta}$ the expression

$$\tilde{\vartheta} = -\frac{1}{R_1}\left(\frac{d\tilde{w}}{d\theta} - \tilde{u}\right) = -\frac{1}{E\delta}\frac{R_2}{R_1}\frac{d\tilde{T}}{d\theta}. \tag{54.17}$$

The real part of this function is obviously the angle of rotation of the tangent to the meridian

$$\vartheta = -\frac{1}{R_1}\left(\frac{dw}{d\theta} - u\right).$$

As regards its imaginary part, it will be equal to [cf. (16.6)]

$$I_m(\tilde{\vartheta}) = -\frac{1}{E\delta}\frac{R_2}{R_1}I_m\left(\frac{d\tilde{T}}{d\theta}\right) = \frac{1}{E\delta c}\frac{R_2}{R_1}\frac{1}{1+\mu}\frac{d(M_1+M_2)}{d\theta} = \frac{R_2}{Ec\delta}N_1.$$

Thus, finally,

$$\tilde{\vartheta} = \vartheta + i\frac{R_2}{Ec\delta}N_1 = -\frac{1}{E\delta}\frac{R_2}{R_1}\frac{d\tilde{T}}{d\theta}. \tag{54.18}$$

It will be noted that the formula (54.18) establishes a connection between the present transformation and that of Meissner [155], since $\tilde{\vartheta}$ is the basic complex auxiliary function of the above author.

For the determination of the displacements $\tilde{u}$, $\tilde{w}$ the expression

$$\tilde{\Delta}_z = \tilde{w}\cos\theta - \tilde{u}\sin\theta. \tag{54.19}$$

will first be formed. Differentiating it with respect to θ, one finds

$$\frac{d\tilde{\Delta}_z}{d\theta} = \frac{d\tilde{w}}{d\theta}\cos\theta - \tilde{w}\sin\theta - \frac{d\tilde{u}}{d\theta}\sin\theta - \tilde{u}\cos\theta = -R_1(\tilde{\vartheta}\cos\theta + \tilde{\varepsilon}_1\sin\theta). \tag{54.20}$$

The right hand side of this formula contains quantities which have already been expressed in terms of $\tilde{T}$; this follows from the first of the formulae (54.13) and from (54.18). Thus the right hand side represents a known function of θ.

Integrating (54.20), one finds

$$\tilde{\Delta}_z = \tilde{C}_2 - \int_{\theta''}^{\theta} R_1(\tilde{\vartheta}\cos\theta + \tilde{\varepsilon}_1\sin\theta)\,d\theta. \tag{54.21}$$

By adding to this formula the second of the equations (54.13) one may determine separately $\tilde{u}$ and $\tilde{w}$.

It will be noted that the real parts of the functions $\tilde{\Delta}_x$ and $\tilde{\Delta}_z$

$$\begin{aligned} \Delta_x &= u\cos\theta + w\sin\theta, \\ \Delta_z &= w\cos\theta - u\sin\theta, \end{aligned} \tag{54.22}$$

have the following independent geometric interpretations:

Δ_z is the component of the displacement of an arbitrary point on the meridian in the direction of the axis of the shell,

Δ_x is the component of the displacement in the direction perpendicular to this axis (Fig. 73).

In correspondence with this interpretation of Δ_z, the real part of the constant $\tilde{C}_2$ (C) in the formula (54.21) is equal to the value of the axial displacement on the parallel circle θ''. By choice of this constant one may place the origin for Δ_z at any parallel circle. The value of this constant obviously does not influence the stresses and the strains in the shell.

Introduce now the two complex functions

$$\tilde{Q}_x = \tilde{T}_1 \cos\theta + i\frac{c}{R_1}\frac{d\tilde{T}}{d\theta}\sin\theta,$$
$$\tilde{Q}_z = \tilde{T}_1 \sin\theta - i\frac{c}{R_1}\frac{d\tilde{T}}{d\theta}\cos\theta, \tag{54.23}$$

the real parts of which will be

$$Q_x = T_1 \cos\theta + N_1 \sin\theta,$$
$$Q_z = T_1 \sin\theta - N_1 \cos\theta. \tag{54.24}$$

As follows from Fig. 74, Q_z and Q_x are the components of the forces in

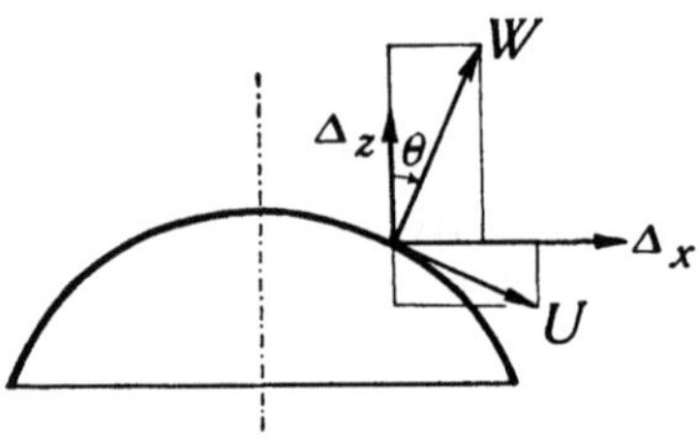

Fig. 73.

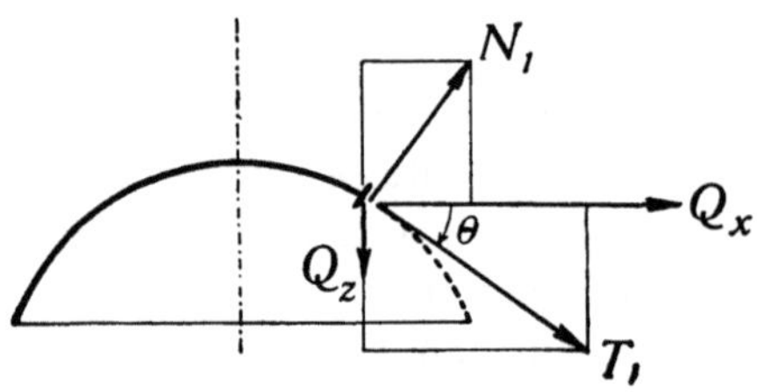

Fig. 74.

the directions of the axis of the shell and perpendicular to this axis which act through the contemplated point of the meridian.

Substituting in (54.23) for $\tilde{T}_1$ its value from (54.12) and (54.14), one finds

$$\tilde{Q}_x = i\frac{c}{R_1}\frac{1}{\sin\theta}\frac{d\tilde{T}}{d\theta} + \frac{\tilde{U}}{R_2}\frac{\cos\theta}{\sin^2\theta},$$
$$\tilde{Q}_z = Q_z = \frac{1}{R_2 \sin\theta}\tilde{U} = T_1{}^* \sin\theta. \tag{54.25}$$

It is seen from the last formula that Q_z does not depend on $\tilde{T}$ and that this formula determines this component exactly apart from an arbitrary constant C_1. It will be noted that as a rule the value of Q_z on one of the parallel circles of the shell may be found from a study of the equilibrium of the shell as a whole. Then the constant C_1 will immediately be determined and thereby Q_z will be known at all sections of the shell. In principle, however, it is possible to have the case when at the two parallel sections bounding the shell the displacement Δ_z will be specified. Then the force Q_z will be statically indeterminate.

The above formulae enable the analysis of symmetrically deformed shells of revolution for any boundary conditions.

In conclusion, some observations will be made regarding the physical implications of the approximation of replacing the particular integral of the equations of the moment theory by a solution of the membrane theory.

For this purpose consider first the case

$$q_1 = q_2 = q_n = U = 0,$$

which will be called the homogeneous problem, since its basic equation (54.6) is homogeneous.

As follows from the second of the formulae (54.25), in this case the axial component of the force Q_z is equal to zero, and hence the homogeneous problem corresponds to the loading shown in Fig. 76 (a shell free from surface loading, but subject to forces Q_x and moments M_1 along the edges).

The general case shown in Fig. 75 will now be considered, where, in addition to symmetrical edge loading, the shell carries also symmetric surface loading.

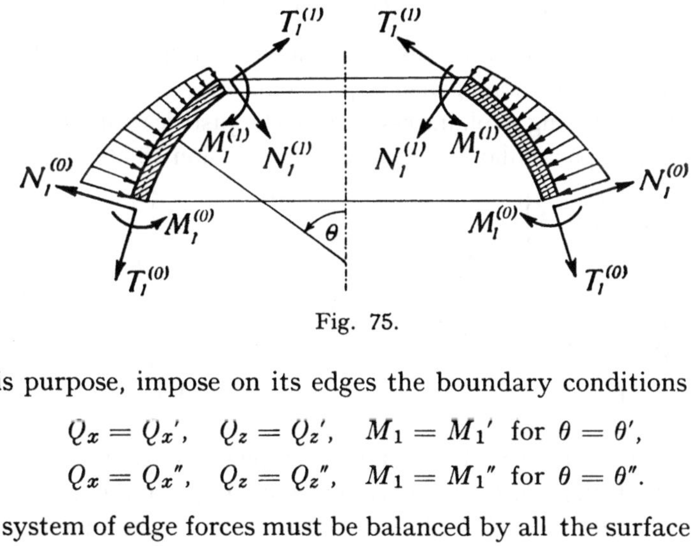

Fig. 75.

For this purpose, impose on its edges the boundary conditions

$$\begin{aligned} Q_x &= Q_x', \quad Q_z = Q_z', \quad M_1 = M_1' \text{ for } \theta = \theta', \\ Q_x &= Q_x'', \quad Q_z = Q_z'', \quad M_1 = M_1'' \text{ for } \theta = \theta''. \end{aligned} \tag{54.26}$$

This system of edge forces must be balanced by all the surface loading, so that the difference

$$2\pi[Q_z'R_2' \sin \theta' - Q_z''R_2'' \sin \theta'']$$

must be equal to the resultant of all the surface forces. It will be assumed that the solution of this problem has been found by means of the membrane theory. To this solution there will correspond boundary conditions which are different from (54.26), namely the conditions

$$Q_x = (Q_x{}^*)', \; Q_z = (Q_z{}^*)', \; M_1 = 0 \text{ for } \theta = \theta', \qquad Q_x = (Q_x{}^*)'', \; Q_z = (Q_z{}^*)'', \; M_1 = 0 \text{ for } \theta = \theta'' \tag{54.27}$$

where the integration constants entering into the formulae (25.3) must be selected in such a way that

$$(Q_z{}^*)' = Q_z', \; (Q_z{}^*)'' = Q_z'', \tag{54.28}$$

i.e., that the boundary conditions (54.26) for the axial components of the force Q_z will be satisfied. This will ensure that the surface and edge loadings will balance.

Subtract now the membrane state of stress from the actual state of stress involving moments; this may be done since all the differential equations are linear and therefore the principle of superposition applies. As a result of this subtraction one arrives at a study of the same shell, but without surface loading, while its edge conditions will be given by

$$Q_z = 0, \; Q_x = Q_x' - (Q_x{}^*)', \; M_1 = M_1' \text{ for } \theta = \theta', \qquad Q_z = 0, \; Q_x = Q_x'' - (Q_x{}^*)'', \; M_1 = M_1'' \text{ for } \theta = \theta''. \tag{54.29}$$

However, this is a loading of the type shown in Fig. 76, i.e., the loading corresponding to the homogeneous problem.

The above reasoning explains essentially the justification of the method of replacing the particular integral of the equation (54.6) by a solution

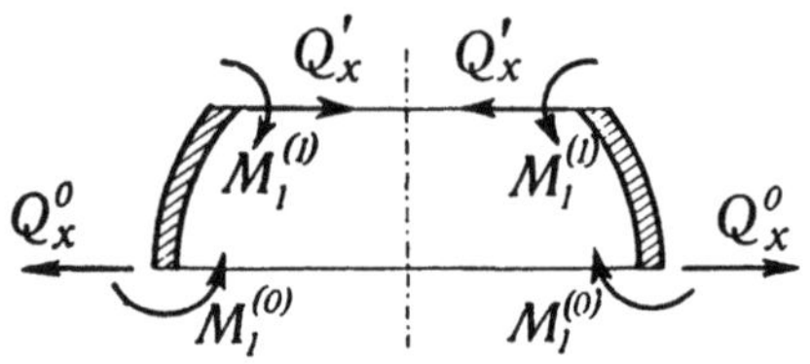

Fig. 76.

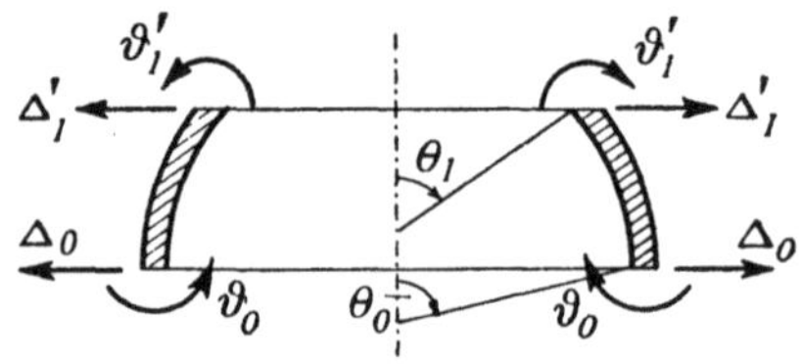

Fig. 77.

of the membrane theory. It should be emphasized that this reasoning applies only if the state of stress corresponding to the given surface loading and edge forces (54.27) will actually be of the membrane type which, as is known from Chapter II, is not always true. This is the limitation imposed on the employment of such an approximate method.

As a result, the following general rule may be formulated by the above reasoning. If a shell of given form and subject to given surface loading, the boundary conditions corresponding to the requirements of the membrane theory, is in a state of stress close to the membrane type, then one may use for the analysis of the same shell for other boundary conditions, as the particular integral of the non-homogeneous equations of the general theory of shells, the solution of the membrane theory.

Fig. 77 shows the positive direction of the displacement Δx and the angle of rotation ϑ for both edges of the shell under consideration.

§ 55. Asymptotic integration of linear second order differential equations

Asymptotic methods of integration of second order ordinary differential equations have been discussed in § 18. Another approach to this problem which extends its possibilities will be discussed in this section.

Consider the linear homogeneous equation of second order

$$\frac{d^2y}{dx^2} + p(x)\frac{dy}{dx} + q(x)y = 0. \tag{55.1}$$

By use of the new unknown

$$z = y \exp\left(\tfrac{1}{2}\int_{x_0}^{x} p\,dx\right), \quad [\exp(x) = e^x] \tag{55.2}$$

(where the lower limit x_0 may be chosen arbitrarily) the above equation may be reduced to the form

$$\frac{d^2z}{dx^2} - F^2 z = 0, \tag{55.3}$$

where

$$F^2 = -q + \frac{1}{2}\frac{dp}{dx} + \frac{1}{4}p^2. \tag{55.4}$$

The equation (55.3) can be transformed to another equation of the same form by means of the following substitutions for the independent and dependent variables:

$$\begin{aligned} dx_1 &= F\,dx, \\ z_1 &= z\sqrt{F}. \end{aligned} \tag{55.5}$$

Introducing these new variables into the equation (55.3), one finds

$$\frac{d^2z_1}{dx_1^2} - F_1^2 z_1 = 0, \tag{55.6}$$

where

$$F_1^2 = 1 + \frac{1}{2}\frac{1}{F}\frac{d^2F}{dx_1^2} - \frac{1}{4}\frac{1}{F^2}\left(\frac{dF}{dx_1}\right)^2 =$$

$$= 1 + \frac{1}{2F^2}\left[\frac{1}{F}\,\frac{d^2F}{dx^2} - \frac{3}{2}\,\frac{1}{F^2}\left(\frac{dF}{dx}\right)^2\right] = 1 + \Delta_1. \qquad (55.7)$$

In the general case, the transformation of the equation (55.3) into the equation (55.6) does not render any important information, since the coefficient F_1 remains variable and the same difficulties are encountered as with the solution of the equation (55.3).

However, firstly if the function F is given the property that it does not increase on differentiation and, secondly, if it is assumed that its magnitude greatly exceeds unity, then, as is clear from (55.7), the variable terms in this formula which are combined in the term Δ_1 will be small compared with its first term and may be neglected with an error of order $\frac{1}{F^2}$ compared with unity.

Then the equation (55.6) takes the form

$$\frac{d^2z_1}{dx_1{}^2} - z_1 = 0 \qquad (55.8)$$

and its general solution is given by

$$z_1' = C_1 e^{x_1} + C_2 e^{-x_1}. \qquad (55.9)$$

Returning now with the help of the formulae (55.5) to the independent variable x and the dependent variable z, one will have

$$z = \frac{1}{\sqrt{F}}\{C_1 e^{\beta} + C_2 e^{-\beta}\}, \qquad (55.10)$$

where

$$\beta = \int_{x_0}^{x} F\,dx, \qquad (55.11)$$

and the lower integration limit x_0 may be chosen arbitrarily to suit the convenience of the analysis.

As can be seen from the preceding derivation, the above solution satisfies the equation (55.3) exactly apart from neglecting in it coefficients of quantities of order $1/F$ in comparison with unity.

As has already been stated, F is a function which is large, but smooth. Therefore the first factor in (55.10) changes considerably more slowly than the second. This property of the solution makes it possible to simplify (55.10). In fact, on differentiating this formula with respect to x, one can neglect the variability of the factor $\sqrt{1/F}$ compared with that of the terms in brackets. Consider the error resulting from such a simplification and compute for this purpose the derivative of (6.10):

$$\frac{dz}{dx} = \sqrt{F}(C_1 e^{\beta} - C_2 e^{-\beta}) - \frac{1}{2F^{3/2}} \frac{dF}{dx} (C_1 e^{\beta} + C_2 e^{-\beta}) =$$

$$= \sqrt{F}\left[C_1 e^{\beta} - C_2 e^{-\beta} - \frac{1}{2F^2} \frac{dF}{dx} (C_1 e^{\beta} + C_2 e^{-\beta})\right]. \tag{55.12}$$

Neglecting the changes in the coefficient of $(1/2F^2) dF/dx$ is equivalent to omitting the group of terms $(1/2F^2)(\partial F/dx)(c_1 e^{\beta} + C_2 e^{-\beta})$ compared with the group of terms $C_1 e^{\beta} - C_2 e^{-\beta}$. However, the factor of $(1/2F^2) dF/dx$ is of order $1/F$, whence the introduced simplification leads to an error of order $1/F$ compared with unity. To this degree of accuracy, one can include the factor $1/\sqrt{F}$ in the constant of integration, whence

$$z = C_1 e^{\beta} + C_2 e^{-\beta}. \tag{55.13}$$

§ 56. Approximate integration of the homogeneous differential equation for symmetrical deformations

The theory of the last section will now be applied to the equation (54.9)

$$\frac{d^2\mathfrak{T}}{d\theta^2} + p(\theta) \frac{d\mathfrak{T}}{d\theta} + q(\theta) . \mathfrak{T} = 0, \tag{56.1}$$

where

$$\begin{aligned} p(\theta) &= \left(2 \frac{R_1}{R_2} - 1\right) \cotan \theta - \frac{1}{R_1} \frac{dR_1}{d\theta}, \\ q(\theta) &= i \frac{R_1^2}{R_2 c} = i\sqrt{12(1-\mu^2)} \frac{R_1^2}{R_2 \delta}. \end{aligned} \tag{56.2}$$

Introducing the new unknown

$$z = \mathfrak{T} e^{\frac{1}{2}\int\left\{\left(2\frac{R_1}{R_2} - 1\right)\cotan\theta - \frac{1}{R_1}\frac{dR_1}{d\theta}\right\} d\theta} = R_2 \sqrt{\frac{\sin\theta}{R_1}} . \mathfrak{T}, \tag{56.3}$$

the equation reduces to the form,

$$\frac{d^2 z}{d\theta^2} - F^2 z = 0, \tag{56.4}$$

where

$$F^2 = \frac{1}{2} \frac{dp}{d\theta} + \frac{1}{4} p^2 - q = -i \frac{R_1^2}{R_2 c} - \frac{1}{2R_1} \frac{d^2 R_1}{d\theta^2} + \frac{3}{4} \frac{1}{R_1^2}\left(\frac{dR_1}{d\theta}\right)^2 +$$

$$+ \frac{\cotan\theta}{2R_1} \frac{dR_1}{d\theta} - \frac{R_1}{R_2} \frac{1}{\sin^2\theta} + \frac{2 + \cos^2\theta}{4\sin^2\theta}. \tag{56.5}$$

It will now be required to find a first approximation to the solution of the equation (56.4) in the same sense as this term has been used in the preceding section. In other words, quantities of order $\frac{1}{F}$ will be neglected on comparison with unity. Further, the approximate solution will not be required to retain its accuracy near the singular points of the equation (56.1), such points being $\theta = 0$ and $\theta = \pi$, and it will be assumed that the radii of curvature R_1 and R_2 are sufficiently smooth functions of θ. With these stipulations all the terms in the formula (56.5), with the exclusion of the first, may be assumed to be quantities of order unity.

As regards the first term

$$-i\frac{R_1^2}{R_2 c},$$

it will have an absolute value of order $\frac{R}{\delta}$. Hence, in correspondence with the above stated principle for the construction of an approximate solution, one may put

$$F^2 \approx -i\frac{R_1^2}{R_2 c}. \tag{56.6}$$

Further, one finds directly on the basis of the formulae (55.10) and (55.11) that

$$z \approx \frac{\sqrt[4]{R_2 c}}{\sqrt{R_1}}\{\bar{C}_1' e^{-(1-i)\beta} + \bar{C}_2' e^{(1-i)\beta}\}, \tag{56.7}$$

where

$$\beta = \frac{1}{\sqrt{2}}\int_{\theta_0}^{\theta}\frac{R_1}{\sqrt{R_2 c}}\,d\theta = \sqrt[4]{3(1-\mu^2)}\int_{\theta_0}^{\theta}\frac{R_1}{\sqrt{R_2\delta}}\,d\theta, \tag{56.8}$$

and by θ_0 will in future be understood an angle determining one of the edges of the shell.

Reverting now from z to the original dependent variable $\mathfrak{T}$ and introducing instead of C_1', C_2' new integration constants, one obtains

$$\mathfrak{T} = \frac{1}{R_2^{3/4}\sin^{1/2}\theta}[\bar{C}_1'' e^{-(1-i)\beta} + \bar{C}_2'' e^{(1-i)\beta}]. \tag{56.9}$$

As has been shown in the last section, this solution satisfies the equation (56.1) exactly apart from neglecting in the coefficients of the equation quantities of order of $1/F^2 \sim \delta/R$ compared with unity. Hence it is everywhere, except near singularities, equivalent to the mathematically exact solution of the differential equation (54.9). This follows from the fact that (54.9), having been based on the initial assumptions of thin shell theory, involves an error of order δ/R compared with unity.

Employment of the solution in the form (56.9) leads to somewhat involved computations. For practical purposes, a simpler solution will be sufficient which can be obtained from (56.9) by including the slowly varying term $(R_2^{3/4} \sin^{1/2} \theta)^{-1}$ in the constant of integration. The solution thus obtained

$$\mathfrak{T} = \tilde{C}_1 e^{-(1-i)\beta} + \tilde{C}_2 e^{(1-i)\beta} \tag{56.10}$$

is related to (56.9) in the same manner as (55.13) to (55.10). It follows from this that the simplified form of the solution (56.10) satisfies (54.9) exactly apart from the omission of terms of order $1/F \sim \sqrt{\delta/R}$ compared with unity.

Certain authors assume that, if in the formula (56.9) one neglects the variation of the factor

$$\frac{1}{R_2^{3/4} \sin^{1/2} \theta},$$

one must also neglect the variation of the quantity

$$\frac{R_1}{\sqrt{R_2 c}}$$

in the formula (56.8), which would lead to the expression

$$\beta = \frac{1}{\sqrt{2}} \frac{R_1^0}{\sqrt{R_2^0 c}} (\theta - \theta_0), \tag{56.11}$$

where by R_1^0, R_2^0 are understood the edge values of the radii of curvature. However, although at a first glance these two simplifications seem to be of the same type, this is actually not quite true.

It is essential that β enters into the formula (56.9) as an exponent, so that its admissible error has a magnified effect on the final result. Practical design calculations show that sometimes corrections such as are given by (56.8) in comparison with (56.11) may be rather significant (for example, in the analysis of elliptical or box-like end sections). Therefore the recommendation to use the formula (56.8) and not (56.11) seems to be better justified.

§ 57. Approximate expressions for the forces, moments and displacements of symmetrically deformed shells of revolution

The work in this section will start off from the approximate formulae (56.10) and (56.8). Substituting (56.10) into the formulae for the complex forces (54.12), one finds

$$\tilde{T}_1 = -\frac{1+i}{\sqrt{2}}\sqrt{\frac{c}{R_2}}\,\mathrm{cotan}\,\theta[\tilde{A}_1 e^{-(1-i)\beta} - \tilde{A}_2 e^{(1-i)\beta}] + \tilde{t}_1 =$$

$$= \frac{1+i}{\sqrt{2}}\sqrt{\frac{c}{R_2}}\,\mathrm{cotan}\,\theta[\tilde{B}_1 \cosh(1-i)\beta + \tilde{B}_2 \sinh(1-i)\beta] + \tilde{t}_1,$$

$$\tilde{T}_2 = \tilde{A}_1\left[1 + \underline{\frac{1+i}{\sqrt{2}}\sqrt{\frac{c}{R_2}}\,\mathrm{cotan}\,\theta}\right]e^{-(1-i)\beta} +$$

$$+ \tilde{A}_2\left[1 - \underline{\frac{1+i}{\sqrt{2}}\sqrt{\frac{c}{R_2}}\,\mathrm{cotan}\,\theta}\right]e^{(1-i)\beta} + \tilde{t}_2 = \quad (57.1)$$

$$= \left[\tilde{B}_1 - \underline{\frac{1+i}{\sqrt{2}}\sqrt{\frac{c}{R_2}}\,\mathrm{cotan}\,\theta.\tilde{B}_2}\right]\sinh(1-i)\beta +$$

$$+ \left[\tilde{B}_2 - \underline{\frac{1+i}{\sqrt{2}}\sqrt{\frac{c}{R_2}}\,\mathrm{cotan}\,\theta.\tilde{B}_1}\right]\cosh(1-i)\beta + \tilde{t}_2.$$

Following the principles underlying the approximate solution studied in the preceding section, the underlined terms in the last formula must be neglected as quantities of order $\sqrt{\frac{\delta}{R}}$. In what follows, similar terms will be neglected without stipulation.

The terms in (57.1) which correspond to the particular integral will be

$$\tilde{t}_1 = T_1^* + i\frac{c}{R_1}\,\mathrm{cotan}\,\theta\,\frac{d\tilde{t}}{d\theta},$$

$$\tilde{t}_2 = \tilde{t} - T_1^* - i\frac{c}{R_1}\,\mathrm{cotan}\,\theta\,\frac{d\tilde{t}}{d\theta}. \quad (57.2)$$

Further, using the formulae (54.17) and (54.13), one finds

$$\tilde{\vartheta} \approx \frac{1-i}{\sqrt{2}}\frac{1}{E\delta}\sqrt{\frac{R_2}{c}}\{\tilde{A}_1 e^{-(1-i)\beta} - \tilde{A}_2 e^{(1-i)\beta}\} - \frac{1}{E\delta}\cdot\frac{R_2}{R_1}\frac{d\tilde{t}}{d\theta} =$$

$$= -\frac{1-i}{\sqrt{2}}\frac{1}{E\delta}\sqrt{\frac{R_2}{c}}\{\tilde{B}_1\cosh(1-i)\beta + \tilde{B}_2\sinh(1-i)\beta\} - \frac{1}{E\delta}\frac{R_2}{R_1}\frac{d\tilde{t}}{d\theta},$$

$$\tilde{\Delta}_x \approx \frac{R_2\sin\theta}{E\delta}\{\tilde{A}_1 e^{-(1-i)\beta} + \tilde{A}_2 e^{(1-i)\beta} + \tilde{t} - (1+\mu)T_1^*\} =$$

$$= \frac{R_2\sin\theta}{E\delta}\{\tilde{B}_2\cosh(1-i)\beta + \tilde{B}_1\sinh(1-i)\beta + \tilde{t} - (1+\mu)T_1^*\}, \quad (57.3)$$

$$R_1\tilde{\varepsilon}_1 = \frac{R_1}{E\delta}\{-\mu[\tilde{A}_1 e^{-(1-i)\beta} + \tilde{A}_2 e^{(1-i)\beta}] - \mu\tilde{t} + (1+\mu)T_1^*\}.$$

Finally, on the basis of the formulae (54.25), one will have

$$\tilde{Q}_x \approx -\frac{1+i}{\sqrt{2}}\sqrt{\frac{c}{R_2}}\frac{1}{\sin\theta}[\tilde{A}_1 e^{-(1-i)\beta} - \tilde{A}_2 e^{(1-i)\beta}] + \tilde{q}_x =$$

$$= \frac{1+i}{\sqrt{2}}\sqrt{\frac{c}{R_2}}\frac{1}{\sin\theta}[\tilde{B}_1 \cosh(1-i)\beta + \tilde{B}_2 \sinh(1-i)\beta] + \tilde{q}_x, \quad (57.4)$$

$$\tilde{Q}_z = Q_z = T_1{}^* \sin\theta,$$

where

$$\tilde{q}_x = T_1{}^* \cos\theta + i\frac{c}{R_1}\frac{1}{\sin\theta}\frac{d\tilde{t}}{d\theta}. \quad (57.5)$$

If the surface loading is sufficiently smooth, one has in these formulae

$$\tilde{t} \approx F(\theta) = T_1{}^* + T_2{}^*, \quad \tilde{t}_1 \approx T_1{}^*, \quad \tilde{t}_2 \approx T_2{}^*, \quad \tilde{q}_x \approx T_1{}^* \cos\theta \quad (57.6)$$

and general expressions for the forces, moments, angles of rotation and displacements may be obtained by separating real and imaginary parts in the formulae (57.1), (57.3), (57.4).

The corresponding computations will only be performed for one of the two forms of solutions deduced above, namely for that which follows from the expression

$$\tilde{T} \approx \tilde{A}_1 e^{-(1-i)\beta} + \tilde{A}_2 e^{(1-i)\beta}. \quad (57.7)$$

For this purpose, the terms corresponding to the particular integral will be omitted, i.e., only the solution of the homogeneous problem will be obtained, since all the aspects of the particular integral have already been sufficiently explained.

Letting

$$\tilde{A}_1 = A_1' + iA_1'', \quad \tilde{A}_2 = A_2' + iA_2'', \quad (57.8)$$

where A_1', A_1'', A_2', A_2'' are arbitrary real constants, one arrives on the basis of (57.1) at the following expressions for the forces and moments:

$$T_1 = -\frac{1}{\sqrt{2}}\sqrt{\frac{c}{R_2}}\operatorname{cotan}\theta\{[(A_1'-A_1'')\cos\beta-(A_1'+A_1'')\sin\beta]e^{-\beta}+$$

$$+ [(A_2''-A_2')\cos\beta - (A_2'+A_2'')\sin\beta]e^{\beta}\},$$

$$T_2 = (A_1'\cos\beta - A_1''\sin\beta)e^{-\beta} + (A_2'\cos\beta + A_2''\sin\beta)e^{\beta}, \quad (57.9)$$

$$\frac{1}{c}\frac{M_2-\mu M_1}{1-\mu^2} = \frac{1}{\sqrt{2}}\sqrt{\frac{c}{R_2}}\operatorname{cotan}\theta\{[(A_1'+A_1'')\cos\beta+(A_1'-A_1'')\sin\beta]e^{-\beta}+$$

$$+ [-(A_2' + A_2'')\cos\beta + (A_2' - A_2'')\sin\beta]e^{\beta}\}, \tag{57.9}$$

$$M_1+M_2 = -(1+\mu)c[(A_1''\cos\beta+A_1'\sin\beta)e^{-\beta}+(A_2''\cos\beta-A_2'\sin\beta)e^{\beta}].$$

It follows from the last two formulae that to an accuracy of order $\sqrt{\frac{\delta}{R}}$ compared with unity

$$M_2 \approx \mu M_1 \tag{57.10}$$

and, therefore,

$$M_1 = -c[(A_1''\cos\beta + A_1'\sin\beta)e^{-\beta} + (A_2''\cos\beta - A_2'\sin\beta)e^{\beta}]. \tag{57.11}$$

Further, using the first of the formulae (57.3), one finds

$$N_1 = -\frac{1}{\sqrt{2}}\sqrt{\frac{c}{R_2}}\{[A_1' - A_1'')\cos\beta - (A_1' + A_1'')\sin\beta]e^{-\beta} +$$
$$+ [(A_2'' - A_1')\cos\beta - (A_2' + A_2'')\sin\beta]e^{+\beta}\},$$
$$\vartheta = \frac{1}{\sqrt{2}}\frac{1}{E\delta}\sqrt{\frac{R_2}{c}}\{[(A_1' + A_1'')\cos\beta + (A_1' - A_2'')\sin\beta]e^{-\beta} - \tag{57.12}$$
$$- [(A_2' + A_2'')\cos\beta - (A_2' - A_2'')\sin\beta]e^{\beta}\}.$$

Now, the formulae (54.25) for $U = 0$ give

$$Q_z = 0, \quad Q_x = \frac{1}{\sin\theta}N_1. \tag{57.13}$$

Finally, it follows from (57.3), (54.18), (54.21) that

$$\Delta_z \approx C + \frac{R_2}{E\delta}\cos\theta(T_1 + T_2) \approx C + \frac{R_2\cos\theta}{E\delta}T_2,$$
$$\Delta_x \approx \frac{R_2\sin\theta}{E\delta}(T_2 - \mu T_1) \approx \frac{R_2\sin\theta}{E\delta}T_2, \tag{57.14}$$

whence

$$u \approx C\sin\theta,$$
$$w \approx -C\cos\theta + \frac{R_2}{E\delta}T_2. \tag{57.15}$$

As has already been indicated, the constant C fixes the origin from which the axial displacement Δ_z is measured.

Summarizing all the obtained results, one arrives at the following set of formulae:

$$
\begin{aligned}
w &\approx \frac{R_2}{E\delta}[(A_1' \cos\beta - A_1'' \sin\beta)e^{-\beta} + (A_2' \cos\beta + A_2'' \sin\beta)e^{\beta}]\\
\Delta_x &\approx \frac{R_2 \sin\theta}{E\delta} T_2,\\
\vartheta &\approx \frac{1}{E\delta}\frac{1}{\sqrt{2}}\sqrt{\frac{R_2}{c}}\{[(A_1' + A_1'')\cos\beta + (A_1' - A_1'')\sin\beta]e^{-\beta} -\\
&\qquad - [(A_2' + A_2'')\cos\beta + (A_2'' - A_2')\sin\beta]e^{\beta}\},\\
T_1 &\approx -\frac{1}{\sqrt{2}}\sqrt{\frac{c}{R_2}}\,\text{cotan}\,\theta\{[(A_1' - A_1'')\cos\beta - (A_1' + A_1'')\sin\beta]e^{-\beta} +\\
&\qquad + [(A_2'' - A_2')\cos\beta - (A_2' + A_2'')\sin\beta]e^{\beta}\},\\
N_1 &\approx T_1 \tan\theta,\; Q_x = \frac{N_1}{\sin\theta},\\
T_2 &\approx (A_1' \cos\beta - A_1'' \sin\beta)e^{-\beta} + (A_2' \cos\beta + A_2'' \sin\beta)e^{\beta},\\
M_1 &\approx -c[(A_1'' \cos\beta + A_1' \sin\beta)e^{-\beta} + (A_2'' \cos\beta - A_2' \sin\beta)e^{\beta}],\\
M_2 &\approx \mu M_1.
\end{aligned}
\tag{57.16}
$$

These formulae are for the general case when the shell of revolution has two edges determined by the angles $\theta = \theta_0$ and $\theta = \theta_1$ (Figs. 76, 77), i.e., when it is a dome or end-panel with a cut-out. If the shell has only one edge $\theta = \theta_0$, being closed at the apex, then the constants A_1', A_1'' in the above solutions must be set equal to zero. Unfortunately, this does not follow directly from the formulae (57.16), since, due to their method of deduction, and as has also been indicated earlier, these formulae are incorrect near the singular points of the differential equation (54.9).

However, if they had been defined on the basis of the second approximation, it would be clear that the functions in the formulae (57.16) with the constants A_1' and A_1'' tend to infinity at $\theta = 0$. Hence, if the point $\theta = 0$ belongs to the middle surface, then, for the solution to remain finite at this point, it is necessary to assume that

$$A_1' = A_1'' = 0. \tag{57.17}$$

Thus, the number of arbitrary constants in the formulae (57.16) is halved and this is quite natural, since in the presence of only one edge the number of boundary conditions is likewise halved.

If the middle surface were to contain the point $\theta = \pi$ rather than $\theta = 0$, then one would have to put instead $A_2' = A_2'' = 0$.

The solution given by (57.16) will now be further analyzed. For this purpose, along with $\beta = \beta_1$, introduce the new variable

$$\beta_2 = \frac{1}{\sqrt{2}} \int\limits_{\theta_1}^{\theta} \frac{R_1 d\theta}{\sqrt{R_2 c}} = \frac{1}{\sqrt{2}} \int\limits_{\theta_0}^{\theta} \frac{R_1 d\theta}{\sqrt{R_2 c}} + \frac{1}{\sqrt{2}} \int\limits_{\theta_1}^{\theta_0} \frac{R_1 d\theta}{\sqrt{R_2 c}} = \beta_1 + \alpha_0, \quad (57.18)$$

where

$$\alpha_0 = \frac{1}{\sqrt{2}} \int\limits_{\theta_1}^{\theta_0} \frac{R_1 d\theta}{\sqrt{R_2 c}} \quad (57.19)$$

is a constant.

Replacing the parameter β by β_2, in the formulae (57.16) in those terms which are multiplied by the constants A_1', A_1'' one will have

$$w \approx \frac{R_2}{E\delta} [(B_1 \cos \beta_2 - B_2 \sin \beta_2) e^{-\beta_2} + (C_1 \cos \beta_1 + C_2 \sin \beta_1) e^{\beta_1}],$$

$$\Delta_x \approx \frac{R_2 \sin \theta}{E\delta} T_2, \quad (57.20)$$

$$\vartheta \approx \frac{1}{E\delta} \frac{1}{\sqrt{2}} \sqrt{\frac{R_2}{c}} \{[(B_1+B_2) \cos \beta_2 + (B_1-B_2) \sin \beta_2] e^{-\beta_2} -$$
$$- [(C_1 + C_2) \cos \beta_1 + (C_2 - C_1) \sin \beta_1] e^{\beta_1}\}.$$

$$T_1 \approx - \frac{1}{\sqrt{2}} \sqrt{\frac{c}{R_2}} \operatorname{cotan} \theta \{[(B_1-B_2) \cos \beta_2 - (B_1+B_2) \sin \beta_2] e^{-\beta_2} +$$
$$+ [(C_2 - C_1) \cos \beta_1 - (C_1 + C_2) \sin \beta_1] e^{\beta_1}\},$$

$$N_1 = T_1 \tan \theta, \quad Q_x = \frac{1}{\sin \theta} N_1 = \frac{1}{\cos \theta} T_1, \quad (57.21)$$

$$T_2 \approx (B_1 \cos \beta_2 - B_2 \sin \beta_2) e^{-\beta_2} + (C_1 \cos \beta_1 + C_2 \sin \beta_1) e^{\beta_1},$$

$$M_1 \approx - c[(B_2 \cos \beta_2 + B_1 \sin \beta_2) e^{-\beta_2} + (C_2 \cos \beta_1 - C_1 \sin \beta_1) e^{\beta_1}],$$

$$M_2 \approx \mu M_1.$$

Here B_1, B_2, C_1, C_2 are new arbitrary constants which are related to the previous ones by the equalities

$$\begin{aligned} B_1 &= (A_1' \cos \alpha_0 + A_1'' \sin \alpha_0) e^{\alpha_0}, \\ B_2 &= (A_1'' \cos \alpha_0 - A_1' \sin \alpha_0) e^{\alpha_0}, \\ C_1 &= A_2', \\ C_2 &= A_2''. \end{aligned} \quad (57.22)$$

It will be noted that the parameter

$$\beta_1 = \frac{1}{\sqrt{2}} \int\limits_{\theta_0}^{\theta} \frac{R_1 d\theta}{\sqrt{R_2 c}} \tag{57.23}$$

vanishes at the edge $\theta = \theta_0$ and that away from this edge it becomes a negative quantity of order $\sqrt{\dfrac{R}{\delta}}$.

Hence it is clear that the functions appearing in the formulae (57.20), (57.21) with the constants C_1, C_2 decrease rapidly as the point under consideration moves away from the edge $\theta = \theta_0$.

As regards the parameter

$$\beta_2 = \frac{1}{\sqrt{2}} \int\limits_{\theta_1}^{\theta} \frac{R_1 d\theta}{\sqrt{R_2 c}}, \tag{57.24}$$

it will vanish at the edge $\theta = \theta_1$ and become, at a short distance from this edge, a positive quantity of order $\sqrt{\dfrac{R}{\delta}}$. Hence it is clear that the functions occurring in the formulae (57.20), (57.21) with the constants B_1 and B_2 decrease rapidly as the point under consideration leaves the edge $\theta = \theta_1$.

Thus the conclusion is reached that the formulae (57.20), (57.21) or, what is the same thing, the formulae (57.16), because the first set were obtained from the second by replacing one set of integration constants by another, give a state of stress which is localized near the edge of the shell.

If the distance between the edges is sufficiently large, the symmetrical loading Q_x^0, $M_1^{(0)}$, applied to the edge $\theta = \theta_0$, does not, for any practical purpose, cause stresses and displacements near the other edge. The same may be said with regard to the loading Q_x^1, $M_1^{(1)}$ applied to the edge $\theta = \theta_1$.

Thanks to this situation, one may in many problems with edge conditions at two boundaries of a shell of revolution consider these independently from each other, i.e., when considering the boundary condition at the edge $\theta = \theta_0$, one may put $B_1 = B_2 = 0$, and for those at the edge $\theta = \theta_1$, one can put $C_1 = C_2 = 0$. This procedure greatly simplifies the calculations. In fact, only in those cases when the distance between the edges of the shell is small so that it has the form of a comparatively narrow belt is it necessary to consider the boundary conditions simultaneously.

The functions in the solutions (57.20) and (57.21) coincide with those functions in terms of which the bending problem of beams on elastic foundations may be solved. This coincidence is not accidental, because there exists an analogy between the problem under consideration and the problem of bending of a beam on an elastic foundation which enabled I. Shtaerman [143] to give an unusual and very clear study of the theory of symmetrically deformed shells of revolution. If one were to aim only at an acquaintance with this particular problem, it would be impossible to design a better approach to it than that given by the above author. This problem has been studied in a different manner here only for the reason that interest in this book is directed towards general rather than particular methods which are to apply to the greatest possible range of problems of the theory of thin shells.

§ 58. Influence coefficients

Consider a shell of revolution bounded by a single parallel circle (a dome without an opening or a cap) and loaded along its edge by uniformly distributed forces $Q_x{}^0 = Q_0$ and moments $M_1{}^0 = M_0$.

Then it will depend on whether the external normal to the edge section of the shell is directed towards the side of increasing or decreasing θ, i.e., on whether one has the case shown in Fig. 78(*a*) or in 78(*b*); whether in the formulae (57.20), (57.21) one has to put $B_1 = B_2 = 0$, or $C_1 = C_2 = 0$, as has been discussed above.

For this purpose, one obtains from these formulae the expressions

$$\begin{aligned}
w &= \frac{R_2}{E\delta}(C_1\cos\beta - C_2\sin\beta)\,e^{-\beta},\\
\Delta_x &= w\sin\theta,\\
\vartheta &= \mp\frac{1}{\sqrt{2}}\sqrt{\frac{R_2}{c}}\,\frac{1}{E\delta}\,[(C_1+C_2)\cos\beta + (C_1-C_2)\sin\beta]\,e^{-\beta},\\
T_1 &= \mp\frac{1}{\sqrt{2}}\sqrt{\frac{c}{R_2}}\,\operatorname{cotan}\theta[(C_2-C_1)\cos\beta+(C_1+C_2)\sin\beta]\,e^{-\beta}, \qquad (58.1)\\
N_1 &= T_1\tan\theta,\quad Q_x = \frac{N_1}{\sin\theta},\\
T_2 &= (C_1\cos\beta - C_2\sin\beta)\,e^{-\beta},\\
M_1 &= -\,c(C_2\cos\beta + C_1\sin\beta)\,e^{-\beta},
\end{aligned}$$

in which the upper signs relate to the case shown in Fig. 78(*a*) and the lower ones to the case shown in Fig. 78(*b*).

In the formulae (58.1), the parameter β is given by

$$\beta = -\beta_1 = -\frac{1}{\sqrt{2}} \int_{\theta_0}^{\theta} \frac{R_1 d\theta}{\sqrt{R_2 c}}, \tag{58.2}$$

if the external normal to the edge section of the shell is directed towards the side of increasing θ, and by

$$\beta = \beta_2 = \frac{1}{\sqrt{2}} \int_{\theta_1}^{\theta} \frac{R_1 d\theta}{\sqrt{R_2 c}}, \tag{58.3}$$

if it points towards the side of decreasing θ.

It will be noted that if one takes as independent variable instead of the angle θ the arc length of the meridian of the shell s, measuring

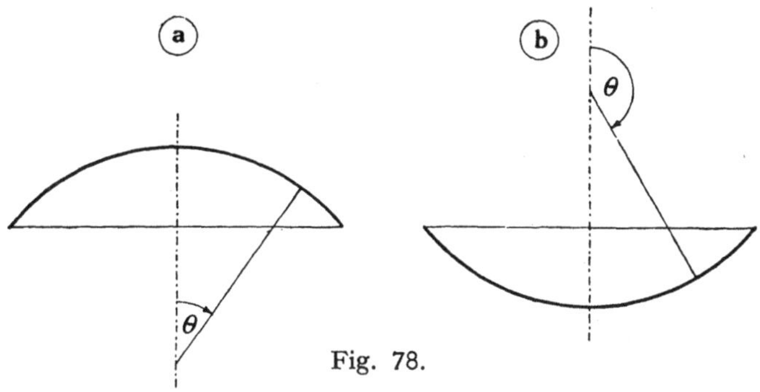

Fig. 78.

it each time from the edge, then the two formulae (58.2) and (58.3) reduce to the single formula

$$\beta = \frac{1}{\sqrt{2}} \int_{0}^{s} \frac{ds}{\sqrt{R_2 c}} = \sqrt[4]{3(1-\mu^2)} \int_{0}^{s} \frac{ds}{\sqrt{R_2 \delta}}, \tag{58.4}$$

since in the first case $R_1 d\theta = -ds$ and in the second case $R_1 d\theta = ds$.

Certain authors [147] neglect in the expression (58.4) the variation of R_2 as a result of which it takes the form

$$\beta = \frac{1}{\sqrt{2}} \frac{s}{\sqrt{R_2^0 c}}, \tag{58.5}$$

where R_2^0 is the edge value of the radius R_2. Normally, employment of this simplification does not raise any objections, although it gives rise to more serious errors than the simplification introduced in the preceding section.

In order to obtain from the formulae (58.1) the general solution of the problem under consideration, one has to add to them the solution of the equations of the membrane theory which, as before, will be provided with an asterisk.

Let the edge of the shell be subjected to the uniformly distributed force Q_0 and moment M_0. Then, on the basis of the formulae (58.1) and (54.29), one may write

$$
\begin{aligned}
Q_0 &= \mp \frac{1}{\sqrt{2}} \sqrt{\frac{c}{R_2}} \frac{1}{\sin\theta_0} (C_2 - C_1) + Q_0^*, \\
M_0 &= -cC_2 + M_0^* = -cC_2,
\end{aligned}
\tag{58.6}
$$

where Q_0^* and $M_0^* = 0$ are the edge values of the force Q_x and the moment M_1, as given by the membrane theory. Hence

$$
C_2 = -\frac{M_0}{c}, \quad C_1 = \pm \sqrt{\frac{2R_2}{c}} \sin\theta_0 (Q_0 - Q_0^*) - \frac{1}{c} M_0. \tag{58.7}
$$

Introducing these values for the constants into the formulae (58.1) and adding there terms corresponding to the particular integral, one finds

$$
\begin{aligned}
w &= \frac{R_2\sqrt{2}}{E\delta}\left[\pm \sqrt{\frac{R_2}{c}} (Q_0 - Q_0^*) \sin\theta_0 \cos\beta - \frac{M_0}{c} \cos\left(\beta + \frac{\pi}{4}\right)\right] e^{-\beta} + w^*, \\
\Delta &= \frac{R_2\sqrt{2}}{E\delta} \sin\theta_0 \left[\pm \sqrt{\frac{R_2}{c}} (Q_0 - Q_0^*) \sin\theta_0 \cos\beta - \right. \\
&\qquad\qquad \left. - \frac{\sqrt{2}}{c} M_0 \cos\left(\beta + \frac{\pi}{4}\right)\right] e^{-\beta} + \Delta^*, \\
\vartheta &= -\sqrt{\frac{2R_2}{c}} \frac{1}{E\delta} \left[\sqrt{\frac{R_2}{c}} \sin\theta_0 (Q_0 - Q_0^*) \cos\left(\beta - \frac{\pi}{4}\right) \pm \right. \\
&\qquad\qquad \left. \pm \frac{1}{c} M_0 \cos\beta\right] e^{-\beta} + \vartheta^*, \\
T_2 &= \left[\pm \sqrt{\frac{2R_2}{c}} \sin\theta_0 (Q_0 - Q_0^*) \cos\beta - \frac{M_0}{c} \cos\left(\beta + \frac{\pi}{4}\right)\right] e^{-\beta} + T_2^*, \\
M_1 &= \left[M_0\sqrt{2} \cos\left(\beta - \frac{\pi}{4}\right) \mp \sqrt{2R_2 c} \sin\theta_0 (Q_0 - Q_0^*) \sin\beta\right] e^{-\beta}.
\end{aligned}
\tag{58.8}
$$

These last formulae determine quantities which characterize the stresses and strains in the shell in terms of the given boundary loading.

By the way, it follows from these formulae that

$$\begin{aligned} \Delta_0 &= \alpha_{11}Q_0 + \alpha_{12}M_0 - \alpha_{11}Q_0^* + \Delta_0^*, \\ \vartheta_0 &= \alpha_{21}Q_0 + \alpha_{22}M_0 - \alpha_{21}Q_0^* + \vartheta_0^*, \end{aligned} \tag{58.9}$$

where

$$\alpha_{11} = \pm \frac{2\sqrt[4]{3(1-\mu^2)}}{E}\left(\frac{R_2{}^0}{\delta}\right)^{3/2}\sin^2\theta_0,$$

$$\alpha_{12} = \alpha_{21} = -\frac{\sqrt{12(1-\mu^2)}}{E\delta^2}R_2{}^0\sin\theta_0, \tag{58.10}$$

$$\alpha_{22} = \pm \frac{4[3(1-\mu^2)]^{3/4}}{E\delta^2}\sqrt{\frac{R_2{}^0}{\delta}},$$

and again the upper sign, as in all the preceding formulae, refers to the case shown in Fig. 78(*a*) and the lower sign to the case shown in Fig. 78(*b*). For the application of the formulae (58.8), it is useful to have the auxiliary functions in Fig. 79.

The quantities α_{11}, $\alpha_{12} = \alpha_{21}$, α_{22} may be called influence coefficients for the edge of a symmetrically loaded shell of revolution, since they characterize completely the deformation of the edge under the influence of the force $Q_0 = 1$ and the moment $M_0 = 1$.

Employment of the formulae (58.9), (58.10) simplifies the statical indeterminateness of a roof section between the shell and its supporting contours.

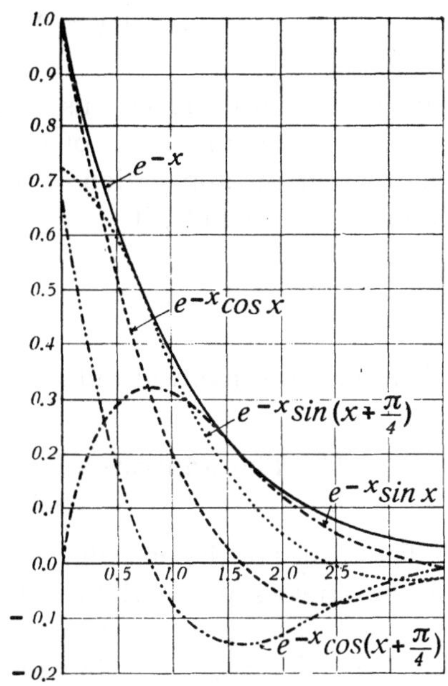

Fig. 79.

It will be noted that in the problem under consideration the greatest stress is provided by the hoop force T_2 and the bending moment M_1. For these stresses, on the basis of formulae (58.8), one may write down the expressions

$$\begin{aligned} \sigma_2 = \frac{T_2}{\delta} = \Big[&\pm 2\sqrt[4]{3(1-\mu^2)}\sqrt{\frac{R_2}{\delta}}\,\frac{\sin\theta_0}{\delta}(Q_0 - Q_0^*)\cos\beta - \\ &- \sqrt{2}\sqrt{12(1-\mu^2)}\,\frac{M_0}{\delta^2}\cos\left(\beta + \frac{\pi}{4}\right)\Big]e^{-\beta} + \sigma_2^*, \end{aligned} \tag{58.11}$$

$$\sigma_M = \frac{6M_1}{\delta^2} = \left[\frac{6\sqrt{2}M_0}{\delta^2}\cos\left(\beta - \frac{\pi}{4}\right) \mp \right.$$
$$\left. \mp \frac{6}{\sqrt{3(1-\mu^2)}}\sqrt{\frac{R_2}{\delta}}\,\frac{\sin\theta_0}{\delta}(Q_0 - Q_0{}^*)\sin\beta\right]e^{-\beta}. \tag{58.11}$$

All the above formulae were obtained on the basis of the assumption that the shell of revolution has only one edge. However, these same formulae may also be used in the case when the shell has not one but two edges, provided the arc length s_0 between these edges is sufficiently large. The errors incurred in this process may be estimated by comparing the quantity

$$\exp\left(-\sqrt{3(1-\mu^2)}\,\frac{s_0}{\sqrt{R_2{}^0\delta}}\right)$$

with unity, since, in fact, this type of error is introduced by the process of separate determination of the constants in these formulae. Hence it follows that the application of the formulae of this section to shells with two edges is possible, if

$$\frac{s_0}{\sqrt{R_2\delta}} \geqslant 2,$$

when the error involved will not be greater than 10%.

§ 59. The analysis of cylindrical pressure vessels and their end-panels

The theory of the preceding section will now be applied to the analysis of the stresses in a cylindrical pressure-vessel with caps of arbitrary form and subject to uniform normal pressure. In order to simplify the calculations, it will be assumed that the end-panel has the same thickness as the cylindrical part of the vessel and that they are connected by a weld joint. Further, it will be assumed that the length of the vessel is sufficiently large, so that one may consider the stresses in the region of the cylinder near one end cap independently of those near the other end. Fig. 80 shows that part of the vessel which will be studied. It is seen from this drawing that

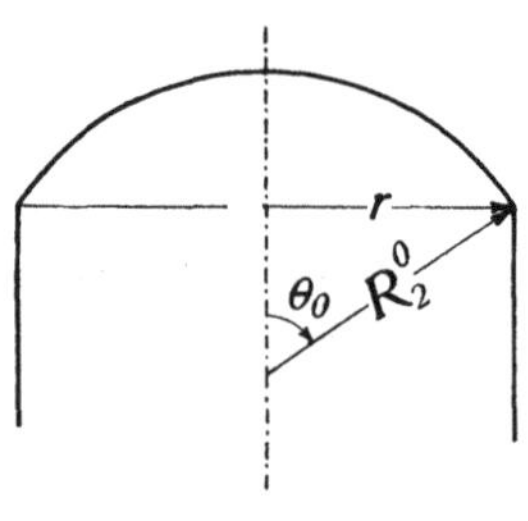

Fig. 80.

$$R_2{}^0 \sin\theta_0 = r.$$

where r is the radius of the cylinder and θ_0 is the angle corresponding to the edge of the end-panel.

On the basis of the formulae of the preceding section, one finds:

a) The influence coefficients of the edge of the cylindrical shell

$$\alpha_{11}' = -\frac{2\sqrt[4]{3(1-\mu^2)}}{E}\left(\frac{r}{\delta}\right)^{3/2},$$
$$\alpha_{12}' = \alpha_{21}' = -\frac{2\sqrt{3(1-\mu^2)}}{E\delta}\,\frac{r}{\delta}, \tag{59.1}$$
$$\alpha_{22}' = -\frac{4[3(1-\mu^2)]^{3/4}}{E\delta^2}\sqrt{\frac{r}{\delta}};$$

b) the influence coefficients of the edge of the end panel

$$\alpha_{11}'' = -\alpha_{11}'\sqrt{\sin\theta_0},$$
$$\alpha_{21}'' = \alpha_{12}'' = \alpha_{12}', \tag{59.2}$$
$$\alpha_{22}'' = -\frac{\alpha_{22}'}{\sqrt{\sin\theta_0}}.$$

When writing down the formulae (59.1) and (59.2), account has been taken of the fact that the external normals to the edge sections of the cylinder and the end-panel are directed towards the sides of decrease and increase of the curvilinear coordinate respectively.

For the cylinder, loaded by uniform normal pressure, the membrane theory gives (§ 25)

$$T_1^* = \tfrac{1}{2}pr,\ T_2^* = pr,\ Q_x^* = T_1^*\cos\theta = 0,$$
$$w^* = \varepsilon_2 r = \frac{r}{E\delta}(T_2^* - \mu T_1^*) = \left(1-\frac{\mu}{2}\right)\frac{pr^2}{E\delta}, \tag{59.3}$$
$$\Delta_x^* = w^*\sin\theta + u^*\cos\theta = w^*,\ \vartheta^* = -\frac{dw^*}{dx} = 0.$$

Correspondingly, for the end-panel, one finds

$$T_1^* = \frac{pR_2}{2},\ T_2^* = \frac{pR_2}{2}\left(2-\frac{R_2}{R_1}\right),\ Q_x^* = \frac{pR_2}{2}\cos\theta,$$
$$\Delta_x^* = \varepsilon_2 R_2\sin\theta = \frac{R_2\sin\theta}{E\delta}(T_2^* - \mu T_1^*) = \tag{59.4}$$
$$= \frac{pR_2^2\sin\theta}{2E\delta}\left(2-\mu-\frac{R_2}{R_1}\right),$$
$$\vartheta^* \approx 0.$$

Now, on the basis of the formulae (58.9), one may construct the following expressions for the displacement Δ_0 and the angle of rotation ϑ_0 at the edge of each of the two shells under consideration:

a) For the cylinder

$$\begin{aligned} \Delta_0{}' &= \alpha_{11}{}'Q_0{}' + \alpha_{12}{}'M_0{}' + \left(1 - \frac{\mu}{2}\right)\frac{pr^2}{E\delta}, \\ \vartheta_0{}^1 &= \alpha_{21}{}'Q_0{}' + \alpha_{22}{}'M_0{}'; \end{aligned} \tag{59.5}$$

b) for the end-panel

$$\begin{aligned} \Delta_0{}'' = &- \alpha_{11}{}'Q_0{}''\sqrt{\sin\theta_0} + \alpha_{12}{}'M_0{}'' + \\ &+ \alpha_{11}{}'\frac{pr}{2}\sqrt{\sin\theta_0}\operatorname{cotan}\theta_0 + \frac{1}{2}\frac{pr^2}{E\delta}\left[2 - \mu - \left(\frac{R_2}{R_1}\right)_0\right]\frac{1}{\sin\theta_0}, \\ \vartheta_0{}'' &= \alpha_{12}{}'Q_0{}'' - \alpha_{22}{}'\frac{M_0{}''}{\sqrt{\sin\theta_0}} - \alpha_{21}{}'\frac{pr}{2}\operatorname{cotan}\theta_0. \end{aligned} \tag{59.6}$$

Since the edges of the shells are rigidly connected, they satisfy at the joint the equalities

$$\Delta_0{}' = \Delta_0{}'', \ \vartheta_0{}' = \vartheta_0{}'', \ Q_0{}' = Q_0{}'', \ M_0{}' = M_0{}'', \tag{59.7}$$

the form of which takes into consideration that the positive directions of the corresponding quantities are as shown in Fig. 81.

If the end cap were inverted with

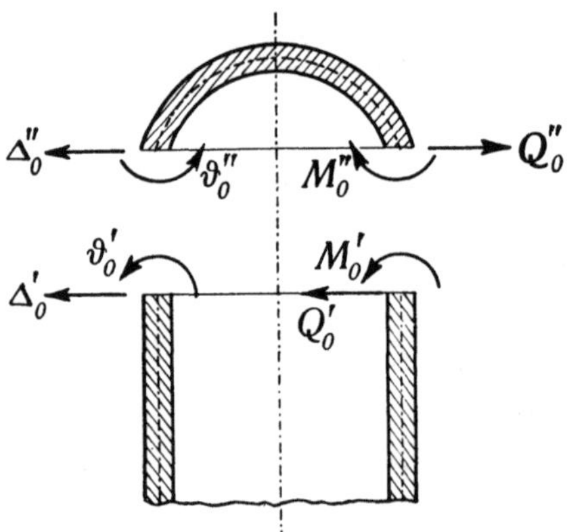

Fig. 81.

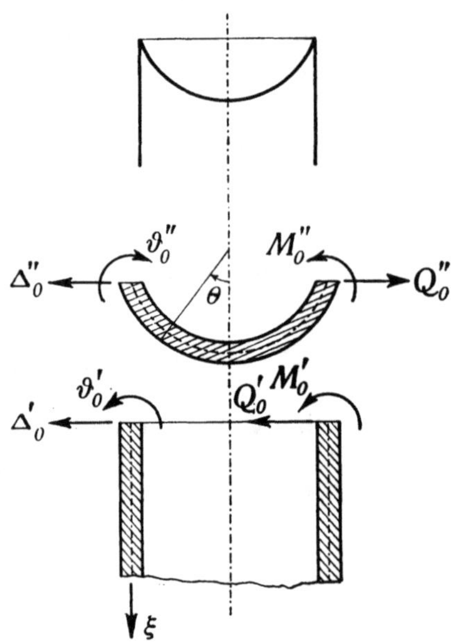

Fig. 82.

respect to the vessel (Fig. 82), then the conditions at the joint of the shells would be different, namely:

$$\Delta_0' = \Delta_0'', \quad \vartheta_0 = -\vartheta_0'', \qquad Q_0' = Q_0'', \quad M_0' = -M_0''. \tag{59.8}$$

With the help of the equalities (59.7), the formulae (59.5) and (59.6) transform into two equations in the two unknowns

$$Q_0 = Q_0' = Q_0'' \text{ and } M_0 = M_0' = M_0'',$$

and one of these equations will lack terms depending on Q_0, the other terms depending on M_0, so that one obtains directly the formulae

$$Q_0 = \frac{pr}{2} \frac{\sqrt{\sin\theta_0}}{1+\sqrt{\sin\theta_0}} \left\{ \cotan\theta_0 - \frac{1}{2\sqrt[4]{3(1-\mu^2)}} \sqrt{\frac{\delta}{r}} \times \right.$$
$$\left. \times \frac{1}{\sin^{3/2}\theta_0} \left[2 - \mu - \frac{R_2{}^0}{R_1{}^0} - (2-\mu)\sin\theta_0 \right] \right\}, \tag{59.9}$$
$$M_0 = -\frac{1}{4} pr \frac{\sqrt{r\delta}}{\sqrt[4]{3(1-\mu^2)}} \frac{\sqrt{\sin\theta_0}}{1+\sqrt{\sin\theta_0}} \cotan\theta_0.$$

Thus, it follows that in the general case when $\theta_0 \neq \frac{\pi}{2}$ the edge force Q_0 is a quantity of order pr and the edge moment a quantity of order $pr\sqrt{r\delta}$.

However, in the particular case $\theta_0 = \frac{\pi}{2}$, the formulae (59.9) give

$$Q_0 = \frac{p\sqrt{r\delta}}{8\sqrt[4]{3(1-\mu^2)}} \left(\frac{R_2}{R_1}\right)_0, \qquad M_0 = 0. \tag{59.10}$$

Thus, when the meridian of the end-panel goes smoothly over into the generator of the cylinder, the moment M_0 vanishes and the force Q_0 decreases by a factor $\sqrt{\frac{r}{\delta}}$ in comparison with the general case, when the meridian of the end-panel joins the cylinder at some angle. Hence one may conclude that the stresses in the pressure vessel will in the first case be essentially smaller than in the second and that, therefore, it is rational to ensure a smooth junction of the end cap and the cylinder with $\theta_0 = \frac{\pi}{2}$. This deduction which has been obtained under the assumption that the thickness of the end cap and the cylinder are the same, remains true also when the thicknesses are different.

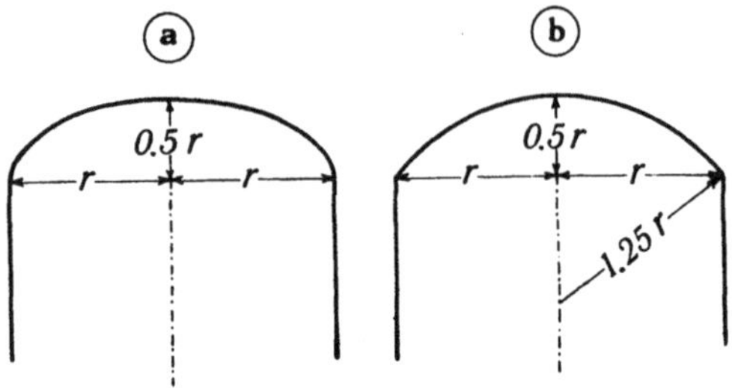

Fig. 83.

In order to gain not only qualitative, but also quantitative impressions of this analysis, two examples will now be considered. In the first of these, the end-panel is to be half an ellipsoid of revolution with a ratio of the semi-axes $\frac{b}{a} = \frac{b}{r} = 0.5$, and in the, second, a spherical segment the height of which is equal to the smaller semi-axis of the ellipsoid of the first example (Fig. 83).

In this case, the trigonometric functions of the edge angle of the spherical shell will be

$$\sin \theta_0 = 0.8, \quad \cos \theta_0 = 0.6,$$
$$\operatorname{cotan} \theta_0 = 0.75,$$

and the radius of the sphere

$$R = 1.25r.$$

Correspondingly, one finds for the radii of curvature of the ellipsoidal end-panel with the given ratio of the semi-axes the expressions

$$R_1 = \frac{2r}{(1 + 3 \sin^2 \theta)^{3/2}}, \quad R_2 = \frac{2r}{(1 + 3 \sin^2 \theta)^{1/2}}, \tag{59.11}$$

and at the edge of the shell

$$\frac{R_2{}^0}{R_1{}^0} = 4.$$

Taking these details into consideration and using the formulae (59.10) and (59.11), one finds the following values for the edge loading:

for the first example (ellipsoidal end-panel)

$$Q_0 = 0.389 p \sqrt{r\delta},$$
$$M_0 = 0;$$

for the second example (spherical end-panel)

$$Q_0 = 0.177pr\left(1 + 0.48\sqrt{\frac{\delta}{r}}\right) \approx 0.177pr,$$

$$M_0 = -\,0.069pr\sqrt{r\delta}.$$

The edge loadings being known, it is not difficult to obtain, on the basis of formulae (58.8), expressions for all the displacements, forces and moments, the most important of which will now be given:

First example

a) Cylindrical shell:

$$w = \frac{pr^2}{E\delta}\left(1 - \tfrac{1}{2}\mu - e^{-b\xi}\cos b\xi\right),$$

$$T_2 = pr\left(1 - e^{-b\xi}\cos b\xi\right),$$

$$M_1 = \frac{pr\delta}{\sqrt{12(1-\mu^2)}}\,e^{-b\xi}\sin b\xi,$$

where $\xi = x/r$ (x being the distance from the edge of the cylinder) and

$$b = \sqrt[4]{3(1-\mu^2)}\sqrt{\frac{r}{\delta}}.$$

b) Ellipsoidal end-panel:

$$\Delta x = \frac{pr^2}{E\delta}\,e^{-\beta}\cos\beta + \Delta^*,$$

$$T_2 = pre^{-\beta}\cos\beta + \frac{pR_2}{2}\left(2 - \frac{R_2}{R_1}\right),$$

$$M_1 = -\,\frac{pr\delta}{\sqrt{12(1-\mu^2)}}\,e^{-\beta}\sin\beta,$$

where, in agreement with the formulae (58.2), (59.11),

$$\beta = \sqrt[4]{12(1-\mu^2)}\sqrt{\frac{r}{\delta}}\int_{\theta}^{\pi/2}\frac{d\theta}{(1+3\sin^2\theta)^{5/4}} = \sqrt{2}.b\int_{\theta}^{\pi/2}\frac{d\theta}{(1+3\sin^2\theta)^{5/4}}.$$

Second example

a) Cylindrical shell:

$$w = \frac{pr^2}{E\delta}\left\{0.85 - 0.322\sqrt{\frac{r}{\delta}}\, e^{-b\xi}\cos\left(b\xi - \frac{\pi}{4}\right)\right\},$$

$$T_2 = pr\left\{1 - 0.322\sqrt{\frac{r}{\delta}}\, e^{-b\xi}\cos\left(b\xi - \frac{\pi}{4}\right)\right\},$$

$$M_1 = -\,0.0975 pr\sqrt{r\delta}\, e^{-b\xi}\cos\left(b\xi + \frac{\pi}{4}\right);$$

b) Spherical end-panel:

$$w = \frac{pr^2}{E\delta}\left\{0.547 - 0.402\sqrt{\frac{r}{\delta}}\, e^{-\beta}\cos\left(\beta - \frac{\pi}{4}\right)\right\},$$

$$T_2 = pr\left\{0.625 - 0.322\sqrt{\frac{r}{\delta}}\, e^{-\beta}\cos\left(\beta - \frac{\pi}{4}\right)\right\},$$

$$M_1 = -\,0.975 pr\sqrt{r\delta}\, e^{-\beta}\cos\left(\beta + \frac{\pi}{4}\right),$$

where $\beta = \sqrt{1.25}b\alpha$ and α is the angle measured from the edge of the spherical shell.

Figs. 84 and 85 show the distributions of the normal deflection w, the hoop stress

$$\sigma_2 = \frac{T_2}{\delta},$$

the bending stress

$$\sigma_M = \frac{6M_1}{\delta^2}$$

and of the sum of the meridional normal stress in the extreme fibre of the shell $\sigma = \sigma_M + \dfrac{T_1}{\delta}$ evaluated on the basis of the above derived results with $r/\delta = 100$.

It will be discovered from a study of these curves that in the first example (the ellipsoidal end-panel) the maximum resultant stress in the pressure vessel exceeds only by 10% the stress

$$\sigma^* = \frac{pr}{\delta}$$

(i.e., the largest stress in the structure in agreement with the membrane theory). Thus, although the pressure vessel with the ellipsoidal end-panels is not in a membrane state of stress, the stresses in it differ only insignificantly from those given by that theory.

In contrast, in the second example (pressure vessel with spherical head), the hoop, bending and resultant meridional stresses significantly exceed the membrane stress $\sigma = \frac{pr}{\delta}$ (being one and a half, four and fives times as large respectively). Hence the advantages which attach to the first mode of construction in comparison with the second, are completely clear and it becomes also quite obvious why end-panels of spherical form usually have to be reinforced in the regions where they are joined to the cylindrical parts of a pressure vessel. In such cases one may use reinforcements (for example, rounded angles) and in those cases where these are not sufficient, ribs may be attached by means of rivets or weld joints to the spherical and cylindrical shells in the neighbourhood of their edges. The theory of the analysis of these reinforcements may be found in the well-known book by P. F. Papkovich [107].

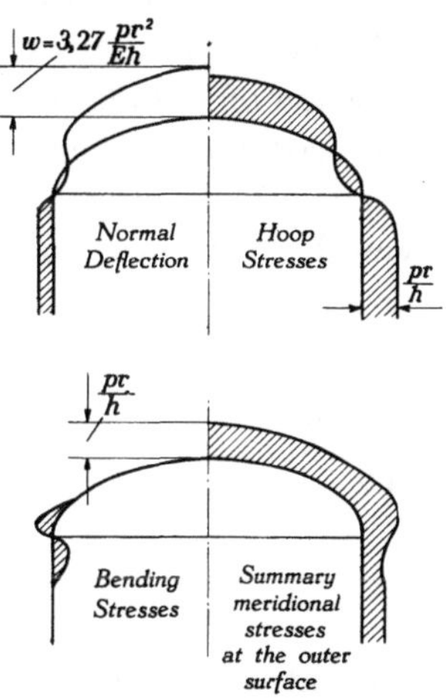

Fig. 84.

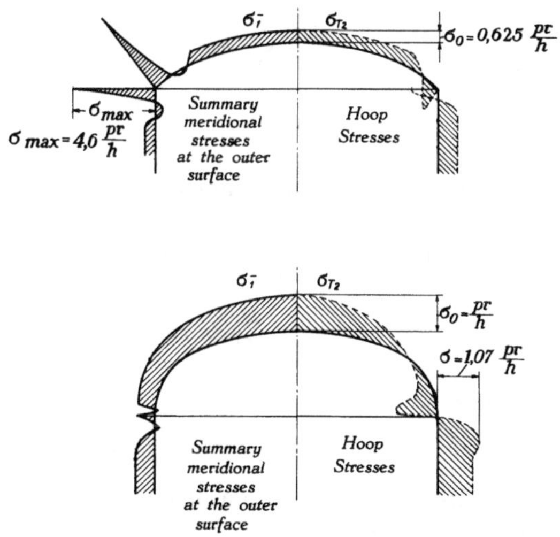

Fig. 85.

In conclusion, it will be noted that the above difference between end-panels which are joined smoothly to the cylinder and end-panels which are joined to the cylinder at an angle diminishes with increasing shell thickness. In particular, if in the above examples one were to take

$r/\delta = 25$ instead of $r/\delta = 100$, then the bending stresses in the spherical cap would not be four times, but only twice as large as the membrane stresses in the cylindrical shell.

§ 60. Symmetrical deformation of toroidal shells

The sequence of calculations studied in the last section for symmetrically deformed shells of revolution was based on the assumptions that the particular integral may be taken from the membrane theory and that the solution of the homogeneous problem may be determined in a first approximation by the given asymptotic method. This approach to the analysis applies to a majority of problems of the type which are encountered in practice. However, it is possible to indicate certain problems to which this method does not apply and which require use of specially adapted methods. As an example, the theory of symmetrical deformation of toroidal shells will be considered (since they have a wide range of applications in the manufacture of pressure vessels, turbines and other machinery).

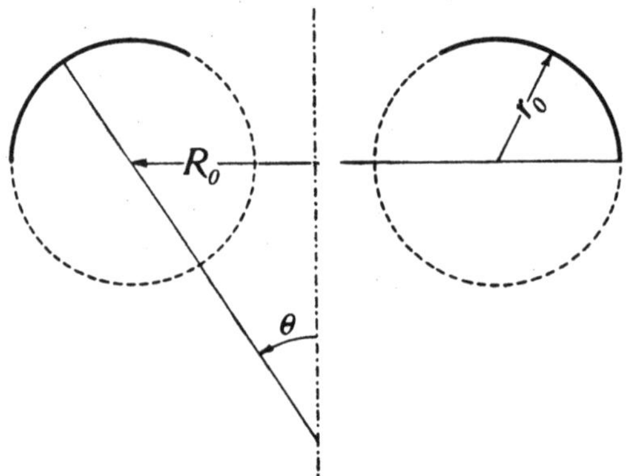

Fig. 86.

For these shells (see Fig. 86)

$$R_1 = r_0, \ R_2 = R_0 \frac{1 + \alpha \sin\theta}{\sin\theta}, \tag{60.1}$$

where

$$\alpha = r_0/R_0.$$

Substituting these values for the radii of curvature into the equation (54.6), one finds

$$\frac{1}{1 + \alpha \sin\theta} \frac{d}{d\theta}\left[\frac{(1 + \alpha\sin\theta)^2}{\sin\theta} \frac{d\tilde{T}}{d\theta}\right] + i2d^2\tilde{T} = 2id^2(T_1{}^* + T_2{}^*), \tag{60.2}$$

where

$$2d^2 = \sqrt{12(1 - \mu^2)} \frac{r_0{}^2}{R_0\delta}. \tag{60.3}$$

It will now be assumed that the torus is loaded by uniform normal pressure p. Then (cf. § 27)

$$T_1^* = \frac{pr_0}{2}\,\frac{2+\alpha\sin\theta}{1+\alpha\sin\theta} - \frac{\alpha A}{(1+\alpha\sin\theta)\sin\theta},$$
$$T_2^* = \frac{pr_0}{2} + \frac{A}{\sin^2\theta}, \tag{60.4}$$

where A is an integration constant which is determined from static considerations, since on one of the parallel circles of the toroidal shell the axial force Q_z is known (§ 54). Substituting (60.4) in (60.2), one obtains

$$\frac{1}{1+\alpha\sin\theta}\,\frac{d}{d\theta}\left[\frac{(1+\alpha\sin\theta)^2}{\sin\theta}\,\frac{d\tilde{T}}{d\theta}\right] + i2d^2\tilde{T} = i2d^2\left[\frac{pr_0}{2}\,\frac{3+2\alpha\sin\theta}{1+\alpha\sin\theta} + \right.$$
$$\left. + \frac{A}{(1+\alpha\sin\theta)\sin^2\theta}\right]. \tag{60.5}$$

Differentiating this equation with respect to θ and introducing the new variable

$$\tilde{Z} = \frac{(1+\alpha\sin\theta)^2}{\sin\theta}\,\frac{d\tilde{T}}{d\theta}, \tag{60.6}$$

one finds

$$(1+\alpha\sin\theta)^2\frac{d}{d\theta}\left[\frac{1}{(1+\alpha\sin\theta)}\,\frac{d\tilde{Z}}{d\theta}\right] + i2d^2\sin\theta\tilde{Z} =$$
$$= -\,i2d^2\cos\theta\left[\alpha\frac{pr_0}{2} + A\,\frac{2+3\alpha\sin\theta}{\sin^3\theta}\right]. \tag{60.7}$$

Compared with (60.5) this equation has the advantage that its coefficients are finite for all values of the variable θ. However, the right hand side of the equation becomes infinite for $\theta = 0$ and $\theta = \pm\,\pi$. In order to remove these singularities, introduce into (60.7) the new variable

$$\tilde{Z} = V - i2d^2A\,\mathrm{cotan}\,\theta, \tag{60.8}$$

after which one arrives at the following differential equation for the auxiliary function V:

$$(1+\alpha\sin\theta)\,\frac{d^2V}{d\theta^2} - \alpha\cos\theta\frac{dV}{d\theta} + i2d^2\sin\theta V = 2d^2\tilde{D}\cos\theta, \tag{60.9}$$

where

$$\tilde{D} = -\left(2d^2A + i\alpha\frac{pr_0}{2}\right). \tag{60.10}$$

The analysis of toroidal shells has thus been reduced to the solution of the differential equation (60.9), since after determination of the auxiliary function V one may find by use of the results of § 54 all the quantities characterizing the deformation of the shell. Omitting the intermediate elementary manipulations, the final results will be given here:

$$\begin{aligned}
T_1 &= -\frac{\alpha\cos\theta}{(1+\alpha\sin\theta)}\,\frac{1}{2d^2} I_m(V) + \frac{pr_0}{2}\,\frac{2+\alpha\sin\theta}{1+\alpha\sin\theta} - \alpha A\,\frac{\alpha+\sin\theta}{(1+\alpha\sin\theta)^2},\\
T_2 &= -\frac{1}{2d^2} I_m\left[\frac{d}{d\theta}\left(\frac{1}{1+\alpha\sin\theta}V\right)\right] + \frac{pr_0}{2} + \alpha A\,\frac{\alpha+\sin\theta}{(1+\alpha\sin\theta)^2},\\
M_1 &= -\frac{c}{2d^2}\left\{\mu\frac{\alpha\cos\theta}{(1+\alpha\sin\theta)^2} R_e(V) + R_e\left[\frac{d}{d\theta}\left(\frac{1}{1+\alpha\sin\theta}V\right)\right]\right\}, \qquad (60.11)\\
N_1 &= -\frac{c}{r_0}\left\{\frac{\sin\theta}{(1+\alpha\sin\theta)^2} I_m(V) + \frac{2d^2A\cos\theta}{(1+\alpha\sin\theta)^2}\right\},\\
\vartheta &= -\frac{1}{E\delta}\,\frac{1}{\alpha(1+\alpha\sin\theta)} R_e(V);
\end{aligned}$$

$$\begin{aligned}
\Delta_x &= \frac{R_0}{E\delta}(1+\alpha\sin\theta)(T_2-\mu T_1),\\
\Delta_z &= -r_0\int_0^\theta \vartheta\cos\theta\, d\theta + C.
\end{aligned} \qquad (60.12)$$

A study of the equation (60.9) readily reveals that the ordinary methods of approximate solution of the equations for symmetrical deformations of shells of revolution do not apply. For example, an attempt to substitute as its particular integral a solution of the membrane theory leads to the formula

$$V_{\text{part.}} \approx -i\tilde{D}\cot\theta, \qquad (60.13)$$

according to which the particular integral tends to infinity at the points $\theta = 0$, $\theta = \pm\pi$. Between these limits, a periodic particular integral of the equation (60.9) must obviously be finite, since (for all values of θ) the coefficients of the equation as well as its right hand side are finite.

Further, it is seen from the equation (60.9) that at the points $\theta = 0$, $\theta = \pm\pi$ the coefficient of the last term on the left hand side of the equation vanishes, whereas the method of asymptotic integration is based on the assumption that this coefficient greatly exceeds unity (§ 55).

It is clear from these observations that, in fact, the ordinary methods of approximate solutions do not apply to toroidal shells (at least not near the points $\theta = 0$ and $\theta = \pm\pi$), so that the integration of the equation (60.9) necessitates a special study which will be presented next.

I. *The homogeneous problem*

Consider the homogeneous equation

$$(1 + \alpha \sin \theta)^2 \frac{d}{d\theta}\left[\frac{1}{1 + \alpha \sin \theta} \frac{dV}{d\theta}\right] + i2d^2 \sin \theta V = 0 \qquad (60.14)$$

and let it be required to determine in a sufficiently convenient manner its integrals in the range $0 \leqslant \theta \leqslant \pi/2$. The equation (60.14) may be written in the form

$$\frac{d^2V}{d\chi^2} + F^2V = 0, \qquad (60.15)$$

where

$$d\chi = (1 + \alpha \sin \theta)\, d\theta, \quad F^2 = 2id^2 \frac{\sin \theta}{(1 + \alpha \sin \theta)^3}. \qquad (60.16)$$

Next apply to (60.15) the transformation studied in § 55, i.e., introduce the new dependent variable

$$W = V\sqrt{F} \qquad (60.17)$$

and the new independent variable

$$dx = F\, d\chi = F(1 + \alpha \sin \theta)\, d\theta = \sqrt{2i}\, d \frac{\sqrt{\sin \theta}}{\sqrt{1 + \alpha \sin \theta}}\, d\theta. \qquad (60.18)$$

Then (60.15) takes the form

$$\frac{d^2W}{dx^2} + (1 + \Delta)W = 0, \qquad (60.19)$$

where in the given particular case

$$\Delta = -\frac{1}{2F}\frac{d^2F}{dx^2} + \frac{1}{4F^2}\left(\frac{dF}{dx}\right)^2 = -\frac{1}{2id^2}\left[-\frac{5}{16 \sin^3 \theta} + \frac{1 + 7\alpha \sin \theta}{16 \sin \theta} + \right.$$

$$\left. + \frac{\alpha}{16 \sin^2 \theta} - \frac{4\alpha \cos^2 \theta}{16 \sin^2 \theta(1 + \alpha \sin \theta)} + \frac{11}{16}\, \frac{\alpha^2 \cos^2 \theta}{\sin \theta(1 + \alpha \sin \theta)}\right]. \qquad (60.20)$$

It is seen from the formula (60.20) that $\Delta \to \infty$ as $\theta \to 0$ and therefore it is impossible to neglect in the equation (60.19) the quantity Δ, as this was done in § 55. However, it may now be concluded that the expression (60.20) may be replaced $\left(\text{in the interval } 0 \leqslant \theta \leqslant \frac{\pi}{2}\right)$ by the simplified expression

$$\Delta \approx \frac{1}{2id^2} \frac{5}{16} \frac{1}{\theta^3}. \tag{60.21}$$

At a first glance, such a simplification seems to be unjustified, since, if the formula (60.21) gives in fact a good approximation to Δ near $\theta = 0$, then it will not even give a rough approximation to Δ near the other end of the interval $\left(\theta = \frac{\pi}{2}\right)$. However, taking into consideration that $\frac{32}{5}d^2 \gg 1$, it becomes clear that only for $\theta \ll 1$ will the quantity Δ be large or have an absolute value comparable with unity $\left(\text{near } \theta = \frac{\pi}{2}, \text{ however, the quantity } |\Delta| \text{ will be much smaller than unity}\right)$. Hence it follows that in determining Δ in the equation (60.19) from the formula (60.21), one does not incur a substantial error, since in that region where the value of Δ is large the formula (60.21) is a good approximation, while in the region where this formula deviates greatly, the quantity Δ is generally negligible.

Next the quantity Δ must be expressed in terms of x. On the basis of (60.18), one has for x

$$x = \sqrt{2i}\xi = \sqrt{2i}d \int_0^\theta \frac{\sqrt{\sin\theta}}{\sqrt{1 + \alpha\sin\theta}}\, d\theta \tag{60.22}$$

or

$$\begin{aligned} \xi = d \int_0^\theta \frac{\sqrt{\sin\theta}}{\sqrt{1+\alpha\sin\theta}}\, d\theta = \frac{2}{3} d\theta^{3/2} \Big\{ & 1 - \frac{1}{28}\theta^2 + \frac{1}{5280}\theta^4 - \dots \\ & - \frac{3}{10}\alpha\theta \left[1 - \frac{5}{36}\theta^2 + \frac{11}{1170}\theta^4 - \dots\right] + \\ & + \frac{9}{56}\alpha^2\theta^2 \left[1 - \frac{35}{132}\theta^2 + \frac{91}{2700}\theta^4 - \dots\right] + \dots \Big\}. \end{aligned} \tag{60.23}$$

The formulae (60.22) and (60.23) may be used for the practical evaluation of the variable x. It follows from them that for $\theta \to 0$

$$x \approx \sqrt{2i}\tfrac{2}{3}d\theta^{3/2}. \tag{60.24}$$

But, as has been stated above, the evaluation of Δ introduces substantial errors into the equation (60.19) only for small θ, i.e., in the region where it is admissible to employ the formula (60.24). Taking this into consideration and substituting (60.24) into (60.21) and (60.21) into (60.19), one finds

$$\frac{d^2W}{dx^2} + \left(1 + \frac{5}{36}\frac{1}{x^2}\right)W = 0. \tag{60.25}$$

This last equation belongs to the group of equations which can be integrated by use of Bessel functions. In fact, its general solution may be written in the form

$$W = \tilde{C}_1 h_{1/3}{}^{(1)}(x) + \tilde{C}_2 h_{1/3}{}^{(2)}(x), \tag{60.26}$$

where $\tilde{C}_1$, $\tilde{C}_2$ are two complex integration constants and

$$\begin{aligned} h_{1/3}{}^{(1)}(x) &= i^{5/6}\sqrt{0.5\pi x}\,H_{1/3}{}^{(1)}(x), \\ h_{1/3}{}^{(2)}(x) &= i^{-5/6}\sqrt{0.5\pi x}\,H_{1/3}{}^{(2)}(x), \end{aligned} \tag{60.27}$$

$H_{1/3}{}^{(1)}(x)$ and $H_{1/3}{}^{(2)}(x)$ being Hankel-functions of the first and second kinds (of order $\frac{1}{3}$). The constants in the formulae (60.27) have been chosen to suit the convenience of the following manipulations. The reduction of the problem to Bessel functions has in this case a definite advantage, since it permits the use of the well-known series expansions of these functions (for small arguments as well as asymptotic expansions). The employment of Hankel functions here is likewise not accidental, since then one of the two solutions in (60.26) has an absolute value which decreases exponentially, while that of the other increases exponentially as ξ increases. In this way it becomes possible (for different problems of the analysis of tori) to satisfy various boundary conditions (§ 57).

Changing from W to V by use of (60.16), (60.17), one finds

$$V = \left(\frac{\eta}{\sin\theta}\right)^{1/4}(1 + \alpha\sin\theta)^{3/4}[\tilde{C}_1 V_1(\sqrt{2i}\xi) + \tilde{C}_2 V_2(\sqrt{2i}\xi)], \tag{60.28}$$

where the following notation has been introduced:

$$\begin{aligned} V_1(\sqrt{2i}\xi) &= i^{5/6}\sqrt{\frac{\pi}{2}}\,(\sqrt{2i}\xi)^{1/3}H_{1/3}^{(1)}(\sqrt{2i}\xi), \\ V_2(\sqrt{2i}\xi) &= i^{-5/6}\sqrt{\frac{\pi}{2}}\,(\sqrt{2i}\xi)^{1/3}H_{1/3}^{(2)}(\sqrt{2i}\xi), \end{aligned} \tag{60.29}$$

$$\eta^{3/2} = \frac{3}{2d}\,\xi = \theta^{3/2}\left\{1 - \frac{1}{28}\,\theta^2 + \ldots\right\}. \tag{60.30}$$

As is seen from (60.30) and (60.23), the ratio $\left(\dfrac{\eta}{\sin\theta}\right)^{1/4}$ varies little from unity in the interval $-\dfrac{\pi}{2} \leqslant \theta \leqslant \dfrac{\pi}{2}$, the variation being not more than a few percent. On this basis, in what follows, will the function V be determined by use of the formula

$$V = (1 + \alpha \sin\theta)^{3/4}[\tilde{C}_1 V_1 + \tilde{C}_2 V_2]. \tag{60.31}$$

For the functions V_1 and V_2, one may give the series

$$\begin{aligned} V_1(\sqrt{2i}\xi) &= i\sqrt{\frac{\pi}{2}}\,\{1.293e^{\frac{\pi}{4}i}\,\xi^{2/3}[f_1(\xi) - if_2(\xi)] - \\ &\qquad - 1.079e^{\frac{5\pi}{12}i}\,[g_1(\xi) - ig_2(\xi)]\}, \\ V_2(\sqrt{2i}\xi) &= -i\sqrt{\frac{\pi}{2}}\,\{1.293e^{\frac{\pi}{12}i}\,\xi^{2/3}[f_1(\xi) - if_2(\xi)] - \\ &\qquad - 1.079e^{-\frac{5\pi}{12}i}\,[g_1(\xi) - ig_2(\xi)]\}, \end{aligned} \tag{60.32}$$

in which

$$\begin{aligned} f_1(\xi) - if_2(\xi) &= (\tfrac{1}{3})!\sum_{k=0}^{\infty}\frac{1}{k!}\,\frac{1}{(k+\frac{1}{3})!}\,(-\tfrac{1}{2}i\xi^2)^k = \\ &= 1 - \frac{i\xi^2}{1!2(1+\frac{1}{3})} - \frac{\xi^4}{2!2^2(1+\frac{1}{3})(2+\frac{1}{3})} + \\ &\quad + \frac{i\xi^6}{3!2^3(1+\frac{1}{3})(2+\frac{1}{3})(3+\frac{1}{3})} - \ldots, \\ g_1(\xi) - ig_2(\xi) &= (-\tfrac{1}{3})!\sum_{k=0}^{\infty}\frac{1}{k!}\,\frac{1}{(k-\frac{1}{3})!}\,(-\tfrac{1}{2}i\xi^2)^k = \\ &= 1 - \frac{i\xi^2}{1!2(1-\frac{1}{3})} - \frac{\xi^4}{2!2^2(1-\frac{1}{3})(2-\frac{1}{3})} + \\ &\quad + \frac{i\xi^6}{3!2^3(1-\frac{1}{3})(2-\frac{1}{3})(3-\frac{1}{3})} + \ldots. \end{aligned} \tag{60.33}$$

The series (60.33) converge quickly only for relatively small values of ξ ($\xi \ll 1$), while for sufficiently large ξ they are inconvenient. Therefore, for the evaluation of V_1 and V_2 in the region of large ξ ($\xi > 1$), one has to use the asymptotic expansions for the Bessel functions which (retaining in them only the first terms) lead to the following formulae:

$$V_1(\sqrt{2i}\xi) = \frac{1}{2^{1/12}\xi^{1/6}} e^{-\xi}\left[\cos\left(\xi - \frac{\pi}{24}\right) + i\sin\left(\xi - \frac{\pi}{24}\right)\right],$$
$$V_2(\sqrt{2i}\xi) = \frac{1}{2^{1/12}\xi^{1/6}} e^{\xi}\left[\cos\left(\xi + \frac{\pi}{24}\right) - i\sin\left(\xi + \frac{\pi}{24}\right)\right]. \tag{60.34}$$

Next, the obtained solutions will be extended into the region of negative θ $\left(-\frac{\pi}{2} \leqslant \theta < 0\right)$. For this purpose attention will first be directed to the fact that the equation (60.14) on replacing θ by $-\psi$ takes the form

$$(1 - \alpha \sin\psi)^2 \frac{d}{d\psi}\left[\frac{1}{1 - \alpha\sin\psi}\frac{dV}{d\psi}\right] - i2d^2 \sin\psi V = 0. \tag{60.35}$$

Comparing (60.35) with (60.14), it follows immediately that the integrals of the first equation may be taken to be conjugate complex with respect to the integrals of the second equation, provided the independent variable ξ is replaced by

$$\zeta = d\int_0^{\psi} \frac{\sqrt{\sin\psi}}{\sqrt{1 - \alpha\sin\psi}}\, d\psi. \tag{60.36}$$

(The last result is clear from the fact that the equations (60.35) and (60.14) differ in the signs with the imaginary terms and likewise in the signs with α).

Thus the integrals of the equation (60.14) in the region $-\frac{\pi}{2} \leqslant \theta \leqslant 0$ may be determined from the expressions

$$\overline{V}_1(\sqrt{2i}\zeta) = -i\sqrt{\frac{\pi}{2}}\{1.293^{-\frac{\pi}{4}i}\zeta^{2/3}[f_1(\zeta) + if_2(\zeta)] -$$
$$- 1.079e^{-\frac{5\pi}{12}i}[g_1(\zeta) + ig_2(\zeta)]\}, \tag{60.37}$$
$$\overline{V}_2(\sqrt{2i}\zeta) = i\sqrt{\frac{\pi}{2}}\{1.293e^{-\frac{\pi}{12}i}\zeta^{2/3}[f_1(\zeta) + if_2(\zeta)] -$$
$$- 1.079e^{\frac{5\pi}{12}i}[g_1(\zeta) + ig_2(\zeta)]\},$$

with the asymptotic expansions

$$\begin{aligned} \bar{V}_1 &= \frac{1}{2^{1/12}\zeta^{1/6}} e^{-\zeta}\left[\cos\left(\zeta - \frac{\pi}{24}\right) - i\sin\left(\zeta - \frac{\pi}{24}\right)\right], \\ \bar{V}_2 &= \frac{1}{2^{1/12}\zeta^{1/6}} e^{\zeta}\left[\cos\left(\zeta + \frac{\pi}{24}\right) + i\sin\left(\zeta + \frac{\pi}{24}\right)\right]. \end{aligned} \tag{60.38}$$

However, it is essential to note that the integrals $\bar{V}_1(\sqrt{2i}\zeta)$, $\bar{V}_2(\sqrt{2i}\zeta)$ are not the analytic continuations of the integrals $V_1(\sqrt{2i}\xi)$, $V_2(\sqrt{2i}\xi)$ in the region of negative θ. These are given by other functions, namely:

$$V_1' = -i\sqrt{\frac{\pi}{2}}\{1.293e^{+\frac{\pi}{4}i}\zeta^{2/3}[f_1(\zeta) + if_2(\zeta)] + \\ + 1.079e^{\frac{5}{12}\pi i}[g_1(\zeta) + ig_2(\zeta)]\}, \tag{60.39}$$

$$V_2' = i\sqrt{\frac{\pi}{2}}\{1.293^{\frac{\pi}{12}i}\zeta^{2/3}[f_1(\zeta) + if_2(\zeta)] + 1.079e^{-\frac{5}{12}\pi i}[g_1(\zeta) + ig_2(\zeta)]\},$$

which are linear combinations of $\bar{V}_1$ and $\bar{V}_2$, as a consequence of which one may write down an expression for the auxiliary function V in the entire region $-\frac{\pi}{2} \leqslant \theta \leqslant \frac{\pi}{2}$ in the form

$$V = (1 + \alpha \sin\theta)^{3/4}[\tilde{C}_1 V_1(\theta) + \tilde{C}_2 V_2(\theta)], \tag{60.40}$$

where instead of V_1 and V_2 one has to substitute

$$V_1(\sqrt{2i}\xi),\quad V_2(\sqrt{2i}\xi) \quad \text{in the region } \theta > 0,$$

$$V_1'(\sqrt{2i}\zeta),\quad V_2'(\sqrt{2i}\zeta) \quad \text{in the region } \theta < 0.$$

The functions V_1 and V_2 which have been defined in this manner will be continuous (together with all their derivatives) at the point $\theta = 0$.

II. *The particular integral*

Since the right hand side of the equation (60.9) is a periodic function of θ (with period 2π), the particular integral of this equation may be sought in the form of the trigonometric series

$$\mathfrak{B} = a_0 + a_1\cos\theta + a_2\cos 2\theta + \ldots + a_n\cos n\theta + \ldots \\ + b_1\sin\theta + b_2\sin 2\theta + \ldots + b_n\sin n\theta + \ldots, \tag{60.41}$$

Substituting this series into (60.9), one obtains the following set of recurrence relations between the coefficients:

$$\begin{aligned} &- n^2 a_n + \tfrac{1}{2}[2id^2 - \alpha(n+1)(n+2)]b_{n+1} - \\ &\qquad - \tfrac{1}{2}[2id^2 - \alpha(n-1)(n-2)]b_{n-1} = \beta_n, \\ &- n^2 b_n + \tfrac{1}{2}[2id^2 - \alpha(n-1)(n-2)]a_{n-1} - \\ &\qquad - \tfrac{1}{2}[2id^2 - \alpha(n+1)(n+2)]a_{n+2} = 0, \end{aligned} \tag{60.42}$$

where

$$\beta_n = 2d^2\tilde{D} \text{ for } n = 1,$$
$$\beta_n = 0 \qquad \text{for } n > 1 \text{ and } n = 0.$$

These relations may be written in the following expanded form:

$$\begin{aligned} &- a_1 + \tfrac{1}{2}(2id^2 - 2.3\alpha)b_2 = 2d^2\tilde{D}, \\ &- b_1 + id^2 a_0 - \tfrac{1}{2}(i2d^2 - 2.3\alpha)a_2 = 0, \\ &- 4a_2 + \tfrac{1}{2}(i2d^2 - 3.4\alpha)b_3 - \tfrac{1}{2}i2d^2 b_1 = 0, \\ &- 4b_2 + \tfrac{1}{2}i2d^2 a_1 - \tfrac{1}{2}(i2d^2 - 3.4\alpha)a_3 = 0, \\ &- 9a_3 + \tfrac{1}{2}(i2d^2 - 4.5\alpha)b_4 - \tfrac{1}{2}(i2d^2 - 2.1.\alpha)b_2 = 0, \\ &- 9b_3 + \tfrac{1}{2}(i2d^2 - 2.1\alpha)a_2 - \tfrac{1}{2}(i2d^2 - 4.5\alpha)a_4 = 0, \\ &\ldots\ldots\ldots\ldots\ldots\ldots\ldots\ldots\ldots \\ &\ldots\ldots\ldots\ldots\ldots\ldots\ldots\ldots\ldots \end{aligned} \tag{60.43}$$

A study of (60.43) shows that one is essentially dealing with two independent infinite algebraic systems one of which involves the a_i with odd and the b_i with even indices while the other contains the remaining a_i, b_i. In addition, the second system will be homogeneous, and therefore (since only a particular integral is required) all the unknowns in it may be taken equal to zero:

$$a_{2n} = b_{2n+1} = 0, \quad n = 0, 1, 2, \ldots$$

The first system, however, may be reduced to the form

$$\begin{aligned} &1^2 \frac{ia_1}{d^2} + \left[1 + i\frac{2.3}{2d^2}\alpha\right] b_2 = -2i\tilde{D}, \\ &2^2 \frac{ib_2}{d^2} + a_1 - \left(1 + i\frac{3.4}{2d^2}\alpha\right) a_3 = 0, \\ &3^2 \frac{ia_3}{d^2} + \left(1 + i\frac{4.5}{2d^2}\alpha\right) b_4 - \left(1 + i\frac{1.2}{2d^2}\alpha\right) b_2 = 0, \end{aligned} \tag{60.44}$$

$$
\begin{aligned}
&4^2 i\,\frac{b_4}{d^2}+\left(1+i\,\frac{3.2}{2d^2}\,\alpha\right)a_3-\left(1+i\,\frac{5.4}{2d^2}\,\alpha\right)a_5=0,\\
&5^2 i\,\frac{a_5}{d^2}+\left(1+i\,\frac{6.7}{2d^2}\,\alpha\right)b_6-\left(1+i\,\frac{4.3}{2d^2}\,\alpha\right)b_4=0,\\
&\dots\dots\dots\dots\dots\dots\dots\dots\dots\dots
\end{aligned}
\tag{60.44}
$$

The solution of this last system may be expressed in the form of an infinite continued fraction which is obtained, if the first equation (60.44) is divided by ia_1, the second by ib_2, etc., and if successively the unknowns (the ratios a_i and b_i) are eliminated. As a result of this operation, one obtains the formula

$$
-\frac{2\tilde{D}}{a_1}=\frac{1}{d^2}+\cfrac{1+i\,\frac{2.3}{2d^2}\,\alpha}{\frac{2^2}{d^2}+\cfrac{\left(1+i\,\frac{3.4}{2d^2}\,\alpha\right)\left(1+i\,\frac{1.2}{2d^2}\,\alpha\right)}{\frac{3^2}{d^2}+\cfrac{\left(1+i\,\frac{4.5}{2d^2}\,\alpha\right)\left(1+i\,\frac{2.3}{2d^2}\,\alpha\right)}{\frac{4^2}{d^2}+\cfrac{\left(1+i\,\frac{5.6}{2d^2}\,\alpha\right)\left(1+i\,\frac{3.4}{2d^2}\,\alpha\right)}{\frac{5^2}{d^2}+\dots,}}}}
\tag{60.45}
$$

for the direct evaluation of a_1.

All the other coefficients in the trigonometric series (60.41) may also be determined in the form of continued fractions in terms of the coefficients preceding them. This last result follows clearly from the fact that, if one omits in the system (60.43) the first few equations and then assumes that the coefficient with the smallest index in the residual system is a known quantity, all the above reasoning will again apply to this last system.

Hence one arrives at the following formulae:

$$
\frac{a_n}{b_{n-1}}\,i=\cfrac{1+i\,\frac{(n-1)(n-2)}{2d^2}\,\alpha}{\frac{n^2}{d^2}+\cfrac{\left[1+i\,\frac{n(n-1)}{2d^2}\,\alpha\right]\left[1+i\,\frac{(n+1)(n+2)}{2d^2}\,\alpha\right]}{\frac{(n+1)^2}{d^2}+\cfrac{\left[1+i\,\frac{(n+1)n}{2d^2}\,\alpha\right]\left[1+i\,\frac{(n+2)(n+3)}{2d^2}\,\alpha\right]}{\frac{(n+2)^2}{d^2}+\dots}}},
\tag{60.46}
$$

$$-\frac{b_{n+1}}{a_n}i = \cfrac{1+i\frac{n(n-1)}{2d^2}\alpha}{\cfrac{(n+1)^2}{d^2}+\cfrac{\left[1+i\frac{(n+1)n}{2d^2}\alpha\right]\left[1+i\frac{(n+2)(n+3)}{2d^2}\alpha\right]}{\cfrac{(n+2)^2}{d^2}+\cfrac{\left[1+i\frac{(n+2)(n+1)}{2d^2}\alpha\right]\left[1+i\frac{(n+3)(n+4)}{2d^2}\alpha\right]}{\cfrac{(n+3)^2}{d^2}+\dots}}},$$

$$n = 1, 3, 5 \dots$$

The expression (60.45) is included in the first of the formulae (60.46) and is obtained from it for $n = 1$ and $b_0 = -i2\tilde{D}$.

It follows from the above that a periodic particular integral of the equation (60.9) is determined by the series

$$\mathfrak{B} = a_1 \cos\theta + b_2 \sin 2\theta + a_3 \cos 3\theta + b_4 \sin 4\theta + \dots, \qquad (60.47)$$

the coefficients of which can be evaluated from the formulae (60.45) (60.46). It may be shown that this periodic solution corresponds to the

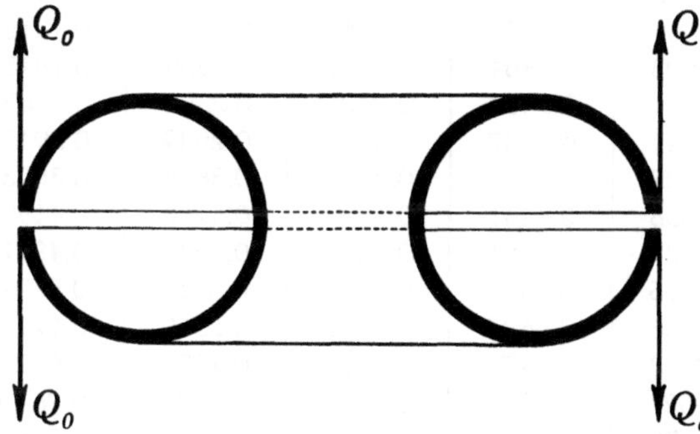

Fig. 87.

problem shown in Fig. 87, i.e., to the case of the complete toroidal shell, loaded by uniform normal pressure, cut along its parallel $\theta = \frac{\pi}{2}$ $\left(\text{or } \theta = -\frac{\pi}{2}\right)$ and subjected along this edge to forces Q_z^0. In addition, it is assumed that on the edges of the shell $\left(\text{i.e., for } \theta = \frac{\pi}{2}, \theta = -\frac{3}{2}\pi\right)$ the angle of rotation and shearing forces vanish. Actually, substituting (60.47) in the expression (60.11), it may be verified that the above particular integral satisfies all the formulated boundary conditions.

It must be emphasized that, while the solution under consideration is periodic for the forces, the moments and the angle of rotation, it will not be periodic for the displacements. Namely, the displacement Δ_z will not be single-valued for $\theta = \frac{\pi}{2}$ and $\theta = -3\frac{\pi}{2}$ which is already clear from Fig. 87 and which may be confirmed by computations, after substituting (60.47) into the third of the formulae (60.12).

Table 23

$\int_0^\theta \sqrt{\frac{\sin\theta}{1+\alpha\sin\theta}}\,d\theta$						
$\theta°$	$\alpha = 0$	$\alpha = 0.1$	$\alpha = 0.2$	$\alpha = 0.3$	$\alpha = 0.4$	$\alpha = 0.5$
0	0	0	0	0	0	0
5	0.0172	0.0171	0.0171	0.0170	0.0170	0.0170
10	0.0486	0.0483	0.0481	0.0478	0.0476	0.0473
15	0.0892	0.0885	0.0878	0.0872	0.0866	0.0859
20	0.1369	0.1355	0.1342	0.1329	0.1316	0.1305
25	0.1908	0.1884	0.1861	0.1839	0.1818	0.1798
30	0.2501	0.2463	0.2428	0.2394	0.2362	0.2332
35	0.3141	0.3087	0.3036	0.2989	0.2943	0.2900
40	0.3822	0.3748	0.3679	0.3614	0.3553	0.3497
45	0.4538	0.4441	0.4352	0.4268	0.4190	0.4116
50	0.5287	0.5164	0.5050	0.4945	0.4847	0.4756
55	0.6065	0.5913	0.5773	0.5644	0.5524	0.5415
60	0.6866	0.6682	0.6515	0.6360	0.6218	0.6086
65	0.7688	0.7470	0.7274	0.7092	0.6924	0.6771
70	0.8527	0.8273	0.8042	0.7834	0.7640	0.7464
75	0.9378	0.9086	0.8823	0.8584	0.8364	0.8168
80	1.0240	0.9909	0.9611	0.9341	0.9095	0.8873
85	1.1109	1.0737	1.0404	1.0103	0.9830	0.9584
90	1.1981	1.1570	1.1202	1.0870	1.0568	1.0294

Employment of the above relations in practical problems leads to comparatively extensive computations. In order to ease this task, use should be made of Tables 23, 24 and the graphs 87a–h. Thus, Tables 23, 24 present the values of the integrals

$$\xi/d = \int_0^\theta \sqrt{\frac{\sin\theta}{1+\alpha\sin\theta}}\,d\theta, \quad \zeta/d = \int_0^\varphi \sqrt{\frac{\sin\varphi}{1-\alpha\sin\varphi}}\,d\varphi.$$

Table 24

$$\int_0^{\varphi} \sqrt{\frac{\sin \varphi}{1 - \alpha \sin \varphi}}\, d\varphi$$

$\varphi°$	$\alpha = 0.1$	$\alpha = 0.2$	$\alpha = 0.3$	$\alpha = 0.4$	$\alpha = 0.5$
0	0	0	0	0	0
5	0.0172	0.0173	0.0173	0.0174	0.0174
10	0.0488	0.0490	0.0493	0.0496	0.0499
15	0.0898	0.0907	0.0915	0.0921	0.0929
20	0.1383	0.1398	0.1415	0.1430	0.1446
25	0.1932	0.1959	0.1987	0.2016	0.2046
30	0.2540	0.2581	0.2625	0.2672	0.2722
35	0.3199	0.3259	0.3325	0.3395	0.3471
40	0.3901	0.3985	0.4077	0.4177	0.4286
45	0.4640	0.4756	0.4880	0.5016	0.5167
50	0.5420	0.5567	0.5728	0.5908	0.6111
55	0.6231	0.6414	0.6619	0.6849	0.7111
60	0.7068	0.7293	0.7545	0.7833	0.8163
65	0.7930	0.8200	0.8505	0.8856	0.9264
70	0.8809	0.9128	0.9492	0.9912	1.0407
75	0.9705	1.0075	1.0499	1.0993	1.1580
80	1.0613	1.1034	1.1526	1.2096	1.2782
85	1.1527	1.2004	1.2558	1.3210	1.3997
90	1.2488	1.2982	1.3603	1.4338	1.5224

Scale:

Quantity	Per subdivision					
	$\theta < 0$			$\theta > 0$		
	No.	$0 < \xi < 1$	$1 < \xi < 3$	No.	$0 < \xi < 1$	$1 < \xi < 3$
$Re\ V_1$	*1*	0.2	1.0	*3*	0.2	0.02
$Im\ V_1$	*2*	0.2	1.0	*4*	0.2	0.02
$Re\ \frac{dV_1}{d\xi}$	*5*	0.2	4.0	*7*	0.2	0.1
$Im\ \frac{dV_1}{d\xi}$	*6*	0.5	1.0	*8*	0.1	0.05

Figs. 87a, b present the functions $V_1(\xi)$, $V_2(\xi)$ and their first derivatives. Finally, Figs. 87c–h permit rapid computation of the first six

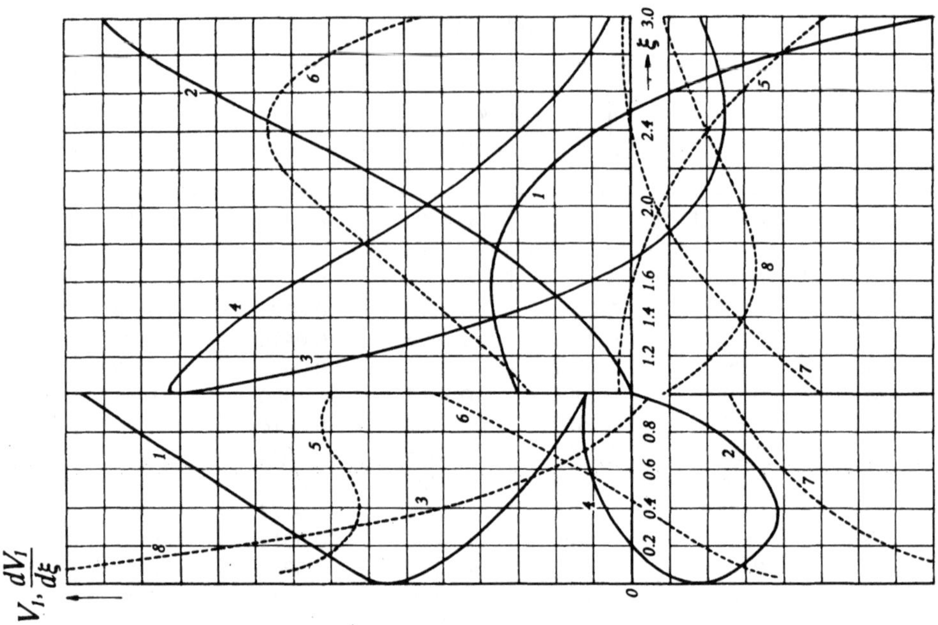
V₁, dV₁/dξ
0
0.2 0.4 0.6 0.8 1.2 1.4 1.6 2.0 2.4 3.0
ξ

Fig. 87a

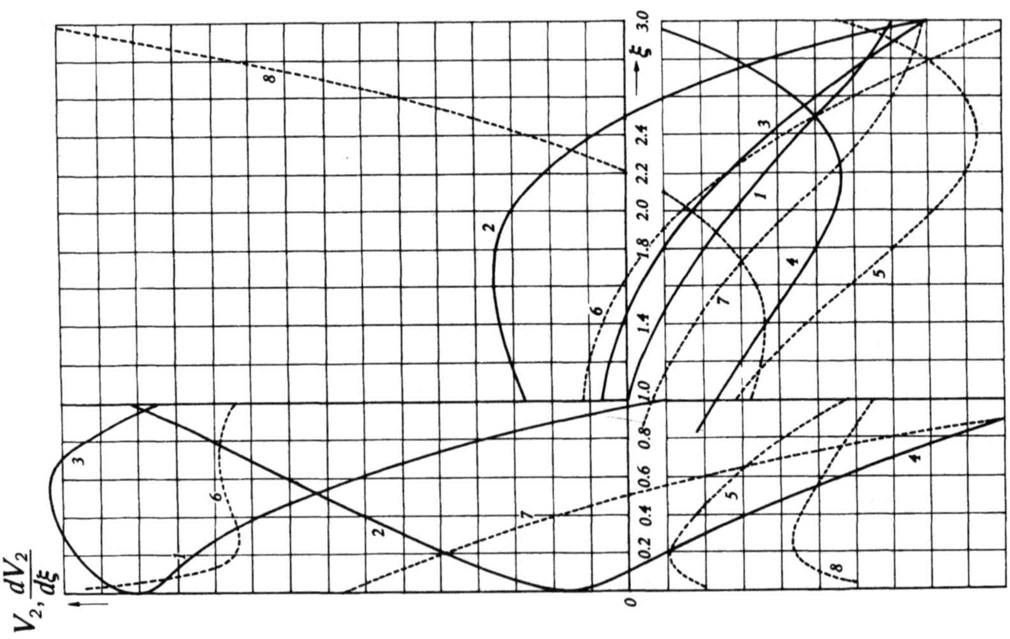
V₂, dV₂/dξ
0
0.2 0.4 0.6 0.8 1.0 1.4 1.8 2.0 2.2 2.4 3.0
ξ

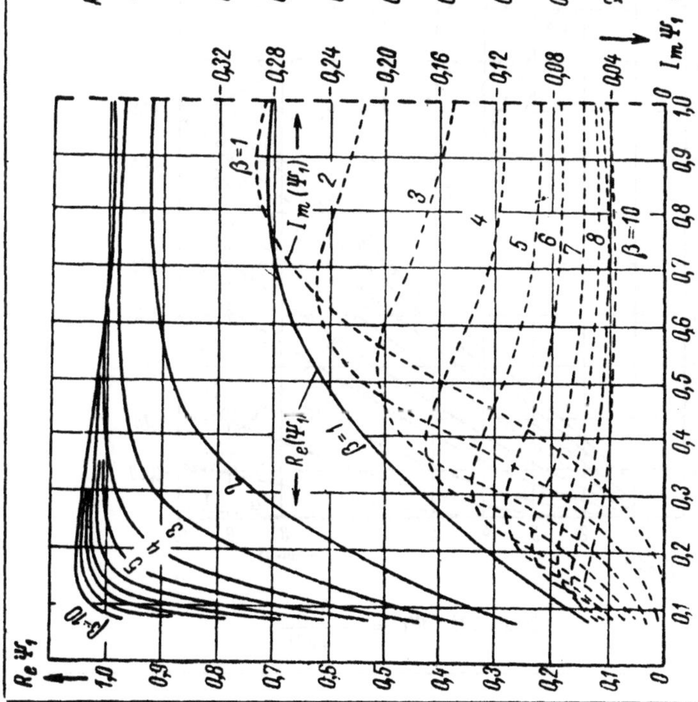

Fig. 87c. $\Psi_1 = \dfrac{a_1}{-2D}$

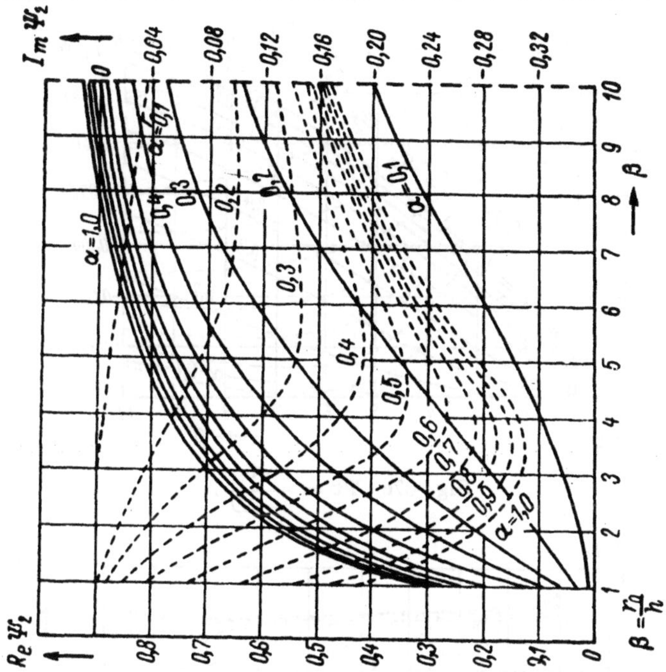

Fig. 87d. $\Psi_2 = \dfrac{b_2}{-i2D}$

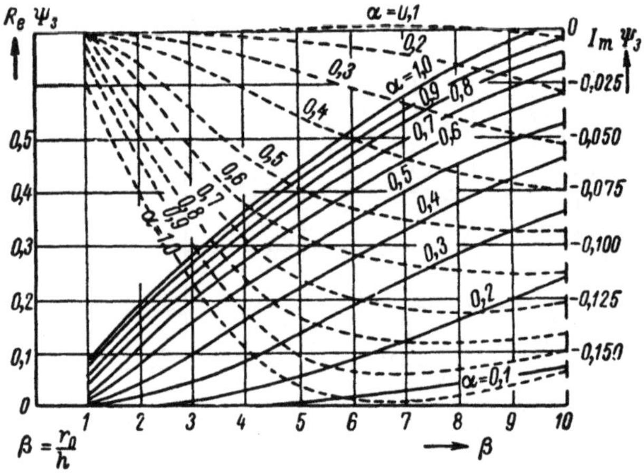

Fig. 87e. $\Psi t = \dfrac{at}{-2D}$.

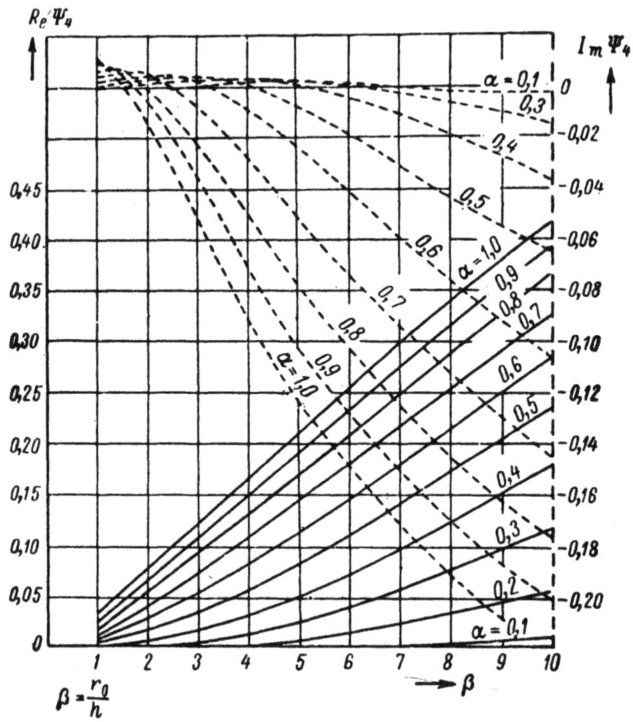

Fig. 87f. $\Psi t = \dfrac{bt}{-i2D}$.

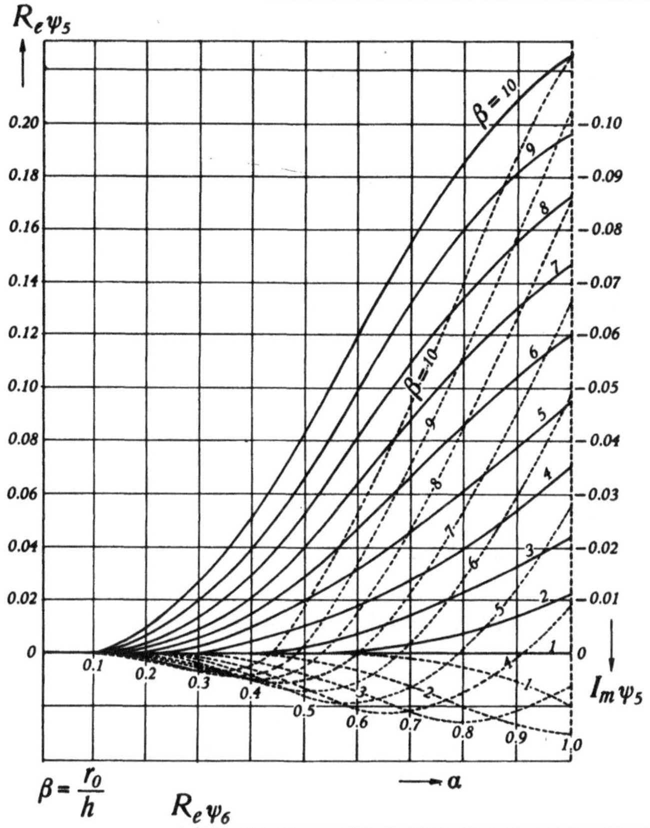

Fig. 87g

$$\psi_5 = \frac{a_5}{-2D}$$

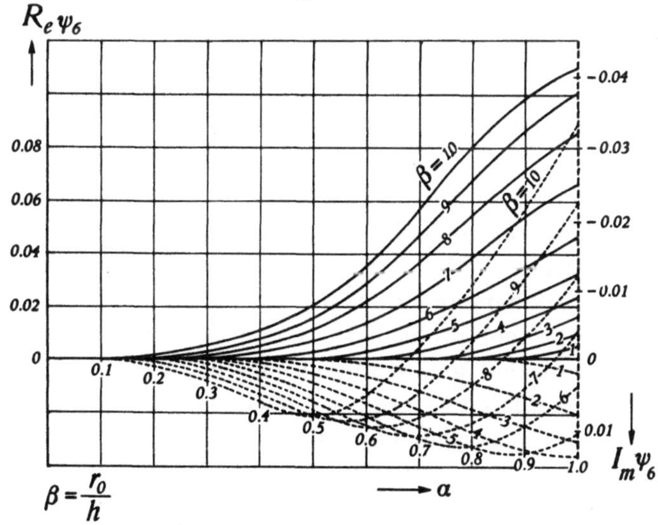

Fig. 87h

$$\psi_6 = \frac{b_6}{-i2D}$$

coefficients of the series (60.47), determining the particular solution. The curves have been reproduced from a paper by V. K. Naumov [104a], devoted to the analysis of toroidal shells applied in turbine design.

Scale:

Quantity	Per subdivision					
	$\theta < 0$			$\theta < 0$		
	No,	$0<\xi<1$	$1<\xi<3$	No,	$0<\xi<1$	$1<\xi<3$
$Re\ V_2$	1	0.1	2.0	3	0.1	2.0
$Im\ V_2$	2	0.2	1.0	4	0.2	1.0
$Re\ \frac{dV_2}{d\xi}$	5	0.5	1.0	7	0.1	2.0
$Im\ \frac{dV_2}{d\xi}$	6	0.2	2.0	8	0.5	1.0

§ 61. Analysis of stress in arched end-panels

Consider an end-panel formed by joining smoothly a spherical shell to part of a torus bounded by two parallel circles.

Depending on the value of the angle θ_0, this end-panel may be convex ($\theta_0 > 0$) or convex-concave ($\theta_0 < 0$). In the second case, the radius of curvature of the meridian of the cap changes its sign along the joint between the sphere and the torus (Fig. 88).

The stresses in the arched end-panel, which will be considered joined to the cylindrical part of a pressure vessel (by means of a weld), will be analyzed for the case of uniform normal pressure. The cylindrical, toroidal and spherical shells will be assumed to have the same thickness. This permits one to fit the solution to the boundary conditions (in this case the conditions at the joints between the shells) without separating the complex auxiliary functions into their real and imaginary parts.

In the present problem the conditions at these joints reduce to the requirement that at a common edge of two shells the quantities Δ_x, ϑ, Q_x, M_1, evaluated for the one shell must be equal (with appropriate sign) to the corresponding quantities, evaluated for the other shell.

The above four equalities indicated may be replaced by the two

$$|\tilde{T}|_{\mathrm{I}} = |\tilde{T}|_{\mathrm{II}}, \qquad \left|\frac{d\tilde{T}}{ds}\right|_{\mathrm{I}} = \left|\frac{d\tilde{T}}{ds}\right|_{\mathrm{II}}, \tag{61.1}$$

where $ds = R_1 d\theta$ is the differential arc length of the meridian of the shell.

Indeed, on the basis of the formulae (51.5) and (54.18), one finds

$$\tilde{T} = T_2 - \mu T_1 + (1 + \mu) T_1 - i \frac{M_1 + M_2}{1(+\mu)c},$$
$$\frac{d\tilde{T}}{ds} = -\frac{E\delta}{R_2}\tilde{\vartheta} = -\frac{E\delta}{R_2}\vartheta - i\frac{N_1}{c}. \tag{61.2}$$

But in the problem under consideration

1) $M_2 \approx \mu M_1$ (since $\kappa_2 \ll \kappa_1$);

2) from the equality $|N_1|_{\text{I}} = |N_1|_{\text{II}}$ follows inevitably $|T_1|_{\text{I}} = |T_1|_{\text{II}}$ (since statics require that $|Q_z|_{\text{I}} = |Q_z|_{\text{II}}$);

3) $T_2 - \mu T_1 = \dfrac{E\delta}{R_2 \sin\theta} \Delta_x.$

Taking all these results into consideration, one arrives at the conclusion that the equalities (61.1) are equivalent to the equalities

$$|\Delta_x|_{\text{I}} = |\Delta_x|_{\text{II}},\ |\vartheta|_{\text{I}} = |\vartheta|_{\text{II}},\ |Q_x|_{\text{I}} = |Q_x|_{\text{II}},\ |M_1|_{\text{I}} = |M_1|_{\text{II}}, \tag{61.3}$$

i.e., equivalent to conditions of compatibility. Thus the conditions holding at the junction of two shells may actually be fulfilled in complex form, using the equalities (61.1).

It will be noted that the above argument applies only to the case when the tangent to the meridian of the compound shell is continuous (since otherwise R_2 will undergo a discontinuous change at the joint). In addition, the proof assumes that the component shells have the same

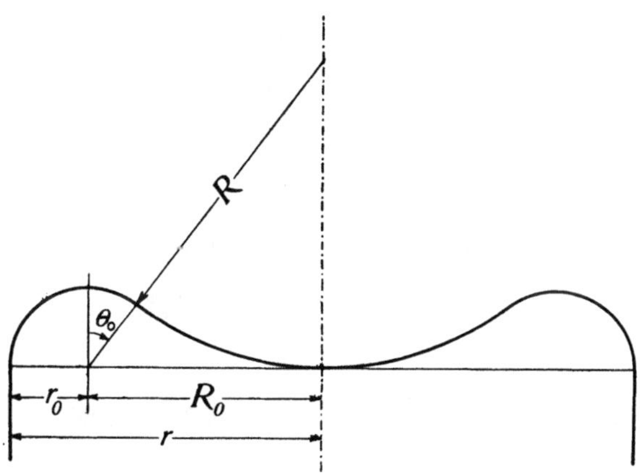

Fig. 88.

thickness (which otherwise would undergo a discontinuity c at the joint). By the way, the first of these limitations may be removed by a certain modification of the boundary conditions (61.1); however, this point will not be discussed further here and it will be left to the reader as an exercise.

Example

The end-panel shown in Fig. 88 will now be analyzed.

1) Given data:

$$r_0 = 0.25r,\ R = 1.026r,\ \delta/r = 1/25,\ \theta_0 = -36°.$$

The thicknesses of the cylindrical, toroidal and spherical shells are the same. The loading is the uniform normal pressure p.

2) General solution for the cylinder:

The length of the cylinder will be assumed to be sufficiently large so that the interaction between the states of stress near the two edges of the cylinder may be neglected.

Then

$$\tilde{T}_c = \tilde{A}e^{-b(1-i)\xi} + T_c^*, \tag{61.4}$$

where $\tilde{A}$ is a constant of integration,

$$b = \sqrt[4]{3(1-\mu^2)}\sqrt{\frac{r}{\delta}} = 6.415,$$

$$T_c^* = T_1^* + T_2^* = 0.5pr + pr = 1.5pr, \tag{61.5}$$

$$\frac{d\tilde{T}_c}{ds} = \frac{1}{r}\frac{d\tilde{T}_c}{d\xi} = -\tilde{A}\frac{b}{r}(1-i)e^{-b(1-i)\xi}.$$

At the joint of the cylinder and the torus

$$\tilde{T}_c = \tilde{A} + 1.5pr,$$

$$\frac{d\tilde{T}_c}{ds} = -\tilde{A}\frac{b}{r}(1-i). \tag{61.6}$$

3) General solution for the spherical segment:

$$T_s = \tilde{B}e^{a(1+i)(\theta_0-\theta)} + T_s^*, \tag{61.7}$$

where

$$a = \sqrt[4]{3(1-\mu^2)}\sqrt{\frac{R}{\delta}} = b\sqrt{1.026} = 6.50,$$

$$T_s^* = T_1^* + T_2^* = -0.5pR - 0.5pR = -1.026pr, \tag{61.8}$$

$$\frac{d\tilde{T}_s}{ds} = -\frac{1}{R}\frac{d\tilde{T}_s}{d\theta} = \frac{a(1+i)}{1.026r}\tilde{B}e^{a(1+i)(\theta_0-\theta)}.$$

At the joint of the torus and the sphere

$$\tilde{T}_s = \tilde{B} - 1.026pr,$$
$$\frac{d\tilde{T}_s}{ds} = \frac{6.33}{r}(1+i)\tilde{B}. \qquad (61.9)$$

In (61.7), the solution has been taken in a form which differs somewhat from that in § 57. This is due to the fact that the end panel, which consists of parts of toroidal and spherical shells, is considered as a single shell with continuously varying normal. Then, assuming the radius of curvature of the meridian of the torus $R_1 = r_0$ to be positive, one must assume both radii of curvature of the spherical shell to be negative. Then one has for the spherical shell

$$-i\frac{R_1^2}{cR_2} = i\frac{R}{c} = [(1+i)\sqrt{R/2c}]^2.$$

4) General solution for the toroidal segment:

$$r_0 = \tfrac{1}{4}r, \ \alpha = r_0/R_0 = \tfrac{1}{3}, \ R_0 = r - r_0 = \tfrac{3}{4}r,$$
$$2d^2 = \sqrt{12(1-\mu^2)}\frac{r_0^2}{R_0\delta} = \sqrt{12(1-0.09)}\frac{25}{4.3} = 6.88.$$

The constant A is determined from the condition

$$Q_z = \left[\frac{pr_0}{2}\frac{2+\alpha\sin\theta}{1+\alpha\sin\theta} - \frac{\alpha A}{(1+\alpha\sin\theta)\sin\theta}\right]\sin\theta = \frac{1}{2}pr \text{ for } \theta = \frac{\pi}{2}. \qquad (61.10)$$

Hence

$$A = -1.125pr.$$

On the basis of the formulae (60.6), (60.8) and (60.2), one has

$$\tilde{T} = \frac{i}{2d^2}\frac{1}{1+\alpha\sin\theta}\frac{dV}{d\theta} + \frac{3+2\alpha\sin\theta}{1+\alpha\sin\theta}\frac{pr_0}{2},$$
$$\frac{1}{r_0}\frac{d\tilde{T}}{d\theta} = \frac{d\tilde{T}}{ds} = \frac{1}{r_0}\left[\frac{\sin\theta}{(1+\alpha\sin\theta)^2}V - i\frac{2d^2\cos\theta}{(1+\alpha\sin\theta)^2}A\right], \qquad (61.11)$$

and by (60.31)

$$V = (1+\alpha\sin\theta)^{3/4}[\tilde{C}_1V_1 + \tilde{C}_2V_2] + \mathfrak{B},$$
$$\frac{dV}{d\theta} = (1+\alpha\sin\theta)^{3/4}\left[\tilde{C}_1\frac{dV_1}{d\theta} + \tilde{C}_2\frac{dV_2}{d\theta}\right] - \qquad (61.12)$$
$$- \frac{\frac{3}{4}\alpha\cos\theta}{(1+\alpha\sin\theta)^{1/4}}(\tilde{C}_1V_1 + \tilde{C}_2V_2) + \frac{d\mathfrak{B}}{d\theta}.$$

a) A particular integral will be

$$\mathfrak{B} = a_1 \cos\theta + b_2 \sin 2\theta + a_3 \cos 3\theta + b_4 \sin 4\theta + \dots \quad (61.13)$$

With the aid of Figs. 87c–h, one determines the first five coefficients of the series:

$$\begin{aligned} a_1 &= (-15.9 + 1.55i)\ pr, \\ b_2 &= -(2.48 + 10.1i)\ pr, \\ a_3 &= (-3.48 + 0.52i)\ pr, \\ b_4 &= -(0.031 + 0.851i) pr, \\ a_5 &= -(0.155 + 0.62i)\ pr. \end{aligned} \quad (61.14)$$

Hence the values of the particular solution at the edges of the torus, i.e., for $\theta = 90°$, $\theta = -36°$ are

$$\begin{aligned} \mathfrak{B}(90°) &= 0, \\ \mathfrak{B}(-36°) &= (-9.29 + 11.2i) pr. \end{aligned} \quad (61.15)$$

b) Calculation of the functions V_1, V_2.

One finds from Figs. 87a, b the values of the functions

$$V_1, \quad V_2, \quad dV_1/d\theta, \quad dV_2/d\theta,$$

in particular, their values at the edges, i.e., for $\theta = 90°$, $\theta = -36°$. Corresponding to these values of θ, one has

$$\xi = d \int_0^{\pi/2} \frac{\sqrt{\sin\theta}\, d\theta}{\sqrt{1 + 0.333 \sin\theta}} \approx 2.01, \quad \zeta = d \int_0^{0.628} \frac{\sqrt{\sin\varphi}\, d\varphi}{\sqrt{1 - 0.333 \sin\varphi}} \approx 0.644. \quad (61.16)$$

For these values of ξ and ζ, one has

$$\left.\begin{aligned} V_1 &= -0.0312 + 0.115i \\ V_2 &= -3.45 - 5.74i \end{aligned}\right\} \text{ for } \theta = \pi/2, \qquad \left.\begin{aligned} V_1 &= 2.54 - 0.676i \\ V_2 &= 0.683 + 2.02i \end{aligned}\right\} \text{ for } \theta = 0.628, \quad (61.17)$$

and hence

$$\left.\begin{aligned} \frac{\partial V_1}{\partial\theta} &= -0.0876 - 0.231i \\ \frac{\partial V_2}{\partial\theta} &= -14.7 - 2.8i \end{aligned}\right\} \text{for } \theta = \pi/2, \qquad \left.\begin{aligned} \frac{\partial V_1}{\partial\theta} &= -2.59 - 1.44i \\ \frac{\partial V_2}{\partial\theta} &= 2.68 - 3.62i \end{aligned}\right\} \text{ for } \theta = -0.628. \quad (61.18)$$

5) The joints between the cylinder, toroid and sphere.

On the basis of the statements at the beginning of this section, the linking up of the solutions for the cylinder, toroid and sphere may be achieved by use of the equalities (61.1).

In the present particular case, this leads to the following system of four, linear, algebraic equations:

$$\begin{aligned}
&\tilde{C}_1(0.0311 - 0.0118i) + \tilde{C}_2(0.378 - 1.98i) + \\
&\qquad + pr(-1.51 + 1.21i) = \tilde{A} + 1.5pr, \\
&\tilde{C}_1(-0.0872 + 0.322i) + \tilde{C}_2(-9.63 - 16.0i) = \\
&\qquad = -6.43(1 - i)\tilde{A}, \\
&\tilde{C}_1(0.247 - 0.299i) + \tilde{C}_2(0.477 + 0.437i) + \\
&\qquad + (0.601 - 3.75i)pr = \tilde{B} - 1.026pr, \\
&\tilde{C}_1(-9.25 + 2.45i) + \tilde{C}_2(-2.48 - 7.35i) + \\
&\qquad + (33.7 - 2.08i)pr = 6.36(1 + i)\tilde{B},
\end{aligned} \tag{61.19}$$

yielding the values

$$\begin{aligned}
&\tilde{C}_1 = (0.614 + 0.293i)pr, \quad \tilde{C}_2 = (0.446 + 0.656i)pr, \\
&\tilde{A} = (-1.52 + 0.571i)pr, \quad \tilde{B} = (1.87 - 3.16i)pr,
\end{aligned} \tag{61.20}$$

thus the problem is essentially solved, since, once the auxiliary function $\tilde{V}$ for the toroid and the auxiliary functions $\tilde{T}_c$ and $\tilde{T}_s$ are known, all the forces, moments and displacements in the three shells comprising the end-panel under consideration may be evaluated (using the formulae of §§ 57, 60).

Table 25 gives the values of the two stresses in the end-panel which are of greatest practical interest, namely $\sigma_2 = \dfrac{T_2}{\delta}$, $\sigma_1 = \dfrac{6M_1}{\delta^2}$, as functions of the angle θ.

Figs. 89 and 90 illustrate the results of this table. Attention should be drawn to the fact that the hoop stress σ_2 reaches a maximum value in the region of the toroid for $\theta = -10°$, where

$$\sigma_{2\max} \approx 3.6pr.$$

As regards the bending stresses, they attain for $\theta = -36°$ their maximum value

$$\sigma_{1\max} \approx 5.5pr.$$

As it has been seen, the investigated end-panel is not a rational design, since the stresses in it are distributed in a very uneven manner and considerable stress concentrations occur in the region of the toroidal shell. In this connection, the greatest danger does not arise from the bending stresses σ_1, although they are greater, but from the stress σ_2 distributed uniformly through the thickness of the shell. In spite of the above stated unfavourable assessment, end-panels of this type find application in exceptional cases, since their profiles may be found to be convenient.

Table 25

Torus			Sphere			Cylinder		
$\theta°$	$\sigma_1 : \frac{pr}{h}$	$\sigma_2 : \frac{pr}{h}$	$\theta°$	$\sigma_1 : \frac{pr}{h}$	$\sigma_2 : \frac{pr}{h}$	ξ	$\sigma_1 : \frac{pr}{h}$	$\sigma_2 : \frac{pr}{h}$
90	−1.04	−0.515	−36	5.52	1.49	0	−1.04	−0.518
75	−2.48	−0.879	−30	3.38	−0.494	0.2	0.633	0.732
60	−7.78	−0.759	−25	1.43	−1.07	0.4	0.181	1.08
45	−4.66	−0.0576	−20	0.736	−0.921	0.6	−0.02	1.03
30	−4.61	1.14				0.8	−0.0179	1
15	−3.22	2.56				1	−0.00104	1
10	−2.14	2.97						
0	0.0775	3.53						
−10	2.43	3.61						
−15	3.5	3.43						
−25	5.21	2.68						
−36	5.52	1.4						

Since there is in the literature a lack of data on concave-convex type arched end-panels and since their analysis is more complex than that of

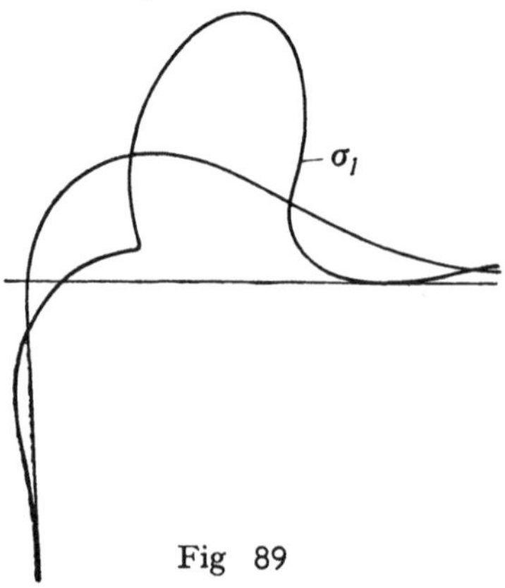

Fig 89

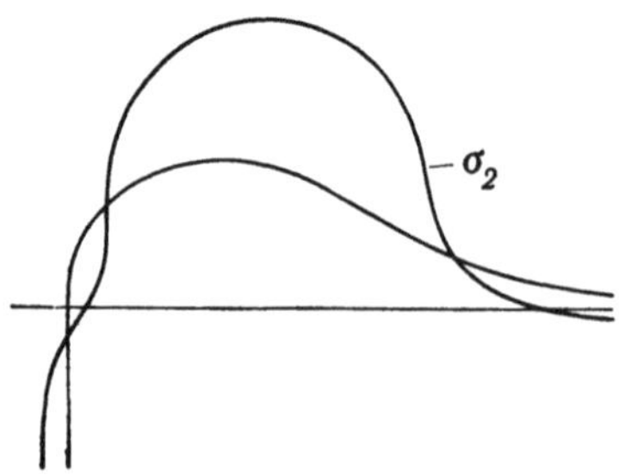

Fig. 90.

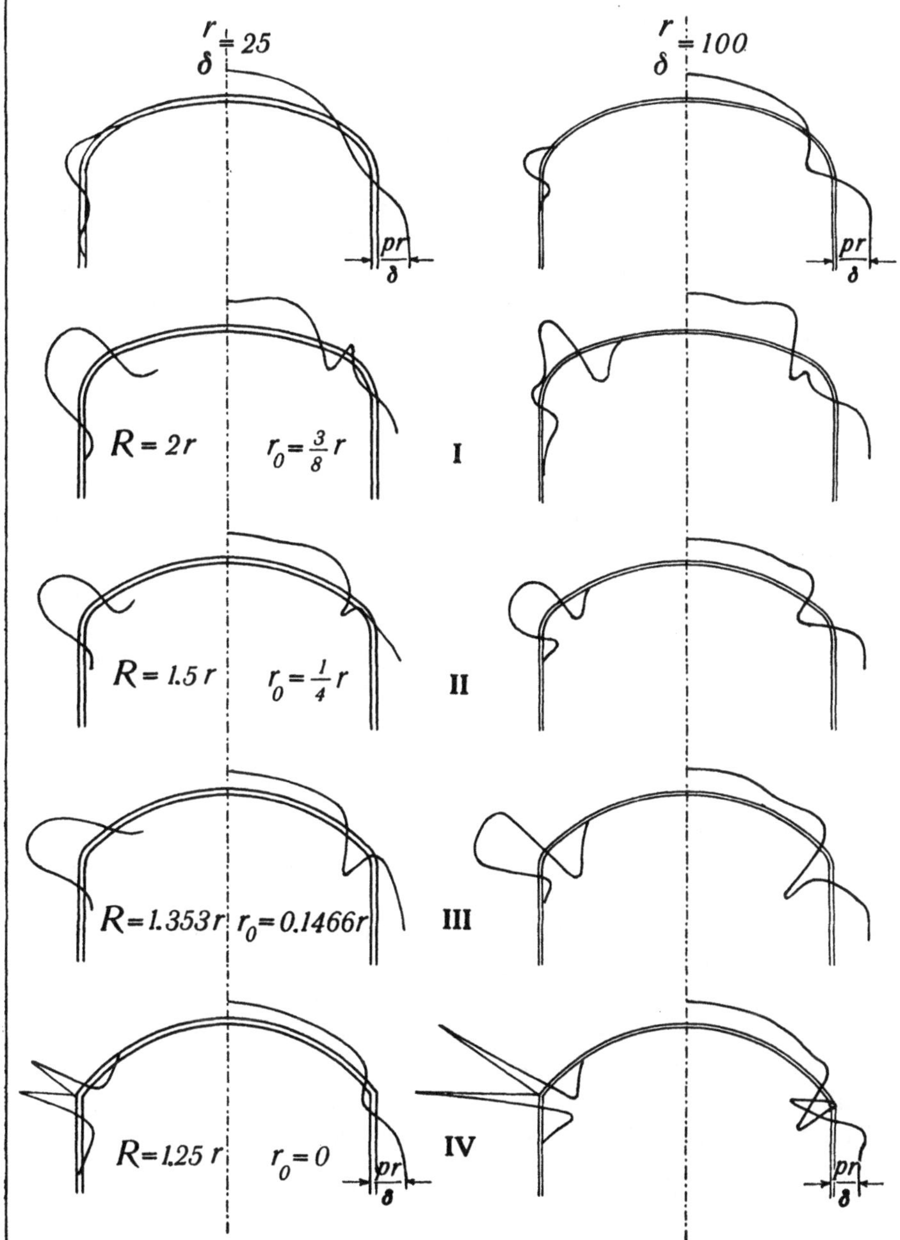

$\frac{r}{\delta} = 25$
$\frac{r}{\delta} = 100$
$\frac{pr}{\delta}$
$R = 2r$
$r_0 = \frac{3}{8} r$
I
$R = 1.5\, r$
$r_0 = \frac{1}{4} r$
II
$R = 1.353 r$
$r_0 = 0.1466 r$
III
$R = 1.25\, r$
$r_0 = 0$
IV

Fig. 91.

ordinary arched end-panels (in view of the necessity to consider the solution for the toroid in the region $\theta > 0$ as well as in the region $\theta < 0$), this type of end-panel has been chosen as an example.

Fig. 91 shows the distributions of the stresses

$$\sigma_1 = \frac{6M_1}{\delta^2} \text{ and } \sigma_2 = \frac{T_2}{\delta}$$

for a number of ordinary arched end-panels, differing from each other by the values of α and r/δ. For the purpose of comparison, the same sketches show the curves for the stresses σ_1 and σ_2 for elliptical ends, the heights of which are equal to those of the arched end-panels (the two upper sketches). The corresponding calculations were performed by the Author in collaboration with B. I. Slepov.

§ 62. Shells of revolution subject to 'wind' loading

As has already been mentioned, the name "wind" loading attaches to that loading system which is obtained from the formulae (52.1) for $k = 1$.

It has been proved in Chapter II that for this loading the equations of the membrane theory may be integrated by quadrature, whatever may be the form of the shell of revolution. It is not difficult to understand that in the case of wind loading the equations of the general theory of shells of revolution, i.e., the differential equations (51.17) may likewise be integrated by quadrature. This follows from the fact that the loading

$$q_1 = q_{1,1} \cos \varphi, \; q_2 = q_{2,1} \sin \varphi, \; q_n = q_{n,1} \cos \varphi \tag{62.1}$$

is not self-equilibrating, so that there must exist not only relations between the derivatives of the forces and moments, but likewise certain linear relations between the same forces and moments expressing the equilibrium conditions of a finite part of the shell of revolution severed by cuts perpendicular to its axis.

Further, by the number of the equations of statics, it may be verified that there must be two such linear relations between the forces and moments which may be written down on the basis of a study of the equilibrium of a part of the shell, cut from it in the above manner.

Hence the system (52.3) for $k = 1$, i.e., the system

$$\begin{gathered} G_1[\tilde{U}^{(1)}] + \left[1 - ic\left(\frac{1}{R_1} - \frac{1}{R_2}\right)\frac{1}{\sin^2\theta}\right]\tilde{T}^{(1)} = f_1(\theta), \\ -icG_1[\tilde{T}^{(1)}] + \tilde{T}^{(1)} + \left(\frac{1}{R_1} - \frac{1}{R_2}\right)\frac{1}{\sin^2\theta}\,\tilde{U}^{(1)} = R_2 q_{n,1}, \end{gathered} \tag{62.2}$$

where

$$f_{1,1}(\theta) = \frac{1}{R_1R_2 \sin\theta}\left\{\frac{d}{d\theta}\,[(q_{n,1}\cos\theta - q_{1,1}\sin\theta)R_2{}^3\sin^2\theta] + \right.$$

$$\left. + \, q_{2,1}R_1R_2{}^2\sin^2\theta\right\}, \qquad (62.3)$$

$$G_1(\ldots) = \frac{1}{R_1R_2\sin\theta}\,\frac{d}{d\theta}\left[\frac{R_2{}^2\sin\theta}{R_1}\,\frac{d(\ldots)}{d\theta}\right] - \frac{1}{R_2\sin^2\theta}\,(\ldots), \qquad (62.4)$$

must have at least two quadratures.

In order to find these quadratures the first equation (62.2) will be subtracted from the second. Then one arrives at the second order differential equation

$$G_1(\tilde{W}) - \left(\frac{1}{R_1} - \frac{1}{R_2}\right)\frac{1}{\sin^2\theta}\,\tilde{W} = f_1(\theta) - R_2q_{n,1} \qquad (62.5)$$

with one unknown, the new auxiliary function $\tilde{W}$

$$\tilde{W} = \tilde{U}^{(1)} + ic\tilde{T}^{(1)}. \qquad (62.6)$$

Taking into consideration that the operator $G_1(\ldots)$ is defined by the formula (62.4), the equation (62.5) may be written in the form

$$\frac{1}{R_1R_2\sin\theta}\,\frac{d}{d\theta}\left[\frac{R_2{}^2\sin\theta}{R_1}\,\frac{d\tilde{W}}{d\theta}\right] - \frac{1}{R_1\sin^2\theta}\,\tilde{W} = f_1(\theta) - R_2q_{n,1}, \qquad (62.7)$$

whence it is clear that it differs from equation (29.2) only in the way that, in the membrane theory, this equation is in terms of the function

$$U = T_1R_2\sin^2\theta, \qquad (62.8)$$

whereas in the theory of shells of revolution, which includes bending, it is in terms of the function

$$\tilde{W} = \tilde{U}^{(1)} + ic\tilde{T}^{(1)} = \tilde{T}_1{}^{(1)}R_2\sin^2\theta - ic\left[\frac{R_2}{R_1}\sin\theta\cos\theta\,\frac{d\tilde{T}^{(1)}}{d\theta} - \tilde{T}^{(1)}\right], \qquad (62.9)$$

which for $c \to 0$ tends to the preceding function.

In § 29, the differential equation (62.7) has been integrated by quadrature. The corresponding computations need not be repeated here and it will be sufficient that the final results are given by the expression

$$\tilde{W} = \frac{1}{R_2\sin\theta}\,[C_1 + C_2\int_{\theta^{*}}^{\theta} R_1\sin\theta d\theta + \int_{\theta^{*}}^{\theta} \Phi R_1\sin\theta d\theta], \qquad (62.10)$$

in which

$$\Phi(\theta) = (q_{n,1}\cos\theta - q_{1,1}\sin\theta)R_2{}^2\sin\theta - \\ - \int_{\theta''}^{\theta} (q_{n,1}\sin\theta + q_{1,1}\cos\theta - q_{2,1})R_1R_2\sin\theta d\theta, \quad (62.11)$$

and C_1 and C_2 are two real arbitrary constants.

The lower limit of integration θ'' may be chosen arbitrarily to suit computational convenience. In what follows, θ'' will be identified with the angle corresponding to the upper edge of the shell of revolution (Fig. 75).

Having found $\tilde{W}$, by use of the formula (62.6) one may now establish the connection between $\tilde{U}^{(1)}$ and $\tilde{T}^{(1)}$. On the basis of this formula, one finds

$$\tilde{U}^{(1)} = \tilde{W} - ic\tilde{T}^{(1)} = - ic\tilde{T}^{(1)} + \frac{1}{R_2\sin\theta}[C_1 + C_2\int_{\theta''}^{\theta} R_1\sin\theta d\theta + \\ + \int_{\theta''}^{\theta} \Phi R_1\sin\theta d\theta]. \quad (62.12)$$

Introducing this value of $\tilde{U}^{(1)}$ into the second of the equations of the system (62.2), one arrives at the following differential equation in the single unknown $\tilde{T}^{(1)}$:

$$G_1[\tilde{T}^{(1)}] + \left(\frac{1}{R_1} - \frac{1}{R_2}\right)\frac{1}{\sin^2\theta}\tilde{T}^{(1)} + \frac{i}{c}\tilde{T}^{(1)} = \\ = \frac{i}{c}\left[R_2q_{n,1} - \left(\frac{1}{R_1} - \frac{1}{R_2}\right)\frac{1}{\sin^2\theta}\tilde{W}\right], \quad (62.13)$$

which may be rewritten in expanded form

$$\frac{d^2\tilde{T}^{(1)}}{d\theta^2} + \left[\left(2\frac{R_1}{R_2} - 1\right)\operatorname{cotan}\theta - \frac{1}{R_1}\frac{dR_1}{d\theta}\right]\frac{d\tilde{T}^{(1)}}{d\theta} + \frac{R_1}{R_2}\left(1 - 2\frac{R_1}{R_2}\right)\frac{1}{\sin^2\theta}\tilde{T}^{(1)} + \\ + i\frac{R_1{}^2}{R_2c}\tilde{T}^{(1)} = i\frac{R_1{}^2}{R_2c}F_1(\theta), \quad (62.14)$$

where

$$F_1(\theta) = R_2 q_{n,1} - \left(\frac{1}{R_1} - \frac{1}{R_2}\right)\frac{1}{R_2 \sin^3\theta}[C_1 + C_2 \int_{\theta''}^{\theta} R_1 \sin\theta\, d\theta +$$

$$+ \int_{\theta''}^{\theta} \Phi R_1 \sin\theta\, d\theta] = T^*_{1,1} + T^*_{2,1}. \quad (62.15)$$

The analysis of shells of revolution of arbitrary shape subject to wind loading has thus been reduced to the integration of the above linear second order differential equation.

Comparing this equation with that deduced earlier for the case of symmetrical deformation, i.e., with (54.6), it is impossible to fail to note the evident similarity of these two equations. As regards their left hand sides, it may be stated that they differ only in one term which has been underlined in equation (62.14). However, as is clear from the analysis of the equation (54.6) which is given in § 55, terms of that type are of secondary importance and are negligibly small (provided the solution is not required near the singular points $\theta = 0$ and $\theta = \pi$). Thus, the equations (54.6) and (62.14) are practically identical from the point of view of methods of solution and all the results in § 53, 54 and 55 referring to the theory of the approximate integration of the equation (54.6) may be carried over to the equation (62.14) almost without modification.

The general solution of the equation (62.14) consists of the sum of any of its particular integrals and the general solution of the homogeneous equation

$$\frac{d^2\mathfrak{T}^{(1)}}{d\theta^2} + \left[\left(2\frac{R_1}{R_2} - 1\right)\cot\theta - \frac{1}{R_1}\frac{dR_1}{d\theta}\right]\frac{d\mathfrak{T}^{(1)}}{d\theta} +$$

$$+ \frac{R_1}{R_2}\left(1 - 2\frac{R_1}{R_2}\right)\frac{1}{\sin^2\theta}\mathfrak{T}^{(1)} + \frac{iR_1^2}{R_2 c}\mathfrak{T}^{(1)} = 0, \quad (62.16)$$

and, provided the function F_1 defined by the formula (62.15) does not essentially increase on differentiation, one may take as particular integral of the equation (62.14)

$$\tilde{\mathfrak{t}}^{(1)} = F_1(\theta) = T^*_{1,1} + T^*_{2,1}. \quad (62.17)$$

This method based on the omission on the left hand side of the equation (62.14) of all terms except the last is, in essence, equivalent to an identification of the particular integral of this equation with the solution of the equation of the membrane theory of shells of revolution for the contemplated type of surface loading. As regards the reasoning, accuracy and limitations of its application, this method is completely analogous to that of the determination of the particular integral of the equation (54.6), which has been discussed in § 54.

Having found in this manner the particular integral, the integration of the equation (62.14) has been reduced to the integration of the homogeneous equation (62.16). Introducing there instead of $\mathfrak{T}_{(1)}$ the new dependent variable

$$Z = R_2 \sqrt{\frac{\sin\theta}{R_1}}\,\mathfrak{T}^{(1)}, \tag{62.18}$$

the equation takes the form

$$\frac{d^2Z}{d\theta^2} - F^2 Z = 0, \tag{62.19}$$

where

$$F^2 = \frac{1}{2}\frac{dp}{d\theta} + \frac{1}{4}p^2 - q = -i\frac{R_1^2}{R_2 c} - \frac{1}{2R_1}\frac{d^2R_1}{d\theta^2} + \frac{3}{4}\frac{1}{R_1^2}\left(\frac{dR_1}{d\theta}\right)^2 +$$
$$+ \frac{\cot\mathrm{an}\,\theta}{2R_1}\frac{dR_1}{d\theta} - \frac{R_1}{R_2}\frac{2}{\sin^2\theta}\left(1 - \frac{R_1}{R_2}\right) + \frac{2+\cos^2\theta}{4\sin^2\theta}. \tag{62.20}$$

As in § 55, the solution will be constructed here only for points lying some distance from the points $\theta = 0$ and $\theta = \pi$ and it will be assumed for this purpose that R_1 and R_2 do not increase on differentiation. Then all terms in the formula (62.20) will be quantities of order unity except for the first term which has an absolute value of order $\frac{R}{\delta}$. Hence, consistent with the accuracy of the first approximation, one may write

$$F^2 \approx -i\frac{R_1^2}{R_2 c}; \tag{62.21}$$

this leads to exactly the same approximate expression for the auxiliary function $\mathfrak{T}^{(1)}$ which was obtained in § 56 for $\mathfrak{T}$,namely

$$\mathfrak{T}^{(1)} = \frac{1}{R_2^{3/4}\sin^{1/2}\theta}\left[\tilde{C}_1 e^{-(1-i)\beta} + \tilde{C}_2 e^{(1-i)\beta}\right], \tag{62.22}$$

where

$$\beta = \frac{1}{\sqrt{2}}\int_{\theta_0}^{\theta}\frac{R_1\,d\theta}{\sqrt{R_2 c}} = \sqrt[4]{3(1-\mu^2)}\int_{\theta_0}^{\theta}\frac{R_1\,d\theta}{\sqrt{R_2\delta}}. \tag{62.23}$$

Since the accuracy of deduction of the result (62.22) does not ensure that all terms of order $\sqrt{\frac{\delta}{R_2}}$ are retained, the factor $\frac{1}{R_2^{3/4}\sin^{1/4}\theta}$ may be assumed to be constant for the differentiation of the expression (62.22) (§ 54, 55). Including this factor with the constant of integration, one obtains for $\mathfrak{T}^{(1)}$ the following approximate expression which coincides with (56.10):

$$\mathfrak{T}^{(1)} = \tilde{C}_1 e^{-(1-i)\beta} + \tilde{C}_2 e^{(1-i)\beta}, \tag{62.24}$$

which will be accepted as the final result.

With a knowledge of $\mathfrak{T}^{(1)}$, all the complex forces are readily found and this will be demonstrated in the following section.

§ 63. Design formulae for 'wind' loading

First, the general expressions for the displacements, forces and moments in shells of revolution subject to wind loading will be deduced.

Consideration will be limited to the homogeneous problem which corresponds to the differential equation (62.16) obtained for

$$q_1 = q_2 = q_n = \tilde{W} = 0. \tag{63.1}$$

The corresponding non-homogeneous problem has been considered in [29a].

It follows from this and the formulae (62.12) that

$$\tilde{U}^{(1)} = -ic\mathfrak{T}^{(1)}, \tag{63.2}$$

whence, taking into consideration (51.6) and (52.2), one obtains for the complex force $\tilde{T}_1$ the expression

$$\tilde{T}_1 = \tilde{T}_1^{(1)} \cos\varphi = ic\left[\frac{\cot an\,\theta}{R_1}\frac{d\mathfrak{T}^{(1)}}{d\theta} - \frac{1}{R_2\sin^2\theta}\mathfrak{T}^{(1)}\right]\cos\varphi. \tag{63.3}$$

However, since $\frac{d\mathfrak{T}^{(1)}}{d\theta}$ in comparison with $\mathfrak{T}^{(1)}$ will be a quantity of order $\sqrt{\frac{R}{\delta}}$, this being a property of the integrals of the equation (62.16), one must neglect in the formula (63.3) the second term, after which (63.3) takes the form

$$\tilde{T}_1 = \tilde{T}_1^{(1)} \cos\varphi = ic\frac{\cot an\,\theta}{R_1}\frac{d\mathfrak{T}^{(1)}}{d\theta}\cos\varphi. \tag{63.4}$$

Further, by (51.7),

$$\tilde{T}_2 = \tilde{T} - \tilde{T}_1 = \left(\mathfrak{T}^{(1)} - ic\frac{\cot\theta}{R_1}\frac{d\mathfrak{T}^{(1)}}{d\theta}\right)\cos\varphi \approx \mathfrak{T}^{(1)}\cos\varphi, \quad (63.5)$$

and, finally, on the basis of the equation (51.11), one finds by taking account of the fact that the problem is homogeneous

$$R_1 \sin\theta \frac{d\tilde{S}}{d\varphi} = -\left[\frac{d\tilde{U}^{(1)}}{d\theta} + ic\cot\theta\frac{R_1}{R_2}\mathfrak{T}^{(1)}\right]\cos\varphi =$$

$$= ic\left[\frac{d\mathfrak{T}^{(1)}}{d\theta} - \cot\theta\frac{R_1}{R_2}\mathfrak{T}^{(1)}\right]\cos\varphi \approx ic\frac{d\mathfrak{T}^{(1)}}{d\theta}\cos\varphi. \quad (63.6)$$

Hence

$$\tilde{S} = \frac{ic}{R_1\sin\theta}\frac{d\mathfrak{T}^{(1)}}{d\theta}\sin\varphi = \tilde{S}^{(1)}\sin\varphi. \quad (63.7)$$

By the way, it follows from these formulae that, when the problem is homogeneous, there exist between the complex forces the relations

$$\tilde{T}_1\sin\varphi - \tilde{S}\cos\varphi\cos\theta = 0,$$

$$\tilde{T}_1\cos\theta\sin\varphi - \tilde{S}\cos\varphi + \frac{ic}{R_1}\frac{d\mathfrak{T}^{(1)}}{d\theta}\sin\theta\sin\varphi = 0, \quad (63.8)$$

which hold, of course, only for 'wind' loading.

Substituting in (63.3), (63.4) and (63.7) for $\mathfrak{T}^{(1)}$ its value (62.24) and separating real and imaginary parts (as was done similarly in § 57), one arrives at the following general expressions for the forces and moments:

$$T_1 = -\frac{1}{\sqrt{2}}\sqrt{\frac{c}{R_2}}\cot\theta\{[(C_1'-C_1'')\cos\beta-(C_1'+C_1'')\sin\beta]e^{-\beta}+$$
$$+[(C_2''-C_2')\cos\beta-(C_2'+C_2'')\sin\beta]e^{\beta}\}\cos\varphi,$$
$$T_2 = \{(C_1'\cos\beta - C_1''\sin\beta)e^{-\beta} + (C_2'\cos\beta + C_2''\sin\beta)e^{\beta}\}\cos\varphi, \quad (63.9)$$
$$S = -\frac{1}{\sqrt{2}}\sqrt{\frac{c}{R_2}}\cdot\frac{1}{\sin\theta}\{([C_1'-C_1'')\cos\beta-(C_1'+C_1'')\sin\beta]e^{-\beta}+$$
$$+[(C_2''-C_2')\cos\beta-(C_2'+C_2'')\sin\beta]e^{\beta}\}\sin\varphi,$$
$$M_2 \approx \mu M_1,$$
$$M_1 = -c[(C_1''\cos\beta+C_1'\sin\beta)e^{-\beta}+(C_2''\cos\beta-C_2'\sin\beta)e^{\beta}]\cos\varphi.$$

The normal shear force N_1 may now be determined as the real part of the function

$$\frac{ic}{R_1}\frac{d\mathfrak{T}}{d\theta}$$

(§ 56), whence, as in the case of symmetrical loading,

$$N_1 = T_1 \tan\theta. \tag{63.10}$$

There remains to evaluate the displacements and the angle of rotation of the meridian of the shell

$$\vartheta = -\frac{1}{R_1}\left(\frac{\partial w}{\partial\theta} - u\right).$$

Putting in the formulae (53.5) $k = 1$, one obtains for the complex displacements the system of differential equations

$$\frac{d\tilde{u}_1}{d\theta} + \tilde{w}_1 = \frac{R_1}{E\delta}(\tilde{T}_1^{(1)} - \mu\tilde{T}_2^{(1)}),$$

$$\tilde{v}_1 + \tilde{u}_1\cos\theta + \tilde{w}_1\sin\theta = \frac{R_2\sin\theta}{E\delta}[\tilde{T}_2^{(1)} - \mu\tilde{T}_1^{(1)}], \tag{63.11}$$

$$\frac{d\tilde{v}_1}{d\theta} - \left(\tilde{v}_1\operatorname{cotan}\theta + \frac{1}{\sin\theta}\tilde{u}_1\right)\frac{R_1}{R_2} = \frac{2(1+\mu)R_1}{E\delta}\tilde{S}^{(1)},$$

with regard to which .t may be stated that it can be integrated by quadrature (on the basis of the results, obtained in §§ 21, 27).

Differentiating the second of the equations (63.11) and combining it with the two other equations of the same system, one finds

$$\tilde{\vartheta}_1 + \frac{\tilde{\Delta}_z}{R_2\sin\theta} = -\frac{R_2}{E\delta}\left\{\left(1 - \frac{ic}{R_1}\right)\frac{1}{R_1}\frac{d\mathfrak{T}^{(1)}}{d\theta} - \frac{2(1+\mu)\tilde{S}^{(1)}}{R_2\sin\theta} + \right.$$

$$\left. + \frac{\tilde{T}_2^{(1)} - \mu\tilde{T}_1^{(1)}}{R_2}\operatorname{cotan}\theta\right\}, \tag{63.12}$$

where

$$\tilde{\Delta}_z = \tilde{w}_1\cos\theta - \tilde{u}_1\sin\theta. \tag{63.13}$$

On the other hand, direct differentiation of Δ_z gives

$$\frac{d\Delta_z}{d\theta} = \frac{d\tilde{w}_1}{d\theta}\cos\theta - \tilde{w}_1\sin\theta - \frac{d\tilde{u}_1}{d\theta}\sin\theta - \tilde{u}_1\cos\theta =$$

$$= -\tilde{\vartheta}_1 R_1\cos\theta - \left(\frac{d\tilde{u}_1}{d\theta} + \tilde{w}_1\right)\sin\theta =$$

$$= -\tilde{\vartheta}_1 R_1\cos\theta - \frac{R_1\sin\theta}{E\delta}(\tilde{T}_1^{(1)} - \mu\tilde{T}_2^{(1)}). \tag{63.14}$$

Hence, if one takes into consideration that $\tilde{T}_1^{(1)}$, $\tilde{S}^{(1)}$ and $\tilde{T}_2^{(1)}$ can be expressed in terms of $\tilde{T}^{(1)}$ by use of the formulae (63.4), (63.5), (63.7), one obtains for Δ_z the first order differential equation

$$\frac{d\Delta_z}{d\theta} - \tilde{\Delta}_z \frac{R_1}{R_2} \operatorname{cotan} \theta = \frac{R_2 \cos\theta}{E\delta} \times$$
$$\times \left\{ \left[1 - \frac{ic}{R_1} \left(1 + 2(1+\mu) \frac{R_1}{R_2} \frac{1+\cos^2\theta}{\sin^2\theta} \right) \right] \frac{d\mathfrak{T}^{(1)}}{d\theta} + \right.$$
$$\left. + (1+\mu) \frac{R_1}{R_2} \operatorname{cotan}\theta \left(1 + \frac{2ic}{R_2 \sin^2\theta} \right) \mathfrak{T}^{(1)} \right\}, \tag{63.15}$$

which, after introduction of the customary simplifications, takes the form

$$\frac{d\tilde{\Delta}_z}{d\theta} - \tilde{\Delta}_z \frac{R_1}{R_2} \operatorname{cotan}\theta = \frac{R_2 \cos\theta}{E\delta} \frac{d\mathfrak{T}^{(1)}}{d\theta}. \tag{63.16}$$

Since it is only necessary to find a particular integral of this equation which for every differentiation will increase in a ratio $\sqrt{\frac{R}{\delta}}$, one may neglect on the left hand side of (63.16) the second term. Further, one may assume for the purpose of the integration that the factor $R_2 \cos\theta$ is constant, neglecting its rate of change in comparison with that of $\mathfrak{T}^{(1)}$.

These considerations lead to the relation

$$\tilde{\Delta}_z \approx \frac{R_2 \cos\theta}{E\delta} \mathfrak{T}^{(1)}, \tag{63.17}$$

which applies in the same region as the remaining approximate formulae of this section. Now, the equation (63.12) permits the writing down of the following expression for $\tilde{\vartheta}_1$:

$$\tilde{\vartheta}_1 = -\frac{1}{E\delta} \left\{ \frac{R_2}{R_1} \left[1 - \frac{ic}{R_1} - (1+\mu) \frac{ic}{R_2} \cdot \frac{2+\cos^2\theta}{\sin^2\theta} \right] \frac{d\mathfrak{T}^{(1)}}{d\theta} + \right.$$
$$\left. + 2\mathfrak{T}^{(1)} \operatorname{cotan}\theta \left[1 + (1+\mu) \frac{ic}{R_2 \sin^2\theta} \right] \right\} \approx -\frac{1}{E\delta} \frac{R_2}{R_1} \frac{d\mathfrak{T}^{(1)}}{d\theta}, \tag{63.18}$$

after which $\tilde{u}_1$, $\tilde{v}_1$ and $\tilde{w}_1$ are readily obtained from the system (63.11).

Without dwelling on these computations which would be similar to the preceding ones, the final results will be given in the form of the following approximate formulae:

$$\tilde{w}_1 = \frac{R_2}{E\delta}\mathfrak{T}^{(1)},$$

$$\tilde{u}_1 = -i\frac{cR_2}{R_1{}^2E\delta}(R_2+\mu R_1)\frac{d\mathfrak{T}^{(1)}}{d\theta} = -i\frac{cR_2}{R_1}\cdot\frac{1}{E\delta}\left(\frac{R_2}{R_1}+\mu\right)\frac{d\mathfrak{T}^{(1)}}{d\theta}, \quad (63.19)$$

$$\tilde{v}_1 = \frac{ic}{E\delta\sin\theta}\left(2-\mu-\frac{R_2}{R_1}\right)\mathfrak{T}^{(1)}.$$

The real parts of the functions (63.18), (63.19) will be the coefficients of the expressions

$$u = u_1\cos\varphi,\ v = v_1\sin\varphi,\ w = w_1\cos\varphi,\ \vartheta = \vartheta_1\cos\varphi \quad (63.20)$$

and, therefore, the formulae (63.18) and (63.19) completely determine the displacements and the angle of rotation of shells of revolution, corresponding to the differential equation (62.16).

In what follows, consideration will be limited to the case when a shell of revolution has only one edge, i.e., when it is a dome without opening. Then one must put in the formulae (63.9)

$$C_1' = C_1'' = 0,$$

since the functions multiplying these constants in the case of a more exact solution of the problem under consideration (for example, in the second approximation) tend to infinity for $\theta = 0$.

Taking this circumstance into consideration, one obtains the following approximate expressions for the displacements, the angle of rotation, the forces and moments in a dome without aperture subject to wind loading:

$$u = \frac{\sqrt{R_2c}}{\sqrt{2}E\delta}\left(\frac{R_2}{R_1}+\mu\right)[(C_2-C_1)\cos\beta+(C_1+C_2)\sin\beta]e^{-\beta}\cos\varphi+u^*,$$

$$v = -\frac{c}{E\delta}\cdot\frac{1}{\sin\theta}\left(2-\mu-\frac{R_2}{R_1}\right)[C_2\cos\beta+C_1\sin\beta]e^{-\beta}\sin\varphi+v^*,$$

$$w = \frac{R_2}{E\delta}(C_1\cos\beta - C_2\sin\beta)e^{-\beta}\cos\varphi + w^*, \quad (63.21)$$

$$\vartheta = -\frac{1}{E\delta}\sqrt{\frac{R_2}{2c}}[(C_1+C_2)\cos\beta+(C_1-C_2)\sin\beta]e^{-\beta}\cos\varphi+\vartheta^*,$$

$$T_1 = \frac{\cot\!\mathrm{an}\,\theta}{\sqrt{2}}\sqrt{\frac{c}{R_2}}[(C_1-C_2)\cos\beta-(C_1+C_2)\sin\beta]e^{-\beta}\cos\varphi+T_1^*,$$

$$S = \frac{1}{\sqrt{2}\sin\theta}\sqrt{\frac{c}{R_2}}\,[(C_1-C_2)\cos\beta-(C_1+C_2)\sin\beta]\,e^{-\beta}\sin\varphi+S^*,$$

$$N_1 = \frac{1}{\sqrt{2}}\sqrt{\frac{c}{R_2}}\,[(C_1-C_2)\cos\beta-(C_1+C_2)\sin\beta]\,e^{-\beta}\cos\varphi,$$

$$T_2 = (C_1\cos\beta - C_2\sin\beta)\,e^{-\beta}\cos\varphi + T_2^*, \qquad (63.22)$$

$$M_2 = \mu M_1,$$

$$M_1 = -\,c(C_2\cos\beta + C_1\sin\beta)\,e^{-\beta}\cos\beta.$$

In the formulae (63.21) and (63.22)

$$\beta = \frac{1}{\sqrt{2}}\int_0^s \frac{ds}{\sqrt{R_2 c}}, \qquad (63.23)$$

where s is the arc length of the meridian of the dome measured from its edge and u^* ..., T_1^* ... are understood to be the solutions of the membrane theory. A method for their determination has been given in § 27.

Some attention will now be given to the case where

$$u^* = v^* = \ldots T_2^* = 0, \qquad (63.24)$$

i.e., where there is no surface loading on the shell, but only edge loading which varies like $\cos\varphi$ or $\sin\varphi$. Obviously the components of such an edge loading (T_1^0, N_1^0, S^0, M_1^0) may not be given in an arbitrary manner, since this would violate the equilibrium conditions of the shell as a whole. Therefore one need only consider systems of edge loading which are self-equilibrating.

It is readily understood that the conditions for the self-equilibration of the edge loadings

$$T_1^0 = t_1^0\cos\varphi,\ N_1^0 = n_0\cos\varphi,\ S^0 = s_0\sin\varphi,\ M_1^0 = m_0\cos\varphi \qquad (63.25)$$

reduce to two equations, whence it follows that in the present case of four components of the edge loading only two may be assigned arbitrarily, the moment M_1^0 and one of the forces. The relation which must hold between the edge forces, in order not to disturb the static equilibrium, will be deduced from the formulae (63.4), (63.7) and (63.10), (63.8). After separating in them real and imaginary parts, one finds

$$\begin{gathered} T_1\sin\varphi - S\cos\varphi\cos\theta = 0, \\ T_1\cos\theta\sin\varphi - S\cos\varphi + N_1\sin\theta\sin\varphi = 0. \end{gathered} \qquad (63.26)$$

Since the problem is homogeneous, these equalities must be fulfilled at all sections of the shell, including the edge regions. They are essentially the equations of static equilibrium for the shell as a whole. Because the equations (63.4) $\sim$ (63.10) were deduced by omitting certain small terms, the formulae (63.26) express the static conditions only in an approximate manner.

In fact, the moment M_1 has been neglected there in comparison with products of forces and moments of order of magnitude of the radius of curvature of the shell, corresponding to an error of order $\sqrt{\frac{\delta}{R}}$.

The question will now be asked what kind of edge loadings will satisfy these formulae. Excluding from them the force S, one finds that

$$Q_z{}^0 = T_1{}^0 \sin\theta - N_1{}^0 \cos\theta = 0, \tag{63.27}$$

whence it follows that the first requirement, imposed on edge loadings of the type (63.25) by the conditions of statics, is that the edge force cannot have a component in the direction of the axis of the shell.

Let Q_0 denote the component of the edge loading, acting normal to the edge and in its plane, so that

$$Q_0 = T_1{}^0 \cos\theta_0 + N_1{}^0 \sin\theta_0 = (t_0 \cos\theta_0 + n_0 \sin\theta_0) \cos\varphi = q_0 \cos\varphi. \tag{63.28}$$

A simultaneous study of the relations (63.27) and (63.28) shows that

$$\begin{aligned} N_1{}^0 &= Q_0 \sin\theta_0, \\ T_1{}^0 &= Q_0 \cos\theta_0. \end{aligned} \tag{63.29}$$

Substituting the second of these formulae into the first of the formulae (63.26), one obtains the following relation between Q_0 and S_0:

$$Q_0 \sin\varphi = S_0 \cos\varphi, \tag{63.30}$$

whence

$$q_0 = s_0. \tag{63.31}$$

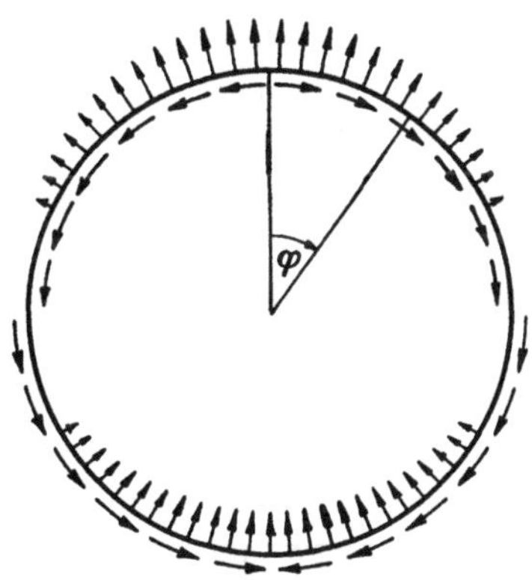

Fig. 92.

Thus, the homogeneous problem, i.e., the solution of the equation (62.16), corresponds to having on the edge of the shell two loadings acting in the plane of the edge (Fig. 92):

a) Normal forces Q_0

$$Q_0 = q_0 \cos \varphi;$$

b) tangential forces S_0

$$S_0 = q_0 \sin \varphi.$$

In addition, there may be given an arbitrary edge moment

$$M_0 = m_0 \cos \varphi.$$

However, its introduction disturbs the static equilibrium; in order to correct this effect, it is sufficient to adjust the edge forces Q_0, S_0 by amounts of the order $\sqrt{\dfrac{\delta}{R}}$ (compared with unity). Since terms of such magnitude lie outside the range of accuracy of the underlying theory, no contradictions will result. Consequently, one will reach the conclusion that in the case of the homogeneous problem only two quantities may be fixed arbitrarily at the edge of the shell, the parameters q_0 and m_0, which may be given directly or indirectly (in terms of the displacements and the angle of rotation). This conclusion also does not change in the general case, when the problem is not homogeneous, since the solutions of the membrane theory $u_1^* \ldots T_1^* \ldots$ do not involve arbitrary constants which might have to be disposed of. Therefore problems of stresses in domes under the action of wind loads are, in general, determined by two arbitrary boundary conditions (instead of the customary four of the theory of shells). This reduction of the number of arbitrary boundary conditions is explained by the fact that in the given case two of the four boundary conditions are compulsory and are dictated by the requirements of the equilibrium of the shell as a whole. In fact, this circumstance relates the problem of wind loading to the earlier studied problem of symmetrical deformation of shells of revolution. Thanks to this circumstance, in certain problems, the differential equations of the theory of shells admit quadrature, and correspondingly their order may be reduced. It is impossible to do so for any other loading included in the formulae (52.1), since for $k \geqslant 2$ these formulae give surface loadings which are self-equilibrating, as a consequence of which all components of the edge loading can be fixed arbitrarily.

Thus, the two particular cases $k = 0$ and $k = 1$ are fortunate exceptions and the physical reasons for this situation should now be completely understood.

§ 64. On the integration of the system (52.3) for shells whose radii of curvature satisfy the relation $\left(\frac{1}{R_1} - \frac{1}{R_2}\right)\frac{1}{\sin^2\theta} = \text{const}$

The two types of surface loading which have been considered above and which correspond to $k = 0$ and $k = 1$ in the formulae (52.1) permit reduction of the solution of the system (52.3) to the integration of differential equations which are only of second order. As has been noted in the previous section, this result cannot be extended to loadings of any other type, so that, as a rule, it is not possible to lower the order of the system (52.3) for $k \geqslant 2$ without introduction of some additional simplifications.

However, there is one family of shells for which the solution of the system (52.3) for any values of k may be reduced to the integration of two independent second order differential equations without any omissions in this system. This property belongs to shells of revolution the radii of curvature of which satisfy the relation

$$\left(\frac{1}{R_1} - \frac{1}{R_2}\right)\frac{1}{\sin^2\theta} = A, \tag{64.1}$$

where A — is a constant parameter.

First of all, the radii of curvature of this family of shells will be determined by supplementing (64.1) by the condition of Codazzi

$$\frac{d(R_2 \sin\theta)}{d\theta} = R_1 \cos\theta. \tag{64.2}$$

Solving (64.1) for R_1 and substituting the value so obtained into the expression (64.2), one arrives at the following differential equation for R_2:

$$\frac{d(R_2 \sin\theta)}{d\theta} = \frac{R_2 \cos\theta}{1 + AR_2 \sin^2\theta}, \tag{64.3}$$

which with

$$x = AR_2 \sin^2\theta \tag{64.4}$$

reduces to the form

$$\frac{1+x}{2+x}\,\frac{1}{x}\,\frac{dx}{d\theta} = \cot\text{an}\,\theta. \tag{64.5}$$

A general integral of this last equation is

$$x = -1 \pm \sqrt{1 + C\sin^2\theta}, \tag{64.6}$$

where C — is a constant of integration. Hence the family of shells under discussion falls into two sub-families the radii of curvature of which are expressed by the formulae:

$$\text{a)}\qquad R_1 = \frac{\sqrt{1 + C\sin^2\theta} - 1}{A\sqrt{1 + C\sin^2\theta}\,\sin^2\theta}, \qquad R_2 = \frac{\sqrt{1 + C\sin^2\theta} - 1}{A\sin^2\theta}; \tag{64.7}$$

$$\text{b)}\qquad R_1 = \frac{1 + \sqrt{1 + C\sin^2\theta}}{A\sqrt{1 + C\sin^2\theta}\,\sin^2\theta}, \qquad R_2 = -\frac{1 + \sqrt{1 + C\sin^2\theta}}{A\sin^2\theta}. \tag{64.8}$$

For the further development it is expedient to introduce instead of A and C two other constants which have clearer geometrical interpretations; in fact,

$$\left.\begin{aligned} \nu &= \left(\frac{R_2}{R_1}\right) \\ r_0 &= R_2 \end{aligned}\right\} \text{ for } \theta = \frac{\pi}{2}. \tag{64.9}$$

The new constant ν is equal to the ratio of the radii of curvature of the middle surface at the equator and the new constant r_0 is equal to the radius of the equatorial section of the shell. Consequently, the first constant characterizes the form of the shell, the second its dimensions.

It follows from the formulae (64.7), (64.8) and (64.9) that

$$C = \nu^2 - 1, \quad A = \frac{\nu - 1}{r_0}, \tag{64.10}$$

whence the formulae (64.7) and (64.8) become

$$\text{a)}\qquad R_1 = \frac{r_0}{\nu - 1}\,\frac{\sqrt{1 + (\nu^2 - 1)\sin^2\theta} - 1}{\sqrt{1 + (\nu^2 - 1)\sin^2\theta}\,\sin^2\theta}, \qquad \nu > 0, \qquad R_2 = \frac{r_0}{\nu - 1}\,\frac{\sqrt{1 + (\nu^2 - 1)\sin^2\theta} - 1}{\sin^2\theta}; \tag{64.11}$$

$$\text{b)}\quad R_1 = -\frac{r_0}{1-\nu}\,\frac{\sqrt{1+(\nu^2-1)\sin^2\theta}+1}{\sqrt{1+(\nu^2-1)\sin^2\theta}\,\sin^2\theta},$$

$$\nu < 0, \tag{64.12}$$

$$R_2 = \frac{r_0}{1-\nu}\,\frac{\sqrt{1+(\nu^2-1)\sin^2\theta}+1}{\sin^2\theta}.$$

Having obtained both radii of curvature as functions of the angle θ, one may now also evaluate the cartesian coordinates of the curves which are the meridians of the shells 'a' and 'b'. Let the origin of the system lie in the equatorial plane, the axis Y coincide with the axis of revolution and the X direction be perpendicular to Y.

The coordinates of the meridian of the middle surface in the above type of cartesian system will be determined by the formulae (Fig. 93)

$$x = R_2 \sin\theta,$$

$$y = -\int_{\frac{\pi}{2}}^{\theta} R_1 \sin\theta\, d\theta. \tag{64.13}$$

Substituting here the above values for R_1 and R_2, one will have

$$x = \frac{r_0}{\nu-1}\,\frac{-1 \pm \sqrt{1+(\nu^2-1)\sin^2\theta}}{\sin\theta}, \tag{64.14}$$

$$y = -\frac{r_0}{\nu-1}\int_{\frac{\pi}{2}}^{\theta} \frac{\mp 1 + \sqrt{1+(\nu^2-1)\sin^2\theta}}{\sin\theta\sqrt{1+(\nu^2-1)\sin^2\theta}}\, d\theta,$$

where the upper and lower signs refer to the sub-families of shells 'a' and 'b' respectively.

The integral in the expression for y may be evaluated to give the expression

$$y = -\frac{r_0}{\nu-1}\left\{\ln\tan\frac{\theta}{2} \mp \frac{1}{2}\ln\left[\frac{\sqrt{1+\nu^2\tan^2\theta}-1}{1+\sqrt{1+\nu^2\tan^2\theta}}\right]\right\}. \tag{64.15}$$

Hence one arrives at the following final formulae for the meridians of the shells 'a' and 'b'

a) $$x = \frac{r_0}{\nu - 1} \frac{\sqrt{1 + (\nu^2 - 1)\sin^2\theta} - 1}{\sin\theta},$$

$$\nu > 0, \qquad (64.16)$$

$$y = \frac{r_0}{\nu - 1} \ln\frac{1}{\nu} \frac{\sqrt{1 + \nu^2 \tan^2\theta} - 1}{\sec\theta - 1};$$

b) $$x = -\frac{r_0}{\nu - 1} \frac{\sqrt{1 + (\nu^2 - 1)\sin^2\theta} + 1}{\sin\theta},$$

$$\nu < 0, \qquad (64.17)$$

$$y = -\frac{r_0}{\nu - 1} \ln\frac{1}{\nu} \frac{1 - \sqrt{1 + \nu^2 \tan^2\theta}}{\sec\theta + 1}.$$

In these formulae, θ is assumed to vary between the limits

$$0 \leqslant \theta \leqslant \frac{\pi}{2}.$$

Fig. 94 shows the meridian curves of the families of surfaces under consideration for different values of ν. All these surfaces have the same equatorial width $2r_0$.

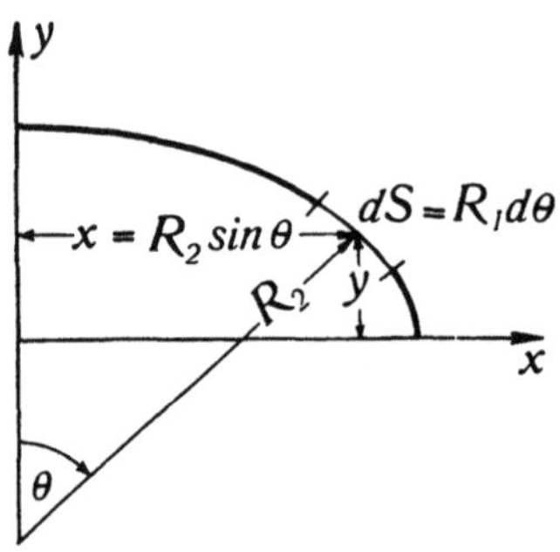

Fig. 93.

It is seen from Fig. 94 and likewise directly from the formulae (64.11) and (64.12) that the sub-family 'a' comprises shells of positive, and the sub-family 'b' shells of negative Gaussian curvature. The shape characteristic for the first group recalls ellipsoids of revolution, the second single-sheeted hyperbóloids. As a particular case, the sub-family 'a' contains for $\nu = 1$ the sphere, and the sub-family 'b' for $\nu = -1$ the catenoid, i.e., the surface, formed by revolution of a catenary the radii of curvature of which are given by the formulae

$$R_1 = -R_2 = -\frac{r_0}{\sin^2\theta}. \qquad (64.18)$$

The limiting surface, belonging to both sub-families, is the surface of zero Gaussian curvature, the circular cylinder which corresponds to $\nu = 0$. Thus, the formula (64.1) covers a wide range of shells of various forms, the study of which is undoubtedly of interest.

The basic system of differential equations (52.3) for all these shells takes the form

$$G_k(\tilde{U}_k) + k^2\left[1 - i\,\frac{c(\nu - 1)}{r_0}\right]\tilde{T}^{(k)} = f_k(\theta),$$
$$G_k(\tilde{T}^{(k)}) + \frac{i}{c}\,\tilde{T}^{(k)} + i\,\frac{(\nu - 1)}{r_0 c}\,\tilde{U}^{(k)} = \frac{i}{c}\,R_2 q_{n,k}, \tag{64.19}$$

where the operator $G_k(\ldots)$ and the functions of the surface loading $f_k(\theta)$ are defined by the formulae (52.4) and (52.5). It is an essential feature that the terms on the left hand sides of the system (64.19) which do not enter into the operator $G_k(\ldots)$ are in this case constant coefficients. Therefore the system (64.19) may be split up into two independent differential equations of second order, as may be shown by the following reasoning.

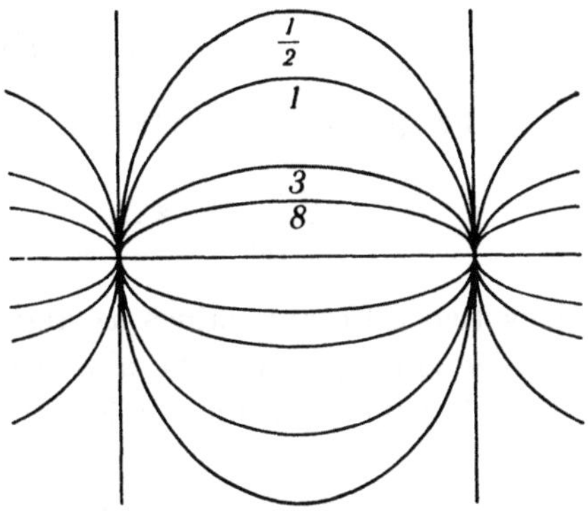

Fig. 94.

Introduce the new auxiliary function

$$\tilde{W} = \tilde{U}_k + \alpha\tilde{T}^{(k)}, \tag{64.20}$$

in which α is a constant but so far undetermined coefficient.

Further, expressing $\tilde{U}^{(k)}$ in terms of $\tilde{W}$ and $\tilde{T}^{(k)}$, using (64.20) and substituting in (65.19), one finds

$$G_k(\tilde{W}) - \alpha G_k(\tilde{T}^{(k)}) + k^2\left[1 - i\,\frac{c(\nu - 1)}{r_0}\right]\tilde{T}^{(k)} = f_k(\theta),$$
$$G_k(\tilde{T}^{(k)}) + \frac{i}{c}\,\tilde{T}^{(k)} + \frac{i(\nu - 1)}{r_0 c}\,\tilde{W} - \alpha\,\frac{i(\nu - 1)}{r_0 c}\,\tilde{T}^{(k)} = \frac{i}{c}\,R_2 q_{n,k}. \tag{64.21}$$

Hence, multiplying the second equation by α and adding it to the first equation, one obtains

$$G_k(\tilde{W}) + i\,\frac{\alpha(\nu - 1)}{r_0 c}\,\tilde{W} + \left\{k^2\left[1 - i\,\frac{c(\nu - 1)}{r_0}\right] + i\frac{\alpha}{c} - \alpha^2\,\frac{i(\nu - 1)}{r_0 c}\right\}\tilde{T}^{(k)} =$$
$$= f_k(\theta) + i\,\frac{\alpha}{c}\,R_2 q_{n,k}. \tag{64.22}$$

If now α is chosen in such a manner that the expression in the braces vanishes, the second unknown $\tilde{T}$ disappears from the equation (64.22). In this way, one obtains for $\tilde{W}$ the second order differential equation

$$G_k(\tilde{W}) + i\frac{\alpha(\nu - 1)}{r_0 c}\tilde{W} = f_k(\theta) + i\frac{\alpha}{c}R_2 q_{n,k}, \tag{64.23}$$

or, more correctly speaking, two such equations for the two auxiliary functions

$$\begin{aligned} \tilde{W}_1 &= \tilde{U}^{(k)} + \alpha_1 \tilde{T}^{(k)}, \\ \tilde{W}_2 &= \tilde{U}^{(k)} + \alpha_2 \tilde{T}^{(k)}, \end{aligned} \tag{64.24}$$

since the quadratic equation

$$i\frac{(\nu - 1)}{r_0 c}\alpha^2 - \frac{i}{c}\alpha - k^2\left[1 - i\frac{c(\nu - 1)}{r_0}\right] = 0 \tag{64.25}$$

gives two different values α_1 and α_2 for the constant α, each of which may be employed for the above transformation.

Solving the equation (64.25), one finds

$$\begin{aligned} \alpha_1 &= \frac{r_0}{2(\nu - 1)}\left\{1 + \sqrt{1 - i4k^2(\nu - 1)\frac{c}{r_0} - 4k^2(\nu - 1)^2\frac{c^2}{r_0^2}}\right\}, \\ \alpha_2 &= \frac{r_0}{2(\nu - 1)}\left\{1 - \sqrt{1 - i4k^2(\nu - 1)\frac{c}{r_0} - 4k^2(\nu - 1)^2\frac{c^2}{r_0^2}}\right\}. \end{aligned} \tag{64.26}$$

Substituting these values for α into the equations (64.23), one arrives at two independent differential equations for the auxiliary functions $\tilde{W}_1$, $\tilde{W}_2$.

It will now be assumed that one has succeeded in integrating these equations. Then one may determine directly the auxiliary functions $\tilde{U}^{(k)}$ and $\tilde{T}^{(k)}$ by solving the linear algebraic system (64.24). However, as has been pointed out already, the determination of these auxiliary functions is equivalent to the complete solution of the problem, since, if they are known, all the complex forces may be evaluated by means of operations involving at the most differentiations (using the formulae of §§ 51, 52). Thus, the analysis for an arbitrary loading of any shell of revolution, the radii of curvature of which satisfy the relation (64.1), has been reduced to the integration of two, independent, linear differential equations of second order of the form (64.23).

It will be noted that the ratio

$$\frac{c}{r_0} = \frac{1}{\sqrt{12(1 - \mu^2)}}\frac{\delta}{r_0}$$

is a quantity of order $\frac{\delta}{r_0}$. Expanding the roots in the formulae (64.26) in power series of this small parameter and neglecting quantities of order $\frac{c^2}{r_0{}^2}$ (and higher) in comparison with unity, one finds

$$\begin{aligned} \alpha_1 &\approx \frac{r_0}{\nu - 1}\left[1 - ik^2(\nu - 1)\frac{c}{r_0}\right], \\ \alpha_2 &\approx ick^2\left[1 + i(k^2 - 1)(\nu - 1)\frac{c}{r_0}\right]. \end{aligned} \tag{64.27}$$

These simpler formulae completely replace the formulae (64.26), provided k and ν are not so large that the parameter

$$k^2(\nu - 1)\frac{c}{r_0} \tag{64.28}$$

is of the same order as unity.

It has already repeatedly been mentioned in this book that in the simplification of the formulae of the theory of shells one has to deal with three factors: the relative thickness $\frac{\delta}{R}$, the degree of smoothness of the loading and the degree of smoothness of the form of the shell. The thinner the shell, the smoother its radii of curvature and the components of the surface loading (as functions of the coordinates of its middle surface), the better are the chances with regard to simplifications. The formulae (64.27) clearly demonstrate the roles of these three factors. Side by side with the relative thickness $\frac{c}{r_0}$, they involve the parameters k and ν, characterizing respectively the loading and the shape of the shell. Thus, the larger the value of k, the less smooth will be the loading, and the larger ν, the less smooth will be the shape of the shell (where the last point is shown by Fig. 94).

A study of the formulae (64.27) will confirm that an increase of k or ν could be equivalent to a thickening of the shell. Hence it follows that for shells of insufficiently smooth shapes which are loaded in an insufficiently smooth manner the errors arising out of the initial assumptions underlying the theory of shells must increase. In this connection, terms which, as a rule, may be neglected $\left(\text{terms of order } \frac{\delta}{r}\right)$ begin to play an essential role.

Let the shape of the shell and the character of its loading be sufficiently smooth, i.e., let the parameter (64.28) be small compared with unity. Then the formulae (64.27) may be further simplified; in fact, the second terms in these formulae may be neglected in comparison with the first terms. Hence one finds

$$\alpha_1 \approx \frac{r_0}{\nu - 1}, \qquad \alpha_2 \approx ick^2. \tag{64.29}$$

These expressions are the first approximations to the parameters α_1 and α_2.

Introducing (64.29) into (64.23), one obtains the two differential equations

$$G_k(\tilde{W}_1) + \frac{i}{c}\tilde{W}_1 = f_k(\theta) + i\frac{r_0}{(\nu - 1)c}R_2 q_{n,k}, \qquad G_k(\tilde{W}_2) - \frac{k^2(\nu - 1)}{r_0}\tilde{W}_2 = f_k(\theta) - k^2 q_{n,k} R_2 \tag{64.30}$$

for the auxiliary functions

$$\tilde{W}_1 = \tilde{U}^{(k)} + \frac{r_0}{\nu - 1}\tilde{T}^{(k)}, \qquad \tilde{W}_2 = \tilde{U}^{(k)} + ick^2\tilde{T}^{(k)}. \tag{64.31}$$

The first of these equations may be written out in full in the following manner:

$$\frac{d^2\tilde{W}_1}{d\theta^2} + \left[\left(2\frac{R_1}{R_2} - 1\right)\cot\theta - \frac{1}{R_1}\frac{dR_1}{d\theta}\right]\frac{d\tilde{W}_1}{d\theta} - \frac{R_1^2}{R_2^2}\frac{k^2}{\sin^2\theta}\tilde{W}_1 + i\frac{R_1^2}{R_2 c}\tilde{W}_1 = \frac{R_1^2}{R_2}\left[f_k(\theta) + i\frac{r_0}{c(\nu - 1)}R_2 q_{n,k}\right], \tag{64.32}$$

whence it becomes clear that it is of the same type as the equations (54.5) and (62.16), i.e., as the equations which had to be integrated for the analysis of arbitrary shells of revolution, subject to symmetrical or wind loading. Consequently, all the methods of approximate solution studied in the preceding sections will also be applicable to the above equation.

As regards the second equation, it takes, on the basis of the equality

$$\frac{\nu - 1}{r_0} = \left(\frac{1}{R_1} - \frac{1}{R_2}\right)\frac{1}{\sin^2\theta}, \tag{64.33}$$

the form

$$\frac{1}{R_1 R_2 \sin\theta}\frac{d}{d\theta}\left[\frac{R_2 \sin^2\theta}{R_1}\frac{d\tilde{W}_2}{d\theta}\right]-\frac{k^2}{R_1 \sin^2\theta}\tilde{W}_2=f_k(\theta)-k^2 R_2 q_{n,k}. \quad (64.34)$$

However, this equation differs in no way from the equation (24.4), i.e., from the basic differential equation of the membrane theory of shells of revolution. Thus, if the parameters α_1 and α_2 in their first approximations are determined from the formulae (64.29), then the analysis of shells of revolution belonging to the family under consideration reduces to the solution of two second order differential equations one of which is analogous to the equation for symmetrical deformation of shells of revolution while the other coincides with the equation of the membrane theory.

When proceeding to the second approximation given by the formulae (64.27), the equations (64.32) and (64.34) undergo a certain change (which increases with the values of ν and k). In that case, the equation (64.34) is found to differ somewhat from the differential equation of the membrane theory, whereas the equation (64.32) preserves its character as an equation of the type for symmetrical deformations of shells of revolution.

Hence it follows that the construction of the general solution of the problem under consideration by superposition of a membrane solution on a solution of the edge effect type is an approximate method which is justified only in those cases when α_1 and α_2 may be determined from the formulae for their first approximations, i.e., from the formulae (64.29). If the radii of curvature of a shell and the components of its loading are sufficiently smooth functions of the curvilinear coordinates of the middle surface, the errors incurred by employment of this method will be insignificant and they will not exceed the limits of accuracy which are customary for the theory of thin shells.

However, this statement requires one reservation. As has been noted by A. L. Gol'denweizer [8], the circular cylindrical shell belonging to the family of shells under consideration exhibits the special feature that the formulae of the first approximations must be used with greater care in this case than in the case of other shells. The fact is that for this shell the radius $R_1 = \infty$ so that the second term on the left hand side of the equation (64.34) vanishes. However, if one determines the formulae from the second, and not from the first approximation, this term will not vanish, although it has a small coefficient. This circumstance is of no importance for short cylindrical shells, while for very long shells it may change the character of the solution significantly.

§ 65. The analysis of spherical shells for arbitrary loading

In the present account of the analysis of spherical shells, it will be assumed that one is dealing with a spherical segment, and no consideration will be given to the study of shells with two edges, i.e., to frusta. As a consequence, one has to exclude from the solution terms which become infinite for $\theta = 0$. In addition, only the homogeneous problem will be considered, i.e., the case when there is no surface loading and the deformations of the shell depend only on the edge loading. For reasons to be explained later this case is of the greatest practical interest.

If the shell has the shape of a spherical segment and its deformations are arbitrary, the complex forces and complex displacements may be represented in the form of trigonometric series in the angle φ, so that it will be sufficient to consider a solution of the form

$$\tilde{T}_1 = \tilde{T}_{1,k} \cos k\varphi, \quad \tilde{T}_2 = \tilde{T}_{2,k} \cos k\varphi, \quad \tilde{S} = \tilde{S}_k \sin k\varphi, \tag{65.1}$$

$$\tilde{u} = \tilde{u}_k \cos k\varphi, \quad \tilde{v} = \tilde{v}_k \sin k\varphi, \quad \tilde{w} = \tilde{w}_k \cos k\varphi, \tag{65.2}$$

in order to make possible a study of the problem in its general form.

Assuming the system (64.19) to be homogeneous and letting $R_1 = R_2 = R$ and $\nu = 1$, one obtains

$$\begin{aligned} \Delta_k[\tilde{U}_k] + k^2 \tilde{T}^{(k)} &= 0, \\ \Delta_k[\tilde{T}^{(k)}] + i2a^2 \tilde{T}^{(k)} &= 0, \end{aligned} \tag{65.3}$$

where

$$\Delta_k(\ldots) = \frac{d^2(\ldots)}{d\theta^2} + \cot\theta \frac{d(\ldots)}{d\theta} - \frac{k^2}{\sin^2\theta}(\ldots), \tag{65.4}$$

$$\tilde{U}_k = \tilde{T}_{1,k} \sin^2\theta - \frac{i}{2a^2} \sin\theta \cos\theta \frac{d\tilde{T}^{(k)}}{d\theta}, \tag{65.5}$$

$$\tilde{T}^{(k)} = \tilde{T}_{1,k} + \tilde{T}_{2,k}, \tag{65.6}$$

$$4a^4 = 12(1 - \mu^2)\frac{R^2}{\delta^2}. \tag{65.7}$$

Thus, for spherical shells, the second of the equations of the system (64.19) is found immediately, without any transformations, to contain only one unknown, namely the auxiliary function $\tilde{T}^{(k)}$.

Written out in full, this equation becomes

$$\frac{d^2\tilde{T}^{(k)}}{d\theta^2} + \cot\theta \frac{d\tilde{T}^{(k)}}{d\theta} - \frac{k^2}{\sin^2\theta} \tilde{T}^{(k)} + i2a^2 \tilde{T}^{(k)} = 0. \tag{65.8}$$

It represents a particular case of the differential equation for spherical harmonics and its general solution is

$$\tilde{T}^{(k)} = \tilde{C}_1 P_n{}^k(\cos\theta) + \tilde{C}_2 Q_n{}^k(\cos\theta), \tag{65.9}$$

where $P_n{}^k$ and $Q_n{}^k$ are spherical harmonics of the first and second kinds and in the present case the lower index is the complex number

$$n(n+1) = i2a^2. \tag{65.10}$$

However, since the spherical harmonics have not been tabulated in the region of interest for the problem under consideration, the above solution is not readily used. Therefore the equation (65.8) will be integrated by another method, namely the asymptotic method studied in §§ 54, 55.

Using the substitution

$$z = \tilde{T}^{(k)}\sqrt{\sin\theta}, \tag{65.11}$$

the equation (65.8) will first be reduced to the form

$$\frac{d^2z}{d\theta^2} - \left(\frac{4k^2-1}{4\sin^2\theta} - \frac{1}{4} - i2a^2\right)z = 0. \tag{65.12}$$

The second term in the bracket may be omitted, since it is clearly small compared with the third term. Introducing this simplification and retaining the accuracy of the first approximation, one obtains

$$z = \tilde{C}_1 e^{\beta} + \tilde{C}_2 e^{-\beta}, \tag{65.13}$$

where

$$\beta = -(1-i)a\int_{\theta_0}^{\theta}\sqrt{1 + i\frac{4k^2-1}{8a^2\sin^2\theta}}\,d\theta, \tag{65.14}$$

and the lower limit in the last integral will be identified with the angle determining the edge of the spherical segment.

In the solution (65.13), the constant C_1 must be put equal to zero, since the function e^{β} becomes infinite for $\theta = 0$. Hence, for the spherical segment,

$$\tilde{T}^{(k)} = \frac{1}{\sqrt{\sin\theta}}z = \frac{1}{\sqrt{\sin\theta}}\tilde{C}_2 e^{-\beta}. \tag{65.15}$$

The deduced solution coincides with that recommended by A. Havers [151] who evaluated the integral in the formula (65.14) and constructed the graph of the function β. However, as was pointed out later on by A. Aas Jakobson [144], for practical purposes sufficient accuracy may be achieved more simply by taking into consideration that the function $e^{-\beta}$ will decay rapidly with the distance from the edge of the shell. On this basis, one may, firstly, neglect in the formulae (65.15) the rate of change of the factor $\frac{1}{\sqrt{\sin\theta}}$ in comparison with that of $e^{-\beta}$, i.e., one may assume that this multiplier is constant and include it in the constant of integration. Secondly, one may neglect in the integrand of (65.14) the variation of $\sin\theta$ and regard it as a constant ($\sin\theta = \sin\theta_0$) within the limits of the narrow strip in which the stresses, corresponding to the solution (65.15), will be essential.

Then one obtains for β the expression

$$\beta = (b_1 - ib_2)a\alpha, \tag{65.16}$$

where

$$\begin{aligned} b_1 &= \frac{1}{\sqrt{2}}\left\{\sqrt{\sqrt{1+\frac{(4k^2-1)^2}{64a^4\sin^4\theta_0}}+1}+\sqrt{\sqrt{1+\frac{(4k^2-1)^2}{64a^4\sin^4\theta_0}}-1}\right\},\\ b_2 &= \frac{1}{\sqrt{2}}\left\{\sqrt{\sqrt{1+\frac{(4k^2-1)^2}{64a^4\sin^4\theta_0}}+1}-\sqrt{\sqrt{1+\frac{(4k^2-1)^2}{64a^4\sin^4\theta_0}}-1}\right\}, \end{aligned} \tag{65.17}$$

and

$$\alpha = \theta_0 - \theta \tag{65.18}$$

is the angle, measured from the edge of the segment (Fig. 95).

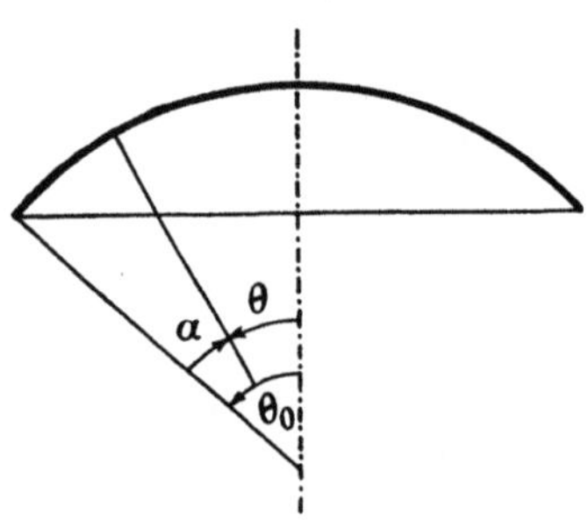

Fig. 95.

Differentiation with respect to this angle α is equivalent to differentiation with respect to $-\theta$.

Thus, one obtains for the auxiliary function $\tilde{T}^{(k)}$ the following final approximate expression

$$\tilde{T}^{(k)} \approx \tilde{C}e^{-(b_1-ib_2)a\alpha} = \tilde{C}e^{-\beta}, \tag{65.19}$$

which behaves similarly to those solutions which decay rapidly with the distance from the edge of the shell and which were considered in the sections devoted to symmetrical and wind loadings.

Introducing this result into the first of equations (65.3), one obtains the differential equation for the auxiliary function $\tilde{U}_k$

$$\frac{d^2\tilde{U}_k}{d\theta^2} + \operatorname{cotan}\theta\frac{d\tilde{U}_k}{d\theta} - \frac{k^2}{\sin^2\theta}\tilde{U}_k = -k^2\tilde{T}^{(k)} = -k^2\tilde{C}e^{-\beta}. \tag{65.20}$$

Since the function $\tilde{T}^{(k)}$ satisfies the equation (65.8), the particular integral of the equation (65.20) will obviously be

$$\tilde{U}_k = -\frac{i}{2a^2} k^2 \tilde{T}^{(k)} \approx -i \frac{k^2}{2a^2} \tilde{C} e^{-\beta}. \tag{65.21}$$

In order to obtain the general solution of the linear equation (65.20), one has to add to this particular integral the solutions of the homogeneous equation

$$\frac{d^2 \tilde{U}_k}{d\theta^2} + \cot\theta \frac{d\tilde{U}_k}{d\theta} - \frac{k^2}{\sin^2\theta} \tilde{U}_k = 0. \tag{65.22}$$

It may be verified by substitution that this equation is satisfied by

$$\tilde{U}_k = \tilde{B}_1 \tan^k \frac{\theta}{2} + \tilde{B}_2 \cot^k \frac{\theta}{2}, \tag{65.23}$$

where the constant $\tilde{B}_2$ must be put equal to zero, since in the particular problem under consideration the solution must remain finite for $\theta = 0$.

Adding (65.21) and (65.23) and taking the last observation into consideration, one arrives at the following general expression for the auxiliary function $\tilde{U}_k$:

$$\tilde{U}_k = \tilde{B} \tan^k \frac{\theta}{2} - i \frac{k^2}{2a^2} \tilde{C} e^{-\beta}. \tag{65.24}$$

Thus, both auxiliary functions $\tilde{U}_k$ and $\tilde{T}_k$ have now been determined and this is equivalent to the determination of all complex forces.

Attention will now be given to the fact that the equation (65.22) coincides with the equation which was obtained from the basic differential equation of the membrane theory for spherical shells, i.e., from an equation of Chapter II, if the last equation is assumed to be homogeneous. Hence it may be said that the general solution for a spherical shell subject to edge loading of the type $\begin{pmatrix}\cos k\varphi \\ \sin k\varphi\end{pmatrix}$ may be obtained by superposition of one or the other of the two solutions:

a)
$$\tilde{T}^{(k)} = 0, \tag{65.25}$$
$$\tilde{U}^{(k)} = \tilde{B} \tan^k \frac{\theta}{2};$$

b)
$$\tilde{T}^{(k)} = \tilde{C} e^{-\beta},$$
$$\tilde{U}^{(k)} = -i \frac{k^2}{2a^2} \tilde{C} e^{-\beta} = -i \frac{k^2}{2a^2} \tilde{T}^{(k)}. \tag{65.26}$$

The first of these coincide exactly with the solutions given for the same problem by the membrane theory, while the second, expressed in terms of a rapidly decaying function, are solutions of the edge effect type.

This result has an important fundamental meaning: it shows that the solution of the homogeneous problem for spherical shells may be obtained by adding the solution of the membrane theory to a solution of the edge effect type. It has been shown in the preceding section that such a subdivision of the general solution is possible for many shells, however, only as an approximation. For spherical shells, it is obtained without any omission and it may be considered a result of the same order of accuracy as that of the initial equations of the theory of thin shells. Thus, having divided the present problem into two parts, these will now be solved separately.

a) *The solution of the form* (65.25)

Substituting the solution (65.25) into the formula (65.5), one finds

$$\tilde{T}_{1,k} = \tilde{B}\frac{\tan^k \frac{\theta}{2}}{\sin^2\theta}. \tag{65.27}$$

After this, the condition

$$\tilde{T}^{(k)} = \tilde{T}_{1,k} + \tilde{T}_{2,k} = 0 \tag{65.28}$$

gives

$$\tilde{T}_{2,k} = -\tilde{B}\frac{1}{\sin^2\theta}\tan^k\frac{\theta}{2}. \tag{65.29}$$

In order to obtain the last complex force $\tilde{S}$, use will be made of the formulae (51.6) and (51.11). The second of these takes for the homogeneous problem of the theory of spherical shells the form

$$k\sin\theta\tilde{S}_k = -\frac{d\tilde{U}_k}{d\theta} - i\frac{k^2}{2a^2}\tilde{T}^{(k)}\cotan\theta. \tag{65.30}$$

Introducing here the values of $\tilde{U}_k$ and $\tilde{T}^{(k)}$ from (65.25), one finds

$$\tilde{S}_k = -\tilde{B}\frac{1}{\sin^2\theta}\tan^k\frac{\theta}{2}. \tag{65.31}$$

Thus, the following values of the coefficients in the formulae for the complex forces correspond to the solution of the form (65.25):

$$\tilde{T}_{1,k} = -\tilde{T}_{2,k} = -\tilde{S}_k = \tilde{B}\frac{1}{\sin^2\theta}\tan^k\frac{\theta}{2}. \tag{65.32}$$

On the basis of this result, the equations for the complex displacements (53.5) and (53.6) for the particular case of spherical shells take the form

$$\frac{d\tilde{u}_k}{d\theta} + \tilde{w}_k = \frac{R(1+\mu)}{E\delta} \tilde{B} \frac{1}{\sin^2\theta} \tan^k \frac{\theta}{2},$$

$$\frac{k}{\sin\theta} \tilde{v}_k + \tilde{u}_k \operatorname{cotan}\theta + \tilde{w}_k = -\frac{R(1+\mu)}{E\delta} \tilde{B} \frac{1}{\sin^2\theta} \tan^k \frac{\theta}{2}; \quad (65.33)$$

$$\frac{d\tilde{v}_k}{d\theta} - \tilde{v}_k \operatorname{cotan}\theta - \frac{k}{\sin\theta} \tilde{u}_k = -\frac{2(1+\mu)R}{E\delta} \tilde{B} \frac{1}{\sin^2\theta} \tan^k \frac{\theta}{2},$$

$$-\frac{d^2\tilde{w}_k}{d\theta^2} + \frac{d\tilde{u}_k}{d\theta} = -\frac{R}{E\delta} i2a^2\tilde{B} \frac{1}{\sin^2\theta} \tan^k \frac{\theta}{2},$$

$$(65.34)$$

$$\frac{k^2}{\sin^2\theta} \tilde{w}_k - \operatorname{cotan}\theta \frac{d\tilde{w}_k}{d\theta} + \frac{k}{\sin\theta} \tilde{v}_k + \tilde{u}_k \operatorname{cotan}\theta = \frac{R}{E\delta} i2a^2\tilde{B} \frac{1}{\sin^2\theta} \tan^k \frac{\theta}{2},$$

$$\frac{d}{d\theta}\left(\frac{k}{\sin\theta}\tilde{w}_k\right) + \frac{d\tilde{v}_k}{d\theta} - \tilde{v}_k \operatorname{cotan}\theta - \frac{k}{\sin\theta}\tilde{u}_k = \frac{R}{E\delta} i2a^2\tilde{B} \frac{1}{\sin^2\theta} \tan^k \frac{\theta}{2}.$$

On the basis of (65.33), the relations (65.34) may be somewhat simplified, after which they become

$$\frac{d^2\tilde{w}_k}{d\theta^2} + \tilde{w}_k = i2a^2 \frac{R}{E\delta}\left(1 - \frac{i(1+\mu)}{2a^2}\right) \tilde{B} \frac{1}{\sin^2\theta} \tan^k \frac{\theta}{2},$$

$$\frac{k^2}{\sin^2\theta} \tilde{w}_k - \operatorname{cotan}\theta \frac{d\tilde{w}_k}{d\theta} - \tilde{w}_k = i2a^2 \frac{R}{E\delta}\left(1 - i\frac{1+\mu}{2a^2}\right) \tilde{B} \frac{1}{\sin^2\theta} \tan^k \frac{\theta}{2}, \quad (65.35)$$

$$k\frac{d}{d\theta}\left(\frac{1}{\sin\theta}\tilde{w}_k\right) = i2a^2 \frac{R}{E\delta}\left(1 - i\frac{1+\mu}{a^2}\right) \tilde{B} \frac{1}{\sin^2\theta} \tan^k \frac{\theta}{2}.$$

Subtracting the second equation from the first, one finds

$$\frac{d^2\tilde{w}_k}{d\theta^2} + \operatorname{cotan}\theta \frac{d\tilde{w}_k}{d\theta} - \frac{k^2}{\sin^2\theta}\tilde{w}_k + 2\tilde{w}_k = 0. \quad (65.36)$$

Hence, as may be verified by substitution,

$$\tilde{w}_k = \tilde{D}_1(k + \cos\theta) \tan^k \frac{\theta}{2} + \tilde{D}_2(k - \cos\theta) \operatorname{cotan}^k \frac{\theta}{2}, \quad (65.37)$$

where $\tilde{D}_1$ and $\tilde{D}_2$ are constants of integration. In order to exclude terms which become infinite for $\theta = 0$, one must put here $\tilde{D}_2 = 0$.

The second constant $\tilde{D}_1$ must be determined in such a manner that the formula (65.37) will conform with all the three equations (65.35). Substituting (65.37) into these equations, one arrives at the following three equalities

$$k(k^2-1)\tilde{D}_1 = i2a^2\frac{R}{E\delta}\left(1-i\frac{1+\mu}{2a^2}\right)\tilde{B},$$

$$k(k^2-1)\tilde{D}_1 = i2a^2\frac{R}{E\delta}\left(1-i\frac{1+\mu}{2a^2}\right)\tilde{B}, \qquad (65.38)$$

$$k(k^2-1)\tilde{D}_1 = i2a^2\frac{R}{E\delta}\left(1-i\frac{1+\mu}{a^2}\right)\tilde{B}.$$

$$k \neq 0, \quad k \neq 1.$$

The cases $k = 0$ and $k = 1$, which correspond to symmetric and wind loading, can be considered with the aid of the earlier general theory.

The first two equalities are seen to be completely consistent, while the third contradicts them to the extent of a term of order $\frac{1}{2a^2}$ in comparison with unity, i.e., a term of order $\frac{\delta}{R}$. This inconsistency does not occur suddenly, since the reader has already been warned (§ 16, § 40) that contradictions of the order $\frac{\delta}{R}$ in comparison with unity would be quite possible in the formulae for the complex displacements as a result of the approximate nature of the reasoning, on the basis of which these formulae were generalized in Chapter I to the case $\mu \neq 0$. Therefore it is impossible to guarantee terms of order $\frac{\delta}{R}$ in (65.38). Hence the second terms on the right hand sides of the formulae (65.38) must be neglected which gives

$$\frac{R}{E\delta}\tilde{B} = -i\frac{k(k^2-1)}{2a^2}\tilde{D}_1 = -i\frac{k(k^2-1)}{2a^2}\tilde{D}. \qquad (65.39)$$

In what follows, the constant $\tilde{D}$ will be used instead of the constant $\tilde{B}$. Substituting the expression for $\tilde{w}_k$ into the formulae (65.33), one finds

$$\frac{d\tilde{u}_k}{d\theta} = -\tilde{D}\left\{i\frac{(1+\mu)k(k^2-1)}{2a^2}\cdot\frac{1}{\sin^2\theta}+k+\cos\theta\right\}\tan^k\frac{\theta}{2},$$

$$\frac{k}{\sin\theta}\tilde{v}_k+\tilde{u}_k\operatorname{cotan}\theta = -\tilde{D}\left\{-i\frac{(1+\mu)k(k^2-1)}{2a^2}\frac{1}{\sin^2\theta}+k+\cos\theta\right\}\tan^k\frac{\theta}{2}, \qquad (65.40)$$

$$\frac{d\tilde{v}_k}{d\theta} - \tilde{v}_k \operatorname{cotan}\theta - \frac{k}{\sin\theta}\tilde{u}_k = i(1+\mu)\frac{k(k^2-1)}{a^2}\tilde{D}\frac{1}{\sin^2\theta}\tan^k\frac{\theta}{2}.$$

The first of these formulae gives

$$\tilde{u}_k = -\tilde{D}\left[i\frac{(1+\mu)k(k+\cos\theta)}{2a^2\sin\theta} + \sin\theta\right]\tan^k\frac{\theta}{2} + \tilde{C}, \qquad (65.41)$$

after which one finds from the second formula

$$\tilde{v}_k = \tilde{D}\left\{i\frac{1+\mu}{2a^2}\left[\frac{k(k+\cos\theta)}{\sin\theta} - \sin\theta\right] - \sin\theta\right\}\tan^k\frac{\theta}{2} - \tilde{C}\frac{\sin\theta}{k}. \qquad (65.42)$$

Substituting these expressions for $\tilde{u}_k$ and $\tilde{v}_k$ into the third of the relations (65.40), one arrives at the relation

$$\tilde{C}\frac{k}{\sin\theta} = 0, \qquad (65.43)$$

whence it follows that the constant $\tilde{C}$ must be put equal to zero in order that the deduced formulae will be consistent with all the three relations (65.40).

Thus, for the complex forces and complex displacements corresponding to the solution of the form (65.25) one may write

$$\begin{aligned}
\tilde{u}_k &= -\tilde{D}\left[\sin\theta + i\frac{1+\mu}{2a^2}\frac{k(k+\cos\theta)}{\sin\theta}\right]\tan^k\frac{\theta}{2},\\
\tilde{v}_k &= -\tilde{D}\left\{\sin\theta - i\frac{1+\mu}{2a^2}\left[\frac{k(k+\cos\theta)}{\sin\theta} - \sin\theta\right]\right\}\tan^k\frac{\theta}{2},\\
\tilde{w}_k &= \tilde{D}(k+\cos\theta)\tan^k\frac{\theta}{2}, \qquad (65.44)\\
\tilde{T}_{1,k} &= -i\frac{E\delta}{R}\frac{k(k^2-1)}{2a^2}\tilde{D}\frac{1}{\sin^2\theta}\tan^k\frac{\theta}{2},\\
\tilde{T}_{2,k} &= \tilde{S}_k = i\frac{E\delta}{R}\frac{k(k^2-1)}{2a^2}\tilde{D}\frac{1}{\sin^2\theta}\tan^k\frac{\theta}{2}.
\end{aligned}$$

Instead of the complex constant $\tilde{D}$ now introduce two real constants B_1 and B_2, defined by

$$\tilde{D} = B_1 - \frac{i2a^2}{1+\mu}B_2.$$

Separating real and imaginary parts, in the formulae (65.44) one obtains the following expressions for the coefficients of $\begin{pmatrix}\sin k\varphi\\ \cos k\varphi\end{pmatrix}$ in the formulae for the displacements, forces and moments:

$$u_k = \operatorname{Re}(\tilde{u}_k) = -\left\{B_1 \sin\theta + B_2 \frac{k(k+\cos\theta)}{\sin\theta}\right\} \tan^k \frac{\theta}{2},$$

$$v_k = \operatorname{Re}(\tilde{v}_k) = -\left\{B_1 \sin\theta - B_2 \left[\frac{k(k+\cos\theta)}{\sin\theta} - \sin\theta\right]\right\} \tan^k \frac{\theta}{2},$$

$$w_k = \operatorname{Re}(\tilde{w}_k) = B_1(k+\cos\theta) \tan^k \frac{\theta}{2}, \tag{65.45}$$

$$T_{1,k} = \operatorname{Re}(\tilde{T}_{1,k}) = -\frac{E\delta}{R(1+\mu)} B_2 \frac{k(k^2-1)}{\sin^2\theta} \tan^k \frac{\theta}{2},$$

$$S_{1,k} = T_{2,k} = -T_{1,k},$$

$$M_{1,k} = -\frac{E\delta^3}{12(1+\mu)R^2} B_1 \frac{k(k^2-1)}{\sin^2\theta} \tan^k \frac{\theta}{2},$$

$$M_{1,k} = -H_k = -M_{2,k}.$$

It follows from these expressions that for $B_1 = 0$

$$M_1 = M_2 = H = 0,$$

and, vice versa, for $B_2 = 0$,

$$T_1 = T_2 = S = 0.$$

Hence it is clear that in (65.45) the terms, multiplied by B_2, represent a pure membrane solution and the terms, multiplied by B_1, a pure moment solution, i.e., solutions which are not accompanied respectively by bending or stretching of the middle surface. As has been stated in Chapter II, the equations of the membrane theory involve side by side with the first also those of the second type, so that the above result is not unexpected.

The above reasoning may be insufficiently logical for the attentive reader, since, in order to satisfy the relations (65.35), terms of order $\frac{1}{2a^2}$ have been neglected in comparison with unity, while terms of the same order have been included in the formulae (65.44) for the purpose of satisfying the relations (65.33). Nevertheless, there is no inconsistency in this procedure, because the relations (65.34) may be handled more freely than the relations (65.33), the left hand sides of which, as a rule, contain small differences of almost equal numbers.

If, instead of the formula (65.39), the constant $\tilde{D}_1$ had been determined by use of any of the formulae (16.38) which differ from the former by a term of order $\frac{1}{2a^2}$, then, after substitution of the thus defined expression for $\tilde{w}_k$ into the relations (65.33), one could establish that the formulae for $\tilde{u}_k$ and $\tilde{v}_k$ would only change by terms of order $\frac{1}{4a^4}$, i.e., by terms of order $\frac{\delta^2}{R^2}$.

Thus, small errors admitted into the relations (65.34) enter into the relations (65.33) at an essentially reduced scale. This statement is of fundamental value and may be regarded as a general rule for handling the formulae connecting the complex forces and complex displacements.

In fact, inconsistencies which may occur between the formulae (53.5) and (53.6) must be removed at the expense of modifications to the second group of formulae, and not to the first.

b) *Solution of the form* (65.26)

Substituting (65.26) into the formula (65.5), one finds

$$\tilde{T}_{1,k} = \frac{i}{2a^2}\left[\cot\theta \frac{d\tilde{T}^{(k)}}{d\theta} - \frac{k^2}{\sin^2\theta}\tilde{T}^{(k)}\right], \tag{65.46}$$

whence

$$\tilde{T}_{2,k} = \tilde{T}^{(k)} - \tilde{T}_1^{(k)} = \tilde{T}^{(k)} - \frac{i}{2a^2}\left[\cot\theta \frac{d\tilde{T}^{(k)}}{d\theta} - \frac{k^2}{\sin^2\theta}\tilde{T}^{(k)}\right]. \tag{65.47}$$

Further, the formulae (65.30) and (65.26) give

$$S^{(k)} = \frac{i}{2a^2}\,\frac{k}{\sin\theta}\left[\frac{d\tilde{T}^{(k)}}{d\theta} - \tilde{T}^{(k)}\cot\theta\right]. \tag{65.48}$$

After this, one finds for the determination of the complex displacements corresponding to the solution of the form (65.26) the six relations

$$\frac{d\tilde{u}_k}{d\theta} + \tilde{w}_k = \frac{R}{E\delta}\left[-\mu\tilde{T}^{(k)} + (1+\mu)\frac{i}{2a^2}\left(\cot\theta\frac{d\tilde{T}^{(k)}}{d\theta} - \frac{k^2}{\sin^2\theta}\tilde{T}^{(k)}\right)\right],$$

$$\frac{k}{\sin\theta}\tilde{v}_k + \tilde{u}_k\cot\theta + \tilde{w}_k = \tag{65.49}$$

$$= \frac{R}{E\delta}\left[\tilde{T}^{(k)} - (1+\mu)\frac{i}{2a^2}\left(\cot\theta\frac{d\tilde{T}^{(k)}}{d\theta} - \frac{k^2}{\sin^2\theta}\tilde{T}^{(k)}\right)\right].$$

$$\frac{d\tilde{v}_k}{d\theta}-\tilde{v}_k\cot\theta-\frac{k}{\sin\theta}\tilde{u}_k=\frac{2(1+\mu)R}{E\delta}\frac{i}{2a^2}\frac{k}{\sin\theta}\left[\frac{d\tilde{T}^{(k)}}{d\theta}-\tilde{T}^{(k)}\cot\theta\right],$$

$$-\frac{d^2\tilde{w}_k}{d\theta^2}+\frac{d\tilde{u}_k}{d\theta}=i2a^2\frac{R}{E\delta}\left[\tilde{T}^{(k)}-\frac{i}{2a^2}\left(\cot\theta\frac{d\tilde{T}^{(k)}}{d\theta}-\frac{k^2}{\sin^2\theta}\tilde{T}^{(k)}\right)\right],$$

$$\frac{k^2}{\sin^2\theta}\tilde{w}_k-\cot\theta\frac{d\tilde{w}_k}{d\theta}+\frac{k}{\sin\theta}\tilde{v}_k+\tilde{u}_k\cot\theta= \tag{65.50}$$

$$=-\frac{R}{E\delta}\left(\cot\theta\frac{d\tilde{T}^{(k)}}{d\theta}-\frac{k^2}{\sin^2\theta}\tilde{T}^{(k)}\right),$$

$$k\frac{d}{d\theta}\left(\frac{\tilde{w}_k}{\sin\theta}\right)+\frac{d\tilde{v}_k}{d\theta}-\tilde{v}_k\cot\theta-\frac{k}{\sin\theta}\tilde{u}_k=$$

$$=\frac{k}{\sin\theta}\frac{R}{E\delta}\left(\frac{d\tilde{T}^{(k)}}{d\theta}-\tilde{T}^{(k)}\cot\theta\right).$$

It will be shown that the first three of these equations are satisfied exactly and the remaining three approximately by letting

$$\tilde{u}_k=-\frac{i(1+\mu)}{2a^2}\frac{R}{E\delta}\frac{d\tilde{T}^{(k)}}{d\theta},\qquad \tilde{v}_k=\frac{i(1+\mu)}{2a^2}\frac{R}{E\delta}\frac{k}{\sin\theta}\tilde{T}^{(k)},$$

$$\tilde{w}_k=\frac{R}{E\delta}\tilde{T}^{(k)}. \tag{65.51}$$

Actually, substituting (65.51) into the formulae (65.49), it is found that the second and third of these equations become identities, while the first is converted into an equation for $\tilde{T}^{(k)}$ which coincides with the earlier equation (65.8) for the same function.

The relations (65.50), on substitution from (65.51), take the form

$$-\left(1+i\frac{1+\mu}{2a^2}\right)\frac{d^2\tilde{T}^{(k)}}{d\theta^2}=i2a^2\tilde{T}^{(k)}+\cot\theta\frac{d\tilde{T}^{(k)}}{d\theta}-\frac{k^2}{\sin^2\theta}\tilde{T}^{(k)},$$

$$\left(1+i\frac{1+\mu}{2a^2}\right)\left(\cot\theta\frac{d\tilde{T}^{(k)}}{d\theta}-\frac{k^2}{\sin^2\theta}\tilde{T}^{(k)}\right)= \tag{65.52}$$

$$=\cot\theta\frac{d\tilde{T}^{(k)}}{d\theta}-\frac{k^2}{\sin^2\theta}\tilde{T}^{(k)},$$

$$\left(1+i\frac{1+\mu}{a^2}\right)\frac{k}{\sin\theta}\left(\frac{d\tilde{T}^{(k)}}{d\theta}-\tilde{T}^{(k)}\cot\theta\right)=\frac{k}{\sin\theta}\left[\frac{d\tilde{T}^{(k)}}{d\theta}-\tilde{T}^{(k)}\cot\theta\right],$$

whence it follows that for the above substitution the first relation coincides with the equation (65.8) exactly to an accuracy of terms of order $\frac{1}{2a^2}$ in comparison with unity, while the other two relations are satisfied identically to the same degree of accuracy. However, as has already been noted, a similar error must be considered to be admissible and even unavoidable in the second group of formulae for the complex displacements.

Now, having expressed all the complex forces and complex displacements in terms of

$$\tilde{T}^{(k)} = \tilde{C}e^{-\beta} = (C_1 - iC_2)\,e^{-(b_1-ib_2)a\alpha}, \tag{65.53}$$

where C_1 and C_2 are two real constants, the following formulae may be obtained for the coefficients $u_k, \ldots, H_k$:

$$u_k = \frac{R}{E\delta}\frac{(1+\mu)}{2a}[-(C_1b_2+C_2b_1)\cos b_2a\alpha+(C_1b_1-C_2b_2)\sin b_2a\alpha]\,e^{-b_1a\alpha},$$

$$v_k = -\frac{R}{E\delta}\frac{1+\mu}{2a^2}\frac{k}{\sin\theta}[-C_2\cos b_2a\alpha + C_1\sin b_2a\alpha]\,e^{-b_1a\alpha},$$

$$w_k = \frac{R}{E\delta}(C_1\cos b_2a\alpha + C_2\sin b_2a\alpha)\,e^{-b_1a\alpha},$$

$$T_{1,k} = \frac{1}{2a}\Big\{[(C_1b_2+C_2b_1)\cos b_2a\alpha-(C_1b_1-C_2b_2)\sin b_2a\alpha]\cot\theta+ + \frac{k^2}{a\sin^2\theta}[-C_2\cos b_2a\alpha+C_1\sin b_2a\alpha]\Big\}e^{-b_1a\alpha},$$

$$T_{2,k} = \Big\{C_1\cos b_2a\alpha+C_2\sin b_2a\alpha+\frac{k^2}{2a^2\sin^2\theta}[C_2\cos b_2a\alpha- -C_1\sin b_2a\alpha]\Big\}e^{-b_1a\alpha}, \tag{65.54}$$

$$S_k = \frac{1}{2a}\frac{k}{\sin\theta}[(C_1b_2+C_2b_1)\cos b_2a\alpha-(C_1b_1-C_2b_2)\sin b_2a\alpha]\,e^{-b_1a\alpha},$$

$$M_{1,k} = \frac{R}{2a^2}\Big\{C_2\cos b_2a\alpha-C_1\sin b_2a\alpha-\frac{1-\mu}{2a^2}\frac{k^2}{\sin^2\theta}(C_1\cos b_2a\alpha+ +C_2\sin b_2a\alpha]\Big\}e^{-b_1a\alpha},$$

$$M_{2,k} = \frac{R}{2a^2}\left\{\mu(C_2\cos b_2a\alpha - C_1\sin b_2a\alpha) + \right.$$
$$\left. + \frac{1-\mu}{2a^2}\frac{k^2}{\sin^2\theta}(C_1\cos b_2a\alpha + C_2\sin b_2a\alpha)\right\}e^{-b_1a\alpha},$$
$$H_k = \frac{1-\mu}{4a^3}R\frac{k}{\sin\theta}[(C_1b_1 - C_2b_2)\cos b_2a\alpha + \tag{65.54}$$
$$+(C_1b_2+C_2b_1)\sin b_2a\alpha]e^{-b_1a\alpha}.$$

In these formulae, α denotes the angle measured from the edge of the shell and the variations in $\sin\theta$ and $\cot\theta$ must be disregarded by putting $\sin\theta = \sin\theta_0$, $\text{cotan}\,\theta = \text{cotan}\,\theta_0$, since this simplification was employed during the deduction of the formula (65.16). Adding the expressions (65.54) to the expressions (65.45), one obtains the general solution of the homogeneous problem for shells, having the shape of spherical segments and subject to edge loads obeying the law

$$\begin{pmatrix}\cos k\varphi \\ \sin k\varphi\end{pmatrix}.$$

These results are readily generalized to the case of a spherical shell bounded by two parallel circles, i.e., having the form of a spherical frustum. For this purpose, one only has to retain in the formulae (65.13) and also in (65.23) the terms which become infinite for $\theta = 0$ and to repeat the above manipulations.

Similarly, the obtained solution may be extended to the case when, in addition to edge loads, the shell is subject to surface loading. For this purpose one has to add to the sum of the formulae (65.45) and (65.54) terms corresponding to the particular integral. These are readily determined if the components of the surface loading are expanded in series of spherical harmonics. However, this question will not be discussed here further; in practice, surface loadings are either of the symmetrical or of the wind type, and these two particular cases have been studied above in sufficient detail.

As far as edge loads are regarded, they may exhibit a much greater variety. The group of practical problems the solution of which necessitates the study of values "k" > 2 includes, for example, the analysis of spherical domes supported by columns.

§ 66. The design of catenoid shells

As far as it is known, catenoid shells have no definite technical applications. However, they will be discussed here, although only briefly, as an example of shells of revolution of negative Gaussian curvature which, as has already been shown in Chapter II, by this very property differ essentially from shells of positive Gaussian curvature. In view of this circumstance, it will be expedient to carry out the solution even though it is for one of the shells of this class. The reason for selecting the catenoid as such an example is that for this shell the equations for shells of revolution may be integrated in terms of elementary functions.

The catenoid shell belongs to the family considered in § 64, and it is obtained from it for $\nu = -1$. Its radii of curvature, as has already been indicated, can be expressed in terms of θ by means of the formulae

$$R_2 = -R_1 = \frac{r_0}{\sin^2\theta}, \tag{66.1}$$

where r_0 is the radius of curvature at the equator and at the same time its smallest radius of curvature.

For this shell, the differential equations (64.23), after putting in them the components of the surface loading equal to zero, take the form

$$\begin{aligned} G_k(W_1) + i2b^2\alpha_1 W_1 &= 0, \\ G_k(W_2) + i2b^2\alpha_2 W_2 &= 0, \end{aligned} \tag{66.2}$$

where in this case

$$G_k(\ldots) = \sin^3\theta \frac{d}{d\theta}\left[\frac{1}{\sin\theta}\frac{d(\ldots)}{d\theta}\right] - k^2(\ldots), \tag{66.3}$$

$$4b^4 = 12(1-\mu)^2\frac{r_0{}^2}{\delta^2}, \tag{66.4}$$

$$W_1 = \tilde{U}_k + \frac{\alpha_1}{2}\tilde{T}^{(k)}, \quad W_2 = \tilde{U}_k + \frac{\alpha^2}{2}\tilde{T}^{(k)}, \tag{66.5}$$

$$\tilde{U}_k = \tilde{T}_{1,k} + \frac{i}{2b^2}\cos\theta\sin\theta\frac{d\tilde{T}^{(k)}}{d\theta}, \tag{66.6}$$

$$\begin{aligned} \alpha_1 &= \frac{1}{2}\left(1 + \sqrt{1 + i\frac{4k^2}{b^2} - \frac{4k^2}{b^4}}\right), \\ \alpha_2 &= \frac{1}{2}\left(1 - \sqrt{1 + i\frac{4k^2}{b^2} - \frac{4k^2}{b^4}}\right). \end{aligned} \tag{66.7}$$

Since both the equations (66.2) are of the same type, it is sufficient to integrate one of them, in order to be able to write down immediately the solution for the other equation. Therefore, attention will be focussed here on the first of these equations.

Introducing the substitution

$$Y = \frac{W_1}{\sin\theta}, \tag{66.8}$$

one finds after certain manipulations

$$\sin\theta \frac{d}{d\theta}\left(\sin\theta \frac{dY}{d\theta}\right) + (i2b^2\alpha_1 - k^2 - 1)Y = 0. \tag{66.9}$$

Now using the new independent variable

$$x = \ln\left(\tan\frac{\theta}{2}\right), \tag{66.10}$$

this equation will be reduced to the form

$$\frac{d^2Y}{dx^2} + (i2b^2\alpha_1 - k^2 - 1)Y = 0, \tag{66.11}$$

whence it is obvious that

$$Y = \tilde{C}_1\left(\tan\frac{\theta}{2}\right)^{m_1} + \tilde{C}_2\left(\operatorname{cotan}\frac{\theta}{2}\right)^{m_1}, \tag{66.12}$$

where

$$m_1 = \sqrt{k^2 + 1 - i2b^2\alpha_1}. \tag{66.13}$$

Now reverting to the dependent variable W_1, one finds

$$W_1 = \left[\tilde{C}_1\left(\tan\frac{\theta}{2}\right)^{m_1} + \tilde{C}_2\left(\operatorname{cotan}\frac{\theta}{2}\right)^{m_1}\right]\sin\theta. \tag{66.14}$$

By a quite similar reasoning, one may find for W_2

$$W_2 = \left[\tilde{C}_1\left(\tan\frac{\theta}{2}\right)^{m_2} + \tilde{C}_2\left(\operatorname{cotan}\frac{\theta}{2}\right)^{m_2}\right]\sin\theta, \tag{66.15}$$

where

$$m_2 = \sqrt{k^2 + 1 - i2b^2\alpha_2}.$$

The quantities α_1 and α_2 will now be determined from the formulae for the second approximation (§ 64)

$$\begin{aligned} \alpha_1 &= 1 + i\frac{k^2}{b^2}, \\ \alpha_2 &= -i\frac{k^2}{b^2}\left[1 - i\frac{k^2 - 1}{b^2}\right], \end{aligned} \tag{66.16}$$

whence

$$m_1 \approx \sqrt{-i2b^2 + 1 + 3k^2},$$

$$m_2 \approx \sqrt{1 - k^2 + i2\frac{k^2(k^2-1)}{b^2}} \approx \sqrt{1-k^2}\sqrt{1 - i\frac{2k^2}{b^2}}. \tag{66.17}$$

If the ratio $\frac{k^2}{b^2}$ is small compared with unity, a further simplification becomes possible after which the formulae (66.17) take the form

$$\begin{aligned} m_1 &\approx -(1-i)b, \\ m_2 &\approx i\sqrt{k^2-1}. \end{aligned} \tag{66.18}$$

For this smallest degree of accuracy, the integrals corresponding to the exponent m_1 will be the rapidly varying part of the solution (edge effect) and the integrals corresponding to the exponent m_2 will coincide with the solution of the homogeneous problem of the membrane theory for the catenoid. This discussion effectively exhausts the question of the integration of the equations for the catenoid shell, apart from purely technical details.

§ 67. Approximate theory of integration for any shells of revolution subject to arbitrary loading

Reverting to § 64 and considering the formulae for the first approximations to the parameters α_1 and α_2, it is seen that for this degree of accuracy the parameter α_2 is obtained independently of ν, i.e., of the shape of the shell. This fact justifies the statement that not only for shells of the family (64.1) one may arrive at a division of the system into two independent second order equations, but also for shells of revolution of arbitrary shape. However, there will be one difference, namely, that for shells of the family (64.1) this division may be achieved without omitting any terms in the initial system of equations, whereas for arbitrary shells it can only be realized as a first approximation. In § 64, the conditions have been explained under which the first approximation gives good results. These conditions are sufficient smoothness of the shell contour and sufficient smoothness of the loading. In mathematical terms, the last condition reduces to the requirement that the parameter $\frac{ck^2}{R}$ must be small compared with unity. Fulfillment of the two stated conditions guarantees the possibility for the use of the formulae of the first approximation with relatively small errors, except in the special case of the circular cylindrical shell for which a third factor, the length of the cylinder, may play an essential role.

It will now be assumed that these conditions are fulfilled. Then, for shells of revolution of arbitrary shape, one may write down the approximate differential equation

$$B_k(\tilde{W}_2) = f_k(\theta) + ik^2 R_2 q_n, \tag{67.1}$$

where

$$\tilde{W}_2 = \tilde{U}_k + ick^2 \tilde{T}^{(k)}, \tag{67.2}$$

$$B_k(\ldots) = \frac{1}{R_1 R_2 \sin\theta} \frac{d}{d\theta}\left[\frac{R_2{}^2 \sin\theta}{R_1} \frac{d(\ldots)}{d\theta}\right] - \frac{k^2}{R_1 \sin^2\theta}(\ldots). \tag{67.3}$$

The left hand side of this equation agrees exactly with the left hand side of the equation (24.4), i.e., with the left hand side of the basic differential equation of the membrane theory of shells of revolution, in view of which the auxiliary function will be

$$\tilde{W}_2 = \tilde{C}_1 \Phi_1 + \tilde{C}_2 \Phi_2 + \tilde{W}_2{}', \tag{67.4}$$

where Φ_1 and Φ_2 are two linearly independent integrals of the homogeneous membrane equation

$$B_k(\Phi) = 0, \tag{67.5}$$

and $\tilde{W}_2{}'$ is a particular solution of the equation.

The last may always be determined as soon as one knows Φ_1 and Φ_2.

The problem of the solution of the equation (67.5) has been considered in detail in Chapter II, in view of which there will be no need to repeat it here. It will only be recalled that for shells formed by rotation of second order curves the equation may be integrated by quadratures.

It will now be assumed that the general solution of the equation (67.1) has been obtained and that $\tilde{W}_2$ is a known function of θ. Expressing $\tilde{U}_k$ in terms of $\tilde{W}_2$ and the as yet unknown function $\tilde{T}^{(k)}$ and substituting this expression into the second of the equations (52.3), one obtains for $\tilde{T}^{(k)}$ the following linear second order differential equation:

$$G_k[\tilde{T}^{(k)}] + \frac{i}{c}\left[1 - i\frac{ck^2}{\sin^2\theta}\left(\frac{1}{R_1} - \frac{1}{R_2}\right)\right]\tilde{T}^{(k)} =$$

$$= \frac{i}{c}\left[R_2 q_{n,k} - \frac{1}{\sin^2\theta}\left(\frac{1}{R_1} - \frac{1}{R_2}\right)\tilde{W}_2\right], \tag{67.6}$$

or, in expanded form,

$$\frac{d^2\tilde{T}^{(k)}}{d\theta^2}+\left[\left(2\frac{R_1}{R_2}-1\right)\cot\theta-\frac{1}{R_1}\frac{dR_1}{d\theta}\right]\frac{d\tilde{T}^{(k)}}{d\theta}+$$
$$+\frac{k^2}{\sin^2\theta}\frac{R_1}{R_2}\left(1-2\frac{R_1}{R_2}\right)\tilde{T}^{(k)}+i\frac{R_1{}^2}{R_2c}\tilde{T}^{(k)}=$$
$$=i\frac{R_1{}^2}{R_2c}\left[R_2q_{n,k}-\frac{1}{\sin^2\theta}\left(\frac{1}{R_1}-\frac{1}{R_2}\right)\tilde{W}_2\right]. \tag{67.7}$$

For $k = 0$, this equation becomes the equation (54.5), and for $k = 1$ the equation (62.14). In other words, the equation includes as special cases the differential equations of shells of revolution subject to symmetrical or wind loadings. Hence, clearly, the same convenient approximate methods of integration will be completely applicable to this equation, which have been developed above for the equations (54.5) and (62.16).

In particular, if the right hand side of the equation (67.7) is a function which does not increase significantly on differentiation, one may take as its particular integral the expression

$$\tilde{T}_k' = R_2q_{n,k}-\frac{1}{\sin^2\theta}\left(\frac{1}{R_1}-\frac{1}{R_2}\right)\tilde{W}_2. \tag{67.8}$$

In order to obtain its general solution, one has to add to the result (67.8) all linearly independent integrals of the homogeneous equation

$$\frac{d^2\tilde{T}^{(k)}}{d\theta^2}+\left[\left(2\frac{R_1}{R_2}-1\right)\cot\theta-\frac{1}{R_1}\frac{dR_1}{d\theta}\right]\frac{d\tilde{T}^{(k)}}{d\theta}+$$
$$+\frac{k^2}{\sin^2\theta}\frac{R_1}{R_2}\left(1-2\frac{R_1}{R_2}\right)\tilde{T}^{(k)}+i\frac{R_1{}^2}{R_2c}\tilde{T}^{(k)}=0. \tag{67.9}$$

These integrals will be rapidly changing functions of the angle θ so that it will not be out of place to determine them by use of asymptotic methods.

Taking this into consideration, let

$$z=\frac{R_2\sqrt{\sin\theta}}{\sqrt{R_1}}\tilde{T}^{(k)}, \tag{67.10}$$

after which the equation (65.9) takes the form

$$\frac{d^2z}{d\theta^2}+\left\{i\frac{R_1{}^2}{R_2c}+\frac{1}{2R_1}\frac{d^2R_1}{d\theta^2}+\frac{3}{4}\frac{1}{R_1{}^2}\left(\frac{dR_1}{d\theta}\right)^2-\frac{\cot\theta}{2R_1}\frac{dR_1}{d\theta}+\right.$$
$$\left.+\frac{R_1}{R_2}\frac{1}{\sin^2\theta}-\frac{2+\cos^2\theta}{4\sin^2\theta}+\frac{k^2}{\sin^2\theta}\frac{R_1}{R_2}\left(1-2\frac{R_1}{R_2}\right)\right\}z=0. \tag{67.11}$$

Agreeing to neglect terms of order $\sqrt{\dfrac{\delta}{R}}$ compared with unity and not to consider points in the neighbourhood of the points $\theta = 0$ and $\theta = \pi$, one obtains the approximate solution

$$z = \tilde{C}_1 e^{\beta} + \tilde{C}_2 e^{-\beta}, \tag{67.12}$$

where

$$\beta = \int\limits_{\theta_0}^{\theta} \sqrt{-i\frac{R_1{}^2}{R_2 c} - \frac{k^2}{\sin^2\theta}\,\frac{R_1}{R_2}\left(1 - 2\frac{R_1}{R^2}\right)}\,d\theta =$$

$$= -\frac{1}{\sqrt{2}}(1-i)\int\limits_{\theta_0}^{\theta} \frac{R_1}{\sqrt{R_2 c}}\sqrt{1 - i\,\underline{\frac{k^2}{\sin^2\theta}\,\frac{c}{R_1}\left(1 - 2\frac{R_1}{R_2}\right)}}\,d\theta. \tag{67.13}$$

Attention will now be given to the reasons for retaining in the formula (67.13) the underlined term which by virtue of the first of the above two conditions for the applicability of the stated approximate solutions must be small in comparison with unity.

The fact is that, if it is desired to obtain a solution with an error of order not exceeding $\sqrt{\dfrac{\delta}{R}}$ in comparison with unity, one is justified in neglecting only terms of order $\dfrac{\delta}{R}$ in the square bracket of equation (67.1). Meanwhile, the last but one term being small, may all the same (for $k > 2$) be more significant than all the preceding terms in the same bracket. In fact, this is the reason why it must be retained. At the same time, one may neglect in the underlined term of (67.13), on the basis of its true character, the variations of R_1, R_2 and $\sin\theta$, putting these functions equal to their edge values, if the shell has a single edge $\theta = \theta_0$, or equal to the average edge value, if the shell has two edges $\theta = \theta_0$ and $\theta = \theta_1$.

As a result, one obtains for β the following approximate expression:

$$\beta = -\frac{(b_1 - ib_2)}{\sqrt{2}}\int\limits_{\theta_0}^{\theta}\frac{R_1 d\theta}{\sqrt{R_2 c}} = \frac{(b_1 - ib_2)}{\sqrt{2}}\int\limits_{0}^{s}\frac{ds}{\sqrt{R_2 c}}, \tag{67.14}$$

where

$$\begin{aligned} b_1 &\approx 1 - \frac{k^2}{2\sin^2\theta_0}\cdot\frac{c}{R_1{}^0}\left(1 - 2\frac{R_1{}^0}{R_2{}^0}\right),\\ b_2 &\approx 1 + \frac{k^2}{2\sin^2\theta_0}\cdot\frac{c}{R_1{}^0}\left(1 - 2\frac{R_1{}^0}{R_2{}^0}\right), \end{aligned} \tag{67.15}$$

and s is the arc length of the meridian of the shell, measured from one of its edges.

For practical purposes, the formula (67.14) may be written with sufficient accuracy in the form

$$\beta = \frac{b_1 - ib_2}{\sqrt{2R_2c}}\, s. \tag{67.16}$$

If the shell has only one edge and the point $\theta = 0$ belongs to its middle surface, one must put $\tilde{C}_1 = 0$ in the solution (67.12), since the corresponding integral for a more rigorous study of the problem would become infinite for $\theta = 0$.

Now, when the auxiliary functions $\tilde{T}^{(k)}$ and $\tilde{U}^{(k)}$ are known and determined by the formulae

$$\begin{aligned} \tilde{T}^{(k)} &= \tilde{C}_1 e^{\beta} + \tilde{C}_2 e^{-\beta} + \tilde{T}^{(k)\prime} = \tilde{T}^{(k)\prime\prime} + \tilde{T}^{(k)\prime}, \\ \tilde{U}^{(k)} &= W_0 - ick^2 \tilde{T}^{(k)}, \end{aligned} \tag{67.17}$$

the problem is essentially solved, since one may immediately determine from the formulae

$$\begin{aligned} \tilde{T}_{1,k} &= \frac{\tilde{U}^{(k)}}{R_2 \sin^2\theta} + i\,\frac{c}{R_1} \cot\!an\,\theta\, \frac{d\tilde{T}^{(k)}}{d\theta}, \\ \tilde{T}_{2,k} &= \tilde{T}^{(k)} - \tilde{T}_{1,k}, \\ \tilde{S}_k &= -\frac{1}{k}\, \frac{1}{R_1 \sin\theta} \cdot \frac{d\tilde{U}^{(k)}}{d\theta} - \frac{ikc}{R_2}\, \frac{\cos\theta}{\sin^2\theta}\, \tilde{T}^{(k)} + \\ &\qquad + (q_{n,k}\cos\theta - q_{1,k}\sin\theta)\,\frac{R_2}{k} \end{aligned} \tag{67.18}$$

all the complex forces after which the evaluation of the forces and moments reduces to a separation of the real and imaginary parts in the expressions (67.18).

When carrying out the operations in the formulae (67.18), it should be remembered that the formula

$$\tilde{T}^{(k)\prime\prime} = \frac{\sqrt{R_1}}{R_2\sqrt{\sin\theta}}\,(\tilde{C}_1 e^{\beta} + \tilde{C}_2 e^{-\beta}) \tag{67.19}$$

contained an error of order $\sqrt{\dfrac{\delta}{R}}$ in comparison with unity, in correspondence with which, for the differentiation of the expression (67.19), one must assume that the coefficient $\sqrt{R_1} : R_2\sqrt{\sin\theta}$ is constant. This makes it possible to include this coefficient in the integration constant as the result of which one obtains

$$\tilde{T}^{(k)''} = \tilde{C}_1 e^{\beta} + \tilde{C}_2 e^{-\beta}. \tag{67.20}$$

Concluding this study of an approximate method of analysis of shells of revolution of arbitrary shape subject to any loading obeying the law $\begin{pmatrix}\cos k\varphi \\ \sin k\varphi\end{pmatrix}$, it must be noted that in essence this method was known to A. E. H. Love and it has been systematically applied in his book [49]. In fact, the above author considered it to be admissible for the study of shells of revolution bounded by parallel circles, to solve by subdividing the solution into three parts — the membrane, the pure moment and the 'edge effect'. Obviously the above results coincide with this recommendation.

At the same time, the present treatment of the problem differs essentially in two respects from that due to Love. In the first place, Love did not consider the question of the errors involved in the method recommended by him; in the method above, this was obtained as the result of neglecting certain terms, the errors of which may be estimated by comparing the formula

$$\alpha_1 = ick^2 \tag{67.21}$$

with the formula

$$\alpha_1 = ick^2 \left[1 - i \frac{c}{\sin^2 \theta} \left(\frac{1}{R_1} - \frac{1}{R_2} \right) (k^2 - 1) \right]. \tag{67.22}$$

Secondly, when considering equations of the form (67.9), Love as a rule tended to integrate them mathematically exactly, which was an unnecessary complication, since, from the point of view of interest in practical calculations, the first approximation obtained from the asymptotic method gives already sufficient accuracy.

REFERENCES

The numbers in square brackets refer to the following list of journals:

1. Prikl. Mat. Mekh., Akademiya Nauk. S.S.S.R.
2. D.A.N., S.S.S.R.
3. Zbirnik prats' Inst. mat. A.N.U.S.S.R.
4. Trudy L.P.I.
5. Trudy Kaz. aviats. in-ta.
6. Izv. fiz. mat. ob-va pri Kaz. Un-te.
7. Izv. A.N. S.S.R. O.T.N.
8. Journ. für reine und ang. Math.
9. Phil. Trans. Roy. Soc.
10. Exercise de mathématique
11. Math. Annalen
12. Mém. de l'Acad. Sci, Paris.
13. Z.A.M.M.
14. Proekt i standart
15. Perebod s nem. izdan. Len. in-ta grazhd. vozd. f-ta.
16. Stroitel'naia promyshlennocm'
17. Bauingenieur
18. Zeit. für techn. Phys.
19. Uchenye zapuski Gor'kovck. Un-ta.
20. Izv. Elektr.-tekhn. in-ta
21. Beton und Eisen
22. Ingenieur Archiv
23. Werft und Reederei
24. Proc. of Amer. Soc. Civil. Eng.
25. Inzhenernyi sbornik
26. Izv. Kievck. politekhn. i c.-kh. in-tov.
27. Zeitschr. für Math. und Phys.
28. Trans. A.S.M.E.
29. Prikladna Mekhanika, A.N.U.S.S.R.
30. London Math. Soc. Proc.
31. Quart. of Appl. Math.
32. Journal Math. Phys.
33. Uchen. Zap. LGU (Mekhan).
34. Acta Technica Acad. Sci. Hungary.
35. Engineering News Rec.
36. Appl. Mech. Rev.
37. J. Boston Soc. Civ. Eng.
38. Zeitschr. für Bauwesen.
39. Schweiz. Bauz.
40. Nauchn. Zap. Dnepropetrov, Univ. (Mekh.)
41. Zeitschr. V.D.I.
42. Journal of Appl. Mech.
43. Math. Quart.
44. J. Aeronaut. Sci.
45. Der Stahlbau

CHAP. I. *General theory of Shells*

1. VEKUA, I. N., On the theory of thin shallow elastic shells, [1], Vol. XII, 1, 1948.
2. —, On the theory of elastic shells, [2], 68, 3, 1949.
3. VLASOV, V. Z., The fundamental differential equations of the general theory of elastic shells, [1], VIII, 2, 1944. [Also NACA T. M. 1241].
4. —, The general theory of shells and its industrial applications, Gostekhizdat, 1949.
5. GOL'DENWEIZER, A. L., Additions and corrections to the theory of thin shells, "Plates and Shells", Gosstroiizdat, 1939.
6. —, The equations of the theory of shells, [1], IV, 2, 1940.
7. —, On the applicability of the general theorems of the theory of elasticity to thin shells, [1], VIII, 1, 1944.
8. —, A qualitative study of the state of stress in thin shells, [1], IX, 6, 1945.
9. —, Certain examples of the integration of the equations of the theory of thin shells, [1], X, 3, 1946.
10. GOL'DENWEIZER, A. L., LUR'E, A. I., On the mathematical theory of equilibrium of elastic shells, [1], XI, 5, 1947.
11. GOL'DENWEIZER, A. L., Review of V. V. Novozhilov's book "Theory of Thin Shells", [1], XIII, 1, 1949, p. 117.
12. KIL'CHEVSKII, N. A., A generalization of contemporary shell theory, [1], II, 4, 1939.
13. —, Certain methods of integration of the equations of equilibrium of elastic shells, [1], IV, 2, 1940.
14. —, The basic equations of equilibrium of elastic shells and some methods for their integration, [3], 4, 5, 6, 1940; 6, 1941.
15. LUR'E, A. I., A study in the theory of thin shells, [4], 6, 1937.
16. —, On the equations of the general theory of elastic shells, [1], XIV, 5, 1950.
17. —, The general theory of thin elastic shells, [1], IV, 2, 1940.
18. —, The static theory of thin-walled elastic shells, Gostekhizdat, 1947.
19. RABOTNOV, IU. N., The fundamental equations of the theory of shells, [2], New Series, XLVII, 2, 1945.
20. MUSHTARI, KH. M., On the stability of cylindrical shells subject to torsion, [5], 2, 1938.
21. —, Certain generalizations of the theory of thin shells, [6], XI, 8, 1938.
22. —, On the region of applicability of the linear theory of elastic shells, [2], 58, 6, 1947.
23. —, On the region of applicability of the approximate theory of shells due to Kirchhoff-Love, [1], XI, 5, 1947.
24. —, The invariant equations of equilibrium of the boundaries of elastic shells in complex form, [1], XII, 2, 1948.
25. —, A qualitative study of the state of stress of an elastic shell subject to small deformations and arbitrary displacements, [1], XIII, 2, 1949.
26. NOVODVORSKII, V., Shells, Baku, 1932.
27. NOVOZHILOV, V. V., Theory of thin shells, Oborongiz, 1941.
28. —, Certain remarks regarding the theory of shells, [1], V, 3, 1941.
29. —, On the errors involved in one of the assumptions of the theory of shells, [2], 38, 5–6, 1943.
30. NOVOZHILOV, V. V., FINKEL'SHTEIN, R. M., Regarding the errors connected with the Kirchhoff hypotheses in the theory of shells, [1], VII, 1943.
31. NOVOZHILOV, V. V., On the question of the solutions of problems of the theory of thin shells subject to forces and moments, [2], 38, 9, 1943.
32. —, A new method for the analysis of thin shells, [7], 1, 1946.
33. —, A generalization of the method of complex displacements to the non-homogeneous problem of the theory of shells, [2], 53, 6, 1946.
34. RABOTNOV, IU. N., The boundary conditions in the theory of thin shells, [2], XLVII, 5, 1945.
35. TIMOSHENKO, S. P., Theory of elasticity, Part II, 1916.
36. FEINBERG, S., On the problem of the construction of an approximate theory of thin shells of arbitrary shape, "Studies in the Theory of Structures", Gosstroiizdat, 1939.

37. ARON, H., Das Gleichgewicht und die Bewegung einer unendlich dünnen, beliebig gekrümmten, elastischen Schale, [8], 78, 1874.
38. BASSET, A., On the extension and flexure of thin elastic shells, [9], 179, (A), 1890.
39. CAUCHY, A., Sur l'équilibre et le mouvement d'une plaque solide, [10], 3, 1828.
40. CLEBSCH, A., Théorie de l'élasticité des corps solides, Paris, 1883. (Translated from the German, with notes added by B. de St. Venant).
41. DONNELL, L., Stability of thin-walled tubes under Torsion, NACA Rep. No. 479, 1933.
42. —, A discussion of thin shell theory, Proc. Fifth Congress for Appl. Mech., 1939.
43. FLÜGGE, W., Statik und Dynamik der Schalen, Verlag J. Springer, Berlin, 1934.
44. FÖPPL. A., FÖPPL, L., Drang und Zwang, Vol. II.
45. GECKELER, J., Statics of elastic bodies, OGIZ, 1932.
46. KIRCHHOFF, G., Vorlesungen über Mathematische Physik, Vol. 1, Mechanik, 1876.
47. KRAUSS, F., Über die Grundgleichungen der Elastizitätstheorie schwachdeformierter Schalen, [11], 101, 1, 1929.
48. LOVE, A. E. H., On the small free vibrations and deformations of thin elastic shells, [9], 179 (A), 1888.
49. —, Mathematical theory of elasticity, 4th ed., 1927.
50. POISSON, S. D., Mémoire sur l'équilibre et le mouvement des corps solides, [12], 8, 1829.
51. TREFFTZ, E., Ableitung der Schalenbiegungsgleichungen mit dem Castiglianoschen Prinzip, [13], 2, 1935.
52. TIMOSHENKO, S. P., Plates and shells.

CHAP. II. *The Membrane Theory*

53. VLASOV, V. Z., Analysis of shells of revolution for unsymmetrical loading, [14], 3–4, 1937.
54. —, Analysis of shells, formed by second order surfaces, "Plates and Shells", Gosstroiizdat, 1939.
55. —, Membrane theory of thin shells, formed by second order surfaces, [1], XI, 4, 1947.
56. HAAS, R., DITTSIUS, A., The stretching and deformation of the shells of a flexible Air-ship, [15], 1931.
57. GVOZDEV, A. A., Contribution to membrane theory, [16], 1, 1933.
58. GOL'DENWEIZER, A. L., MROSHCHINSKII, A. K., REPMAN, IU. V., Methods of Analysis of spherical domes using membrane theory, "Plates and Shells", 1939.
59. GOL'DENWEIZER, A. L., The membrane theory of shells, bounded by second order surfaces, [1], XI, 2, 1947.
60. NOVODVORSKII, V., On the so-called membrane theory of shells, [16], 1, 1933.
61. RABOTNOV, IU. N., Certain solutions of the membrane theory of shells, [1], X, 5, 6, 1946.
62. RAPOPORT, I. IA., ZARKHI, L. I., MALIEV, A. S., The floor systems of Zeiss -Dywidag, Soviet Inst. of Structures, Gosstroiizdat, Leningrad, 1932.
63. SOKOLOVSKII, V. V., On membrane shells of revolution, [1], I, 3, 1938.
64. —, The equations of equilibrium of membrane shells, [1], VII, 1, 1943.
65. BELTRAMI, E., Sull equilibrio delle supperficie flessibili ed inestendibili, Mem. R. Acad. Sci di Bologna, 1881.
66. DISCHINGER, F., Shells, Gosstroiizdat, 1932.
67. DISCHINGER, F., Die Rotationsschalen mit unsymmetrischer Form und Belastung, [17], 1935.
68. —, Der Spannungszustand in affinen Schalen und Raumfachwerken, [17], 17, 23–27, 1936.
69. FEDERHOFER, K., Über Schalen gleicher Festigkeit, [17], 27–28, 1939.
70. LAMÉ, G. and CLAPEYRON, B. P. E., Mémoire sur l'équilibre interieur des corps solides, Mém. prés. par div. savants 4 1833

71. LAGALLY, M., Über Spannung und elastische Deformation von unebenen Membranen, [13], 4, 1924.
72. LECORNU, Z., Sur l'équilibre des surfaces flexibles et inextendibles., Journ. de l'école Polytechnique, c. XLVIII.
73. MEGAREUS, G., Die Kuppel gleicher Festigkeit, [17], 20, 1939.
74. MILANCOVIČ, M., O membranoma jednakog otpora, Rad. Iugoslavenske Akademije, kn, 175, 1908.
75. PÖSCHL. TH., Die Verwendung der Spannungsfunktion beim statischen Schalenprobleme, [18], Bd. 8, 1921.
76. —, Technische Daten über tropfenförmige Mineralölbehälter, [17], 8, 1927.
77. PÜCHER, A., Die Berechnung von doppeltgekrümmten Schalen mittels Differenzengleichungen, [17], 18, 9/10, 1937.
78. Ueber die Spannungsfunctionen beliebig gekrümmter dünner Schalen, Proc. of the fifth Int. Congr. for Applied Mech., 1939.

CHAP. III. *Analysis of cylindrical shells*

79. ADADUROV, R. A., Stresses and strains in a cylindrical shell with rigid transverse sections, [2], 62, 2, 1948.
80. AMBARTSUMIAN, S. A., On the analysis of shallow shells, [1], XI, 5, 1947.
81. —, On the theory of anisotropic elastic shells, [1], XII, 1, 1948.
82. BYCHKOV, D., MROSHCHINSKII, A. K., The torsion of metal beams, Stroiizdat, 1944.
83. VLASOV, V. Z., A new practical method for the analysis of folded covers and shells, [16], 11–12, 1932.
84. —, The moment theory of cylindrical shells, [14], 10, 1938.
85. VLASOV, V. Z., GOL'DENWEIZER, A. L., KAGAN, M., Graphs for the analysis of thin-walled wooden shells for the determination of bending and twisting moments, [14], 6, 1933.
86. VLASOV, V. Z., A new method for the analysis of thin-walled prismatic covers and shells, Gosstroiizdat, 1933.
87. —, The structural mechanics of shells, O.N.T.I., Stroiizdat, 1936.
88. —, Analysis of rib-stiffened shells and beams of thin-wall construction subject to bending and torsion, [14], 10, 1936.
89. —, Certain problems of the theories of strength of materials, structural mechanics and elasticity, [7], 9, 1950.
90. —, Thin-walled elastic rods (buckling, stability and vibrations), Gosstroiizdat, 1940.
91. —, Certain new problems of the structural mechanics of shells and thin-walled structures, [7], 1, 1947.
92. —, Contact problems in the theory of shells and thin-walled structures, [7], 6, 1949.
93. GALERKIN, B. G., On the theory of elastic cylindrical shells, [2], IV, 5–6, 1934.
94. —, The equilibrium of elastic cylindrical shells, Trudy-Leningrad Institute of Structures, 1935.
95. GVOZDEV, A. A., On the analysis of thin-walled cylindrical shells, [16], 1, 1932.
96. VLASOV, V. Z., Structural mechanics of thin-walled three-dimensional systems, Gosstroiizdat, 1949.
97. GALERKIN, B. G., PEREL'MAN, IA. I., Stresses and displacements in circular cylindrical pipe-line, Izv. Naychono-issledov, Inst., Gidro-techniki, 27, 1940.
98. GOL'DENWEIZER, A. L., Analysis of thin-walled shells and structures, based on rigid diaphragms, taking into account bending and twisting moments, "Plates and Shells", Gosstroiizdat, 1939.
99. —, On approximate methods for the analysis of thin shells of zero Gaussian curvature, [1], XI, 4, 1947.
100. —, On the theory of thin-walled rods, [1], XIII, 6, 1949.
101. DZHANELIDZE, G. IU., On the theory of slender and thin-walled rods, [1], XIII, 6, 1949.
102. LUR'E, A. I., Concentration of stress in the region of an opening in the surface of a circular cylinder, [1], X, 3, 1946.

103. Markov, A. N., On a method of integration of the equations for bending of cylindrical shells, [19], 6, 1938.
104. —, On the theory of bending of cylindrical shells, [19], 6, 1938.
105. Mroshchinskii, A. K., Analysis of cylindrical shells subject to concentrated forces, "Plates and Shells", 1939.
106. Novozhilov, V. V., Analysis of cylindrical shells, [7], 6, 1946.
107. Papkovich, P. F., Structural mechanics of ships, "Sea Transport", II, 1, 1947.
108. Pasternak, P. L., Practical analysis of structures and cylindrical shells allowing for bending moments. Inf. bull. NKTP, 1932.
109. —, Practical analysis of structures and cylindrical shells allowing for bending moments, [14], 2, 1933.
110. Rabotnov, Iu. N., Bending of cylindrical shells under concentrated loads. [2], 52, 4, 1946.
111. Timoshenko, S. P., On the problem of deformation and stability of cylindrical shells, [20], II, 1914.
112. Umanskii, A. A., Torsion and bending of air-frames, Oborongiz, 1939.
113. Feinberg, S., On the problem of the construction of a moment theory of cylindrical shells, [14], 12, 1936.
114. Khachaturian, T. T., Shallow cylindrical shells, A. N. Arm., S.S.R., Erevan, 1949.
115. Aas Jakobsen, A., Zylinderschalen mit veränderlichem Krümmungshalbmesser und veränderlicher Wandstärke, [17], 1937.
116. —, Über das Randstörungsproblem an Kreiszylinderschalen, [17], 20, 1939.
117. Dischinger, F., Die strenge Theorie der Kreiszylinderschale mit ihrer Anwendung auf Zeiss-Dywidag-Schalen, [21], 1935.
118. Finsterwalder, U., Die querversteiften, zylindrischen Schalengewölbe mit kreissegmentförmigem Querschnitt, [22], 1933.
119. Miesel, K., Über die Festigkeit von Kreiszylinderschalen bei nicht achsensymmetrischer Belastung, [22], I, 1929.
120. Reissner, H., Formänderung und Spannungen einer dünnwandigen, an den Randen frei aufliegenden Zylinderschale, [13], 1933.
121. Von Sanden, K., Günter, K., Über das Festigkeitproblem querversteifter Hohlzylinder unter allseitig gleichmässigem Aussendruck, [23], 8, 9, 10, 1920; 17, 1921.
122. Shorrer, H., Line load action on thin cylindrical shells, [24], March, 1935.
123. Schwerin, E., Über die Spannungen in freitragenden gefüllten Rohren, [13], 2, 1922.

Chap. IV. *Analysis of shells of revolution*

124. Burmistrov, E. F., Symmetrical deformation of a nearly cylindrical shell, [1], XIII, 4, 1949.
125. Vekua, I. N., Integration of the equations for spherical shells, [1], X, 5, 1945.
126. —, Some fundamental problems of the theory of thin spherical shells, [1], XI, 5, 1947.
127. Vlasov, V. Z., On two representations of the equations for spherical shells, [1], XI, 5, 1947.
128. Galerkin, B. G., Equilibrium of an elastic spherical shell, [1], VI, 6, 1942.
129. Gol'denweizer, A. L., A study of the state of stress of spherical shells, [1], VIII, 6, 1944.
130. Kantorovich, Z. B., Approximate solution of the edge problem for conical shells, [25], I, 2, 1941.
131. Lur'e, A. I., Equilibrium of an elastic symmetrically loaded spherical shell, [1], VII, 6, 1943.
132. Feodos'ev, V. I., The elastic elements of precision instrument construction, Oborongiz, 1949.
133. Novozhilov, V. V., Analysis of stresses in thin spherical shells for arbitrary loading, [2], XXVII, 6, 1940.
134. —, Analysis of shells of revolution, [7], 7, 1946.

135. REPMAN, IU. V., Analysis of spherical shells by use of the moment theory for arbitrary loading, "Plates and Shells", 1939.
136. SOKOLOVSKII, V. V., Analysis of spherical shells, [2], XVI, 1, 1937.
137. —, On a certain problem of the theory of shells, [2], XVI, 8, 1937.
138. FEODOS'EV, V. I., On the analysis of stiffened box-sections (stressed skins), [25], IV, 1, 1947.
139. SHAPIRO, G. S., On the equilibrium of a cone and conical shells, [1], VIII, 1944.
140. SHTAERMAN, I. IA., On the theory of symmetrical deformation of anisotropic elastic shells, [26], 1924.
141. —, On the integration of the differential equations of equilibrium of elastic shells, [26], 2, 1927.
142. —, Fundamental ideas of the contemporary theory of domes and vaults, Trudy Soviet Conference on Concrete and Reinforced Concrete, 1932.
143. —, Analysis of a dome as an arch on an elastic foundation, [14], 9, 1933.
144. AAS JAKOBSEN, A., Beitrag zur Theorie der Kugelschale auf Einzelstützen, [22], 1937.
145. BLUMENTHAL, O., Über asymptotische Integration von Differentialgleichungen mit Anwendung auf die Berechnung von Spannungen in Kugelschalen, [27], 62, 1914.
146. BOLLE, Z., Festigkeitsberechnung von Kugelschalen, Dissertationsarbeit, Zürich, 1916.
147. COATES, W., The state of stress in full heads of pressure vessels, [28], 52, 1930.
148. DUBOIS, F., Über die Festigkeit der Kugelschale, Promotionsarbeit, Zürich, 1917.
149. GECKELER, J., Über die Festigkeit achsensymmetrischer Schalen, Forsch, Arb. a.d. Geb. Ing. Wes., 276, 1926.
150. —, Die Festigkeit achsensymmetrischer Schalen, [22], I, 1930.
151. HAVERS, A., Asymptotische Biegetheorie der unbelasteten Kugelschalen, [22], 4, 1935.
152. HÖHN, E., Über die Festigkeit der gewölbten Böden und der Zylinderschale, Berlin, Verlag, J. Springer, 1927.
153. MAULBETSCH, J., HETENYI, M., Stresses in pressure vessels, [28], 58, 1936.
154. MEISSNER, E., Über Elastizität und Festigkeit dünner Schalen, Viertelschr. d. natur. Ges., Bd. 60, Zürich, 1915.
155. —, Das Elastizitätsgesetz für dünne Schalen von Ringflächen, Kugel- und Kegelform, Phys. Zeitschr., Bd. 14, 1913.
156. VAN DER NEUT, H., De elastische stabiliteit van den dunwandigen bol., Diss., Delft, 1932.
157. PASTERNAK, P., Die Praktische Berechnung biegefester Kugelschalen kreisrunder Fundamentsplatten auf elastischer Bettung und kreiszylinderscher Wandungen in gegenseitiger monolither Verbindung, [13], 1926.
158. PASTERNAK, P., Verhandl d. 2. Int. Kongr. für techn. Mech., Zürich, 1927.
159. REISSNER, H., Spannungen in Kugelschalen (Kuppeln.), Müller-Breslau,-Festschr., 1912.
160. SCHWERIN, E., Über Spannungen in symmetrisch und unsymmetrisch belasteten Kugelschalen (Kuppeln) insbesondere bei Belastung durch Winddruck, Diss., Berlin, 1918.
161. STANGE, K., Der Spannungszustand einer Kreisringschale, [22], 1931.
162. WISSLER, H., Festigkeitberechnung von Ringflächen, Promotionsarbeit, Zürich, 1916.
163. PONOMAREV, S. D., BIDERMAN, V. L., LIKHAREV, K. K., FEODOS'EV, V. I., Fundamentals of contemporary methods of analysis of the strength of machine designs, Mashgiz, 1950.

SUPPLEMENTARY REFERENCES

CHAP. I. *General Theory of Shells*

1a. VLASOV, V. Z., Thin-walled space systems, Gosstroiizdat, 1958.
2a. —, Some problems of the strength of materials, structural mechanics and the theory of elasticity, [7], 9, 1950.
3a. Second International Congress for thin-walled shell roofs, Oslo, 1957.
4a. GATEWOOD, B. E., Thermal stresses, McGraw-Hill, New York, 1957.
5a. GOL'DENWEIZER, A. L., Theory of thin elastic shells, GITTL, Moscow, 1953.
6a. —, On the application of the solution of the Riemann-Hilbert problem to the membrane analysis of shells, [1], Vol. XV, 2, 1951.
7a. —, A precise definition of the theory of the simple edge effect, [1], Vol. XV, 3, 1951.
8a. —, On Reissner's theory of bending of plates, [7], 4, 1958.
9a. —, A geometric criterion in the membrane theory of thin elastic shells, "Problems of continuum mechanics", SIAM-ANSSSR, Philadelphia-Moscow, 1961.
10a. KIL'CHEVSKII, N. A., Integro-differential and integral equations of equilibrium of thin elastic shells, [1], Vol. XXIII, 1, 1959.
11a. LUR'E, A. I., On the static geometric analogue of shell theory. "Problems of continuum mechanics", SIAM-ANSSSR, Philadelphia-Moscow, 1961.
12a. MIKHLIN, S. G., Error estimates in the analysis of elastic shells as plates, [1], Vol. XVI, 4, 1952.
13a. MISHONOV, M. K., On the theory of shallow shells, [1], Vol. XXII, 5, 1959.
14a. MUSHTARI, KH. M., GALIMOV, K. Z., The nonlinear theory of elastic shells, Tatknigoizdat, 1957.
15a. MUSHTARI, KH. M., TEREGULOV, I. G., The theory of shallow, orthotropic shells of medium thickness, [7], 6, 1959.
16a. NAGHDI, P., On the theory of thin elastic shells. [31], *14*, 4, 369—380, 1957
17a. —, A survey of recent progress in the theory of elastic shells, [36], 9, No. 9.
18a. NAZAROV, A. A., Equilibrium equations of shallow shells and their application, [29], 1956, 2, No. 3.
19a. NOVOZHILOV, V. V., Theory of elasticity, Sudpromgiz, 1958.
20a. ONIASHVILI, O. D., Analysis of shells and other thin-walled space structures, "Structural mechanics in the USSR", (1917–1957), Gostroiizdat, 1957.
21a. PONOMAREV, S. D., BIDERMAN, V. L., LIKHAREV, K. K., MAKUSHIN, V. M., MALININ, N. N., FEODOS'EV, V. I., Strength analysis in mechanical engineering, Vol. 2, Mashgiz, 1958.
22a. RABOTNOV, IU. N., Plates and shells "Mechanics in the USSR over 30 years", Moscow-Leningrad, Gostekhizdat, 1950.
23a. SAVIN, G. N., Stress distribution in a thin shell with an arbitrary hole "Problems in continuum mechanics", SIAM-ANSSSR, Philadelphia-Moscow, 1961.
24a. SURKIN, R. G., On the work of the Kazan school in the theory of shallow shells, Izv. Kazan. Fil. ANSSSR, Iubil. Sbornik, 1957.
25a. TIMOSHENKO, S. P., History of strength of materials, McGraw-Hill, New York, 1953.
26a. CHERNYKH, K. F., On conjugate problems of the theory of thin shells, [2], 117, No. 6, 1957.
27a. –, On a variational principle of the complex theory of shells, [1], Vol. XXII, 2, 1958.

28a. —, The conjugate problems of the theory of thin shells, "Problems of continuum mechanics" SIAM-ANSSSR, Philadelphia-Moscow, 1961.
29a. —, The linear theory of shells, Part I, Izd. Leningrad University, 1961.
30a. FADLE, I., Eine einfache Ableitung der Gleichgewichtsbedingungen des Schalenproblems, [22], XIV, 6, 1944.
31a. FLÜGGE, W., Statik und Dynamik der Schalen, Berlin, Springer, 1957.
32a. GREEN, A. E., ZERNA, W., Theoretical elasticity, Oxford, 1954.
33a. LAMB, H., On the deformation of an elastic shell, [30], 21, 1891.
34a. NAGHDI, P. M., The effect of transverse shear deformation on the bending of elastic shells of revolution, [31], 14, 1, 1957.
35a. NASH, W. A., Bibliography on shells and shell-like structures, U.S. Navy Dept.-David Taylor Model Basin Reports, November, 1954.
36a. PFLÜGER, A., Elementary Schalenstatik, Berlin-Göttingen-Heidelberg, 1957.
37a. REISSNER, E., Stress-strain relations in the theory of thin elastic shells, [32], 31, 1952.

CHAP. II. *The Membrane theory*

38a. VLAZOV, V. Z., On the membrane theory of shells of revolution, [7], 5, 1955.
39a. DISCHINGER, F., Dachbauten, Kragdächer, Schalen und Rippenkuppeln, Handbuch für Eisenbetonbau, 4th ed., Ernst & Sohn, Berlin 1927.
40a. LEBEDEV, V. A., Thin-walled free shells. Gosstroiizdat, 1958.
41a. PAVILAINEN, V. IA., On the equations of the membrane theory of shells, [33], 280, 3, 1960.
42a. RZHANITSYN, A. R., The membrane theory of shallow shells, "Analysis of space structures", 3, M. Gos. Izd-vo Lit. po str-vu i arkh., 1955.
43a. CHERNYKH, K. F., Analysis of a dome on columns, [25], 21, 1955.
44a. AIMOND, F., Einführung in die allgemeine Theorie der biegungsfreien Schalen, Int. Ver. f. Brücken- und Hochbau, II. Kongress. Berlin-München, Vorberichte, 1936.
45a. CSONKA, P., Results on shells of translation, [34], 10, 1, 1955.
46a. —, Calculation of calotte shells over rectangular bases, [34], 11, 3–4, 1955.
47a. COLBERG, O., 40 Jahre Eisenbetonkuppelbau, [21], XL, 1, 2, 3, 1941.
48a. DAY, G. L., New Spheroidal design of large oil tanks, [35], 103, 1929.
49a. DISCHINGER, F., Schalen und Rippenkuppeln, Handbuch für Eisenbetonbau, Vol., 6, 4-th ed., 1928.
50a. DISCHINGER, F., FINSTERWALDER, U., Eisenbetondächersystem Zeiss-Dywedag, [17], g, 1928.
51a. DISCHINGER, F., Theorie der Vieleckkuppeln, [21], 28, 1929.
52a. FLÜGGE, W., Recent trends in the theories of plates and shells, [36], 4, 489–491, 1951.
53a. HAGER, K., Die vierseitiggelagerte Kreisschale mit gleichförmiger Belastung, [17], 31/32, 1934.
54a. LÖBELL, F., Ein Beitrag zur Bestimmung der Deformationen einer elastischen Membran, [13], 7, 1927.
55a. PÖSCHL, TH., Berechnung von Behältern nach neueren analytischen und graphischen Methoden, Berlin, 1926.
56a. PUCHER, A., Beitrag zur Theorie tragender Flächen, Diss. Techn. Hochsch., Graz, 1931.
57a. —, Die Berechnung der Dehnungsspannungen von Rotationsschalen mit Hilfe von Spannungsfunktionen, Int. Ver. f. Brücken- u. Hochbau, Abh. 5, 1937–38.
58a. —, Die Schalenbauweise im Bauwesen, [41], N 17–18, 1913.
59a. REISSNER, E., On some aspects of the theory of thin elastic shells, [37], 42, 100–133, 1955.
60a. SCHWEDLER, Die Konstruktion der Kuppeldächer, [38], 1866.
61a. THOMA, D., Spannungen in dünnen zylindrischen Gefässwänden, Föppl. Festschr., 1924.
62a. TÖLKE, F., Über Rotationschalen gleicher Festigkeit für konstanten Innen- oder Aussendruck, [13], 19, 6, 1939.

63a. WIEDEMANN, E., Berechnung nicht biegungsteifer Rotationsschalen für Winddruck, [39], 1936.
64a. —, Ein Beitrag zur Frage der Formgebung räumlichtragender Tonnenschalen, [22], 4, 1937.
65a. WORCH, G., Die statisch bestimmte Kugelschale unter Winddruck, [21], 1932.

CHAP. III. *Analyses of cylindrical shells*

66a. AMBARTSUMIAN, S. A., On the problem of the construction of an approximate theory for the analysis of shallow cylindrical shells, [1], XVIII, 3, 1954.
67a. —, On the limits of the applicability of certain assumptions of the theory of thin cylindrical shells, [7], 5, 1954.
68a. VLAZOV, V. Z., A new method of analysis of prismatic beams of thin-wall construction under bending and torsion. Vestnik Voenno-inzhenernoi Akadamii RKKA im Kuibysheva, 20, Sb. po stroit. Mekh. II, 1937.
69a. —, Structural mechanics of shells, ONTI, Stroiizdat, 1936.
70a. —, Structural mechanics of thin-walled space systems, Gosstroiizdat, 1949.
71a. GIL'MAN, L. K., On the analysis of isotropic cylindrical shells under arbitrary loading, Tr. Vyssh. Voenno. morsk. inzh. str. uchilishcha RKMF, 1939.
72a. GRIGOLIUK, E. I., On the theory of circular cylindrical shells with rigid ribs, [7], 11, 1954.
73a. DAREVSKII, V. M., The solution of certain problems of the theory of cylindrical shells, [1], XVI, 5, 1952.
74a. MALKINA, R. L., On the analysis of tubes with arbitrary cross-section, [25], 19, 1954.
75a. NIKITIN, V. A., An approximate solution of the problem of concentrated forces acting on a cylindrical shell, [33], 280, 3, 1960.
76a. NOVOZHILOV, V. V., LASHMANOVA, I. A., Torsion of tubes, [33], 217, 1957.
77a. NOVOZHILOV, V. V., KOZHEVNIKOVA, M. K., An approximate theory of constrained torsion of thin-walled rods of closed profile taking into account warping of the cross-section, [7], 9, 1956.
78a. PIROGOV, I. M., Stress distribution around a hole in a cylindrical shell subject to concentrated forces, [7], Ser. Mekh i Mash., 2, 1959.
79a. SHIMANSKII, IU. A., Analysis of local strength of tubes and barrels. Sb., Sudpromgiz 1954.
80a. ESTRIN, M. I., Analysis of cylindrical shells clamped along an oblique contour, [7], ser. mekh. i mash., 2, 1959.
81a. AAS JAKOBSEN, A., Sur le calcul de la voute cylindrique circulaire, Trav. Architect, construct. 21, 1937.
82a. BAZANT, ZD., Théorie éxacte des enveloppes cylindrique épaisses, Int. Ver. f. Brücken- u. Hochbau, 1936.
83a. —, Théorie de l'enveloppe cylindrique sollicitée par une préssion radiale variable, Acad. Tchéque Sci. Bull. int., 39, 1939.
84a. BIJLAARD, Stresses from radial loads and external moments in cylindrical pressure vessels, Weld. Journal, Dec. 1955.
85a. BLEICH, F. und BLEICH, H., Biegung, Drillung und Knickung von Stäben aus dünnen Wänden, Int. Ver. f. Brücken- u. Hochbau, II. Kongress, Berlin-München, Vorberichte, 1936.
86a. BROWN, W., Stress and strain in thin elliptic cylinder under internal pressure, Engineer, vol. 162, 28/VIII, 1936.
87a. ENYEDI, B., Praktische Berechnung der Kreiszylindrischen Behälterwände aus Eisenbeton, Int. Ver. f. Brücken- u. Hochbau, Abh. 2, 1934.
88a. GRÜBER, E., Die Berechnung zylindrischer biegungssteifer Schalen unter beliebigem Lastangriff., Int. Ver. f. Brücken- u. Hochbau, Abh. 2, 1934.
89a. HOFF, N., KEMPNER, J., POHLE, F. V., Line load applied along generators of thin-walled circular cylindrical shells of finite length, [31], II, N 4, 1954.
90a. HOFF, N. I., The Accuracy of Donnell's equations, [42], 22, N 3, 329–334. 1955.
91a. MEISSNER, E., Beanspruchung und Formänderung zylindrischer Gefässe mit linear verändlicher Wandstärke, Vierteljahresschr. der naturforsch, Ges. Zürich, 1917.

92a. Roark, R. I., The strength and stiffness of cylindrical shells under concentrated loading, [42], Dec., 1935.
93a. Shorrer, H., Line load action on thin cylindrical shells, [24], March, 1936.
94a. Volterra, Enrico, Contributo al calcolo delle volte sottili cilindriche, Ann. Lavori Publ., 76, 1938.
95a. Withum, D., Die Kreiszylinderschale mit kreisförmigem Ausschnitt unter Schubbeanspruchung, [22], 26, 6, 1958.
96a. Yuan, S. W., Ting, L., On radial deflections of a cylinder, subjected to equal and opposite concentrated radial loads, [43], 24, 2, 1957.
97a. —, —, On radial deflection of a cylinder of finite length with various end conditions, [44], 25, 4, 1958.

Chap. IV. *Analysis of shells of revolution*

98a. Belova, V. I., Stress distribution in an extended plane sheet with an axisymmetric dent, [33], 217, 1957.
99a. Kantorovich, Z. B., The foundations of the analysis of chemical equipment and apparatus, Mashgiz, 1960.
100a. Clark, R. A., Reissner, E., A problem of finite bending of toroidal shells, [31], 10, Jan. 1953 pp. 321–334.
101a. Kovalenko, A. D., Survey of studies in the theory of plates and shells in rotating turbines, [29], 1, 2, 1955.
102a. Korolevich, Iu. S., Asymptotic solution of problems of axi-symmetric deformation of conical shells with linearly varying thickness, (Ukrainian) [29], V, 1, 1959.
103a. Lakhtin, A. A., On the analysis of certain details including toroidal parts, [25], 26, 1958.
104a. Naumov, V. K., Graphs for the analysis of toroidal elements, Leningrad metallurgical works, A study of parts of boilers and gas turbines and axial compensators, Moscow-Leningrad, 6, 1960.
105a. Palatnikov, E. A., Analysis of axial compensators in pipelines, Oborongiz, 1957.
106a. Razletov, B. K., Determination of stress fields in a plane sheet with an axisymmetric dent, [25] 17, 1953.
107a. Timoshenko, S. P., On a problem of the analysis of spherical shells, Vestn. o-va tekhnologov, 1914.
108a. Tumarkin, S. A., The asymptotic solution of a linear, nonhomogeneous, second order differential equation with a singular point and its application to the analysis of toroidal shells and blades, [1], XXIII, 6, 1959.
109a. Feodos'ev, V. I., Elastic elements of precision instrument making, Oborongiz, 1949.
110a. Chernykh, K. F., Meissner's equation: the case of skew-symmetric loading, [7], 6, 1959.
111a. —, St. Venant's problem for thin-walled tubes with a circular axis, [1], XXIV, 3, 1960.
112a. Chernykh, K. F., Chistiakova, K. N., On the analysis of a circular plate reinforced by radial ribs, [33], 280, 3, 1960.
113a. Shevliakov, Iu. A., Bespal'ko, L. A., Analysis of conical shells rigidly fixed to a base, [40], 73, 10, 1959.
114a. Becker, H., Load distribution at the intersection of several coaxial axisymmetric shells, [42], 22, 2, 1955.
115a. Bijlaard, P. P., Local stresses in spherical shells from radial or moment loading, Welding I, 36, 5, 1957.
116a. Clark, R. A., On the theory of thin elastic toroidal shells, [32], 29, 1950.
117a. Clark, R. A., Reissner, E., On axially symmetric bending of nearly cylindrical shells of revolution, [42], 23, 1, 1956.
118a. Clark, R. A., Asymptotic solutions of toroidal shell problems, [31], 16, April, 1958.
119a. Ekström, J. E., Studien über dünne Schalen von rotationssymmetrischer Form und Belastung mit konstanter oder verändlicher Wandstärke, Stockholm, 1933 (Ingenieurs Vetensk. Akademien Handl., N 121), 1936.

120a. —, Die allgemeine Lösung des achsensymmetrischen Schalenproqlems mit besonderer Rücksicht auf elliptische Dampfkesselböden, Akad. Ing. Vetensk. Handl., 140, 1936.
121a. FANKHAUSER, E., Experimentelle und theoretische Untersuchungen der Festigkeit von Kegel- und Kugelschalen, Berlin, 1913.
122a. GRANHOLM, H., Massive Kuppeln, Zylindrische Behälter und ähnliche Konstructionen, Int. Ver. f. Brücken- u. Hochbau, II Kongr., Berlin-München, Bd. 1, 1936.
123a. HABEL, A., Beitrag zur Berechnung der Biegungsspannungen in kreiszylindrischen Behältern mit Kuppeldeck und Zugring bei achsensymmetrischer Belastung, Mitt. Hauptver. Deutsch, Ing. Tschechoslav., Rep. 25, 1936.
124a. HAMPL, M., Beanspruchung und Durchbiegung der dünnen Kugelböden durch radialen constanten Druck. Acad. Tchêque, Bull. Inst., 1937.
125a. —, Das Spannungsproblem der achsensymmetrischbelasteten Kugelschale, [45], 11, 1940.
126a. HOFF, N. J., Thin circular conical shells under arbitrary loads, [42], 22, 4, 1955.
127a. HONEGGER, E., Festigkeitberechnung von Kugelschalen mit linear veränderlicher Wandstärke, Dissertationarbeit, Luzern, 1919.
128a. HOLLISTER, S., Stresses in symmetrically loaded hemispherical shells having tapered edges, [42], 4, 1937.
129a. LOHMANN, W., Beitrag zur Integration der Reissner-Meissnerschen Schalengleichungen für Behälter unter konstantem Innendruck, [22], 1935.
130a. MEISSNER, E., Zur Festigkeitsberechnung von Hochdruckkesseltrommeln, [39], 86, 1925.
131a. Proceedings of the third U.S. National Congress of Applied Mechanics, N.Y., 1958.
132a. RYS, K., Stresses in dished heads of pressure vessels, [28], 1931.
133a. SCHULTZ-GRUNOV, F., Die Festigkeitberechnung achsensymmetrischer Böden und Deckel, [22], 1933.
134a. SIEBEL, E., SCHWAIGER, S., Neuere Untersuchungen an Dampfkesseltrommeln und Behältern, Forschung, Ing. Wes., 11, 400, 1940.
135a. SPOTS, M., Analysis of spherical shells of variable thickness, [42], 1939.
136a. STEUERMANN, E., Some considerations of the calculation of elastic shells, Proc. of third Congr. for Appl. Mech., 1930.
137a. —, Berechnung der Kuppeln als Bögen auf elastischer Unterlage, [17], 21–22, 1933.
138a. TÖLKE, F., Zur Integration der Differentialgleichungen der drehsymmetrischen belasteten Rotationschale bei beliebiger Wandstärke, [22], 4, 1938.

AUTHOR INDEX

The numbers in brackets refer to the references listed on p. 407–417.

SUBJECT INDEX

MIX
Papier aus verantwortungsvollen Quellen
Paper from responsible sources
FSC® C105338

If you have any concerns about our products,
you can contact us on
ProductSafety@springernature.com

In case Publisher is established outside the EU,
the EU authorized representative is:
Springer Nature Customer Service Center GmbH
Europaplatz 3, 69115 Heidelberg, Germany

Printed by Libri Plureos GmbH
in Hamburg, Germany